Bioelectrochemistry:
Principles and
Practice

Volume 5

Bioelectrochemistry of Biomacromolecules

Edited by
G. Lenaz
G. Milazzo†

Coordinated by D. Walz

Birkhäuser Verlag
Basel · Boston · Berlin

Editors:

Giorgio Lenaz
Professor of Biochemistry
Faculty of Medicine
University of Bologna
Via Irnerio 48
40126 Bologna
Italy

Giulio Milazzo†
formerly Professor
Istituto Superiore di Sanità
Rome
Italy

Coordinator:
PD Dr. Dieter Walz
Biozentrum
University of Basel
Klingelbergstrasse 70
CH-4056 Basel
Switzerland

A CIP catalogue record for this book is available from the Library of Congress, Washington, D.C., USA.

Deutsche Bibliothek – Cataloging-in-Publication Data

Bioelectrochemistry of Biomacromolecules / ed. by G. Lenaz ;
G. Milazzo. Coordinated by D. Walz. - Basel ; Boston ; Berlin :
Birkhäuser, 1997
(Bioelectrochemistry ; Vol. 5)

ISBN-13: 978-3-0348-9936-9 e-ISBN-13: 978-3-0348-9179-0
DOI: 10.1007/978-3-0348-9179-0

NE: Lenaz, Giorgio [Hrsg.]; GT

Printed on acid-free paper produced from chlorine-free pulp. TCF ∞

9 8 7 6 5 4 3 2 1

GIULIO MILAZZO (1912–1993)

Giulio Milazzo, the father of Bioelectrochemistry, died on January 6 1993 in Rome. He often pointed out that the roots of the subject go back two hundred years to Galvani and Volta, and that he had only resurrected the science. But the Bioelectrochemistry he started was certainly different from the Natural Philosophy of the past, and in many ways quite different from parallel modern developments.

At a time when science is becoming more narrowly focused and scientists more specialized, Giulio Milazzo catalyzed the formation of an interdisciplinary grouping that was broad in scope and inclusive in its organization. He envisaged Bioelectrochemistry as a discipline including all aspects of the overlap of biology and electrochemistry. He believed that science is international and that one should use all scientific means possible to foster cooperation across national barriers. His ideas catalyzed the founding of the *Bioelectrochemical Society* which to this day attempts to follow the high standards set by him.

Giulio Milazzo was convinced that communication between scientists from various fields would be greatly facilitated if a comprehensive textbook written in a common language existed. He therefore initiated the preparation of what he called a *Treatise on Bioelectrochemistry*, but his untimely death prevented him from finishing this enormous task. The Bioelectrochemical Society, under whose auspices the work had been commenced, considers it both an honor and a duty to pursue the project to completion. But with the guiding spirit no longer with us, the project of a comprehensive textbook seemed too ambitious, and the Treatise was therefore converted to the present Series of Texts. May it nevertheless not only serve the purpose envisaged by Giulio Milazzo, but also be a living memory to a great scientist and a dear friend.

Martin Blank   Roy Caplan   Dieter Walz

Contents

Contributors

Fraser A. Armstrong, Inorganic Chemistry Laboratory, University of Oxford, Oxford OX1 3QR, UK

Viktor Brabec, Institute of Biophysics, Academy of Sciences of the Czech Republic, 612 65 Brno, Czech Republic

Benjamin A. Feinberg, Department of Chemistry, University of Wisconsin-Milwaukee, WI 53201, USA

Hiriam F. Gilbert, Verna and Marrs McLean Department of Biochemistry, Baylor College of Medicine, Houston, TX 77036, USA

H. Allen O. Hill, Inorganic Chemistry Laboratory, University of Oxford, Oxford OX1 3QR, UK

Vladimir Kleinwächter, Institute of Biophysics, Academy of Sciences of the Czech Republic, 612 65 Brno, Czech Republic

Chi-Hao Luan, The Laboratory of Molecular Biophysics, School of Medicine, University of Alabama at Birmingham, University Station, Birmingham, Alabama 35294-0019, and Bioelastics Research Ltd., Birmingham, Alabama 35205, USA

C. Bruce Moore, Inorganic Chemisty Laboratory, University of Oxford, Oxford OX1 3QR, UK

David M. A. NabiRahni, Inorganic Chemistry Laboratory, University of Oxford, Oxford OX1 3QR, UK

David A. Pye, Cancer Research Campaign Department of Medical Oncology, and Cancer Research Campaign Department of Drug Development, Paterson Institute for Cancer Research, Christie Hospital NHS Trust, Manchester M20 9BX, UK

Michael D. Ryan, Department of Chemistry, Marquette University, Milwaukee, WI 53233, USA

Dan W. Urry, The Laboratory of Molecular Biophysics, School of Medicine, University of Alabama at Birmingham, University Station, Birmingham, Alabama 35294-0019, USA

Vladimir Vetterl, Institute of Biophysics, Academy of Sciences of the Czech Republic, 612 65 Brno, Czech Republic

Introduction

Volume 5 of this Series deals with biomacromolecules, in particular nucleic acids, proteins, and charged polysaccharides. These macromolecules constitute the building blocks of cell structure and function. They are linear, sometimes also branched, polymers of bifunctional organic molecules. The properties of macromolecules are not simply the sum of the properties of their monomeric components because the monomers are linked by anhydride bonds which mask most of their functional groups. This is particularly evident for the electrical properties of polymers made up of monomers such as amino acids or nucleotides whose linkage is formed from charged functional groups. In addition, the conformation of a given macromolecule is the result of cooperative phenomena which cannot be predicted from the individual units *per se* but depend on their sequence in the polymer.

Among biomacromolecules there are both water-insoluble species, which form supramolecular structures (fibers, membranes, organelles), and water-soluble species, which presumably exist as free components dissolved in the aqueous phase of compartments or are loosely bound to such structures. This volume is restricted to water-soluble species, and particular emphasis is placed on the electrochemical properties which are essential for many biological functions, e.g., electron transfer between redox centers in proteins. Moreover, electrochemical techniques have proven very useful for the investigation of biomacromolecules. In Chapter 1 the structure, chemical reactivity and electromagnetic properties of nucleic acids are extensively reviewed. The structure of proteins as well as their folding and function are discussed in Chapter 2. The electrochemistry of proteins containing redox centers is dealt with in Chapter 3, while Chapter 4 investigates the use of voltammetric methods for studying such proteins. Two special and important redox centers in proteins, i.e. the thiol/disulfide couple and the iron-sulfur clusters are treated in Chapters 5 and 6, respectively. Finally, Chapter 7 describes the properties of glycosaminoglycans which are complex charged polysaccharides occurring on the cell surface and in the extracellular matrix.

This volume is intended not only for students and investigators in bioelectrochemistry but also for biochemists, biologists, and biotechnologists, as well as chemists with an interest in biological aspects of macromolecules. Owing to its comprehensiveness it can also serve as a textbook to be used by graduate students in biochemistry, biophysics, and bioenergetics.

Giorgio Lenaz

Bioelectrochemistry of Biomacromolecules
ed. by G. Lenaz and G. Milazzo†

CHAPTER 1
Structure, chemical reactivity and electromagnetic properties of nucleic acids

Viktor Brabec, Vladimír Kleinwächter and Vladimír Vetterl
Institute of Biophysics, Academy of Sciences of the Czech Republic, Brno, Czech Republic

1. Introduction

Nucleic acids and proteins may be considered the most significant biomacromolecules. While nucleic acids carry genetic information and represent the project of life, proteins provide the means of executing it.

At present, information on nucleic acids is growing enormously. Although much remains to be learned about the function of these biomacromolecules, most textbooks on molecular biology or genetics for the past 30 years have mentioned the "central dogma" of molecular biology as the basis of this scientific discipline, describing the pathway of the flow of genetic information (Fig. 1): the arrows in Figure 1 show the directions proposed for the transfer of genetic information. The arrow around DNA indicates the fact that DNA serves as a template for its self-replication. The other two arrows also show a one-way flow of genetic information: the sequence of bases in RNA is never determined by a protein template, and also it is not suggested that DNA would be synthesized on an RNA template. (The strands of RNA can – in sporadic cases, as in the course of reverse transcription in the life cycle of RNA retroviruses – serve as a template for the synthesis of DNA with a complementary sequence of bases). As regards RNA it is worth mentioning that this nucleic acid can also function as an enzyme that influences the rate of protein synthesis. The RNA molecule can be cleaved at a certain point by means of an RNA enzyme and separate a segment carrying only one piece of information. In connection with this observation a hypothesis has been put forward, according to which both DNA and proteins are unnecessary for the origin of life. RNA alone is sufficient: it gives information about the primary structure of proteins and also performs protein synthesis itself. The biological roles of nucleic acids have been amply detailed in numerous review articles and books (see, e. g., references 1–4).

In this article the primary structure (chemical structure and base sequence) is described first. In the following text the most significant physical properties of nucleic acids, determined by their chemical structure, are discussed. Attention is particularly paid to the conformation, electric and magnetic properties of these biomacromolecules. Understanding the properties underlying the functioning of nucleic acids may have great significance for the study of phenomena observed in bioelectrochemistry.

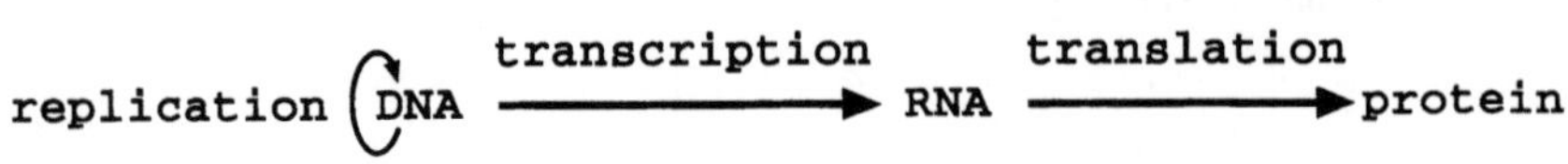

Fig. 1

2. Nucleic acid structure

A century of experimentation has clarified the basic principles of nucleic acid structure, but our knowledge about it is still less complete than it is about protein structure. This difference is largely a result of the higher molecular mass and heterogeneity of nucleic acids. Nucleic acids are of two kinds: deoxyribonucleic acids (DNAs) and ribonucleic acids (RNAs).

2.1. Chemical composition

DNA and RNA can be hydrolyzed to low molecular mass products of three kinds: purine and pyrimidine bases, a pentose sugar and phosphoric acid. The major bases occurring in nucleic acids are adenine, guanine, cytosine, and thymine (in DNA) or uracil (in RNA) (Fig. 2). Equality exists between

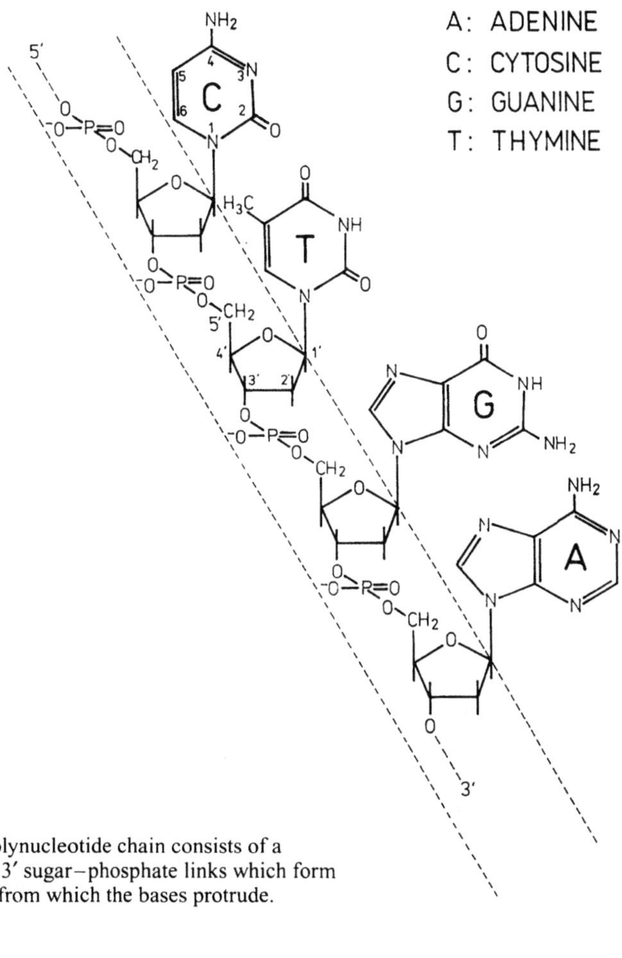

Fig. 2. A polynucleotide chain consists of a series of 5′–3′ sugar–phosphate links which form a backbone from which the bases protrude.

the numbers of purine and pyrimidine bases in DNAs. An important chemical feature of nucleic acids is the position of the hydrogen atoms in the purine and pyrimidine base residues. The nitrogen atoms attached to the purine and pyrimidine rings are usually in the *amino* (NH_2) tautomeric forms and only rarely assume the *imino* (NH) form. Likewise, the oxygen atoms attached to the C(6) atoms of guanine, thymine and uracil are normally in the *keto* (C=O) form and only rarely take up the *enol* (C–OH) form.

The pentose of RNA is always D-ribose and that of DNA is always 2-deoxy-D-ribose. The basic monomeric building blocks of nucleic acid chains are the nucleotides shown in Figure 2. Nucleotides are composed of nucleosides, which contain a base molecule linked to C(1′) of ribose or deoxyribose. The nucleotides of nucleic acids are nucleosides with the 3′ or 5′ hydroxyl esterified with phosphoric acid. The sugar moiety of nucleosides or nucleotides is not planar. The puckering of ribose or deoxyribose, i.e. the displacement of one or two atoms from a main plane, is limited because of steric interactions between the base on C(1′) and the exocyclic CH_2–O–R group at C(5′). Sugar residues exist in two principal

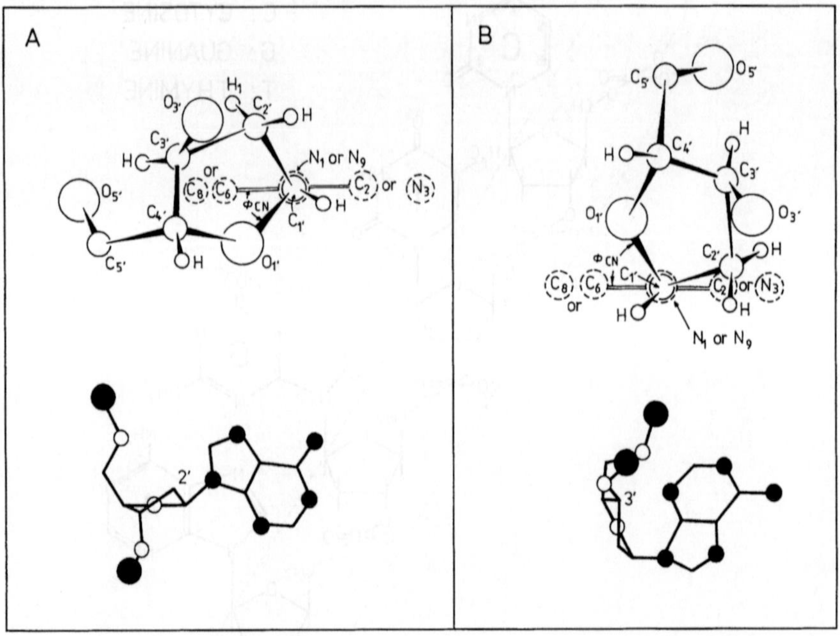

Fig. 3. Schematic illustration of the torsion angle in a pyrimidine and purine nucleoside. In the top pictures the plane of the base is viewed end-on with the glycosyl bond between C′(1) and N(1) (for a pyrimidine) or N(9) (for a purine) perpendicular to the page. The torsion angle ϕ_{CN}, is the dihedral angle between the plane of the base and the plane formed by the C′(1) to N(1) bond. The torsion angle is zero when O′(1) lies directly in the front of C(6) [or C(8) for a purine] and positive angles are measured when C′(1) to O′(1) is rotated in a clockwise direction when viewing from C′(1) to N. When ϕ_{CN} is negative (about −30°), the system is in *anti* conformation (A). When ϕ_{CN} is positive (about +150°), the system is in *syn* conformation (B).

conformations: 3′-*endo*-2′-*exo* or 3′-*exo*-2′-*endo* (Fig. 3). Moreover, the position of the base relative to the sugar is sterically restricted. The conformations of nucleosides and nucleotides can be better understood by considering the sugar–base torsion angle, ϕ_{CN}, and the related *syn* and *anti* conformations which are defined in Figure 3. Extensive X-ray diffraction studies have revealed that pyrimidine nucleosides and nucleotides have negative values of ϕ_{CN} and exist in the *anti* conformation [5, 6]. This is because the energy barrier between *anti* and *syn* conformations in the pyrimidine nucleosides and nucleotides is high (c. 15–25 kJ/mol) [7, 8]. In contrast this energy barrier in purine nucleosides and nucleotides is considerably smaller (5–10 kJ/mol), so that in solution they exist in an equilibrium between the two conformations.

Nucleic acids are formed by polynucleotide chains (Fig. 2) in which the sequence of the purine and pyrimidine nucleotides is highly irregular, varying from one molecule to another. The base sequence that contains the information specifying the structure of proteins is written in the triplet genetic code [1, 2]. The sequence in which successive triplets of the bases are arrayed along a single polynucleotide chain encodes the sequence in which successive amino acids are linked to form a protein chain.

Polynucleotide is a generic term for nucleic acids and their synthetic analogues. All polynucleotide chains have the same basic structure with the nucleotide as the repeating unit (Fig. 2). In the polynucleotide chain, the sugar and phosphate groups are always linked together by the same chemical bond (3′–5′ phosphodiester linkage), so that the sugar-phosphate backbone is extremely regular. The fact that the nucleic acids always occur as the 3′–5′ phosphodiesters (Fig. 2) provides these chains with a polarity, which is of primary importance for the structure and function of nucleic acids.

2.2. Conformation of DNA

DNA encodes not only the genetic message itself but also instructions for the selective expression of the message. Today, almost 40 years after the discovery of the double helical structure of DNA, we realize that in particular the information specifying genetic control is three-dimensional and its interpretation depends on the conformational properties of DNA. Early information about the conformation of DNA was obtained from X-ray diffraction patterns of DNA fibres. Now short oligonucleotides of any desired sequence can be synthesized and crystallized. This makes it possible to study DNA conformation by conventional single-crystal X-ray diffraction methods, which provide more precise data compared to those obtained by investigating DNA fibres.

2.2.1. DNA as double helix: All cellular DNA occurs as right-handed double helices. The two polynucleotide chains in DNA associate by hydro-

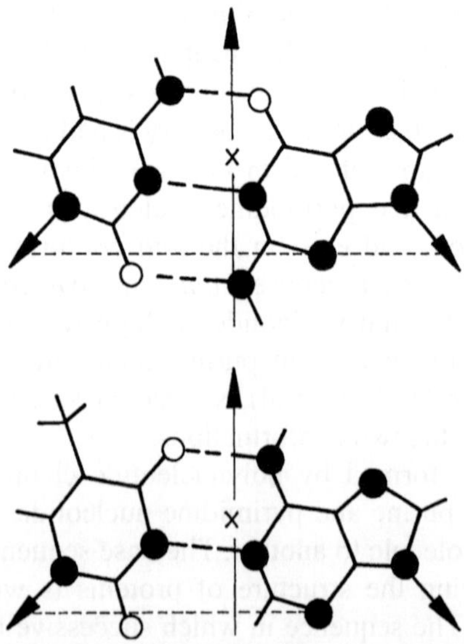

Fig. 4. Complementary base pairing involves the formation of three hydrogen bonds between guanine and cytosine (top), and of two hydrogen bonds between adenine and thymine (bottom).

gen bonding between adenine and thymine and between guanine and cytosine (Fig. 4). This way of base pairing is a means of interaction specific to the structure of DNA and of RNA (if base pairing occurs in RNA, adenine is paired with uracil). The two polynucleotide strands in DNA run in opposite directions so that they have opposite polarity, i. e. are antiparallel. Looking along the helix one strand runs in the 5′–3′ direction, while its partner runs in the 3′–5′ direction. The sugar-phosphate chain is on the outside and carries negative charges on the phosphate groups, giving the nucleic acids their polyanionic character. When DNA is in solution, these charges and their electrostatic repulsion are neutralized by the binding of cations present in the solution. This neutralization contributes to the thermodynamic stability of the double helix. The bases lie inside, forming a hydrophobic core, and are stacked above each other. The base pairs also contribute to the thermodynamic stability of the double helix by the hydrogen bonding between the complementary bases and by base stacking, resulting from interactions between the electron systems of the stacked base pairs. Dipoles, dipole–induced dipole moments and London dispersion forces appear to be important in the vertical base stacking. In addition, evidence has been presented for the contribution of hydrophobic forces to DNA stability in an aqueous solution. Each base pair is rotated around the axis of the helix relative to the next base pair. The twisting of the two strands around one another forms a helical structure with grooves on this surface (Fig. 5). The double helix is right- or, less frequently, left-handed, the turns running clockwise or counterclockwise, respectively, looking along the helical axis.

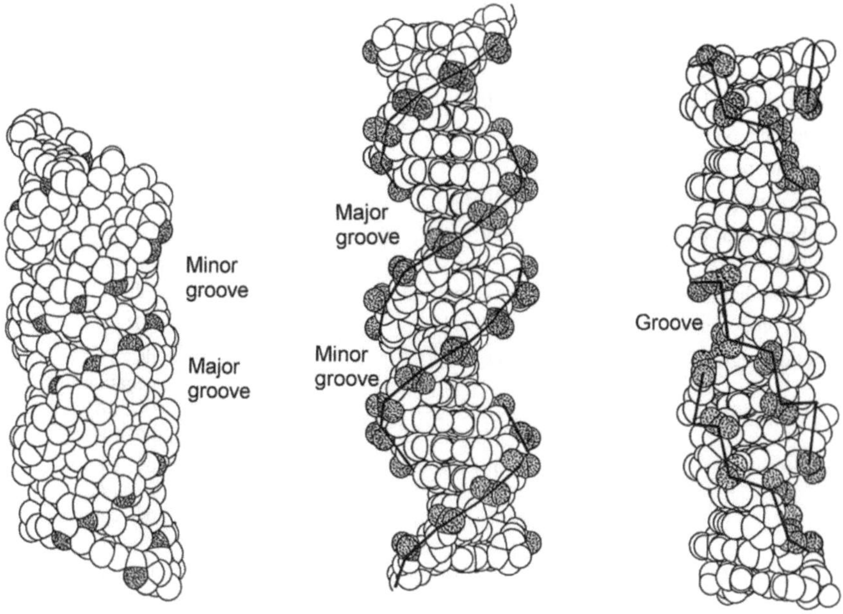

Fig. 5. Nucleic acids can form several types of double helix: A-DNA (left), B-DNA (centre) and Z-DNA (right). The black line traces the sugar-phosphate backbone.

The DNA of certain small viruses is single-stranded. Such single-stranded DNA can be converted into the double helical form after entry into cells. Single-stranded DNAs are more compact than double helices.

2.2.1.1. A-, B- and Z-DNAs: This section illustrates the main structural features of double-stranded DNA helices. DNAs can adopt several conformations depending on environmental conditions such as counter-ion and relative humidity and, in synthetic polynucleotides with defined repetitive oligonucleotide building blocks, on their composition and the sequence of nucleotide residues. DNA double helices can be classified in three groups (Fig. 5): the A-family is characterized by the 3′-*endo* puckering of the sugar, whereas sugar residues in the helices of the B-family have the 2′-*endo* conformation. Both these families are right-handed helices. The third family is the left-handed Z-DNA. The essential structural parameters of various forms of DNAs derived from single-crystal X-ray analysis [9–13] are given in Table 1.

DNA adopts the A-form under conditions of low relative humidity (e.g. in media containing high concentrations of alcohols) and also in its hybrid helices containing one strand of RNA [14]. Classical B-conformations occur in natural DNAs under the usual physiological conditions, i.e. at high relative humidity and the physiological ionic conditions. Single-crystal studies carried out on short DNA molecules of any selected sequence made it

Table 1. A comparison of the structural parameters of various forms of DNA

	A-DNA	B-DNA	Z-DNA
Helix rise per base pair (nm)	0.23	0.332	0.38
Base pairs per helix repeat	1	1	2
Rotation per base pair	33.6°	35.9°	–60°/2
Base pairs per turn of helix	10.7	10.0	12
Inclination of base normals to helix axis	19°	–1.2°	–9°
Pitch per helix turn (nm)	2.46	3.32	4.56
Mean propeller twist [a]	18°	16°	0°
Torsion angle, ϕ_{CN} [b]	negative	negative	dC, dT: negative; dG, dA: positive
Sugar pucker	3′-*endo*	O′-*endo* 2′-*endo*	dC, dT: 2′-*endo*; dG, dA: 2′-*exo*
Major groove	narrow and very deep	wide and deep	flattened out on helix surface
Minor groove	very broad and shallow	narrow and deep	very narrow and very deep

[a] For definition of the base-pair propeller twist see Figure 6.
[b] For definition of the torsion angle ϕ_{CN} see Figure 3.

possible to obtain accurate parameters of various forms of DNA helices [9–13].

A-DNA and B-DNA differ mainly in the positioning of the base pairs around the helix axis and in base inclination (i.e. the tilt of the bases with respect to the helix axis). In B-DNA the inclination is close to zero (the base pairs are oriented nearly perpendicular to the helix axis) and the axis runs through the centre of each base pair. Although the minor groove is narrower than the major one, the grooves are of similar depth. In A-DNA the base pairs are tilted between 13 and 19 degrees from the perpendicular position. Moreover, they are shifted towards the outside of the helix, whose axis goes through the major groove, bypassing the bases. Hence the minor groove is shallow and the major groove is deep. B-DNA in solution has about ten base pairs per turn of the helix, with a spacing of 0.332 nm along the helix axis from one base pair to the next. The A-DNA has close to 11 base pairs per turn and the distance along the axis is only 0.23 nm per base pair. Thus, in contrast to the rather uniform A-DNA, which is short and thick-set, B-DNA looks longer and thinner (Fig. 5). B-DNA exhibits various sequence-dependent conformational variations, which could be of a fundamental significance in the process of recognition of DNA by specific proteins.

DNA containing regularly alternating purine and pyrimidine residues can form either right-handed or left-handed double helices [15, 16]. The left-handed Z-DNA helices can for instance be formed in synthetic polynucleotides such as poly(dG-dC)·poly(dG-dC) and poly(dA-dC)·poly(dG-dT). The agents capable of converting DNA from a right-handed to a left-handed form are 4M NaCl, 60–65% alcohols, di- and trivalent cations, amine complexes of cobalt, platinum, zinc and copper, bromination of deoxyguanosine residues yielding dBr^8Guo, bromination or methylation of deoxycytidine at the C(5) position, N-acetoxy-aminofluorene, or negative supercoiling in closed circular DNA. The basic repeating unit of the Z-DNA double helix is not a single base pair, as it is in A- and B-DNA, but rather a pair of dinucleotides. The orientation of the bases around the glycosidic linkage is *anti* for the pyrimidine nucleoside residue, but *syn* for the purine nucleoside residue. The sugar puckering is 2′-*endo* in pyrimidine nucleoside residues, but 3′-*endo* for purine nucleoside residues. The sugar–phosphate backbone of the Z-DNA double helix is not smooth but follows a zig-zag pattern (hence its designation as Z-DNA) (Fig. 5) to accommodate the unusual glycosidic linkages and puckers with the two sugars facing each other. The purine–pyrimidine sequence is thus not equivalent to the pyrimidine–purine one. Only one deep minor groove is distinguishable in Z-DNA. It should also be noted that under certain conditions even non-alternating sequences of purines and pyrimidines can adopt the conformation of Z-DNA with thymidines in the *syn* conformation [17, 18]. In comparison with A- and B-DNA the molecule of Z-DNA looks elongated and slim (Fig. 5). Until recently, details of the structure of naturally occurring, left-handed DNA have been assumed to conform to the canonical helix of Z-DNA, whose structural parameters and properties are discussed above. Most of the measurements presumed to substantiate

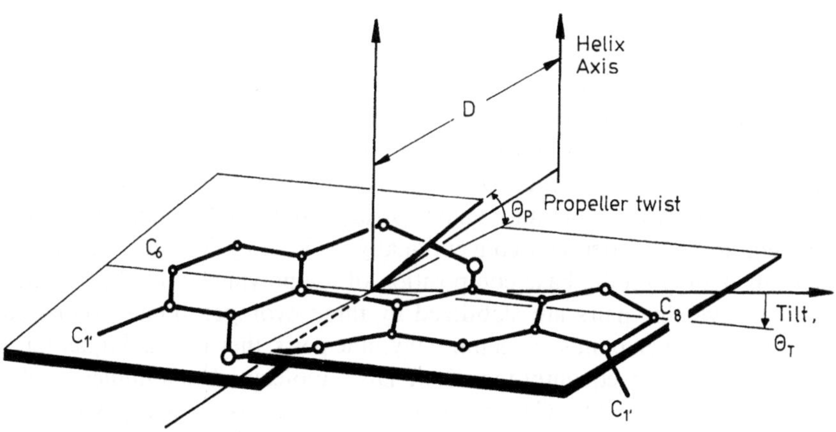

Fig. 6. The propeller twist between the pyrimidine and purine base pairs of a right-handed helix. C′(1) atoms are shifted up and down by the twist and may serve to stabilize the helix by increasing stacking overlap between bases.

the existence of Z-DNA do not provide direct evidence about many of the subtle features of the structure, nor about one of the more radical aspects of the Z-DNA model, namely the reversed chain polarity (direction of the 5′ to 3′ progression in the backbone chains) relative to B-DNA. A new left-handed, zig-zag double helix with B-DNA backbone directions was proposed [19]. Structural features of this new Z-DNA include the presence of an additional hydrogen bond between each guanine N(2)-amino group and an adjacent phosphate oxygen, and the backbone chain polarities that are the same as in B-DNA. As this new Z-DNA model allows a relatively unhindered right- to left-handed transition (without rotation of each base pair around its own horizontal axis), it is assumed to be preferentially generated *in vivo*.

2.2.1.2. Helix–coil transition: The forces that stabilize the double helix of DNA can be disturbed by environmental factors (e.g. elevated temperature, extremely low or extremely high ionic strength, extreme pH values of the medium, or organic solvents). After breaking all hydrogen bonds two separate single polynucleotide chains result, which in solution usually assume an arbitrarily coiled form. The ability of the DNA helix to unwind (and rewind) without breaking covalent bonds is important for its biological functions, since the processes in which DNA is involved, i.e. replication and transcription, take place on single DNA strands.

The formation of helical structures of a polynucleotide is always accompanied by changes in some physico-chemical parameters of the polymer. The method most widely used for monitoring the helix–coil transition in polynucleotides (alternative terms are melting or denaturation) is based on measurement of absorbance at about 260 nm. The polymerization of nucleotide units into a single chain results in an overall decrease in the absorption spectrum compared to that of the corresponding mixture of monomers. A further decrease in absorbance, which is the result of mutual interactions between the chromophoric purine and pyrimidine bases, accompanies the formation of the double helix. This effect, called hypochromism, amounts to a change of 30–35%. Thus, if the helical structure changes into the irregular coil, absorbance at 260 nm increases by this amount [20, 21].

The helix–coil transition (or melting) occurs over a narrow temperature range, which reflects the cooperativity of the process. The transition is characterized by a temperature value corresponding to the midpoint of this temperature range, denoted conventionally as T_m (Fig. 7). The value of T_m depends mainly on the base composition of the polynucleotide [22]. Since guanine · cytosine pairs are stabilized by three hydrogen bonds compared with two hydrogen bonds in adenine · thymine pairs, they melt at higher temperatures, and consequently the whole DNA molecule has a higher T_m the greater the fraction of G · C pairs it contains. T_m increases linearly with G · C content. Another important factor is the ionic strength of the solution. The increase in T_m depends on the logarithm of concentration of monovalent cations [23] (for details see section 4.1.3.5.).

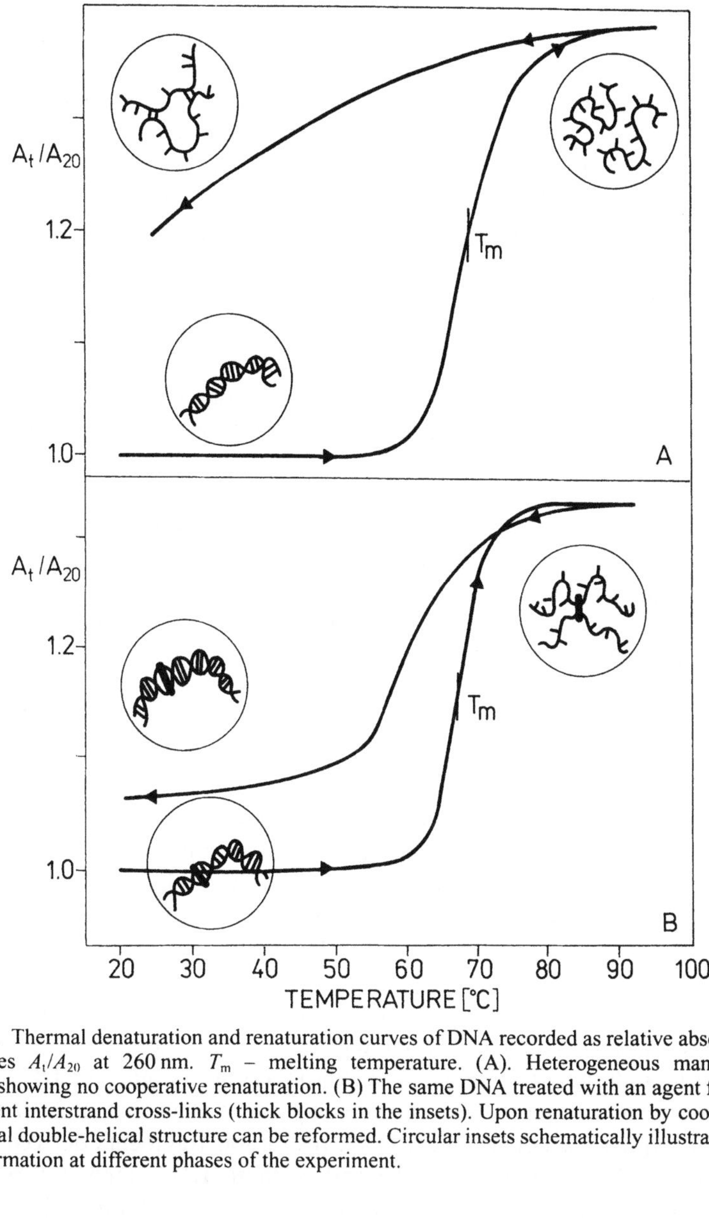

Fig. 7. Thermal denaturation and renaturation curves of DNA recorded as relative absorbance changes A_t/A_{20} at 260 nm. T_m – melting temperature. (A). Heterogeneous mammalian DNA showing no cooperative renaturation. (B) The same DNA treated with an agent forming covalent interstrand cross-links (thick blocks in the insets). Upon renaturation by cooling the original double-helical structure can be reformed. Circular insets schematically illustrate DNA conformation at different phases of the experiment.

Instead of plotting the conventional melting curves (relative absorbance change vs. temperature), plots of dA/dt vs. t, so-called differential melting curves, can be obtained, which can yield information about the melting of individual regions in the polymer characterized by a certain base composition and sequence. The necessary requirement is a homogeneous population of polynucleotide molecules [24].

A reverse process, usually called renaturation, can also be investigated. If, for example, a DNA preparation denatured by heating is slowly cooled down, complementary bases can reform hydrogen bonds, and at sufficiently low temperatures the helical stacked structure can appear. The extent and perfection of renaturation depend on the quality of the polynucleotide preparation. In a heterogeneous population of molecules the degree of renaturation is very low compared with a homogeneous one. A very high yield of renatured molecules is obtained in polynucleotide preparations modified by a chemical agent that forms cross-links between the two strands. Upon denaturation, when hydrogen bonds are broken, the strands are still held together by cross-links and the original structure is reformed after cooling. The renaturation process is accompanied by a decrease in absorbance at 260 nm and the perfection of renaturation can be characterized by the magnitude of the hypochromic effect and cooperativity of the renaturation transition (Fig. 7) [20].

2.2.2. Unusual and local DNA structures: The significant variations in some of the global parameters dependent on the nucleotide sequence result in local changes along the DNA double helix. The effect of the nucleotide sequence on the fine geometrical features of each DNA form has been demonstrated but not fully elucidated. Moreover, supercoiling of closed circular DNA influences the structure of the double helix so that many unusual forms of DNA have been identified.

2.2.2.1. DNA curvature: The change in linearity of the DNA helical axis is also a conformational feature that can be modulated by base sequence. The DNA axis can become nonlinear in space due to intrinsic physico-chemical properties without applying an external force. This DNA is called "curved" in contrast to DNA which is bent by external force (for example by a protein or by a drug–DNA interaction) [25–28]. This section deals only with curved DNA.

The discovery of curved DNA molecules was made through examination of DNA fragments that displayed unusual electrophoretic behaviour on polyacrylamide gels [29–31]. The electrophoretic mobility of DNA with certain repeated sequences is considerably lower than that of DNA with random sequences. For instance, DNA restriction fragments derived from trypanosome kinetoplast minicircles and synthetic DNA fragments, both of which contain short runs of adenine · thymine pairs, exhibit such electrophoretic behaviour [31–34]. The anomalous electrophoretic behaviour of these fragments is attributed to the presence of short tracts of adenine · thymine pairs which are able to induce the stable axial curvature of DNA fragments [35, 36]. Locally curved adenine tracts must be aligned in phase with the helix repeat in order to produce the macroscopic curvature observable in electrophoretic mobility experiments. When the tracts of adenine · thymine pairs are separated by a half of the number of base pairs per helix turn, the DNA fragments display normal electrophoretic mobility. These in-phase and out-of-phase effects indicate that the curvature occurs in a fixed plane. As a rule, DNA fragments containing double-stranded homopolymeric (deoxyadenosine)$_n$ sequences (where in general n = 4), repeated in phase with the helix turn, show anomalous migration in polyacrylamide gels [37] with six deoxyadenosine residues giving the maximum effect. The greatest degree of curvature is observed if the adenine stretches are flanked by thymine on the 3′ side and by cytosine on the 5′ side.

The conformational features of DNA containing a deoxyadenosine tract that gives rise to the curvature have not yet been completely defined. Several structural models have been proposed for the intrinsic curvature of the DNA double helix. The model that conforms with most experimental data is the "wedge model" [37] exploiting a suggestion that the angle between base planes is sequence-dependent. Thus the smooth deformation of DNA by a series of small tilt and roll components between planes of adjacent base pairs could be responsible for the curvature.

2.2.2.2. Supercoiled DNA: Double-helical DNA was initially considered to be a linear molecule with two free ends. After isolation procedures had been improved so that undegraded samples could be analysed, it became evident that many DNA molecules have a closed, circular structure, which lacks free ends [38–43]. As shown below, this form of organization imposes an additional constraint on the DNA double helix.

Linear DNA molecules can freely rotate to accommodate changes in the number of times the two chains of the double helix twist about each other. However, once the two ends become linked to form a loop or a covalently closed circular molecule, the absolute number of times one strand winds about the other – the *linking number* (L) – becomes invariant. In relaxed circular DNA L is equal to the number of double helical turns (twists), given by the *twisting number* (T). For right-handed DNA both L and T have positive values. If the double helix in circular DNA is underwound or overwound, L is changed and supercoiling results (Fig. 8). The supercoiling cannot occur in linear molecules, because an open molecule can release the torsion simply by free rotation. Supercoiling is characterized by the *writhing number* (W), which represents the turning of the duplex axis in space.

The linking number can then be expressed as equal to the sum of the writhe and twist,

$$L = W + T \tag{1}$$

If the DNA duplex was able to under- or overwind easily, L would always be equal to T (as it is in the case of relaxed DNA) and no supercoiling would arise. However, DNA tends to retain its B-form conformation characterized by approximately 10.5 base pairs per unit twist, thus limiting T within a narrow range. Changes in L can then be compensated for by supercoiling or by disruption of base pairs. Underwinding of the double helix in circular

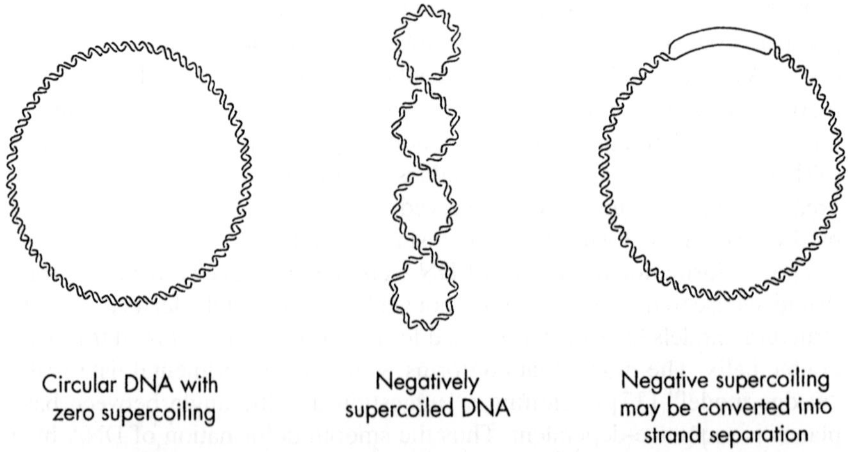

Fig. 8. Supercoiling causes a DNA duplex to be twisted about itself in space; negative supercoiling can be relieved by disrupting base pairs.

DNA results in *negative supercoiling* (characterized by a negative value of W), which corresponds to a right-handed supercoil, while overwinding results in *positive supercoiling* (with a positive value of W), yielding a left-handed supercoil.

Most of the natural supercoiled DNAs are underwound. They are thus negatively supercoiled and form a right-handed superhelix. A typical superhelix density (defined as W/T) of an isolated plasmid or viral circular DNA is around −0.05 (corresponding to about one negative turn per 200 base pairs). A break even in one strand of a circular molecule allows unwinding. Thus nicking or cutting of even one strand of the circular DNA instantly converts a supercoiled molecule into a relaxed form.

The linking number is an invariant property of any individual closed DNA molecule. It is related to the actual enzymatic events by which changes are made in the topology of DNA. The linking number of a particular closed DNA molecule can be changed only by breaking one or both DNA strands, winding or unwinding the molecule, and rejoining the broken ends. The enzymes that catalyse these conversions are called *topoisomerases:* type I makes a transient single-strand break, relaxing the circular DNA molecule; type II acts on relaxed molecules, making a double-strand break, followed by DNA writhing and resealing. Isomeric DNA forms which result from topological differences are called *topoisomers*.

The introduction of negative supercoiling requires energy so that the supercoiled molecule has more energy than the relaxed form owing to its possession of supercoils. Thus, less energy may be needed to induce a conformational alteration in a negatively supercoiled DNA than in its relaxed form. In other words, supercoiled DNA may undergo structural changes under conditions in which relaxed DNA would be unable to do so.

The stress present within negatively supercoiled DNA molecules sometimes leads to localized denaturation. When this happens the energy lost by breaking the hydrogen bonds is lower than that gained by the unwinding of the double helix. The segments rich in adenine · thymine pairs are usually the first to denature, while the regions that denature last are usually rich in much stronger guanine · cytosine pairs.

This local denaturation should be especially favoured in DNA segments containing self-complementary regions or a nucleotide sequence with a local twofold symmetry, usually represented by a sufficiently long inverted repeat capable of forming hydrogen-bonded hairpin structures called *cruciform loops* [39, 44–47]. In such cases, relatively few hydrogen bonds are lost and considerable energy is gained from the loss of one or more supercoils. Two structural features are characteristic of the cruciform, a

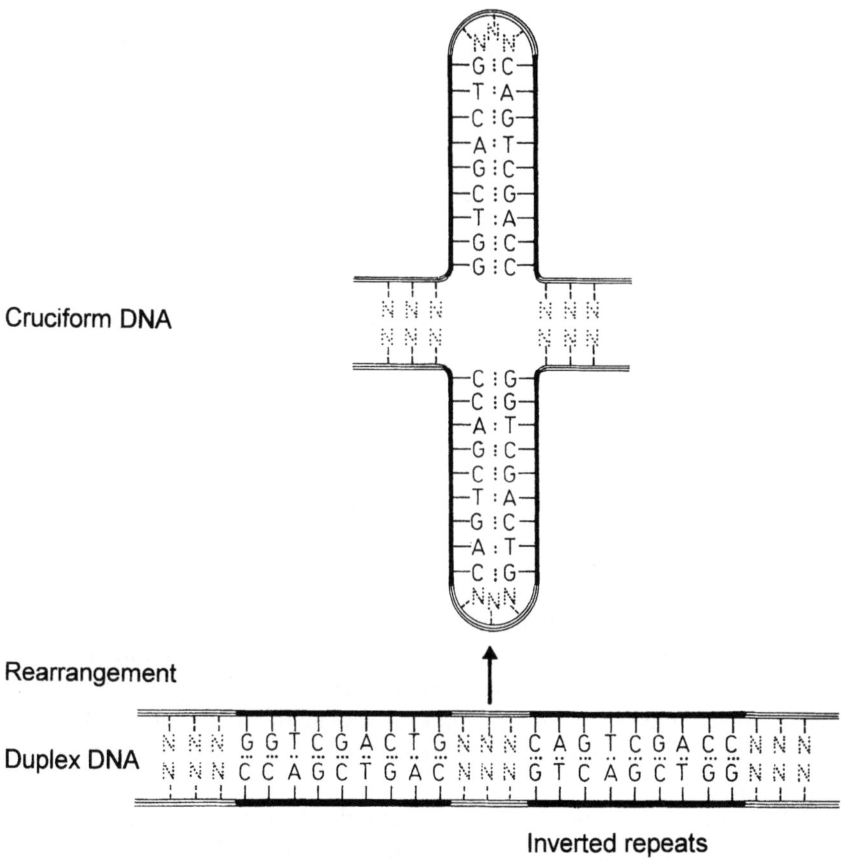

Fig. 9. Inverted repeats can generate cruciform in DNA. The double-helical DNA in the centre shows the sequences of the inverted repeats. A cruciform is generated if the repeats rearrange so that each repeat pairs with the complementary sequence on its own strand instead of with the complement on the other strand.

formally unpaired loop and a four-way junction (Fig. 9) [48]. The optimum size of the formally single-stranded loop lies between four and six bases. The loop adopts a structure probably involving base stacking and even non-Watson-Crick base pairing. Hairpin loops may affect the conformation of the double helical stem close to the loop. The four-way junction is normally fully base-paired, asymmetric and X-shaped. It introduces a bend into the molecule, which differs from the intrinsic sequence-directed curvature (see section 2.2.2.1.).

Negative supercoils can also be lost through conversion of any alternating sequence of purine–pyrimidine residues within the DNA molecules from the B-form to the Z-form of the DNA double helix. When supercoils are present, conversion into the Z-form occurs at a much lower concentration of the agent capable of inducing the left-handed conformation in linear DNA (see section 2.2.1.1.).

Some inverted repeats may be composed of alternating purine–pyrimidine sequences and (in principle) can form either a cruciform or a Z-DNA structure. Formation of Z-DNA in purine-pyrimidine sequences of circular DNA requires more energy than the formation of cruciform. Nevertheless, the results so far obtained suggest that equilibria between B, Z, and cruciform structures exist in alternating purine–pyrimidine sequences depending on superhelicity and environmental conditions [49].

Segments of left-handed Z-DNA can be contiguous with B-DNA both *in vitro* and *in vivo*. The boundary between these two structures is known as the *B-Z junction* [50, 51]. Bases in the junction not only show hypersensitivity to a single-strand-selective chemical probe but also represent sites of enhanced sensitivity to double-strand-selective probes (see section 3.1.). Structural models of the B-Z junction suggest that their structure is polymorphic depending on nucleotide sequence, superhelical density, and environmental conditions.

The intramolecular *triplex* represents another unusual supercoil-stabilized structure of DNA. It may occur at weakly acidic and neutral pH values. A new structure called the *H-form* (a protonated triplex H-DNA) was proposed [52, 53] for any homopurine–homopyrimidine sequence containing a mirror repeat in recombinant plasmids. In this DNA the Watson-Crick duplex extends to the centre of the $(dT\text{-}dC)_n \cdot (dG\text{-}dA)_n$ tract (Fig. 10) and the second half of the homopyrimidine strand folds back upon itself, winding down the major groove of the helix. This returning strand forms Hoogsteen pairs with the purines of the Watson-Crick duplex, the cytosines being protonated. The second half of the homopurine strand also folds back, but is probably unstructured.

Also among the local DNA structural changes are the structures of telomeric sequences and parallel-stranded DNA. Telomeres [54, 55], the ends of eukaryotic chromosomes, have tandemly repeated G·C-rich sequences with a pronounced strand-specific base composition asymmetry (e.g. $d(C_4A_2) \cdot d(T_2G_4)$). A novel protonated DNA conformation, a *C, A hair-*

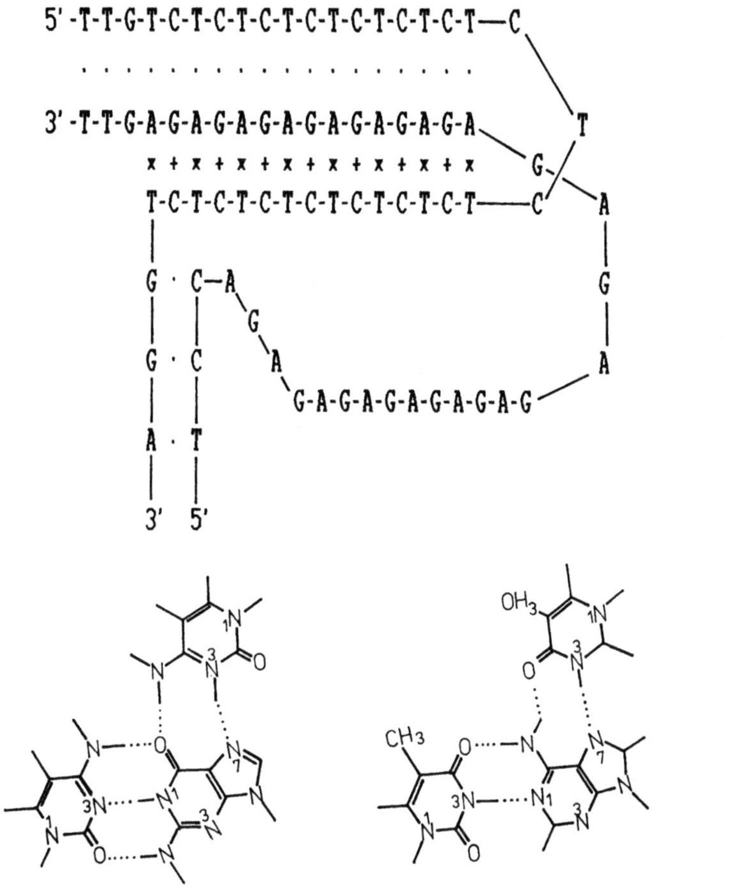

Fig. 10. Top. The triplex H-form. The major element of the structure is the triplex which includes the Watson-Crick (•) duplex associated with a pyrimidine loop through Hoogsteen base pairing (x, +). Bottom. The TAT and CGC^{-} base triads. They consist of the canonical Watson-Crick A·T and G·C base pairs with thymine and cytosine attached by the Hoogsteen pairing. The two base triads are isomorphous. The respective distances between glycosyl bonds are equal.

pin, has been proposed [56] for the telomeric sequence poly[d(C_4A_2)]$_n$·poly[d(T_2G_4)]$_n$ inserted in a plasmid (Figs. 11 and 12). In this structure two independent hairpins are formed which are stabilized by C·C$^+$ and A·A$^+$ base pairs. Moreover, synthetic oligonucleotides containing a telomeric sequence adopt new DNA structures based on non-Watson-Crick base pairing, and the formation of various *quadruplexes* may be observed [57, 58].

Parallel-stranded duplexes can be formed under physiological conditions by oligonucleotides consisting of adenine·thymine pairs [59, 60]. The strands in parallel-stranded DNA have the same polarity and form a right-handed duplex with cross-linked ends. Parallel-stranded DNA inserted into a supercoiled plasmid undergoes transition to a (purine)$_2$·pyrimidine triplex.

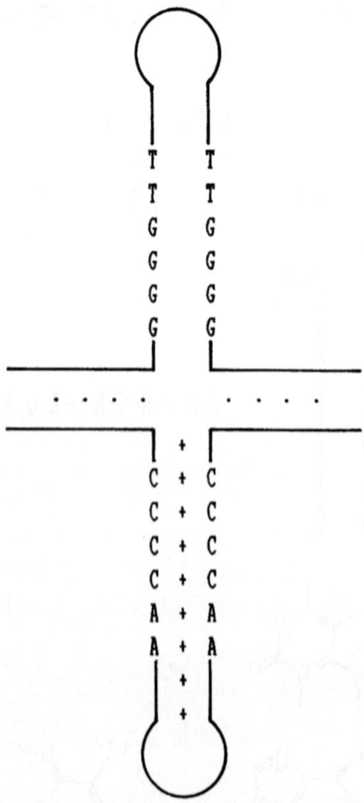

Fig. 11. The (C,A)-hairpin. The C-rich strand forms a hairpin stabilized by the non-canonical C·C⁺ and A·A⁺ base pairs (see Fig. 12). The G-rich strand is probably unstructured.

2.2.3. Higher organization of DNA: If the organization and function of DNA in biological systems are considered, the main problem is the packing of the relatively long double-stranded helical molecule into the much smaller volume of a cell. For instance, in the bacterium *Escherichia coli* a DNA molecule approximately 1.2 mm long has to fit into a cell with a diameter not larger than 2 μm. An even higher degree of packing is required in eukaryotic cells. The DNA corresponding to the human genome is about 1.5 m long and must be compacted about 75,000-fold to get into the nucleus which is only a part of the cell having diameter of 20 μm. Nature solved this problem by folding and packing the DNA fibres, with the participation of proteins or RNA, into chromosomes, which are visible as elongated structures in certain parts of the life cycle of eukaryotic cells, i.e. in the metaphase. In the simplest biological structures, such as viruses or bacteriophages, DNA is probably packed without the help of interaction with other biopolymers.

The latter fact together with the findings that more compact structures can be formed in DNA solutions at higher concentrations, suggest that DNA

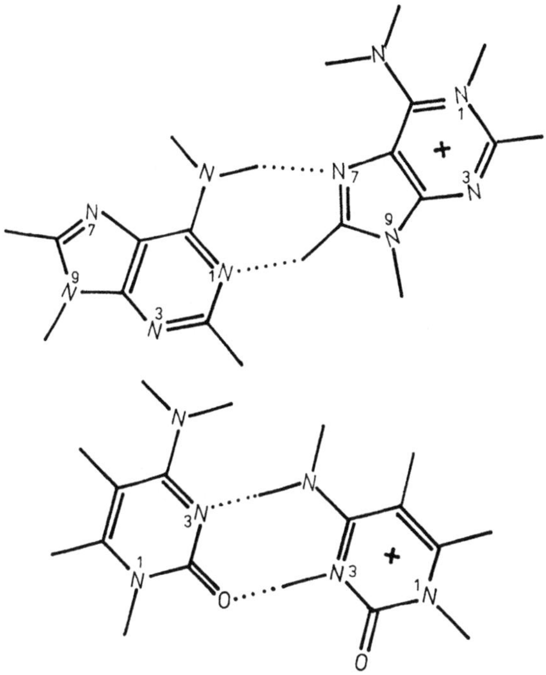

Fig. 12. The C·C⁺ and A·A⁺ base pairs for antiparallel strands.

itself can, at least to some extent, govern its compaction *in vivo*. In *in vitro* experiments condensation of DNA can be achieved by adding excess salt, inert polymers, polyamines, or ethanol to aqueous solutions. Depending on the medium composition, various structures ranging from fibrous precipitates to lamellar microcrystals can be obtained.

In the following sections different types of compact DNA structures will be described.

2.2.3.1. Condensed forms of DNA: DNA is precipitated by ethanol concentrations of between 50 and 95%, and depending on the ionic strength of the solution different types of condensation occur [61].

At low ionic strength and high ethanol concentration (95%) formation of a fibre structure with a diameter of 8–9 nm is observed, in which DNA exhibits a left-handed supercoil. The compaction is relatively low, about nine-fold.

At intermediate ionic strength (about 10 mM) precipitation with ethanol yields fibrous structures with a diameter of about 15 nm, in which DNA is compacted 30 to 35 times. In some instance the fibres show a beaded appearance.

At a higher ionic strength (0.15–0.5 M), 95% ethanol precipitates DNA in rod-like particles of a uniform length of 200 nm. Their diameter depends on the molecular mass of the DNA preparation and it has been suggested that these rods represent lamellar structures with DNA folded back in 200 nm segments. As has been revealed by electron microscopy, the rod-like structures can be changed into toroidal ones by the fusion of the ends of a particle. This transformation depends on the type of cations present in the solution; toroidal shapes are favoured by divalent ions.

2.2.3.2. Liquid crystalline phases of nucleic acids: In concentrated solutions of DNA (up to 10 mg/ml) or in the presence of high salt concentrations, a condensation of DNA can occur which is characterized by the formation of an anisotropic phase. This condensation was interpreted as far back as the early 1960s as being due to the rise of liquid crystalline mesophases (see ref. 62 and references therein). Several authors [63–67] described the formation of liquid crystalline mesophases in concentrated solutions of relatively short DNA molecules (molecular mass $1 \times 10^5 - 1 \times 10^6$). On the basis of typical fingerprint-like textures, i.e. periodically alternating light and dark bands, observed under optical microscope in polarized light and crossed nicols, it was suggested that the DNA mesophase is structurally analogous to the cholesteric phase (Fig. 13).

More insight into the properties of the liquid crystalline mesophases of DNA was obtained in systems in which condensation was induced in a diluted DNA solution by exposing DNA molecules to different chemical environments. These systems were studied in several laboratories and include polyethylene glycol or polyethylene oxide + salt, synthetic polypeptides + salt, polyamines, organic solvents + salts, and transition–metal ions + polylysine (see ref. 68 and references therein).

The macroscopic condensed state of DNA obtained in solutions containing high salt concentration and polyethylene oxide [69] or polyethylene glycol [70] is known under the acronymic name Ψ-DNA (Ψ standing for **p**olymer- and **s**alt-**i**nduced condensation). One of the characteristic features of this condensed form is a large negative band in the circular dichroism (CD) spectra in the region of 265–270 nm, i.e. in the region where the native B- or A-forms of DNA yield a much less intense positive band. In a more recent detailed study examining the properties of DNA in a system containing different concentrations of ethanol and NaCl [68], two different types of condensed DNA, Ψ^- and Ψ^+ state could be transformed into the Ψ^- state by changing the composition of the medium. It was demonstrated that the condensation was not accompanied by a change in the secondary structure of DNA. In fact, similar condensation was also observed for double- or triple-stranded synthetic polyribonucleotides and double-stranded RNA (see refs. 71 and 62).

The properties of condensed DNA described above, together with the characteristic fingerprint-like optical textures, indicate once again the pre-

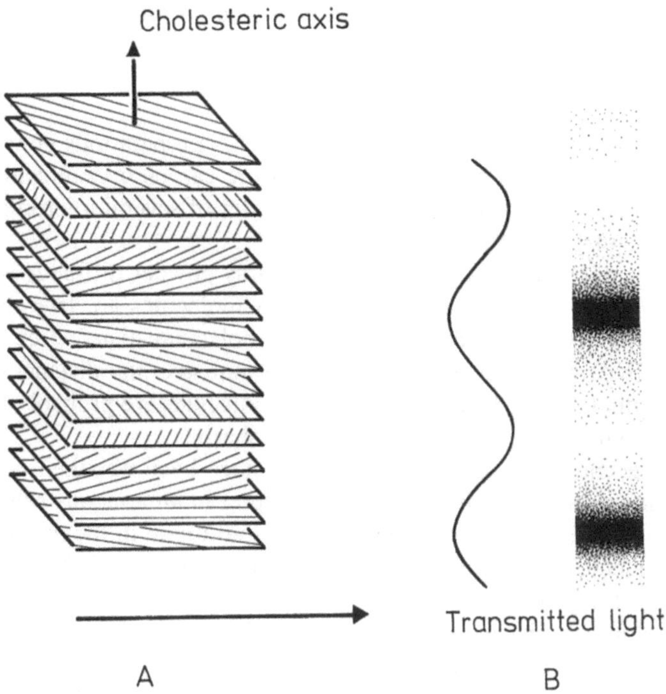

Fig. 13. Schematic illustration of cholesteric liquid-crystalline phase. DNA molecules are located parallel in each plane and their orientation rotates by a constant angle when going from one plane to the next one (A). Viewed between crossed nicols, light is transmitted only by molecules oriented perpendicular to the light beam; molecules rotated by 90° extinguish the light completely. The cholesteric phase appears as a pattern of alternating light and dark bands (B).

sence of the cholesteric liquid crystalline mesophase. DNA molecules are packed parallel in equidistant layers perpendicular to the cholesteric axis, and their orientation changes by a constant angle from one layer to another (Fig. 13). The intermolecular distances and the sense and pitch of the helix depend on environmental conditions (i.e. on the composition of the solution and temperature) and the polynucleotide nature. The lowest intermolecular distances, determined by means of low-angle X-ray scattering, are close to the value of the diameter of the DNA double helix, i.e. about 2.5 nm. The pitch of the helix can vary between 0.5 and 10 μm [62].

Since the DNA molecule contains chromophores, i.e. purine and pyrimidine bases, absorbing in the UV region with a maximum at about 260 nm, the sign of the abnormally intense CD band induced in the UV region depends not only on the sense of the cholesteric helix but also on the orientation of the chromophoric units with respect to the long axis of the DNA helix. The changes in CD spectra can thus be used for monitoring changes in the DNA secondary structure induced by external factors, e.g. by the binding of ligands [72].

2.2.3.3. Lamellar microcrystals of DNA: DNA preparations consisting of molecules fragmented by ultrasound exposure can be crystallized to form hexagonal lamellar microcrystals of a diameter of several micrometers and a thickness of 160–240 nm [73].

Electron microscopic observation and X-ray powder diffraction analysis of the microcrystals have shown that DNA molecules are laterally close-packed in a hexagonal array with an intermolecular distance of 2.39 nm. Shorter DNA molecules are nearly straight and span the distance between the hexagonal crystal faces. Longer DNA molecules are probably folded to fit the crystal thickness, but the bend is not sharp.

More recently, liquid crystalline phases have been obtained with DNA preparations containing molecules 50 nm long in an aqueous salt solution [74], which exhibited a structure similar to that found in the microcrystals. DNA molecules were aligned in columns with their axes parallel, showing a hexagonal array on cross-section. X-ray diffraction analysis of the hexagonal columnar phase yielded a value of 2.97 nm for the interhelical distance. No longitudinal order of the DNA helices was found, which excluded the possibility of a smectic-type layer arrangement.

These experiments indicate that DNA can form a continuum of condensed structures depending on environmental conditions and the length of the DNA molecules, which can serve as a reasonable model for condensed DNA structures found *in vivo*.

2.2.3.4. Chromatin and chromosomes: As has already been mentioned, packing of the long DNA molecule representing the bacterial genome into a cell of much smaller dimensions requires considerable compaction. Prokaryotes, including bacteria, represent the simplest case of the higher-order DNA structure in a cell (see ref. 75 and references therein).

In bacteria, DNA exists in a superhelical form and is divided into several domains, which behave independently as regards topological parameters. All topological alterations occurring during cell life, e.g. DNA reduplication, repair or recombination, are confined to a particular domain. *In vivo* there are 43 ± 10 domains per bacterial chromosome, which is equivalent to the DNA content in a cell, each domain containing 30–100 kilobases.

Early results indicated that the domain boundaries in bacterial nucleoids were delimited by interactions of DNA with RNA. However, it was shown later [76] that DNA-bound proteins are more likely to be responsible for the stabilization of the condensed domain structure. Even though no direct evidence exists, it has been suggested that the enzyme topoisomerase II might be at least one of the candidates for this function.

Analysis of the superhelical properties of bacterial DNA showed that only a part of the negatively supercoiled DNA in the domains is unrestrained, the other part being restrained by interactions with proteins, probably in nucleosome-like structures (see below). A group of small, histone-like basic proteins which could be involved in these interactions was discovered in bacteria [77, 78]. Of these, the protein HU is the most abundant in bacterial cells. It binds to DNA reversibly without any base specificity and induces formation of beaded structures on the DNA fibre, resulting in negative supercoiling. Studies of binding of the HU protein to DNA *in vitro* led to the conclusion that DNA wrapping can occur on the surface of a protein core formed of several HU dimers and that the DNA helix pitch is changed from the usual 10.4 to 8.7 base pairs per turn.

In the nuclei of eukaryotic cells DNA is always found associated with proteins that are either directly involved in the organization of the higher-order structures of DNA or participate in processes of preservation or expression of genetic information. In the basic nucleoprotein complex, which can be isolated from eukaryotic cell nuclei, DNA is organized into a condensed regular structure composed of DNA and histones, called chromatin. Chromatin fibres are then packed into higher-order structures, so-called thin (10 nm) and thick (30 nm) fibres, from which chromosomes are constituted (Fig. 14). The appearance of chromatin varies according to the phase of the cell cycle. Whereas in interphase, i.e. the phase in which the processes connected with DNA reduplication and transcription occur, chromatin is present in the nucleus as finely dispersed material, in metaphase it becomes more condensed, and elongated particles called chromosomes are formed. Eukaryotic cells of different species contain variable amounts of these particles, form several to several tens of chromosomes. During mitosis chromosomes can be observed under the optical microscope and characterized as to shape, possible deformations and irregularities indicating genetic defects, and even internal structure (by means of staining techniques producing chromosome banding).

In metaphase chromosomes the densely packed chromatin forms loops, which are attached at adjacent sites to an elongated structure consisting of proteins, termed the scaffold. The loops contain 50–100 kilobases and represent, similar to the organization of the bacterial chromosome, topologically constrained domains.

The system of topologically constrained loops is preserved in the interphase, when the chromosome structures disappear and chromatin is more or less homogeneously dispersed inside the cell nucleus. The loops are anchored to the nuclear matrix, a net-like protein structure, or to the nuclear envelope. The base sequences that interact with the nuclear matrix are probably identical to those interacting with the scaffold in metaphase chromosomes.

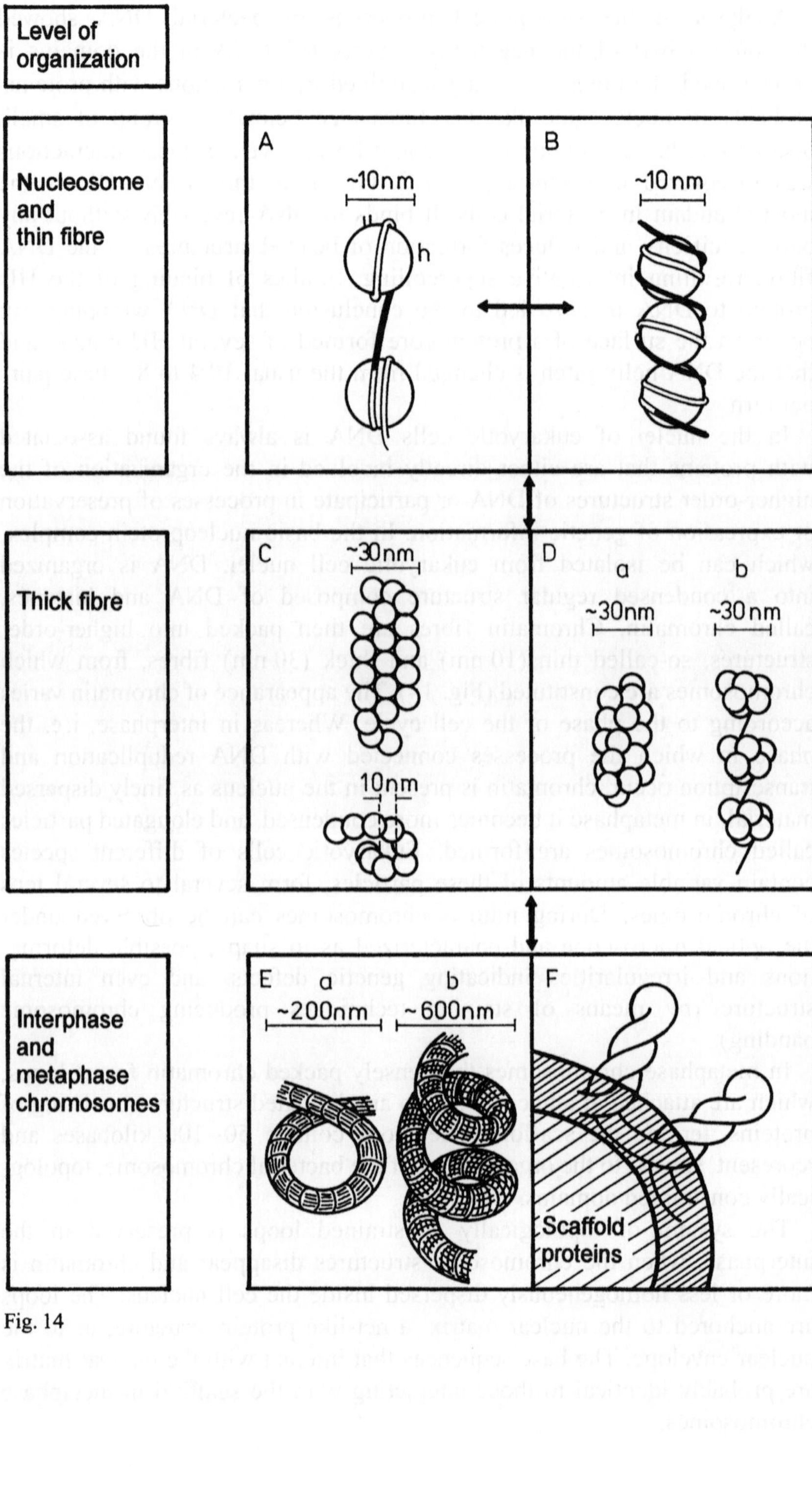
Level of organization
Nucleosome and thin fibre
Thick fibre
Interphase and metaphase chromosomes
A
~10nm
h
B
~10 nm
C
~30nm
10nm
D
a
b
~30nm
~30nm
E
a
b
~200nm
~600nm
F
Scaffold proteins

Fig. 14

The lowest structural unit of chromatin appears under the electron microscope as a fibre of a diameter of about 10.0 nm. In a medium of low ionic strength the fibre becomes extended and a linear array of globular particles called nucleosomes linked by a filamentous structure is observed (Fig. 14A).

The nucleosome particle consists of an octamer aggregate containing pairs of four histones, H2A, H2B, H3 and H4 [79, 80], around which the DNA double helix is wound in supercoil turns. The fifth histone, H1, is associated with the linker regions of DNA. The length of DNA corresponding to one nucleosome subunit is 180–250 base pairs. After thorough digestion of chromatin with nuclease the linker regions with histone H1 are removed and so-called nucleosome core particles are obtained.

The nucleosome core represents the basic unit of chromatin. Its structure was investigated by means of electron microscopy and X-ray and neutron scattering crystallographic methods [81]. These analyses show that the nucleosome core has a wedge-like flat cylindrical shape with a dyad symmetry, about 5.7 nm high and 11.0 nm in diameter. DNA is wrapped around the histone octamer in $1^3/_4$ turns of left-handed supercoil with a helix pitch of 2.7 nm. The length of DNA associated with a nucleosome core is 146 ± 1 base pairs. Energy calculations show that the B-helix of DNA can be smoothly bent into the superhelical coil. It has been suggested that the bending is facilitated by interactions with histones. Histones interact with DNA mainly in its major groove and neutralize phosphate charges on one side of the DNA molecule. Repulsive forces between phosphate negative charges on the other side may then contribute to the bending [82].

Experiments with DNase I digestion led to the conclusion that the number of base pairs per turn in the DNA double helix wound around the histone octamer is 10.0, whereas straight DNA has 11.7 base pairs per turn [81]. Thus, DNA in the nucleosome is overwound with respect to free DNA in solution.

Fig. 14. Schematic representation of various levels of structural organization of chromatin. (A) Nucleosomes in extended state forming „beads on string". Nucleosome cores are connected by segments of linker DNA. (B) Compact arrangement of nucleosomes generating the thin chromatin fibre with a diameter of about 10 nm. h, histone octamer core; –, linker region; =, superhelical stretch of DNA wound on the nucleosome core. (C) Solenoidal model of the thick fibre (30.0 nm). Helically wound thin fibre forms a cylindrical structure with a central hole. (D) Close-packed models of the thick fibre. (a) Continuous arrangement of nucleosomes with helical symmetry. (b) Discontinuous arrangement of nucleosomes in clusters, which can have different types of point group symmetry or be without symmetry. (E) Helical models of interphase and metaphase chromosomes. (a) Thick fibre is wound to form an interphase chromosome with a diameter of ~ 200 nm. (b) Further winding generates a fibre of metaphase chromatid approximately 600 nm in width. (F) A model of chromosome structure. Each chromatid contains a backbone of non-histone proteins (the scaffold), to which the thick fibre is attached in the form of loops.

The fifth histone H1 participates in the packing of nucleosomes into the thin chromatin fibre [83]. Histone H1 interacts with the linker region of DNA and fixes the full turns of DNA in the nucleosome core, inducing formation of the so-called nucleosome filament. An increase in ionic strength induces a denser configuration of the fibre, in which the length of linker segments is reduced and the nucleosomes become close-packed. This structure is termed thin or 10 nm fibre (Fig. 14B).

The thin fibre of chromatin can be further compacted into a thick fibre with a diameter of 25–30 nm. Several models have been proposed to explain how further compaction of the thin fibre is achieved (Fig. 14C, D):

(a) In *a solenoid model* [84] the thin fibre is wound into a helical structure, either right- or left-handed, with three to six nucleosomes per turn depending on the ionic strength. Its diameter varies between the required 25–30 nm, and the pitch of the helix is 11 nm. In the centre of the solenoid is a hole of approximately 10 nm in diameter in which histones H1 are located. It is assumed that interactions between histone H1 molecules stabilize the solenoid structure.

(b) *The close-packed model* describes the thick fibre as a tight arrangement of nucleosomes without the central hole [85]. The packing can be either continuous, with helical symmetry, or discontinuous. In the latter case the clusters of nucleotides can possess different types of point group symmetry or be without symmetry.

(c) In a more recent model the thick fibre is composed of nucleosome dimers formed by *a zig-zag folding of the thin fibre*. A helical arrangement of these dimers produces a cylindrical structure in which DNA linkers are located in the centre [86]. The fibre is stabilized again by interactions between histones H1 which, however, are located on the surface of the fibre.

It has been proposed that in metaphase chromosomes the thick fibre is further compacted into a helical structure of about 60 nm in diameter [87], which is attached to a protein scaffold forming the central part of the chromosome (Fig. 14E, F).

2.3. Conformation of RNA

The expression of genetic information stored in DNA depends in large part on the closely related nucleic acid RNA. The base sequence on one strand of the DNA helix is transcribed into a single strand of messenger RNA. Transfer RNA molecules read the sequence and bring the specific amino acids to ribosomes to be assembled into a protein chain. The ribosomes themselves are composed partly of ribosomal RNAs, which are major components of the protein synthesis machinery. Apart from a genomic role in some viruses, RNAs can also act as enzymes, i.e. RNAs are themselves formed by an RNA-mediated processing of their primary transcripts [88–90]. This functional complexity of RNA is associated with its large conformational diversity.

2.3.1. RNA double helices: Naturally occurring RNAs are almost always single-stranded; nevertheless, almost every RNA molecule has many short double-helical regions. The two segments of a polyribonucleotide chain within a hairpin fold can be in the correct antiparallel orientation to form complementary base pairs. In such globular RNA molecules (transfer RNA, ribosomal RNA and genes coding coat-proteins of some bacteriophages), short double-helical regions are linked with single-stranded segments. RNAs can also form long double-stranded molecules (viral RNAs and their replicative forms, the complexes of synthetic polyribonucleotides).

2.3.1.1 A- and Z-RNAs. One of the two differences in the chemical composition of DNA and RNA is that the RNA backbone contains ribose sugar rings instead of the deoxyribose rings present in the backbone of DNA. An important consequence of the presence of the extra hydroxyl group attached to C(2′) of the sugar ring in RNA is the inability of RNA to coil into the B-type double helix. In a hypothetical B-form of RNA, the extra hydroxyl group cannot fit into the space allotted for it in the B-type duplex for steric reasons. Long double-stranded RNAs and self-complementary double-helical loops and coils of RNA molecules should therefore be variations of an A- or Z-helix.

At low ionic strength A-RNA prevails, which has the following structural parameters: 11 base pairs per helix turn, a pitch of 3.0 nm, helix rise per base pair of 0.27–0.28 nm, and an inclination of base normals to the helix axis in the range of 16°–19°. If the ionic strength is increased, A′-RNA appears, which has 12 base pairs per helix turn, a pitch of 3.6 nm, helix rise per base pair of 0.30 nm, and an inclination of base normals to helix axis of only 10°. Both A- and A′-RNA helices contain Watson-Crick base pairs and are right-handed. The ribose conformation is always 3′-*endo*. The distance between base pairs and the helix axis is 0.44 nm, so that the major groove is deep and the minor groove is narrow and shallow.

The transition from the right-handed A-form to the left-handed Z-form of linear double-stranded RNA occurs in a synthetic alternating copolymer poly[r(C-G)]·poly[r(C-G)] in 6 M $NaClO_4$ at 45 °C [91, 92].

This Z-RNA has many structural similarities to Z-DNA. Guanosine residues are in *syn* conformation and the purine–pyrimidine and pyrimidine–purine sequences have very different phosphodiester backbone conformations. Bromination stabilizes the Z-RNA at a much lower ionic strength, which makes investigation of Z-RNA under physiological conditions possible [93]. The left-handed Z-RNA is stable and is specifically recognized by proteins at physiological temperature and ionic strength. The comparison of Z-RNA with Z-DNA reveals that there may be differences in the glycosidic torsion angles.

2.3.1.2. DNA–RNA hybrids: Double helices composed of one strand of DNA and one strand of RNA occur *in vivo* when DNA base sequences are transcribed into complementary messenger RNA. New RNA is synthesized from only one strand of DNA after the separation of polydeoxyribonucleotide chains. Similarly, such hybrids occur if DNA is synthesized from a RNA strand during reverse transcription.

A-DNA and A-RNA have very similar conformational parameters so that it is not surprising that both RNA and DNA polynucleotide chains adopt the A-conformation in the hybrids. DNA-RNA hybrids resist general conversion into the B-form. This is very probably because the extra hydroxyl group in the C(2′) position of the ribose residue in the RNA strand can barely fit the space allotted for it in the B-form.

If hybrids of homopolynucleotides are formed, for instance from polydeoxyriboadenylic and polyribouridylic acids, then under special conditions triplexes can be formed. The conformation of the single strands in these triple helices, however, once again adopts, the features of the A-conformation. Periodicity in DNA–RNA hybrids is 10.9 base pairs per helix turn.

2.3.1.3. Pseudoknotted structures and bulged bases: The double helix is not the only type of secondary structure exhibited by RNA. For instance, a pseudoknot can be described as the alternating linear sequence [94–96], (Fig. 15]:

halfstem1 – loop – halfstem2 – halfstem1′ – loop – halfstem2′

The two stems fold in such a way that they form a continuous double helix with stacked base pairs [95].

Bulges (unopposed bases in one strand of a duplex) introduce pronounced kinks into RNA and into DNA–RNA double helices, which bend the helix axis by an angle that depends on the number and type of unpaired bases [97]. Bulge-induced kinks in double-stranded RNAs are significant in the tertiary

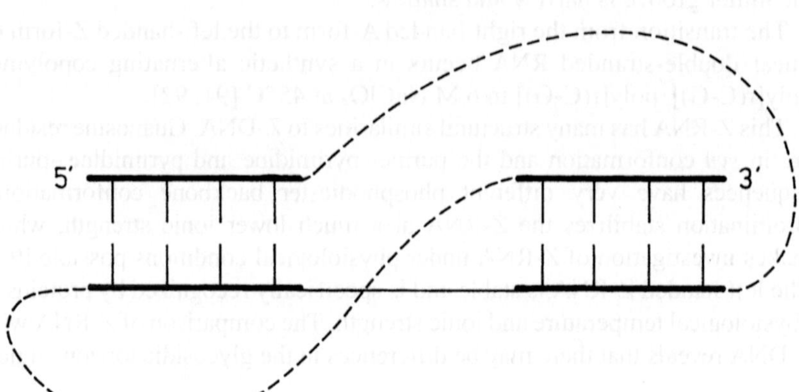

Fig. 15. Pseudoknot secondary structure of RNA.

structure folding of these nucleic acids and function as potential recognition sites for RNA-binding proteins.

2.3.2. Higher organization of RNA: As has already been mentioned, RNA exists in living organisms in a single-stranded form with the strand folded back onto itself, forming helical double-stranded segments separated by loops of non-hydrogen-bonded bases. The functionally active RNAs, which assume specific tertiary structures, are much shorter than DNA molecules even in the simplest organisms. *Transfer RNAs* (abbreviated tRNAs), which are mediators transporting monomeric amino acids to the site of synthesis of a polypeptide chain, are 74–95 bases long. *Ribosomal RNAs* (abbreviated rRNAs), comprise several types of molecules ranging in length from 120 to about 4,700 bases and are constituents of the ribosomes on which protein synthesis takes place on the template of a single strand of *messenger RNA* (abbreviation mRNA).

2.3.2.1. Transfer RNA: tRNAs represent the smallest nucleic acid molecules found *in vivo*. One of them, phenylalanine-transporting tRNA ($tRNA^{Phe}$) of yeast, was the first nucleic acid molecule successfully analysed by X-ray crystallographic techniques at atomic resolution [98, 99]. Structures of other tRNAs have been determined since then and the present knowledge of tRNA structures, physico-chemical properties, and structure-function aspects is quite extensive. It is beyond the scope of this article to discuss the details of tRNA structures. Only a brief outline of tRNA spatial organization will be described and the reader is referred to more exhaustive review articles or books on the subject [100–102].

It follows from the function of tRNA that each cell should contain at least 20 different tRNAs, one type for each amino acid. Usually each amino acid is represented by several tRNAs, which are called isoaccepting tRNAs. tRNAs must fulfill two functions: (i) to serve as an acceptor of a specific amino acid and (ii) to interact with a corresponding codon on mRNA to bring the amino acid to the site of the nascent polypeptide chain. Whereas the second function is ensured by a specific base sequence, which is able to make hydrogen-bonded pairs with a particular codon in mRNA (the socalled anticodon), much less is known about the mechanism by which an enzyme synthetase (specific for each amino acid) recognizes the right tRNA molecule in the process of aminoacylation. It is assumed that the enzyme contacts specific base locations in the tRNA molecule, and thus the tertiary structure of tRNA is also important for the recognition.

The general features of tRNA structures are based on analyses of about 200 tRNA sequences. All sequences can be arranged in a so-called cloverleaf structure (Fig. 16 A). This has four main arms, which are named for their function or characteristic invariant structural feature. The *acceptor arm* is a helical stem of seven base pairs terminated by an unpaired

–C–C–A sequence, which becomes aminoacylated at the free 2′- or 3′-OH group. The other three arms consist of a double-helical stem four base pairs long and a loop of unpaired bases. These are termed *anticodon arm* (its loop contains the anticodon sequence at its centre); the *T arm* (its loop contains ribothymidine); and the *D arm* (its loop contains dihydrouridine). Between the anticodon and T arms lies the so-called *extra arm*, usually a loop of variable length. tRNAs of class 1, which represent abut 75% of known tRNA molecules, have a short extra arm (3–5 bases). tRNAs of class 2 have a long extra arm (13–21 bases). Another variable structure is the D loop, whose length can vary by up to four bases. Besides the two unusual bases mentioned tRNAs contain other modified bases, which can amount to up to 10% of base residues. The modified bases are produced from the four usual bases after their incorporation into the tRNA chain.

The cloverleaf structure is folded into an L-shaped tertiary structure (Fig. 16B, C). The anticodon and D arms form one leg, the acceptor and T arms the other. The structure is stabilized by base-base interactions mainly in the middle part. The interactions include base stacking, base intercalation, unusual base pairing, and formation of base triplets. Bases incorporated into the helical stems adopt the classical A-RNA conformation, whereas nucleotides in the loops often exhibit different, 2′-*endo* puckering. Sharp turns in the folded L-shape structure are realized by base-phosphate interactions. The majority of bases are involved in stacking interactions (with the exception of the nonplanar dihydrouracils, the terminal adenine and a few others), which give a rigid conformation to the loops. This is especially important for the anticodon loop, in which the anticodon sequence can easily form a minihelix with the corresponding codon sequence in mRNA.

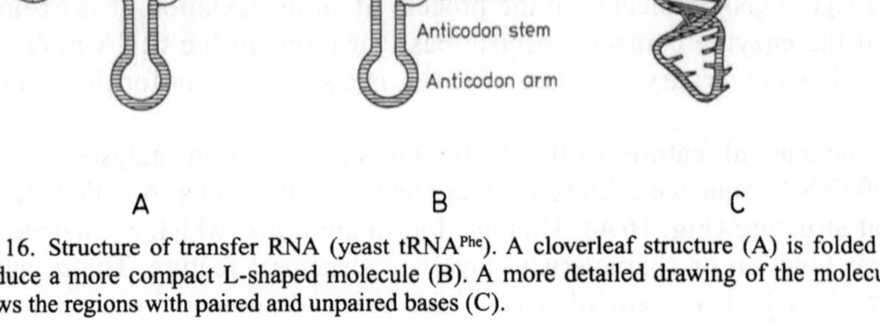

Fig. 16. Structure of transfer RNA (yeast $tRNA^{Phe}$). A cloverleaf structure (A) is folded to produce a more compact L-shaped molecule (B). A more detailed drawing of the molecule shows the regions with paired and unpaired bases (C).

2.3.2.2. Ribosomes: The largest fraction of cellular RNA is organized in compact ribonucleoprotein particles called *ribosomes*. Ribosomes control the transcription of a sequence of codons borne by mRNA into a sequence of amino acid residues in a synthesized protein molecule. The ribosome is attached to mRNA from the 5′ end and moves along it. Two aminoacyl-tRNA molecules associate with the ribosome–mRNA complex, simultaneously bringing together amino acid residues corresponding to two successive codons, and the formation of the peptide bond is realized on the ribosome particle. Several ribosomes move along one mRNA molecule forming a polyribosome (for a review see ref. 103).

The structural organization of ribosomes is preserved in all organisms. Only relatively small differences are observed in the size and composition of cytoplasmic ribosomes. Much greater variability is observed in the ribosomes of cellular organelles, which differ considerably even from the cytoplasmic ribosomes of the same cell. Each ribosome consists of two subunits, a small and a large one. (The size of the subunits and of the rRNA contained in them is standardly characterized by the corresponding sedimentation coefficients expressed in Svedbergs [S].) In total a eukaryotic cytoplasmic ribosome contains 60% RNA (one molecule in the small subunit and three molecules in the large one) and 40% protein (33 molecules in the small subunit and 49 molecules in the large one). For illustration, the composition of a ribosome of the bacterium *E. coli* is given in Table 2. A schematic drawing of a bacterial ribosome particle based on electron microscopic visualization of ribosomes is shown in Figure 17. The details of the interaction of a ribosomal particle with mRNA and tRNA are still a matter of speculation.

Models of the secondary structure of the large 16 S and 23 S rRNA molecules of a bacterial ribosome in free state were constructed on the basis of known base sequences. The structures follow the general pattern for the folding of the single strand of RNA. The complicated structure contains double-helical regions interspersed with numerous loops and bulges. About one half of the bases is paired, but the helical regions are not longer than eight base pairs. A few of the bases are methylated (about 0.6% and 2% in bacterial and mammalian cells, respectively). A much higher amount of base-paired segments is possible in the small 5 S rRNA molecule. Only a little is known about the folding of these structures in the ribosome subunits. There is an indication, however, that the rRNA conformation changes during the processes taking place on the ribosomal particles.

A map of the binding of protein to rRNA was constructed on the basis of nuclease digestion of the ribonucleoprotein complex. It can be said generally that proteins are bound to double-helical regions containing bulges. Each protein binds to a specific region and some could be identified with certain structural parts of the ribosome. Separate active centres were identified on the ribosome, each having specific functions in the process of protein synthesis on the mRNA template.

Table 2. Composition of the *E. coli* ribosome

	Ribosome	Small subunit	Large subunit
Sedimentation coefficient	70 S	30 S	50 S
RNA			
proportion	66%	60%	70%
sedimentation coefficients		16 S	23 S 5 S
Proteins			
proportion	34%	40%	30%
number		21	31

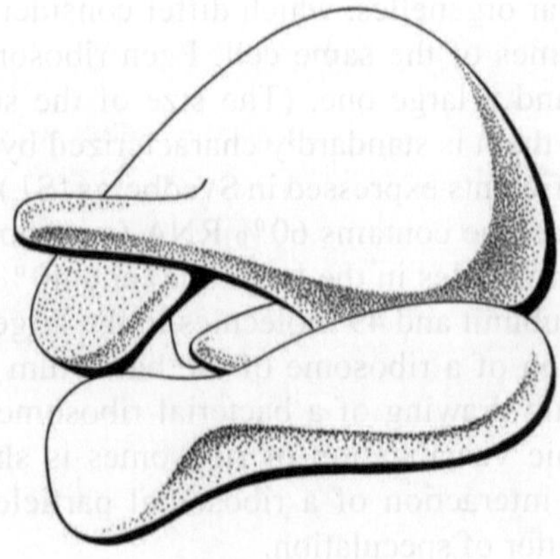

Fig. 17. A schematic drawing of a bacterial 70 S ribosome composed of two subunits (50 S and 30 S). The size is approximately 15 × 20 × 20 nm.

3. Chemical reactivity of nucleic acids

So many chemical reactions of nucleic acids have so far been described that a description of them all would extend the scope of this article too far. Chemical modifications of nucleic acids have already been summarized in several review articles [104–110]. In this section attention is given to two areas of chemical reactions of nucleic acids: (i) chemical probing of the conformational variability of nucleic acids and (ii) the action of so-called chemical nucleases. These areas have been chosen for discussion because their development and application in molecular biology are now particularly intense.

3.1. Chemical probes of nucleic acid structures

Specific reactions of different chemical or, under some circumstances, physical agents with the constituents of nucleic acids, i.e. the heterocyclic bases adenine, guanine, cytosine, thymine, and uracil, and the sugar–phosphate backbone, can be exploited in the analysis of conformational characteristics of polynucleotides. Depending on the chemistry of the reaction, chemical probes can reveal the presence of defects in the regular double-helical structure, e.g. the occurrence of single-stranded regions, so-called unusual structures (loops, cruciforms, B-Z junctions, or triple-stranded helices), or even small distortions of the double helix that need not be connected with the disruption of hydrogen bonds in the base pairs. These conformational irregularities may be either intrinsic properties of the polynucleotide molecule originating, for example, in a specific nucleotide sequence or superhelical winding, or be induced by interactions with external factors such as low-molecular ligands, proteins (or more generally polypeptides), or environmental agents. The information sought includes the topological and sequence specificity of both the intrinsic conformational irregularities and conformational changes induced by the interaction with different types of ligands or other external factors. In favourable cases structural and chemical characterization of these changes can be obtained.

Among the techniques most frequently used in analysing probing experiments are detection of specific probe binding (i.e. the ascertainment of the availability of the particular binding site on the polynucleotide molecule for the reaction); interference experiments (in these experiments competition between a probe binding and ligand interaction with a polynucleotide is investigated); and the footprinting technique, in which a probe acts as a specific "chemical nuclease". Footprinting can, however, be used generally for localization of chemical probe binding to DNA. Chemical nucleases and the footprinting technique will be discussed in section 3.2.

Over the last ten years quite a high number of probes have been designed and exploited in studies of nucleic acid structure. The discussion here will be limited to the chemical probes that are most widely used and that have been demonstrated to have certain advantages in studies on the structure and interactions of nucleic acids.

The probes may be divided into two groups: (i) single strand-selective probes and (ii) probes reacting with double-stranded polynucleotides. However, specificity very often depends on reaction conditions and the single strand-selective probe can, after reacting with a double-stranded polynucleotide, induce a distortion in its regular structure. It should also be borne in mind that single strand selectivity also includes reactivity with the unusual structures mentioned above.

3.1.1. Single-strand selective probes:

Diethyl pyrocarbonate (DEPC, Fig. 18A) carbethoxylates purines in the N(7) positions (Fig. 18B), even though under specific conditions reaction at other sites and with pyrimidine bases can be observed [111]. The reaction of DEPC is more efficient in those polynucleotide structures in which the N(7) of purines is more exposed for carbethoxylation, i.e. in single-stranded DNA and RNA, and in left-handed Z-DNA [112–114]. Intercalative binding of other ligands into the DNA double helix also increases its reactivity towards DEPC, most probably due to DNA unwinding in the vicinity of the intercalation site [115–117].

The carbethoxylation of N(7) of purines results in the ring opening [118], which labilizes the N(9) glycosidic bond. Treatment with alkali leads to strand scission at the reaction site, which makes it possible to detect the modified bases by DNA sequencing or footprinting techniques.

Terbium ion (Tb^{3+}) can be used with a relatively high selectivity as a probe for guanine residues that are present in perturbed regions of DNA molecules. The interaction results in an enhancement of Tb^{3+} fluorescence intensity, which is used for quantification of the binding.

In aqueous solutions Tb^{3+} exhibits a relatively low level of fluorescence, which is enhanced by energy transfer upon formation of complexes with organic ligands. Among the aromatic constituents of nucleic acids the highest effect was produced by guanine [119, 120]. The lanthanide cation interacts primarily in a nonspecific way with negatively charged phosphate groups and the specific interaction with N(7) guanine residues occurs only

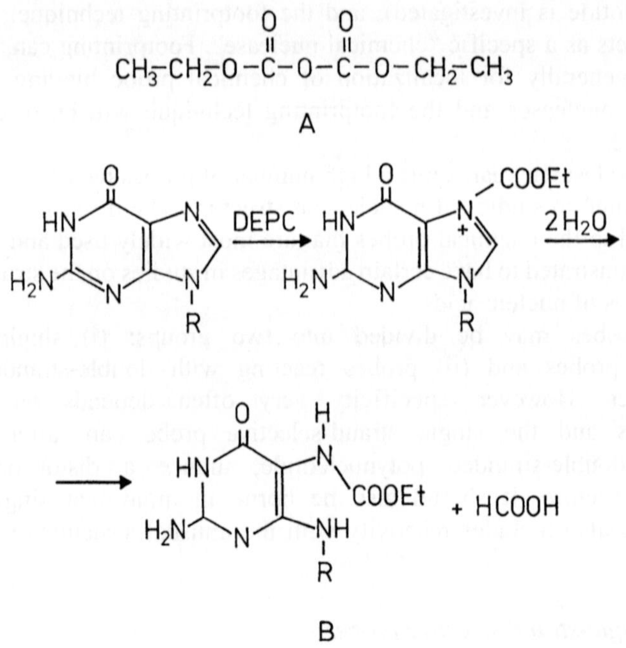

A

B

Fig. 18. Structure of diethylpyrocarbonate (A) and scheme of the carbethoxylation reaction with guanosine (B).

if the base becomes more accessible due to alteration of the secondary structure [121].

Tb^{3+} was used to characterize conformational changes in DNA modified with anticancer platinum compounds, which bind preferentially to guanine bases and distort the DNA structure at these sites [122, 124].

Osmium tetroxide is an effective reagent for *cis* hydroxylation of alkenes. Upon reaction of DNA with OsO_4, Os(VI) esters (Fig. 19 A) are formed by addition to the 5,6 double bond of thymine and, to a lesser extent, of cytosine bases [125, 126]. In probing experiments the reaction is usually carried out in the presence of heterocyclic ligands (e.g. pyridine or bipyridyl), which affect the product structure, hydrolytic stability and kinetics of formation [126].

Osmium tetroxide in the presence of an excess of pyrimidine (Os,py) reacts most readily with thymine residues in those regions of DNA molecules in which pyrimidine 5,6 double bonds became accessible to the rela-

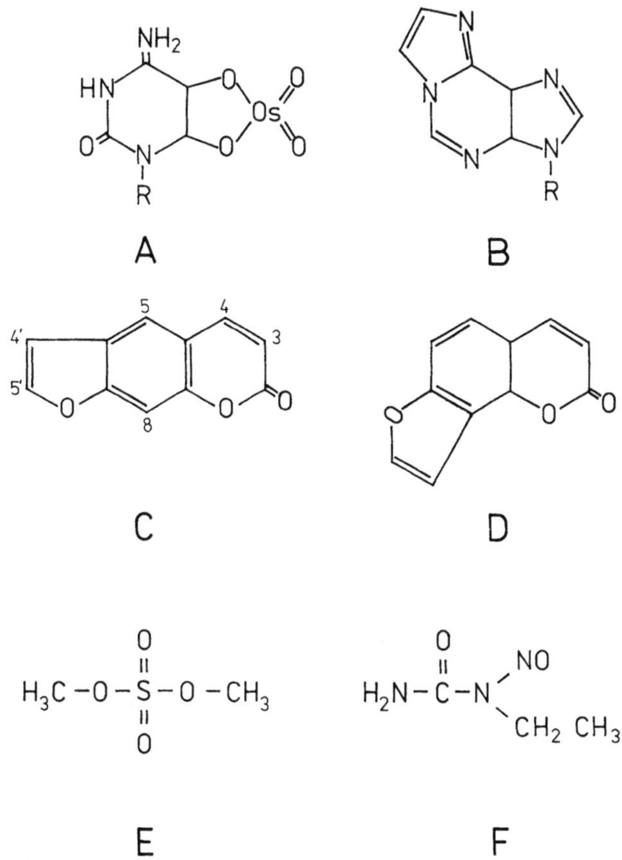

Fig. 19. Structures of thymine Os(VI) ester (A), ethenoadenosine (B), psoralen (C), isopsoralen (D), dimethyl sulphate (E), and N-ethyl-N-nitrosourea (F).

tively bulky Os,py probe due to alteration in the helix structure [127]. These local conformational changes include, for example, helix distortions in the vicinity of single-strand breaks, single-base mismatches, base unstacking in the four-way junction in cruciform structures, premelting changes, formation of local denatured regions, etc.

The binding of osmium can be detected either by means of polarographic techniques [128] or, at single nucleotide resolution, by methods based on labilization of the sugar–phosphate backbone at the binding site to piperidine cleavage [125] or on the ability of the reaction product to terminate transcription of the DNA molecule [129]. In the probing experiments with Os,py the reaction conditions must be chosen properly to prevent induction of secondary structural changes at the site of initial Os,py attachment [127].

Chloroacetaldehyde (CAA) or *bromoacetaldehyde* (BAA) react with N1 and the C6 amino group of adenine, and N3 and the C4 amino group of cytosine, forming cyclic ethenoadenine (Fig. 19B), which exhibits relatively strong fluorescence at about 410 nm [130]. Since these groups are involved in the formation of hydrogen bonds in base pairs, the reaction can take place only if hydrogen bonding is disturbed. Reactions with CAA or BAA have been used for detection of, for example, junctions between B-DNA and Z-DNA [131] and cruciform [132] or triplex structures [133].

3.2.1. Probes reacting with double-stranded polynucleotides:

Psoralens represent a group of photochemical reagents capable of detecting the presence of double-helical structure in nucleic acids [134–136]. Psoralens (or furocoumarins) are natural substances derived from the basic tricyclic structure of psoralen (Fig. 19C). 8-Methoxypsoralen is the most frequently used derivative. Isopsoralens (Fig. 19D) are psoralen stereoisomers with angular geometry.

Both groups of psoralens bind initially to double-helical nucleic acids by intercalation (so-called dark reaction). Irradiation of this complex by long-wavelength ultraviolet light (320–400 nm, the UVA region) induces a photoaddition of the 4′,5′ or 3,4 double bond of psoralens preferentially with the 5,6 double bond of thymine or uracil residues of nucleic acids. Binding of the two reactive sites of psoralen to adjacent thymines in opposite DNA strands results in the formation of a covalent interstrand cross-link [137]. Isopsoralens cannot form cross-links in DNA in diluted solutions due to the angular structure. However, cross-linking with isopsoralens was reported in folded DNA structures [138]. Both the cross-links and monoadducts can be reversed by irradiation with short-wavelength ultraviolet light.

The photochemistry of psoralens makes it possible to probe for double-helical structures in both DNA and RNA molecules. The covalent cross-links are stable under a broad range of conditions and their positions can be mapped even after DNA or RNA isolation, if the cross-linking reaction was carried out *in vivo*. The position of the cross-links can be determined under conditions where the two strands are separated except at the cross-linked sites. Either electron microscopy of nucleic acid molecules under denaturing conditions [139] or enzymatic techniques, which can yield the sequence resolution [140], can be exploited.

The use of psoralens for probing the secondary structure of nucleic acids is very wide. For instance, psoralens were used in investigating chromatin structure [141–143]. It was found that psoralens only cross-link DNA regions not complexed with proteins. By means of psoralen cross-linking double-helical regions were detected under certain conditions in single-stranded DNAs of phage *fd* or some viruses and the double-stranded regions in various RNAs were mapped (for a review see references 135 and 136). Psoralen binding was also used in the determination of DNA torsional strain since, as intercalators, psoralens have an enhanced binding affinity to negatively supercoiled DNA [144].

Dimethyl sulphate (DMS, Fig. 19E) methylates nitrogen atoms of bases not involved in hydrogen bonding. In double-stranded B-DNA DMS reacts with N(7) of guanine and N(3) of adenine [145], while in single-stranded DNA N(1) of adenine and, to a lesser extent, N(3) of cytosine can also be methylated [104]. In the Hogsteen type of base pairing (for an example see Fig. 10), guanine N(7) is involved in hydrogen bonding and is not accessible for methylation. DMS can thus be exploited for probing structures involving Hogsteen base pairs, e.g. in protonated triplex structures (Fig. 10) [146]. DMS is often used in interference studies and can identify guanine – ligand contacts in the major groove of DNA [116, 147].

N-ethyl-N-nitrosourea (ENU, Fig. 19F) forms phosphotriesters upon reaction with the phosphates of the DNA backbone. The ethylation sites can be identified by alkaline hydrolysis of the relatively labile phosphotriester [147].

The ethylation reaction is used in interference studies of ligand and protein binding [147, 148]. Introduction of the ethyl group accompanied by the neutralization of the negative charge of the phosphate group decreases the probability of interaction of a cationic ligand with the DNA backbone, and represents a steric hindrance to binding. In the converse type of experiment, the proximity of bound ligands can prevent ethylation of a particular phosphate group. Probing with ENU thus cannot map direct contacts in the DNA–ligand complex, but detects rather the phosphates which become inaccessible to the reaction due to the ligand binding.

3.2. Chemical nucleases

The term nuclease is usually reserved for enzymes that accomplish the scission of the phosphodiester backbone of polynucleotide chains. Nucleases are divided into two fundamental groups according to the mechanism by which they cleave the phosphodiester backbone. Exonucleases clip off terminal nucleotides, while endonucleases preferentially cut internal bonds. It had been initially assumed that endonucleases are generally unspecific, i.e. that they cut phosphodiesteric bonds in polynucleotides regardless of the surrounding nucleotide sequence. However, endonucleases also exist that cleave only at specific nucleotide sequences. These endonucleases are referred to as restriction endonucleases and make nicks only within specific sequences exhibiting twofold symmetry around a given point. An example is given in Figure 20. As the same sequence is found on both strands, restriction nucleases bring about double-strand breaks. Restriction nucleases usually recognize a sequence of four to six nucleotides and the sites of cleavage usually lie within the recognition sequence.

Over the last decade attention has also been focused on synthetic nucleolytic agents that cleave the phosphodiesteric backbone by oxidation of ribose or deoxyribose residues. Ferrous-EDTA free or linked to DNA-binding ligands, a 2:1 1,10-phenanthroline-copper complex or uranyl acetate (Fig. 21) are among the most commonly exploited agents of this kind. Moreover, there are also natural low-molecular mass substances – bleomycin, calicheamycin γ_1^I and neocarzinostatin – that cleave the phosphodiesteric backbone of polynucleotides without hydrolysis. Synthetic nucleolytic agents are referred to as chemical nucleases in spite of the fact that the mechanism of their scission does not involve a catalysis. As shown below, all chemical nucleases exhibit the endonuclease cleaving activity. Their reaction with nucleic acids and their utilization as footprinting reagents and site-specific nucleolytic agents are described in this section. Some agents with nucleolytic activity can also be used as probes of nucleic acid conformation. This application is discussed in section, 3.1., dealing specifically with chemical probes of nucleic acid conformation.

The nuclease function of *ferrous-EDTA* (Fig. 21 A) consists of DNA cutting by hydroxyl radical generated by this complex and H_2O_2 [149]. The

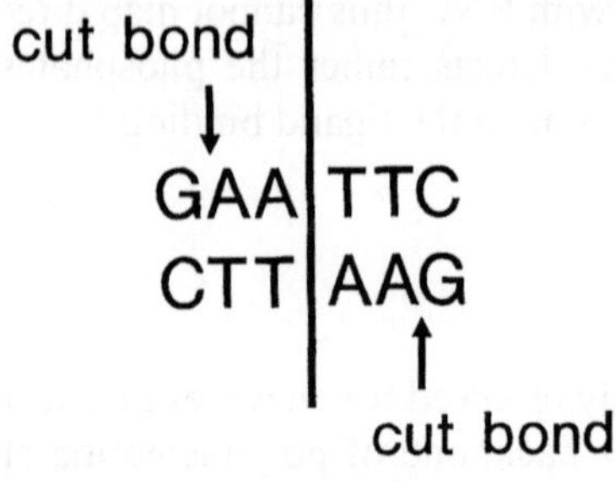

Fig. 20. Recognition sequence of *EcoRI* restriction enzyme.

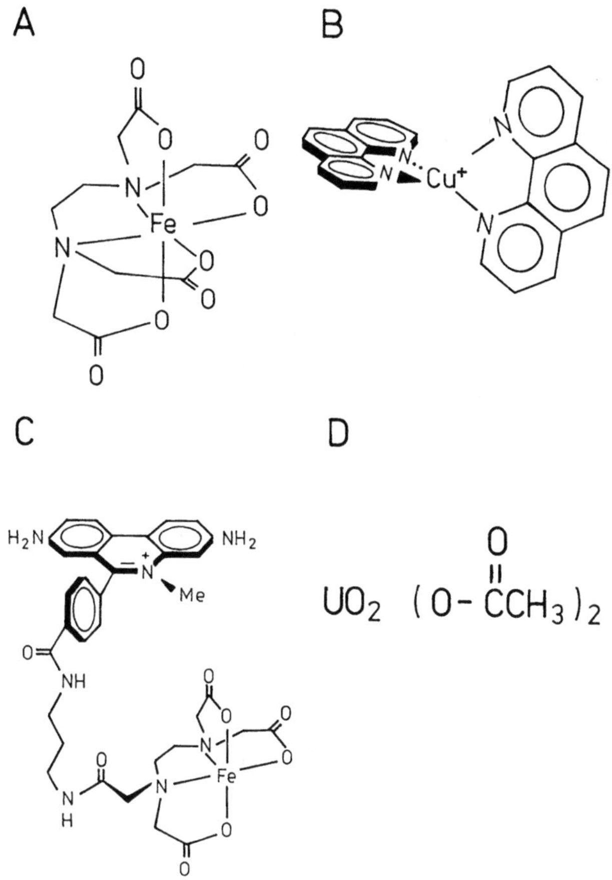

Fig. 21. Structures of synthetic chemical nucleases. (A) Ferrous-EDTA; (B) 2:1 1,10-phenanthroline-cuprous ion; (C) Methidium propyl-EDTA-iron(II); (D) Uranyl acetate.

reaction involves one-electron reduction of hydrogen peroxide by Fe^{2+}, to give $\cdot OH$ and OH^-:

$$Fe^{2+} + H_2O_2 \rightarrow Fe^{3+} + \cdot OH + OH^-$$

The hydroxyl radical makes a single-strand break in DNA by extracting a hydrogen atom from a deoxyribose in the backbone, which then breaks down by secondary reactions, cleaving the chain. The products of the reaction are DNA strands terminated by 5′ and 3′ phosphates flanking the deoxyribose that was attacked. The likely position of the attack is the C(4′) hydrogen of deoxyribose. As ferrous-EDTA is negatively charged, it does not bind to polyanionic DNA. Hence the hydroxyl radical is the only species that interacts with DNA.

Ferrous-EDTA bound to methidium propyl (*methidium propyl-EDTA-Fe^{2+}*, MPE), (Fig. 21C), has also been used as a chemical nuclease [150,

151]. The mechanism of DNA cleavage has the same background as in the case of ferrous-EDTA, i.e. the cleavage results from the hydroxyl radical generated by the ferrous-EDTA moiety in the presence of H_2O_2. The single-strand breaks occur at the sites of double-helical DNA where MPE is bound. A complication in the application of MPE as a chemical nuclease may arise from its sequence-specific binding to DNA. Moreover, the rate of cleavage of DNA with MPE is low so that relatively high concentrations of the nuclease have to be used.

The *2:1 1,10-phenanthroline-copper complex* (Ph-Cu^+), (Fig. 21B), also cleaves DNA through an attack on the deoxyribose [152–154] so that the Ph-Cu^+ reaction should not be specific for the nucleotide at the site of scission. In spite of this fact the scission pattern exhibits sequence-dependent reactivity [155, 156]. The sequence-dependent scission is probably associated with the structural variation of the minor groove. Ph-Cu^+ binds in this groove, where it is oxidized by hydrogen peroxide to form a copper-oxo species. The latter species appears to be responsible for oxidation of the deoxyribose resulting in the cleavage of the polynucleotide strand (Fig. 22).

The products of the reaction are (i) DNA strands that end with the 5′ and 3′ phosphates flanking the deoxyribose that was attacked; (ii) the free base originally linked to the deoxyribose that was attacked; and (iii) 5-methylenefuranone and 3′-phosphoglycolates [152]. The efficiency of Ph-Cu^+ is dependent on the secondary structure of DNA [157]. Denatured DNA and Z-DNA are not cleaved under conditions where B-DNA is cleaved easily. A-DNA is also cleaved but less readily than B-DNA [158].

Light-activated *uranyl acetate* (Fig. 21D) also efficiently cleaves polynucleotide chains [159]. Its use as a chemical nuclease is advantageous in those cases where traces of glycerol quench the scission by ferrous-EDTA (glycerol is commonly used in the isolation and storage of nucleases and other DNA-binding proteins). The photocleavage of DNA by the uranyl ion, UO_2^{2+}, can be accomplished with long-wavelength ultraviolet radiation (300–420 nm). The photocleavage involves oxidation of deoxyribose residues proximal to uranyl ions complexed to phosphates of the DNA backbone.

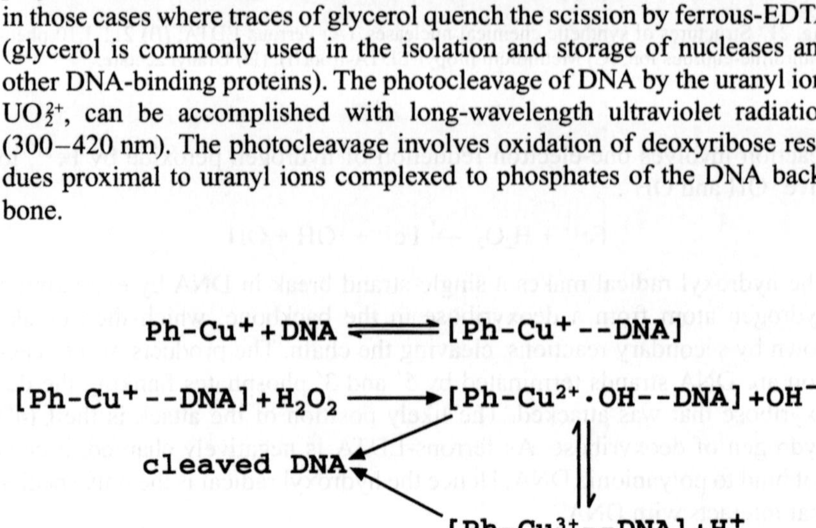

Fig. 22

Detailed information on the sequence specificity of the binding of proteins and drugs to nucleic acids can be obtained by *footprinting techniques* [160, 161]. Nucleic acid footprinting exploits an enzymatic or chemical cleavage of nucleic acid fragments combined with the analysis of the resulting products on the gels used for nucleic acid sequencing. The segments of the nucleic acid molecule to which proteins or drugs are bound can be protected from the scission by a footprinting nuclease so that the corresponding bands in the sequencing ladder are missing. The apparent requirement of the footprinting agents is that they have to react with and cleave all sites of nucleic acids with approximately the same efficiency. In order to demonstrate the advantage of the use of chemical nucleases as footprinting agents, we shall describe first the footprinting procedure exploiting a classical footprinting agent – deoxyribonuclease I (DNase I).

DNase I is a glycoprotein with a molecular mass of 31,000, which cleaves the O(3′)-P bond in polydeoxyribonucleotide chains. The specificity of the cleavage is determined by the conformation of DNA at the site of its cleavage. DNase I recognizes the minor groove in DNA. The most effective cleavage of DNA by this nuclease occurs if the width of this groove is that of DNA in B-Form, i.e. where the base residues are oriented perpendicular to the duplex axis. Deviations from the base tilting result in changes of the minor groove width, which consequently lowers the hydrolytic activity of DNase I. The segments of DNA containing longer sequences of adenine · thymine or guanine · cytosine base pairs are cleaved by DNase I less effectively because these segments have a change in the width of the minor groove. Moreover, many proteins or drugs may be specifically bound to A·T- or G · C-rich segments or their binding may induce conformational alterations not only at the site of their intimate contact with DNA but beyond. Thus the use as footprinting agents of DNase I or other nucleases, even chemical ones, that exhibit some sequential or conformational preference of binding or cleavage, is limited. From this point of view the use of hydroxyl radicals produced by ferrous-EDTA appears to be one of the most suitable techniques for footprinting experiments. This hydroxyl-generating compound cannot itself interact with nucleic acids and the hydroxyl radicals have apparently no sequential or conformational preference for the attack on deoxyribose, which is the primary reaction responsible for polynucleotide chain scission. The use of any of the chemical nucleases as a footprinting reagent presents an undisputed advantage over enzymatic nucleases: because of their small size, chemical nucleases permit scission closer to the edge of a nucleic acid sequence protected by protein or drug binding and therefore give a more precise definition of the binding site.

Synthetic nucleolytic agents can also be used as site-specific endonucleases. In this respect they resemble restriction enzymes. However, none of the *site-specific chemical nucleases* are yet sufficiently efficient or specific to challenge restriction enzymes as reagents to manipulate and analyse DNA. A frequent approach in preparing synthetic nucleolytic agents with site-specific nuclease activities is to link these agents to peptides or proteins [162, 163], oligonucleotides [162, 164], and drugs binding to nucleic acids [165, 166], to target the cutting. The most widely used synthetic nucleolytic agents for site-specific chemical scission of DNA have been ferrous-EDTA and 1,10-phenanthroline-copper.

4. Electric and magnetic properties of nucleic acids

4.1. Electric properties

4.1.1. Electronic charge distribution

4.1.1.1. Electrons σ and π, delocalization: Electromagnetic forces maintain the structure of molecules and govern the processes of chemistry and biochemistry. Because of the long-range nature of coulombic forces, electrostatics play a vital role in intra- and intermolecular interactions. An understanding of these interactions is important because the encounter of molecules is the initial step in the biomolecular processes [167] and requires a knowledge of the electric properties of biologically important molecules.

Those physical properties of nucleic acids that involve electric phenomena are given primarily by the electric charge distribution of bases, nucleosides and nucleotides. Quantum mechanics has made it possible to calculate the electronic structure of molecules with a high degree of accuracy. Every particle or collection of particles has a quantum mechanical wave function that describes the state of the system. The product of the wave function Ψ with its complex conjugate Ψ^* is proportional to the probability density ρ associated with the particle. The probability density ρ is defined so that the probability that a particle is located in a small volume $dx\,dy\,dz$ is represented by $\rho dx\,dy\,dz$. The total molecular wave function Ψ may be resolved into an electronic wave function Ψ_e and a nuclear wave function Ψ_N (Born-Oppenheimer approximation, see pp. 397, 455 in ref. 168):

$$\Psi = \Psi_e \Psi_N \tag{2}$$

This approximation supposes that nuclei, being much heavier than electrons, move much more slowly. Therefore the Schrödinger equation may be solved for the electronic wave function with fixed nuclei. Electron density distributions are obtained directly from the electronic molecular wave functions by integration

$$\rho(r) = \int\int \ldots \Psi_e^* \Psi_e dt_1\, dt_2 \ldots dt_n \tag{3}$$

where dt_i is the volume element for electron i consisting of a space part and a spin part: $dt_i = dx_i\, dy_1\, dz_i\, ds_i$ [169]. Since $\rho(r)$ is a three-dimensional function, it can be tabulated but not easily presented by a graphical plot. In planar molecules the two-dimensional sections containing the molecular plane provide usually the information needed.

When two atomic orbitals of two nuclei are combined two molecular orbitals are formed – a bonding orbital and an antibonding orbital. Bonding between the two nuclei occurs when atomic orbitals overlap. The greater the overlap the greater is the strength of the bond. It is the buildup of electron density between the nuclei that is responsible for the bonding (see p. 467 in ref. 168).

There are two different ways of forming bonding molecular orbitals from atomic *p* orbitals. If the lobes of the *p* orbitals are pointed at each other along the internuclear axis two σ-orbitals are formed. If the lobes of the *p* orbitals are perpendicular to the internuclear axis they can overlap sideways to form π-orbitals (Fig. 23).

The angular momentum of the electron about the internuclear axis is given by $\lambda\hbar$. The bonding molecular orbitals σ are symmetrical around the

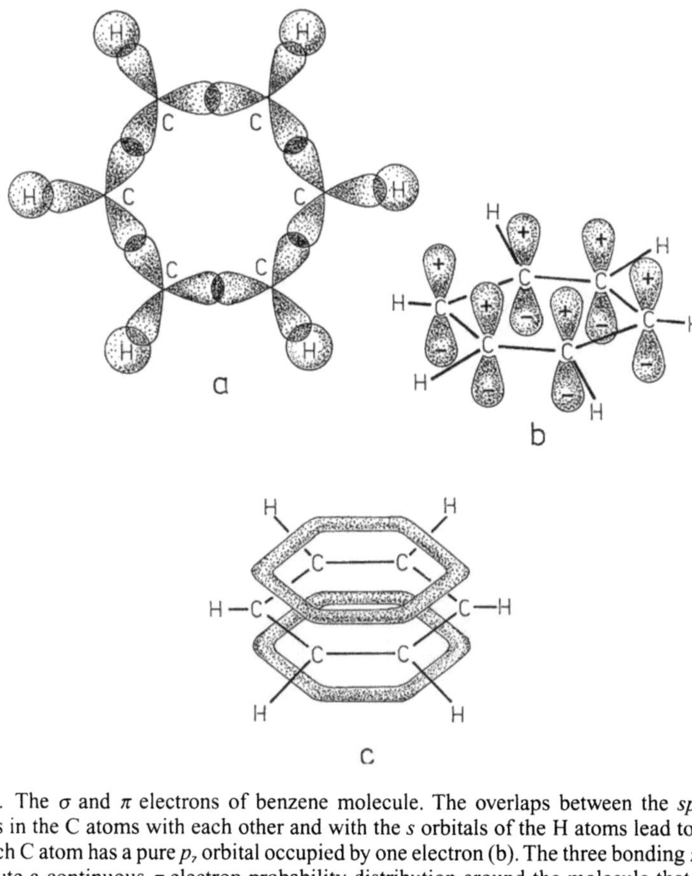

Fig. 23. The σ and π electrons of benzene molecule. The overlaps between the sp^2 hybrid orbitals in the C atoms with each other and with the *s* orbitals of the H atoms lead to σ bonds (a). Each C atom has a pure p_z orbital occupied by one electron (b). The three bonding π orbitals constitute a continuous π electron probability distribution around the molecule that contains six delocalized π electrons (see p. 314 in ref. 170).

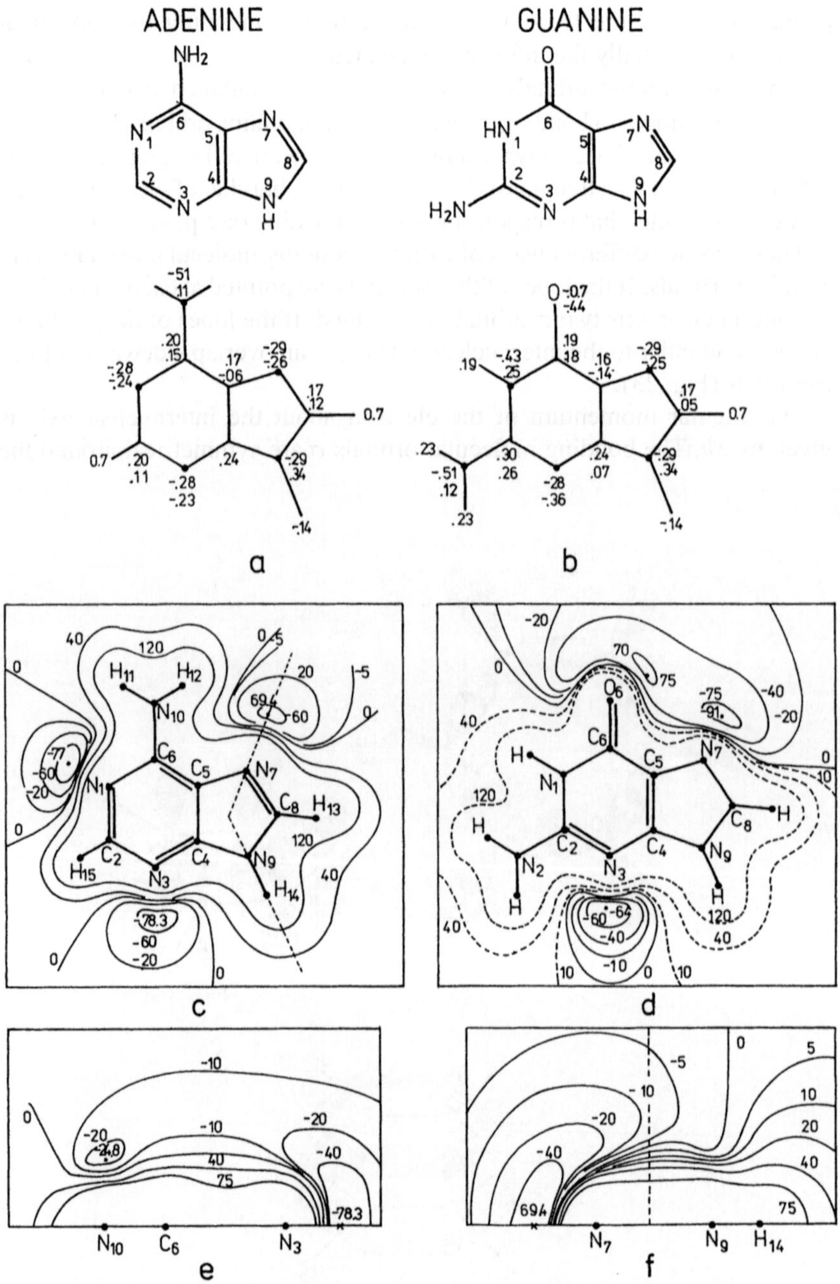

Fig. 24. Adenine and guanine. (a, b) Electronic charge densities (σ charge, lower numbers π charge) (see p. 106 in ref. 40). (c, d) Molecular electrostatic interaction energies (in kcal · mol^{-1}), cut in the plane of the base [185, 186]. (e, f) Molecular electrostatic interaction energies (in kcal · mol^{-1}), cut in the planes perpendicular to the base [186].

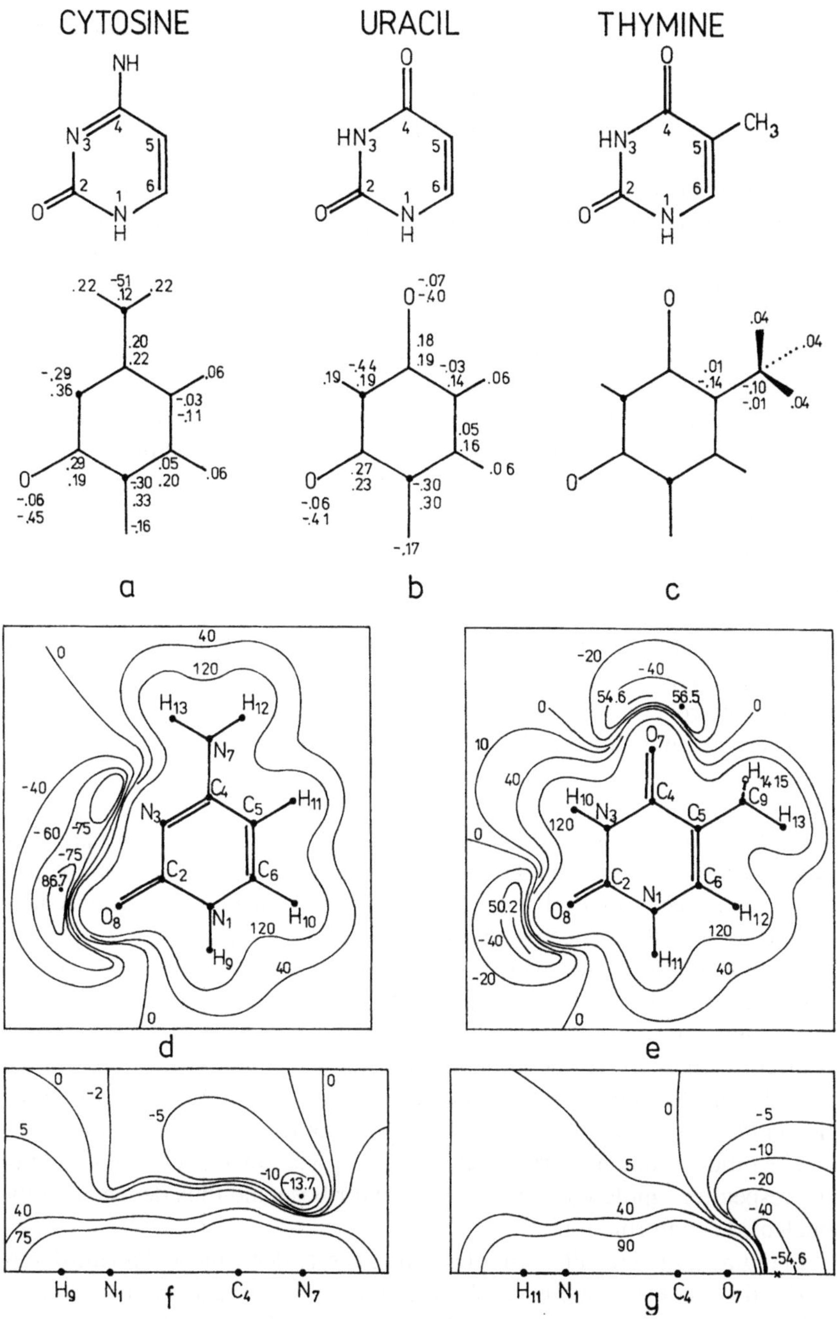

Fig. 25. Cytosine, uracil, and thymine. (a, b, c) Electronic charge densities (σ charge, lower numbers π charge) (see p. 106 in ref. 40). (d, e) Molecular electrostatic interaction energies (in $kcal \cdot mol^{-1}$), cut in the plane of the base [186]. (f, g) Molecular electrostatic interaction energies (in $kcal \cdot mol^{-1}$), cut in the planes perpendicular to the base [186].

internuclear axis and $\lambda = 0$. π orbitals change sign upon a 180° rotation about the bond axis and $\lambda = 1$ (see p. 305 in ref. 170). The bonding molecular orbitals with $\lambda = 2$ are referred to as the δ orbitals (see p. 460 in ref. 168).

The σ electrons form the basic rigid skeleton of localized bonds. The π electrons form a more fluid, single system spreading over the σ skeleton. We cannot speak, in this case, of a strict localization of the π electrons between any pair of adjacent nuclei, the π electrons are "delocalized" (see p. 63 in ref. 171). Figure 23 shows the benzene molecule. The bonding orbitals are occupied by the available six electrons. The result is a continuous electron probability distribution above and below the plane of the ring whose six electrons belong to the molecule as a whole and not to any particular pair of atoms (see p. 326 in ref. 170).

The existence of delocalized π electrons is the most important property of conjugated molecules. The principal physico-chemical, chemical and biochemical properties of such systems are determined by their π electrons. The reason for this is that these electrons are much more mobile than the σ electrons and therefore participate more readily in chemical and biochemical processes (see p. 63 in ref. 171).

Electronic charge densities have been calculated for a number of bases, nucleosides and nucleotides applying either the method of Del Re and Hückel to obtain partitioned σ and π charge contributions (see p. 105 in ref. 40 and references 172 and 173) or PCILO (**P**erturbative **C**onfiguration **I**nteraction using **L**ocalized **O**rbitals) to produce total charges [174–176] (Figs. 24 and 25).

4.1.1.2. Electronic polarizability and dipole moment: The information on charge density distribution contained in Figures 24 and 25 can be summarised into a single vector for each individual base. The orientation and magnitude of this vector give the respective permanent electric dipole moment. In general, the dipole moment lies within the molecular plane of the base and its orientation is indicated by an angle Θ relative to a reference line in the molecule. Figure 26 shows the direction of the dipole moment in the bases, with respect to the C(4)-C(5) bond in purines and the N(1)-C(4) axis in pyrimidines. The angle Θ is positive for anticlockwise and negative for clockwise rotation [177]. The absolute values of the dipole moments are given in Table 3.

The dipole moment of ribonucleosides is given by the vector addition of the respective dipole moments of base and ribose and thus depends on the nucleoside conformation.

In Figure 27 the resultant dipole moments for adenosine and guanosine in *anti* and *syn* conformations are shown. The results of the quantum-mechanical calculations PCILO are presented in Table 3 [177]. The dipole moments of nucleosides are calculated for the glycosidic torsion angle ϕ_{CN}, indicated as the most stable one (see p. 21 in ref. 40). Theoretical calculations, in agreement with experimental evidence [177], point to the predominance of *anti* forms (pp. 69 and 71 in ref. 40). The nucleosides of adenine, cytosine,

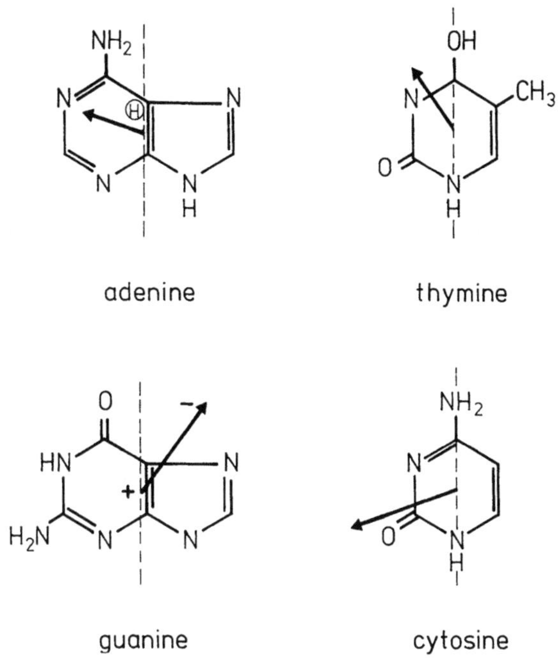

Fig. 26. Permanent electric dipole moments of bases.

uracil, purine and xanthine have a higher dipole moment in the *anti* than in the *syn* conformation. The opposite is true for guanosine and inosine [177]. Theoretical data can be tested by measurements that yield the magnitude of the moment, but its orientation can only be derived if moments of a series of differently substituted compounds of the same molecule are at hand (see p. 106 in ref. 40).

The electronic charge density distribution can be changed by an external field so that the induced dipole moment can arise. The magnitude of the induced dipole moment μ_i is (at low field strengths) a linear function of electric field strength E

$$\mu_i = \alpha_e E \tag{4}$$

where α_e is the electronic polarizability.

π electrons of nucleic acid bases are much more mobile than the σ electrons. Thus the polarizability $\alpha_\perp$ of bases in the direction perpendicular to the plane of the heterocyclic ring is much lower than the polarizability $\alpha_\parallel$ in the direction parallel to the plane of the molecule (see section 4.3.1.). The polarizability anisotropy is defined by

$$\Delta\alpha_e = \alpha_\parallel - \alpha_\perp \tag{5}$$

The average group polarizability (isotropic polarizability) α_e is usually estimated from atomic refractions [178] or by the use of additivity rules [184].

Table 3. Dipole moments and polarizabilities of bases and nucleosides

	Dipole moment				Polarizability volume		
	Calculated				Measured		
	Magnitude [debye]	Direction [deg]	Ref.	[debye]	Ref.	[Å³]	Ref.
Adenine	2.9	64°	[40]	3.0 [a]	[40]	13.9	[177]
	3.1	75°	[177]	3.2	[177]		
	2.8 [b]	88° [b]	[178]				
	2.4		[179]				
Uracil	4.6	36°	[40]	3.9 [a]	[40]	10.2	[1779
	4.8		[179]				
	4.6	38°	[180]	4.16	[180]		
Thymine	3.5	39°	[177]	3.6	[177]	12.0	[177]
	3.5 [b]	33° [b]	[178]				
	4.7		[179]				
	4.4	40°	[180]	4.13	[180]		
Guanine	7.5	–31°	[40]	6.8	[177]	14.4	[177]
	6.9	–36°	[177]				
Cytosine	7.6	102°	[40]	7.2	[177]	11.0	[177]
	7.1	97°	[177]				
	8.1	108°	[178]				
	7.65		[181]	6.5 [c]	[181]		
Adenosine							
anti	5.6		[182]				
syn (ϕ_{CN} = 241°)	1.9						
syn (ϕ_{CN} = 280°)	2.8						
Uridine *anti*	7.0						
Uridine *syn*	3.4		[182]	3.9	[183]	10.2	[183]
Thymidine				3.6	[183]	12.0	[183]
Guanosine *anti*	4.6		[182]				
Guanosine *syn*	10.0		[182]				
Cytidine *anti*	8.8		[182]	7.2	[182]	11.0	[182]
Cytidine *syn*	10.0		[182]				
Deoxycytidine				10.0	[177]		
Inosine				5.2	[183]	13.0	[183]

[a] For 9-methyladenine and 1,3-dimethyluracil, ref. [40], p. 107; [177].
[b] For methylderivatives [178].
[c] For methylderivatives [181].

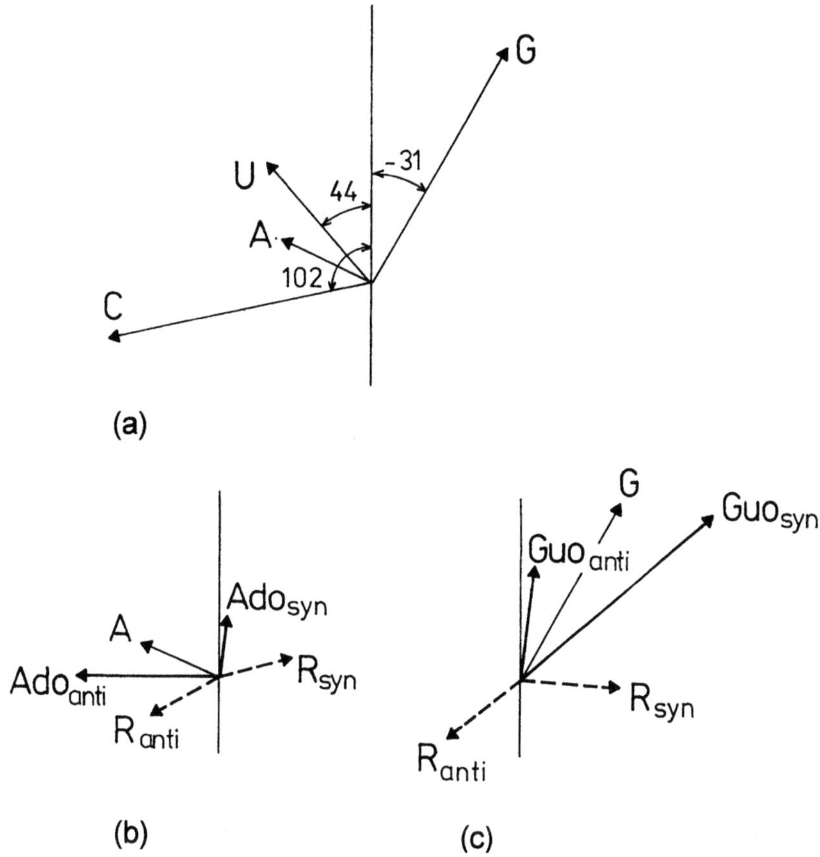

Fig. 27. Permanent electric dipole moments of bases and nucleosides [182]. (a) Dipole moments of adenine (A), cytosine (C), guanine (G) and uracil (U). (b) Dipole moments of adenine (A) and adenosine (Ado) in *anti* and *syn* conformations; component of the dipole moment of ribose in the plane of the base (R). (c) Dipole moments of guanine (G) and guanosine (Guo) in *anti* and *syn* conformations.

The polarizability α is usually converted to a polarizability volume α' [173] by the relation

$$\alpha' = \frac{\alpha}{4_{\|}E_0} \tag{6}$$

The range of the values for α' so obtained for uracil, cytosine and thymine is from 10 to 12 Å^3 and about 14 Å^3 for adenine and guanine, (see p. 465 in ref. 5 and references 178 and 184).

4.1.1.3. Molecular electrostatic potential: The information enclosed in a molecular wave function can be used in the form of the electrostatic potential created in the neighbouring space by the nuclear charges and the electronic distribution. For a given wave function with the corresponding electronic charge density function the average value of such a potential $V(r_i)$ at a given point i in space can be calculated [185].

Since the electrostatic potential is a three-dimensional function, the two-dimensional sections usually in the plane of the base or perpendicular to it use to be presented in the form of maps with equipotential (isopotential) and/or isoenergetic curves. Maps of electrostatic interaction energies with a unit positive charge have been drawn for nucleic acid bases as well as for nucleosides and nucleotides. Figures 24 and 25 show molecular electrostatic interaction energies of some nucleic acid bases [185]. Isoenergetic values are given in kcal/mol. The values of the potential minima at specific important reactive sites of the base pairs were evaluated. The potential and/or energetic minima occur principally in the plane of the bases in the vicinity of their ring nitrogens and carbonyl oxygens, the deepest minimum being generally associated with N(7) of guanine. Out-of-plane minima are also present and they are associated with the amino groups of the bases, carbons C(8) of the purines, and the C(5)-C(6) bonds of pyrimidines (see p. 395 of ref. 186). Electrostatic potential maps have been used successfully in interpreting the reactivity of nucleic acid bases towards a number of electrophilic agents (see p. 382 in ref. 186). While protonating and alkylating agents are charged, the coulombic part of the interaction energy would be dominant [187] and the electrostatic molecular potential can be used for the evaluation of the affinities of the individual sites in DNA towards protonation or alkylation.

The electrostatic molecular potential of a complete double-helical turn of B-DNA at any point in space was obtained by the addition of the molecular potentials at that point produced by all constituent units placed in the appropriate geometry (see p. 389 in ref. 186).

The absolute values of the potential created solely by the sugar-phosphate double-helical backbone without the inner base pairs are very large compared to the corresponding values of a single phosphate unit; this is due to the accumulation of the effects of all the phosphates of the complete double-helical turn. The molecular potentials are quite appreciable at the large distances from the helical axis; the numerical values reflect the inverse distance dependence (Table 4).

As the outer surface of the helix a cylinder with a radius of 1.7 nm was chosen, because at distances larger than 16–17 nm from the axis of the double helix isopotential surfaces are almost perfect cylinders. At shorter distances (e.g. for the isopotential surface of –2,000 kJ/mol at distances between 0.4 nm and 1.02 nm from the helix axis; the 1.02 nm cylinder passes through the most remote phosphate oxygen) the isopotential surfaces present fluctuations so that on any given cylinder variations in potential may reach about 20%, and the surfaces have characteristics of a wreathed column presenting major (M) and minor (m) grooves of different widths and depths (Fig. 28) (see p. 392 in ref. 186). One must come closer to the axis in the major groove to encounter as strong a potential as in the minor groove,

Table 4. Sugar–phosphate backbone potential energy w in plane perpendicular to the helix axis and containing the central (sixth) phosphorus atom of one strand at the distance d from the outer surface of the helix, ref. [186], p. 391

d [Å]	−w [kJ · mol^{-1}]
10	1410
30	753
50	506
70	381
90	305
150	192
350	84

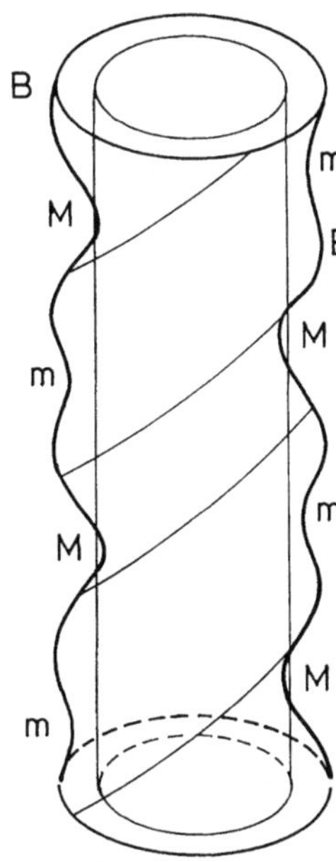

Fig. 28. Isopotential surface of DNA. *B* – sugar-phosphate backbone, *M* – major groove, *m* – minor groove (according to Pullman [186].

i.e. a given value of the potential is encountered at larger distances from the helix axis in the minor groove than in the major groove.

The most interesting effect, associated with the incorporation of the two sugar–phosphate backbones of the helix, on the potential around the complementary base pairs in planes of the base pairs is a very strong increase in the absolute values of the potential minima, indicating a parallel increase in the affinity of the bases towards electrophilic agents. This obviously corresponds to the penetration of the strong potential generated by the phosphates into the vicinity of the bases. Thus there appears to be a general constant, progressive increase in the absolute value of the potential minima in the series: free bases < nucleosides < nucleotides < single helix < double helix. For example the following values (in kJ/mol) were determined for the electrostatic molecular potential energy minima at the N(7) site of guanine (see p. 388, 395 in ref. 186):

guanine	guanosine	guanylic acid	single helix	double helix
−369	−368	−613	−1760	−2856

As has already been mentioned, the negative electrostatic potential minima represent obvious centres of attraction for electrophiles: protons and other positive ions or uncharged molecules with reactive centres at their electron-deficient atoms. In this respect attention was drawn particularly to chemical carcinogens regarded as being active *per se* (several alkylating agents such as diethylsulphate or N-methyl-N-nitrosourea) or to those that require a metabolic activation (polycyclic aromatic hydrocarbons, aromatic amines, aflatoxins, etc.). The possible occurrence of intercalation or of partial stacking with the bases of some of the large carcinogens, and the associated deformation of the DNA double helix, are expected to influence the ease of the reaction (see p. 397 in ref. 186).

The supermolecular approach, which also takes into account other components of interactions – polarization, charge transfer, and exchange repulsion – is more sophisticated than calculations using molecular electrostatic potential [188, 189].

4.1.2. Interaction of bases, nucleosides and nucleotides with ions: The electronic charge distributions and the electrostatic molecular potential maps are the characteristics closely related to what a charged reactant (proton, cation, anion) "feels" upon approaching the substrate, or at least what it feels at not too close a distance (see p. 381 in ref. 186). They can help to predict the most probable sites of protonation or the most probable site of interaction (association) with metal cations and anions of nucleic acid bases, nucleosides and nucleotides.

4.1.2.1. Protonation and deprotonation, p*K values:* The location of the minimum electrostatic molecular potential of the nucleic acid bases correlates very satisfactorily with the position of the most favourable sites of protonation in these molecules, which show the preference of N(1) and N(3) over N(7) in adenine [189, 190]; the preference of N(7) in guanine; of N(3) in cytosine; of O(4) in uracil and thymine; and a general dominance of the ring nitrogens over the oxygens and over the amino or imino nitrogens in all these bases. The preference for protonation is determined by the protonation energy. The protonation energies of N(1), N(3) and N(7) nitrogens of adenine are −1010, −1004, and −971 kJ/mol, respectively [191]. Thus the order of preference is N(1) > N(3) > N(7) [190]. The N(7) site can be protonated only at very low pH values [190]. The relative protonation energy ordering for the double protonated adenine is N(1)N(7) > N(3)N(7) > N(1)N(3), i.e. the most probable sites of protonation of adenine dication are N(1) and N(7) [190]. In nonaqueous acids HCF_3CO_2 and HSO_3F [192] or in a very strong hydrochloric acid (5 M HCl) [193] adenine trication can exist with protonated N(1), N(7) and N(3) nitrogens [194]. The preferred protonation sites of nucleic acid bases are summarized in Table 5. The subscripts 1, 2, 3 and 4 are used with pK to denote the first, second, and subsequent ionization steps. Thus at pH = pK_2 there are 50% neutral and 50% cationic molecules in the solution. The preferred protonation sites do not generally coincide with the preferred hydrogen bonding sites of these bases with water [191]. The preferred protonation site (in other words, the most basic site of the molecule) is not necessarily determined by the highest electron density in the ground state, but rather by the conditions in the transition state (see p. 107 in ref. 40 and ref. 195). The most favourable site of deprotonation (the more acidic atoms in the molecule) correlates with the position of the lowest σ electron charge densities ("amide-like" hydrogen atoms at uracil N(3) and at guanine N(1), which have a positive σ charge of +0.19 e); (see p. 105 in ref. 40).

The probability of protonation or deprotonation of the respective favourable site depends on the pH. At slightly acidic pH (around pH 3) only the most favourable sites of protonation are protonated (N(1) of adenine or N(3) of cytosine). At more acidic conditions (around pH 2) the less favourable sites (N(7) of adenine, N(7) of guanine, and C(4)O of uracil) are also protonated. At pH 2 adenine is thus doubly-protonated; it exists as a dication. At

Table 5. Sites of protonation and deprotonation of bases, ref. [40], p. 105; [190, 191, 194–197], ref. [198], p. 262; [199]

Base	Dication	pK_1	Cation	pK_2	Neutral
Ade	N1, N7	−0.35	N1	4.19	
Gua		0	N7	3.2	
Cyt		–	N3	4.5	
Thy		–	C4O	0	
Ura		–	C4O	0.5	
	Neutral	pK_3	Anion	pK_4	Dianion
Ade		9.8	N9	–	
Gua		9.6	N1/C6O	12.5	N1/C6O, N9
Cyt		12.2	N1/C2O	–	
Thy		9.8	N1/C2O or N3/C4O	13	N1/C2O and N3/C4O
Ura		9.5	N1/C2O or N3/C4O	13	N1/C2O and N3/C4O

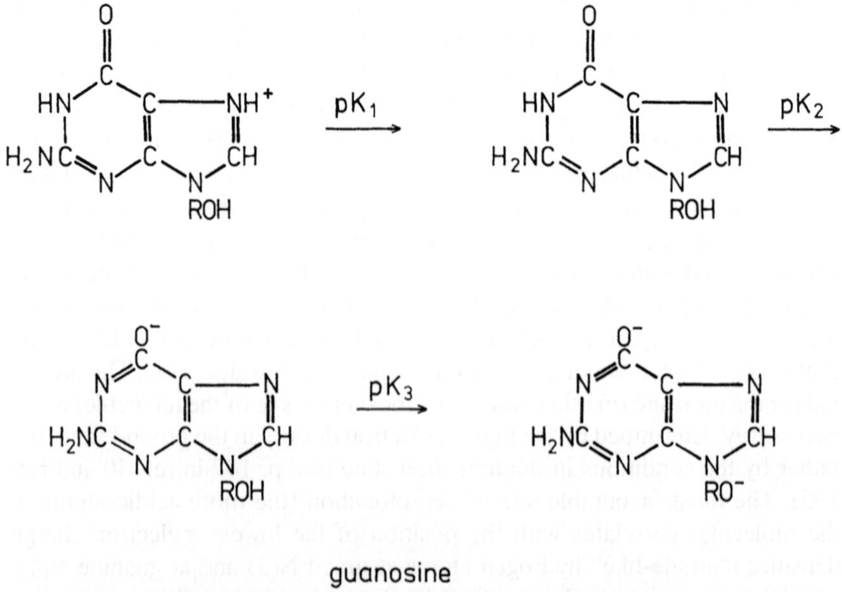

Fig. 29. The ionization steps for guanosine; $pK_1 = 1.6$, $pK_2 = 9.16$, $pK_3 = 12.3$ [196].

alkaline pH deprotonation occurs. The hydrogens at N(3) of uracil and thymine, and at N(1) of adenine and guanine are removed and the bases become anions with the negative charge situated at N(9) (adenine), C(6)O (guanine), N(1)/C(2)O (cytosine), and N(1)/C(2)O or N(3)/C(4)O (uracil, thymine), (Table 5) [195–199].

The ionization steps for guanosine are represented in Figure 29 [196]. The subscripts 1, 2 and 3 are used with pK to denote the first, second and third ionization step, but the numbering does not correspond to that given in Table 5 (for example, pK_1 in Figure 29 is the ionization step between cation and neutral molecule and not between dication and cation); (see p. 108 in ref. 40). Under the influence of stronger alkali, the secondary hydroxyls in riboses are deprotonated. The corresponding pK values are around 12.5.

The phosphate groups in nucleoside mono-, di-, and triphosphates display two main pK values, pK_1 around 1–2 and pK_2 around 7.

At neutral pH the nucleoside monophosphate is mainly in the dianion form and the base is neutral [200]. As the pH is decreased to values below the phosphate pK_2 (pH 6.6 at 25 °C), the phosphate becomes protonated and a monoanion results. Upon further lowering of the pH the base is protonated, the nucleotide is now a zwitterion which possesses a large dipole moment (the calculated dipole moment of neutral GMP is 8.04 debye, of GMP zwitterion 17.40 debye, and of AMP zwitterion 22–27 debye [200]). The zwitterionic structure is at equilibrium with a definite, even though unknown, amount of neutral nucleotide molecules. Upon lowering of the pH below the phosphate pK_1 the negative charges at phosphate groups of ribose

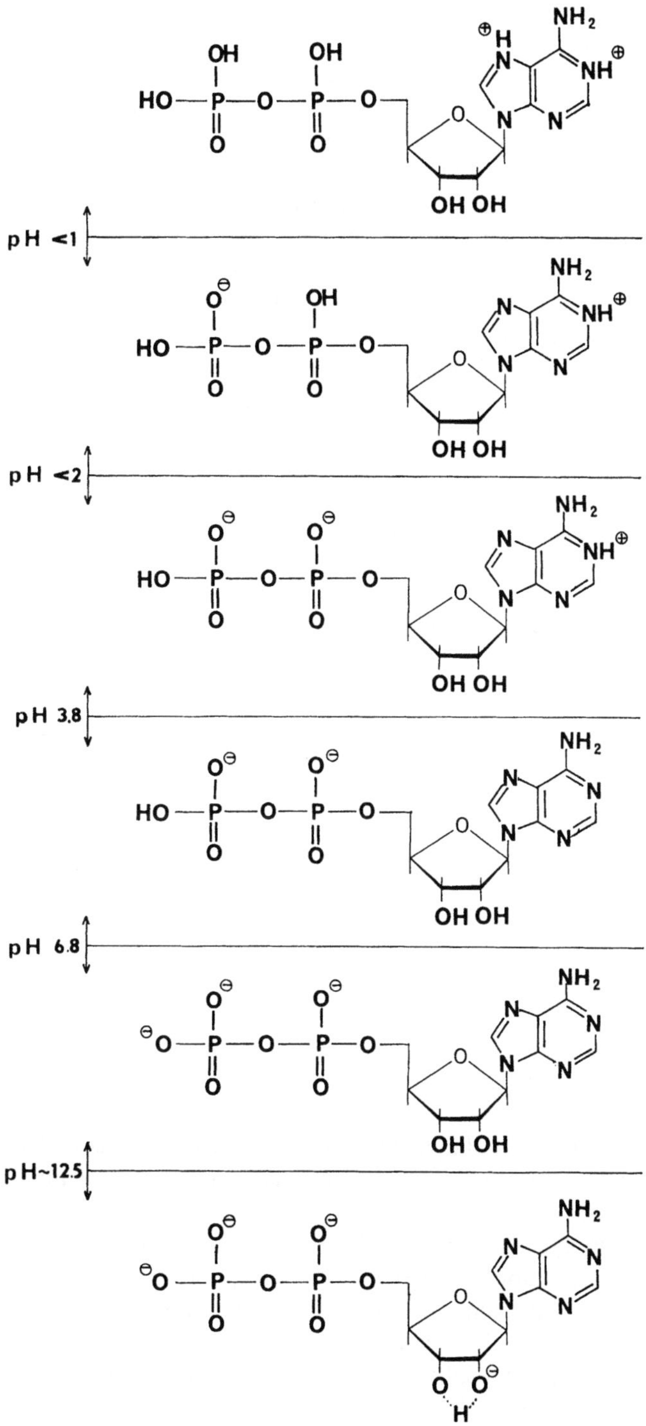

Fig. 30. Sites of ionization in adenosine diphosphate (see ref. [40]).

monophosphates are further neutralized by protonation (i.e. there is a zero charge on phosphate) and the base can pick up one or two other protons depending on the kind of base (dications can be formed by adenine and guanine), Figure 30 (see p. 109 in ref. 40). In ribose di- and triphosphates the negative charges at phosphate groups are neutralized at pH < pK_1 only in part, resulting in average charge/phosphate ratios of 1/2 in diphosphates and 1/3 or 2/3 in triphosphates. Figure 30 shows the ionization steps of ADP. In the primary ionization step (pK_1) one proton per phosphate is removed, resulting in a single negative charge per phosphate group (a charge/phosphate ratio of 1.0). This ionization state prevails up to a pH around 7 where the last proton is removed from the terminal phosphate group in the secondary ionization step (pK_2) (see p. 109 in ref. [40]).

Protonation affects all atomic electron densities (both σ and π) and increases the ionization potential. Guanine protonation increases the *syn* population in 2′- and 3′-guanosine monophosphates [200].

4.1.2.2. Metal ions: The potential sites for the binding of metal cations to nucleic acids are the same as the potential sites of protonation: the base ring nitrogens and the exocyclic base keto groups (oxygens), the ribose hydroxyls, and the negatively charged phosphate oxygen atoms (see p. 201 in ref. 40). However, the preferred site of cation attraction to the base may differ from the preferred site of protonation, because the cation carries an electron cloud which, overlapping with that of the base molecule, gives rise to an exchange repulsion and thus prevents the cation from approaching into the molecular skeleton [201].

The base ring nitrogens carry lone electron pairs and are good ligands for alkali (Li^+, Na^+, K^+, Rb^+, Cs^+) and transition metals (Mn^{2+}, Co^{3+}, Co^{2+}, Ni^{2+}, Pt^{2+}, Cu^{2+}, Ag^+, Au^{3+}, Zn^{2+}, Cd^{2+}, Hg^{2+}) as well as for alkaline earth cations (Mg^{2+}, Ca^{2+}, Sr^{2+}, Ba^{2+}); (see p. 203 in ref. 40).

The *keto* oxygens of pyrimidines (C(2)O of cytosine, C(2)O and C(4)O of uracil and thymine), but not C(6)O of purines, can be involved in direct metal binding (see p. 207 in ref. 40).

The amino substituents of bases are never involved in direct metal coordination because the lone electron pair is delocalized over the π-bonding system of the base and is not available for metal ion binding.

In crystals the N(3) of cytosine is a ligand for metal ions with a coordination number lower than 6; N(7) of purines can accommodate all kinds of metal cations.

Sugar hydroxyls interact preferentially with alkali and alkaline earth cations but not with transition metals (Cd^{2+} is an exception; see p. 210 in ref. 40).

Phosphate groups can bind electrostatically to all metal cations.

The monovalent alkali cations Na^+ and K^+ bind nonspecifically (electrostatically) to the phosphate moieties rather than to specific base sites and act as counter-ions [195, 197] (see section 4.1.2.3). Nevertheless, determination of binding energy with the quantum-mechanical supermolecule approach provides some information about the cation binding tendencies of nucleic acid bases [189].

The results of supermolecule computations for the interaction of bases with Na^+ show that the preferred site of Na^+ binding is essentially dictated by electrostatic attraction and is thus the same as the preferred site of protonation (N(1) in adenine, C(4)O in uracil), when the available acceptor sites on the bases are all of the same nature (oxygens of uracil, nitrogens of adenine) [189].

When acceptor sites of a different nature (oxygens and nitrogens) are present simultaneously on the same base (as in cytosine and guanosine), the role of exchange repulsion (repulsion arising from an overlap of the electron distribution of the two interacting species) becomes important. Exchange repulsion is fundamentally larger for nitrogen than for oxygen. The valence shell orbitals of a nitrogen atom extend farther in space than those of an oxygen atom, thus giving rise to a larger overlap with those of the attacking Na^+ ion at a given distance and hence to a larger repulsion. In cytosine the preferred site of attraction of Na^+ is thus C(2)O instead of the preferred protonation site N(3), and in guanine C(6)O instead of N(7).

In crystals the reverse is true: the preferred metal ion coordination sites are the same as the protonation sites for guanine (N(7) > N(3)) and cytosine (N(3)), but different for adenine (coordination N(7) > N(1), protonation N(1) > N(7)) and thymine (coordination C(2)O > C(4)O, protonation C(4)O > C(2)O; see p. 209 in ref. 40).

Divalent cations (alkaline earth metals Mg^{2+}, Ca^{2+}, Sr^{2+}, Ba^{2+} and transition metals Mn^{2+}, Co^{2-}, Ni^{2+}, Cu^{2+}, Zn^{2+}, Cd^{2+}, Hg^{2+}) can interact specifically with bases rather than electrostatically with phosphate groups. The affinity for base complexing relative to phosphate binding increases in the series (see p. 201 in ref. 40)

$$\mathrm{Mg} < \mathrm{Co} < \mathrm{Ni} < \mathrm{Mn} < \mathrm{Zn} < \mathrm{Cd} < \mathrm{Cu}$$

Thus, addition of Mg^{2+} to a DNA solution shifts the helix–coil transition of DNA to higher value of melting temperature T_m, indicating increased stability of the DNA double helix, because repulsive forces between the negatively charged phosphate groups are screened by Mg^{2+} cations in the same way as has been observed with monovalent cations (see section 4.1.3.5.). The addition of Cu^{2+}, however, markedly decreases the T_m due to the interaction with bases, which results in a destabilizing effect on the DNA double helix. Behaviour similar to that of Cu^{2+} is also found for Hg^{2+} and Ag^+ (see p. 201 in ref. 40).

Using the electrostatic molecular potentials and taking into account the distance of the closest possible approach of the Cl^- anion to guanine and/or cytosine, the existence of attraction zones for Cl^- binding in specific regions of bases was found [201]. The values of the interaction energies involved indicate that these interactions are much stronger with protonated bases than with neutral ones.

Binding of Cl^- anions to the protonated bases is not necessarily limited to the vicinity of the nitrogen atom carrying the added proton (N(3) of cytosine and N(7) of guanine, with binding energies of −435 kJ/mol for cytosine and −368 kJ/mol for guanine), but may occur at other, sometimes quite distant sites (between $-NH_2$ and C(5)H for cytosine, binding energy −367 kJ/mol, and between $-NH_2$ and N(1)H for guanine, binding energy −368 kJ/mol). Thus, protonated guanine has two equally strong sites of maximum binding (−368 kJ/mol), one of which is in the vicinity of the protonated N(7), but the other in a position between the N(1)H and the $-NH_2$ groups, very close to the position most effective for anion complex formation in the neutral guanine. Theoretical prediction of the sites of the strongest attraction of Cl^- to cytosine and guanine seems to be in agreement with experimental data, i. e. with the sites of location of Cl^- ions in the crystal structure of cytosine hydrochloride and Br^- ions in 9-methylguanine hydrobromide [202].

4.1.2.3. Counter-ion atmosphere: In nucleotides there are functional groups that may be protonized or deprotonized depending on the pH of the solution, as has been shown in section 4.1.2.1. Nucleic acids thus belong to a large family of electrically charged polymers – polyelectrolytes [203]. In solution at neutral pH the nucleic acids exist in polyanionic form, negative charges being situated at the phosphate groups [204]. The polyanionic character of nucleic acids favours the electrostatic binding of small cations to the negatively charged phosphates. These cations are usually hydrated and form around the fixed phosphate groups an ionic atmosphere composed primarily of counter-ions and, to a smaller extent, of co-ions from low-molecular weight salts which may be added to the nucleic acid solution. There are electrostatic interactions between the charged phosphate groups themselves and interactions of the phosphate groups with the counter-ions and co-ions. The effect of counter-ion binding (or counter-ion association) to the phosphate charges is treated by assuming an effective charge for the phosphates less than unity. The second counter-ion effect is the screening of the phosphate-phosphate repulsion [203–206]. The mutual repulsion of the negatively charged phosphate groups decreases the stability of the double-helical conformation of DNA and has a stretching effect on single-stranded denatured DNA.

The screening of the phosphate–phosphate repulsion can be treated in the framework of the Debye-Hückel theory of simple electrolytes [204]. The charge density of the screening ions is given by the Boltzmann distribution

$$\rho = Ne\,[\exp(-e\psi/kT) - \exp(e\psi/kT)] \qquad (7)$$

and Poisson equation

$$\nabla^2 \psi = (-\rho)/\varepsilon \qquad (8)$$

where e is the electron charge, N is the bulk ionic concentration, ψ the electrostatic potential around the molecule of nucleic acid, and ε dielectric permittivity.

The distribution of the screening ions and the electrostatic potential ψ in the vicinity of the DNA helix (approximated as a discretely charged cylinder) with the atmosphere of the counter-ions can be calculated by solving the Poisson-Boltzmann equation (Fig. 31). Segments of the polymeric chain of DNA can be treated as rods surrounded by cylindrical atmosphere even if the nucleic acid as a whole has a more or less random shape [203].

In Figure 31 the potential ψ versus the distance r from the centre of the cylinder (a rod-like polynucleotide) is plotted. It can be seen that the potential drops rapidly with the distance from the polyion and approaches zero at a distance of 1.5 to 5 nm from the centre of the polyelectrolyte depending on counter-ion concentration [203]. The electric field intensity, which is the gradient of the electrostatic potential given by $\delta\psi/\delta r$, reaches the tremendous value of 3×10^5 kV/cm, i.e. of 3×10^7 kV/m, at the surface of the highly charged macromolecule. This powerful electrostatic field attracts an appreciable fraction of the counter-ions and creates a fairly dense atmosphere of lower mobility.

A direct expression of counter-ion immobilization by the polyion is given by the osmotic pressure of a polyelectrolyte solution. Since generally a polyelectrolyte molecule carries many hundreds of fixed charges, the number of counter-ions is much higher than the number of polyions and practically all the osmotic pressure of a polyelectrolyte solution may be attributed to the small counter-ions. The contribution of the large polyions is negligible [203].

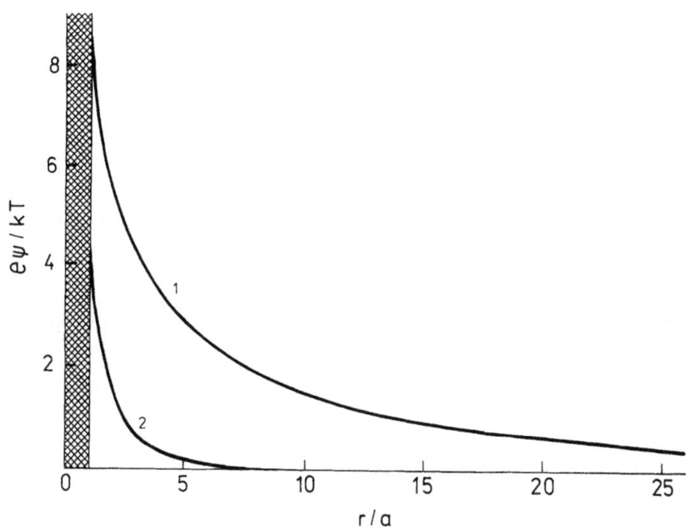

Fig. 31. The electrostatic potential ψ around a rod-like poly-ion with a counter-ion atmosphere; a – radius of a cylindric polyion, e – electron charge. The counter-ion concentrations: 1 – 0.01 M, 2 – 0.1 M [203].

The osmotic pressures of the neutral polyelectrolytes such as nucleic acids should thus be comparable to those of low-molecular weight neutral salts. Experimentally it has been found that the osmotic pressure of the polyelectrolyte solution is lower than that of the low molecular salt. Each polyion carrying γ charges is neutralized by γ counter-ions. This means that a large fraction of the counter-ions is immobilized and does not contribute to the osmotic pressure. For highly charged polyions only 10–20% of the counter-ions are osmotically active and up to 90% are fixed in the neighborhood of the polyion [203].

With divalent ions, specific factors as well as nonspecific electrostatic forces become very prominent, as has been discussed in section 4.1.2.2, and immobilization of the divalent counter-ions is much higher. For the interaction of monovalent ions with alginic acid [203] the osmotic pressure indicates that about 30% of the counter-ions are free. In the case of divalent ions there are about 15% free Mg^{2+}, 3% free Cu^{2+} and 1% free Ca^{2+} in solution. For such a strong degree of ion binding it is appropriate to assume the specific ion association with the formation of neutral ion pairs localized at definite sites along the polyion chain [203].

Divalent ions such as Mg^{2+} and Mn^{2+} interact nearly stoichiometrically with polyribonucleotides [205]. Certain multivalent hydrated cations bind not only to the phosphates, but also to certain functional groups of the bases, in particular to N(7) of the purines [197], as has been mentioned in section 4.1.2.2. At present it cannot be predicted which type of binding dominates for a given cation [197].

4.1.3. Forces stabilizing double-helical conformation of DNA and synthetic polynucleotides: The electronic properties of the bases are important for the theoretical evaluation of the forces involved in interactions between bases (see p. 465 in ref. 5), which stabilize the double-helical conformation of native DNA and synthetic polynucleotides. These interactions are of two kinds (see p. 116 in ref. 40 and ref. 197):

(i) Horizontal interactions in the plane of the bases due to hydrogen bonding, dipole-dipole and dipole-induced dipole interactions. Hydrogen bonds are treated in section 4.1.3.1.

(ii) Vertical interactions perpendicular to the base planes (so-called stacking interactions) stabilized mainly by London dispersion forces and hydrophobic effects. These interactions are discussed in sections 4.1.3.2–4.1.3.4.

The stability of the double-helical conformation depends further on the coulombic interactions between electric charges on bases and phosphates. Such interactions together with the effects of solubility, water structure and other factors on the stability of the double-helical conformation are discussed in section 4.1.3.5.

4.1.3.1. Electronegativity, hydrogen bonds: Electronegativity means the tendency of an atom in a molecule to attract electrons to itself (see p. 472 in ref. 168). Electronegative atoms in a molecule thus display a negative electronic charge density. Fluorine is the most electronegative atom and caesium is the least electronegative one. The atoms become less electronegative as we go down a column of the periodic table because of the increasingly effective screening of the charge of the nucleus by inner electrons with the increase in atomic number. From the electronic charge distribution of bases (Figs. 24 and 25) it can be seen that the most electronegative atoms are nitrogens and oxygens, which display negative charges of about −0.5 e to −0.65 e.

A hydrogen bond results when a proton is shared between two electronegative atoms such as F, O or N which are the right distance apart, e.g.

$$N - H \ldots O$$

In this case the hydrogen atom covalently bound to nitrogen has a positive electronic charge density (the electrons are withdrawn from the H atom to the N atom) and is electrostatically attracted to the oxygen atom with a negative electronic charge density. The hydrogen bond is conventionally marked by dots between H and O. Positively charged hydrogen is a donor and negatively charged oxygen is an acceptor of the hydrogen bond (see p. 475 in ref. 168).

As hydrogen bonds are largely electrostatic in nature, their strength, reflected in the distance between donor and acceptor atoms, depends on partial electronic charges located on N, H, and O. Compared with covalent bonds with well-defined length, strength and orientation, hydrogen bonds are about 20–30 times weaker (bond energy of 12–25 kJ/mol) and only weakly directional (see p. 117 in ref. 40).

The electronic charge density distributions indicated in Figures 24 and 25 suggest that in the bases, hydrogen atoms attached to amino groups in adenine, guanine and cytosine have nearly equivalent positive charges of 0.22 e and therefore exhibit a weak acidity, favouring them as donors in hydrogen bonds (see p. 105 in ref. 40). Similarly, the "amide-like" hydrogen atoms at uracil N(3) and at guanine N(1) have a positive charge of 0.19 e, nearly equal to each other but significantly lower than the charges at amino group hydrogens.

On the other hand, bases can act as hydrogen bond acceptors: N(1), N(3) and N(7) in adenine; N(3), N(7), and C(6)O in guanine; C(2)O and C(4)O in uracil and thymine; N(3) and C(2)O in cytosine. All these atoms display negative charges in the range of −0.47 e (C(2)O of thymine) to −0.65 e (N(3) of cytosine), which render them good acceptors for hydrogen bonds.

The affinity of negatively charged N and O atoms of bases and of phosphate groups for hydrogen bond formation is largely paralleled by their ability to act as ligands for a great variety of metal cations, as has been pointed out in section 4.1.2.2. (see p. 106 in ref. 40).

4.1.3.2. London dispersion forces and stacking forces: As recognized by London [169, 207], the principal contribution to the ever-effective van der Waals attractive forces between atoms in close proximity is given by electrokinetic interactions between the systems. At any instant, the electronic charge distribution within atomic groups is asymmetric due to electron fluctuations. Therefore, dipoles created in one group of atoms polarize the electronic system of neighboring atoms or molecules, thus inducing parallel dipoles that attract each other (see p. 138 in ref. 40). London pointed out that a relationship exists between this effect and optical dispersion in gases, and he named the attractive forces produced by this inherently quantum-mechanical phenomenon "dispersion forces" [187, 206]. These London dispersion forces are additive and decrease with the sixth power of the distance. They are independent of temperature and increase with the product of the polarizabilities of the partner molecules. The corresponding interaction energy W_L between molecules A and B based on the London dipole approximation can be obtained by the following equation (see p. 465 in ref. 5):

$$W_L = -\frac{3}{2}\frac{I_A I_B}{I_A + I_B}\frac{\alpha_A \alpha_B}{r_{AB}^6} \tag{9}$$

where r_{AB} is the distance between points of location of these two dipoles, α is the polarizability, and I ionization potential. The ionization potential represents the energy required to remove an electron from an atom in the gaseous state. The first ionization potential refers to the most weakly bound electron of the molecule (see p. 37 in ref. 171). A measure of the first molecular ionization potential of the π electrons is the energy of the highest occupied molecular orbital (see p. 79 in ref. 171). The values of the ionization potentials of some bases determined by the mass spectroscopic method are quoted on p. 467 in ref. 5.

As the bases also possess a permanent dipole moment, the two electronic effects, London dispersion and permanent dipoles, combine and lead to appreciable attraction between the bases which is more pronounced in purine than pyrimidine bases (see p. 139 in ref. 40).

Electrostatic forces between stacked bases and base pairs are intrinsically related to charge distributions and are therefore accessible to quantum-chemical calculations. The stacking energies for the ten possible combinations of stacks formed by G · C and/or A · T base pairs were calculated within the framework of B-DNA geometry employing monopole–monopole interactions, polarization and dispersion interactions, and repulsion due to the r^{-12} term in the Lennard-Jones approach (see p. 140 in ref. 40). From these calculations it follows that the stack

$$\uparrow \begin{matrix} \mathrm{A} \cdot \mathrm{T} \\ \mathrm{T} \cdot \mathrm{A} \end{matrix} \downarrow$$

is least stable (stacking energy of –16 kJ/mol per dimer), whereas the stack

$$\uparrow \begin{matrix} G \cdot C \\ C \cdot G \end{matrix} \downarrow$$

is most stable (stacking energy of –61 kJ/mol per dimer). (The arrows indicate the direction of the sugar-phosphate chain and point from C(3′) of one sugar unit to C(5′) of the next one, both carbons being attached to the same phosphodiester link.) From studies partitioning the total stabilizing energy of base-paired dimers into horizontal (base-pairing) and vertical (base-stacking) components it becomes clear that horizontal complementary hydrogen bonding is only composition-dependent and is more important than vertical stacking which is influenced by both composition and sequence, [140, 207]. It should be stressed, however, that these calculations consider only molecules in vacuum and do not take into account hydrophobic and base–water hydrogen bonding interactions, which contribute significantly to stacking interactions (see sections 4.1.3.3–4.1.3.4).

4.1.3.3. Hydrophobic interactions: The tendency of a macromolecule to minimize contacts of its nonpolar groups with water by assuming a contact conformation is a phenomenon which has generally come to be considered a hydrophobic interaction [208]. The decrease in the entropy of water observed when a nonpolar solute is introduced implies an increase in the local water structural organization. (An "iceberg structure" with more hydrogen bonding than exists in bulk water is formed.) An increase in the entropy of aqueous solvent is predicted when nonpolar macromolecular groups reduce their solvent contacts. This entropic change contribution represents the hydrophobic interaction.

In native DNA, the purine and pyrimidine bases are largely "stacked" in the interior of the macromolecule with most of their surfaces thus removed from contact with the solvent. Under complete denaturing conditions these groups are generally exposed to complete solvation [209].

In 1959 Kauzmann [209] first pointed out that a major contribution to the stability of the folded form of globular proteins may result from hydrophobic interactions. Thermodynamically this requires that, under conditions favouring the folded form, the free energy of the transfer of nonpolar residues from the interior of the protein into the solvent as a consequence of unfolding is positive. While the enthalpy of transfer for this process is actually negative, the entropy of transfer is also sufficiently negative to make the free energy of the net process positive and thus thermodynamically unfavourable. This negative entropy of transfer for nonpolar compounds arises from a net increase in the order of the aqueous phase as a consequence of the nonpolar groups. Water becomes organized into hydrogen-bonded clusters of "icebergs" around nonpolar groups. Salts that destabilize the folded form should decrease the (positive) free energy of transfer, and salts that stabilize the folded macromolecule should have the opposite effect. Salts that decrease the positive free energy of transfer should also increase the solubility (decrease the activity coefficient) of a model compound in the aqueous phase.

4.1.3.4. Purine – water interactions in base stacking: Experimental data have clearly established that in water the nucleic acid bases are associated by vertical stacking [210–223] rather than by hydrogen bonding as in organic solvents [224–226]. For this type of molecules, association by hydrogen bonds in vacuum is energetically more favourable than association by stacking [227]. Since stacking is thus not the most favourable mode of association of the bases in vacuum, the stacking of the bases in water has been attributed to hydrophobic interactions [211, 212].

According to the theory of hydrophobic interactions [213, 228, 229] the stacked association of the nucleic acid bases should be essentially determined by the water–water restructuring around the solute molecules [227]. However, thermodynamic data on the stacking of the bases [211, 222, 223, 230] and of nucleotides [231] are not in agreement with those expected for stacks held by the standard hydrophobic force [227]. Since stacking occurs only in water, this mode of association is related to some property of water [230].

Quantum-mechanical calculations have shown that the solute–water interaction energies for the hydration of nucleic acid bases [232–234] and base pairs [232, 234, 235] are larger than the water–water interactions [236–238]. A number of molecules of water can thus be "bound" to the bases. These results suggest that the molecules of water might be more intimately involved in the mechanism of stacking of the nucleic acid bases than they are in standard hydrophobic interactions concerning nonpolar molecules [227].

NMR studies on the stacking of the bases, mainly purines, in water have given indications as to the geometrical arrangement of the stacks formed and are in favour of conformations in which the long axes of the two purines are parallel or antiparallel with a large overlap of the bases [212, 213, 219, 221].

The possible mechanism by which water promotes the stacking of the nucleic acid bases was investigated by Langlet *et al.* [227]. They determined the possible conformation of stacked purine dimers in vacuum by minimizing the purine–purine interaction energy. The association energy of two purines by vertical stacking was found to be of the order of 20–25 kJ/mol. The electrostatic contribution determines the relative stability of the different geometrical arrangements of the stacks in spite of the fact that the dispersion contribution is numerically larger by almost one order of magnitude [239–244].

In considering stacked purine dimers in water, Langlet *et al.* [227] examined the possible structure of water around the two stacked purine molecules. Experimental studies [245–246] confirmed the theoretical results on the existence of base–water hydrogen bonds. The first hydration shell around the purine dimer thus contains the water molecules hydrogen-bonded to the purine molecule. The purine has four sites that can give appreciable interactions with water, namely N(1), N(3) and N(7), which are proton acceptors, and N(9)H, which is a proton donor. Thus, four water molecules are placed around each purine molecule of the dimer. In order to optimize the location and orientation of the water molecules bound to each purine with respect to the molecules of water hydrogen-bonded to the other purine of the stack, the interaction energy of the purine dimer with water molecules was minimized. From Figure 32 it can be seen that the water molecules of one base have moved towards the water molecules of the other and rotated in order to form hydrogen bonds between them and thus form "water bridges" between the two purines. The formation of these water–water hydrogen bonds is responsible for an important energy gain (more than 80 kJ/mol), which is obtained for the value of the interaction energy of the purine dimer.

Almost each water molecule of the first shell of one purine is connected to one molecule of water of the first shell of the second purine through a third molecule of water (belonging to the second shell), thus forming what can be called a three-body complex. We may again speak of "water bridges" between the two purines but these are formed by water trimers.

The stabilization of the hydrogen bonds responsible for the bridges originates from the polarization contribution. Stacking thus seems to be due largely to hydrophilic interactions rather than to hydrophobic ones. These hydrophilic interactions appear to influence the conformation of the stacks. The "special properties of water" considered in ref. 224 to explain the stacking of the bases in water are due to the possibility of the water molecule to form hydrogen bonds with the solute and with the other molecules of the solvent. A characteristic of water that probably contributes to the stabilization of the stacks is that the most favourable O...O distance of a water dimer is of the order of 0.28–0.29 nm [236–238], while the molecular planes of the bases are 0.32 nm apart. Thus the promotion of bridges made of

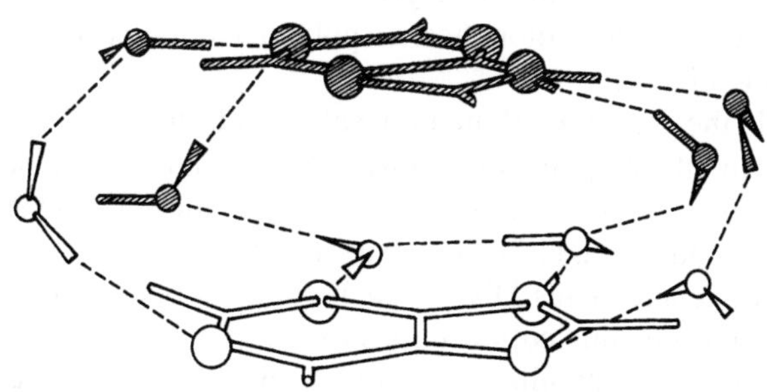

Fig. 32. Water bridges between two stacked purines (according to Langlet *et al.* [227]).

water dimers between the bases requires only a small distortion of the purine–water hydrogen bonds.

All the structures proposed in Figure 32 for the dimers as well as for the molecules of water are not meant as rigid crystal-like aggregates, but are subject to many fluctuations [227].

4.1.3.5. Effect of neutral salts on the stability of the double-helical conformation: The term "neutral salt" is used for a strong electrolyte which is significantly soluble in water without bringing about a major change in solution pH. Such salts can obviously affect electrostatic interactions in macromolecules on a simple charge-shielding basis; they are all capable of serving as simple sources of ionic strength. In this respect they would be expected to differ from one another in effectiveness only on the basis of charge and might be expected to influence macromolecular conformation primarily by weakening attractive or repulsive inter- and intrachain charge–charge interactions (see section 4.1.2.3). However, in addition to such nonspecific effects on electrostatic interactions, neutral salts also exhibit strikingly specific effects on the conformations of biopolymers which seem to have nothing to do with the sign or magnitude of the ionic charge *per se* [247].

The conventional parameter characterizing the stability of the double-helical conformation is the melting temperature T_m (the temperature at which half the base pairs are in helical regions [204], see section 2.2.1.2). Increasing salt concentration in a DNA solution increases the melting temperature. The effect is analogous to the stabilization of DNA double helix by surface-bound metal cations (see section 4.1.2.2) or other cations bound in different manner, e.g. by intercalation [248]. It was found that T_m varies linearly with the logarithm of the activity of monovalent cations. The slope of this linear function is principally due to the difference in linear charge density between the double helix and the much more extended single-strand coils [197, 204].

When protonation of the bases is involved in the formation of an ordered structure, the ionic strength dependence of T_m may be reversed. Thus, the acid form of poly(rA) is destabilized by the addition of a salt. This effect is ascribed to the reduction of electrostatic attraction between protonated bases and negatively charged phosphate groups as a result of screening of the phosphate charges. The protonation of polynucleotide bases is suppressed by the addition of a salt.

In general, the addition of neutral salts tends to stabilize the helical (ordered) form of charged polyelectrolytes in which most or all of the charges have the same sign (thus tending to repel one another) and in which the charges are, on average, further apart in the unfolded than in the folded form of the molecule. Thus DNA, which contains one negatively charged phosphate group per nucleotide unit, is subject to considerable intramolecular electrostatic destabilization. In fact, in many DNAs the helix–coil transition reaches completion spontaneously at room temperature in salt-free water. The addition of small quantities of neutral salt moves the transition temperature to higher levels, approaching 100 °C for DNAs of high G · C (guanine · cytosine) content [209].

Schildkraut and Lifson [23] have demonstrated that the observed dependence on NaCl concentration can be represented by the equation

$$T_m = 16.6 \log c + 0.41\,(G \cdot C) + 81.5 \tag{10}$$

where c is the molar concentration of uni-univalent salt and (G · C) is the mole percentage of guanine plus cytosine bases in the DNA used. The data start to deviate from this equation only at uni-univalent salt concentrations exceeding 1 mol/l. The difference ΔG in free energy between the helix G_H and the coil G_C can be divided into two parts: one (ΔG_0) is an intrinsic component, which is practically independent of salt concentration and is defined as free energy at a very high salt concentration; and the other is an electrostatic component (ΔG_e), which measures the difference in free energy due to interphosphate electrostatic repulsions at a low ionic strength. Thus, ΔG_e represents the reversible work against electrostatic forces necessary to bring two fully charged extended polynucleotide strands from infinite separation into the double-stranded helical arrangement [209]. The calculations showed a linear dependence of T_m on ΔG_e (and thus a logarithmic dependence on c). The results obtained predicted that ΔG_e should go to zero at 3 M NaCl, if the logarithmic relation was valid at salt concentrations in excess of 1 mol/l [209].

The stretching effect of the mutual repulsion of the phosphate groups of nucleic acids can be followed by viscosity measurements. Solutions of spherical macromolecules or coiled linear chains present a lower hydrodynamic resistance to flow so that the viscosity of these solutions is low. If the linear macromolecules stretch and approach a rod-like shape the frictional dissipation of energy accompanying their flow increases and viscosity is high. The addition of a neutral salt screens off the phosphate charges, diminishes the repulsion and the stretching effect is lower, so that viscosity then decreases [203].

An increase in solubility S corresponds to a decrease in activity coefficient γ. The dependence of the solubility or of the activity coefficient on the molar concentration c can be represented by the Setschenow equation

$$\log \gamma = \log C_0/C = K_s c \tag{11}$$

where C_0 is the solubility in water and K_s is the so-called salting-out constant [209]. For salts that increase solubility the value of K_s is negative (the "salting-in" effect); the activity coefficient decreases with increasing salt concentration. Salts that decrease solubility (the "salting-out" effect) have a positive K_s value. K_s is negative for T_m depressors, such as $LiClO_4$, or $NaClO_4$, while it is positive for T_m elevators such as LiCl, $NaCHO_2$ or Na_2SO_4 [209].

Robinson and Grant [249] studied the effect of various neutral salts on the solubility of adenine and thymine. Solubility is increased most in Na_2SO_4, does not change very much in NaBr, and is decreased most in $Cl_3CCOONa$. The salting-out parameters K_s for the more soluble nucleosides, adenosine and deoxyadenosine, were found to be very similar to those for adenine for most salts. Salts that induced the salting out, e. g. NaCl and Na_2SO_4, showed a somewhat larger positive value of K_s for the nucleoside than for the free base. The specific effects of different monovalent cations were found to differ much less than those for the monovalent anions, as was also seen with DNA [209]. Protonation of polynucleotides increases their solubility [250].

The formation of hydrogen bonds in liquid water is a cooperative process. That is, once a hydrogen bond is formed between a pair of water molecules, the formation of a second hydrogen bond between this pair of complexed water molecules and a free water molecule is facilitated. The disruption of a hydrogen bond in such a polymeric complex is also considered to be a cooperative phenomenon [209]. The thermodynamic parameter, entropy, offers a useful index of the degree of structural organization of water in ionic aqueous solutions [209]. The addition of Li^+ and F^- to water results in a decrease in entropy, presumably reflecting some kind of increase in the structural order of the water surrounding these ions. For the other monovalent ions a progressive increase in entropy is connected with increasing ionic radius, suggesting that the entropy gain might arise from steric difficulties attendant upon the introduction of these larger ions into the water framework ("structure-breaking" ions). A decrease in electrostatic ordering of the dipolar water molecules with increasing ion size (larger ionic radius means lower charge density) may also explain the increase in entropy. The most powerful "structure breakers" are Cs^+ ($\Delta S = 23.8$ eu) and ClO_4^- ($\Delta S = 35.2$ eu). The other less powerful structure breakers are, for example, Cl^-, Br^- and I^- [209].

Divalent ions (both cations and anions) all decrease entropy to a considerable degree (at most, for example, Mg^{2+}, $\Delta S = -36.2$ eu and PO_4^{3-}, $\Delta S = -60$ eu). This probably reflects a partial electrostatic ordering of the second water layer around these ions of high charge density ("structure-forming" ions). Other structure-forming ions are, for example, Li^+, F^-, Ca^{2+}, Ba^{2+}, Sr^{2+} or SO_4^{2-} [209].

It has now become clear that ions (such as ClO_4^-, SCN^-, and Ca^{2+}) that are particularly effective in decreasing the stability of the native conformation of a fibrous protein in water (promoting unfolding) are also effective destabilizers of the native conformations of globular proteins and nucleic acids. These same ions are also particularly effective in dissociating non-covalently linked aggregates of proteins and other molecules and in increasing the rate of unfolding (denaturation) of "native" structures. Anions and cations are essentially independently effective in these functions and in their relative effectiveness they may be ranked as first described by Hofmeister in

1888 in a comparative study of the "salting out" of euglobulins form an aqueous solution by various neutral salts. This characteristic ranking of ionic effectiveness in processes such as salting out is called the lyotropic or Hofmeister series. According to this series the relative effectiveness of various ions in destabilizing the native form of DNA in an aqueous solution increases in the order [209]:

$$Cl^-, Br^- < CH_3COO < I^- < ClO_4^- < CNS^- < (CH_3)_4N^+ < K^+ < Na^+ < Li^+ \qquad (12)$$

4.1.4. Electric field effects on DNA

4.1.4.1. Polarization of the ionic atmosphere: relaxation. In section 4.1.2.3, which deals with the atmosphere of small ions surrounding a central polyion, it was shown that the fraction of the counter-ions that is not immobilized by the strong electrostatic field in the neighbourhood of the polyion can be determined by measuring the osmotic pressure of the polyelectrolyte solution, i.e. by an experimental method depending on equilibrium properties only. There is, however, another group of non-equilibrium methods that also allows the evaluation of the degree of ion fixation, e.g. electrophoresis, the measurements of conductivity, diffusion and dielectric permittivity [203].

The fraction of free counter-ions in the solution determined by electric conductivity measurements is larger by a factor of 2 than the fraction determined by osmotic measurements. This discrepancy was clarified by Katchalsky (see p. 32 in ref. 203). The counter-ion fixation in the atmosphere around the polyion consists of a strong immobilization of the counter-ion movement in a direction perpendicular to the axis of the polyion. There remains, however, a relatively free mobility parallel to the axis. In other words, any counter-ion in the atmosphere is free to move along the equipotential surfaces but is restricted in its movements to a direction perpendicular to the equipotentials. In osmotic pressure measurements, both the central polyion and its atmosphere move as a single particle, act osmotically as a unit and do not contribute to osmotic pressure.

On the other hand, in an electric field the counter-ion atmosphere is attracted in an opposite direction to the central polyion and the fixed counter-ions move in relation to the polyion. Due to their relative mobility counter-ions take part in the transport of electricity and they behave as partially free particles [203].

This mobility of the counter-ion atmosphere allows the polyion to become polarized in an electric field (Fig. 33). The conventional polarizability α_e (electron polarizability) is based on the mobility of electrons in relation to the atomic skeleton of the molecule. With polyelectrolytes we encounter a new polarizability α_i (interfacial polarizability) based on the mobility of

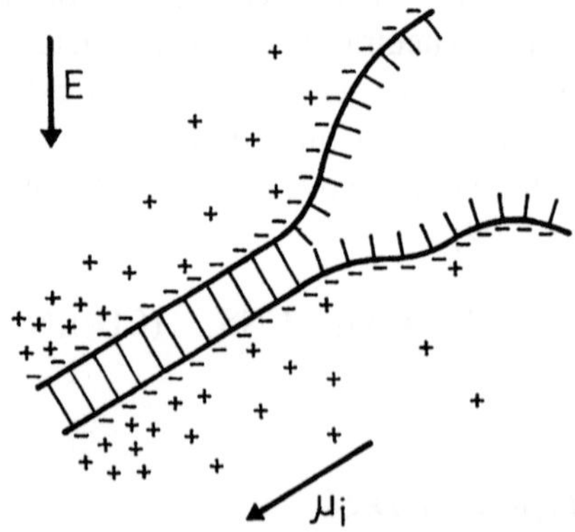

Fig. 33. Polarization of counter-ion atmosphere and denaturation of DNA in an electric field E; μ_i is the induced dipole moment.

counter-ions in relation to the polyion chain. The latter polarizability is about 10^{10} times larger than the electron polarizability of small organic molecules; in a polyphosphate it reaches a value of 10^{-12} cm^3 [203]. The corresponding induced dipole moment in the electric field of 1 kV/cm = 100 kV/m would be 3.3×10^6 debye units. With DNA the resulting induced dipole moment may reach a value of up to 5,000 debye (1 debye = 3.336×10^{-30} C m) [251].

The different effects that can be caused by an electric field in a solution containing polyions are illustrated in Figure 33. The effect of a weak field will be essentially restricted to the displacement of all particles carrying an electric charge with respect to the centre of the polyion. Thus the effect of a weak field will be represented by transference of ions (expressed by the real part of conductivity) and some polarization of the polyion (measurable as dielectric increment). At high frequencies relaxation effects occur due to the fact that the ions diffuse with finite velocity (see section 4.1.5).

4.1.4.2. Orientation: At field strengths higher than about 100 V/cm the orienting force of the field acting on the induced dipole moment can overcome the disorienting tendency of the thermal motion, and a partial orientation of DNA molecules with their long axis parallel to the electric field strength E occurs [251]. At field strengths of about 1,000 V/cm a complete orientation of molecules with a length of several hundred nanometers takes place [252]. The high degree of DNA orientation observed at already low electric field strengths is attributed to a high polarizability of the counter-ion atmosphere around the DNA. The orientation is a consequence of polarization and is slower than polarization. The time constant of the polarization of the ionic atmosphere of DNA is around 10 ns and the time constants of orientation are in the range of 50–1,000 ns depending upon the electric field strength and the chain length of DNA [253].

The degree of orientation ϕ (orientation function) depends on the ratio between the electrostatic potential energy and the thermal energy of the molecule. For molecules with a permanent dipole moment μ_p this ratio is $\beta = \mu_p E/kT$ and the orientation is given by [254]

$$\phi = 1 - \frac{3(\coth\beta - 1/\beta)}{\beta} \tag{13}$$

For molecules with an induced dipole moment $\mu_i = \alpha E$ the ratio between electrostatic and thermal energies is

$$\gamma = \alpha E^2/2kT \tag{14}$$

where α is the polarizability and the orientation function ϕ is given by [253]

$$\phi = \frac{3}{4}\left\{\left(e^{\gamma}\gamma^{-1/2}\int e^{x^2}dx\right) - 1/\gamma\right\} - \frac{1}{2} \tag{15}$$

A quantitative analysis of the experimentally determined dependence of the orientation ϕ of DNA on the electric field strength E shows that the orientation follows the expected induced-dipole mechanism (Eqn. 15) only for DNA molecules of short chain lengths (up to about 50 base pairs, i.e. about 17 nm [251]), and/or at low electric field strengths (with DNA of 258 base pairs up to 4 kV/cm). With high chain lengths and/or high electric field strengths the orientation data suggest the existence of a permanent dipole moment, because the dependence of ϕ on E follows equation 14. However, the symmetry of the native DNA chains in the double helix rules out a permanent dipole directed along the helix axis [255]. Therefore it has been postulated that the induced dipole moment reaches saturated values μ_s at relatively low field strengths E_s [253].

With DNA fragments of intermediate chain lengths the orientation follows an induced-dipole mechanism (Eqn. 15) at low field strengths and changes to a permanent dipole mechanism (Eqn. 14) at higher field strengths. For a DNA fragment with 118 base pairs the transition between the two mechanisms of orientation was observed at field strengths between 11 and 25 kV/cm (the midpoint of the transition E_0 = 17 kV/cm, the transition range E_m = 8 kV/cm).

Both the field strength E_0 at the midpoint of the transition and the width of the transition E_m decrease with increasing chain length [251] (Table 6).

The orientation will cause an anisotropy of the electrical conductivity, i.e. an increase in the direction parallel to the field, and a corresponding decrease in the perpendicular directions. This is mainly due to the fact that the polyion has a much higher mobility parallel to its long axis than perpendicular to it. However, an anisotropy of the internal field may also contribute to this effect. The anisotropy again shows a characteristic times dependence, which is determined by the rotational movement of the whole particle. For the electrolytes under consideration the orientation relaxation times range from seconds to milliseconds. The absolute magnitude of the anisotropy depends upon the completeness of orientation and will decrease

Table 6. Electric properties of DNA restriction fragments

Number of DNA base pairs	E_0 [kV · cm^{-1}]	E_m [kV · cm^{-1}]	$\mu_s \times 10^{-30}$ C.m
69	35.5 ± 1	10.8 ± 0.5	7910 ± 400
118	17.0 ± 1	8.0 ± 1	11730 ± 850
194	9.9 ± 1	6.0 ± 1	13860 ± 1660
258	<7	<4	15100 ± 300
603	–	–	15070 ± 300
4362	–	–	11000 ± 2000

The dipole moment of DNA fragments which is induced at low field strengths is saturated at a field strength E_0 within a transition range E_m to give a constant dipole moment μ_s at high field strengths [251]. The dipole moments are given currently in debye units. 1 debye = 3.33 × 10^{-30} C.m. [173].

if polarization becomes incomplete. This is the case at high frequencies of the polarizing field. As the polarization effect is proportional to the square of the field intensity and therefore independent of its sign, the anisotropy will occur in alternating fields just as well as in d.c. fields, if the frequencies are not too high to prevent the displacement of the counter-ions. The relaxation of polarization at high frequencies mentioned in section 4.1.4.1 will thus cause a decrease in the dipole moment and a dispersion of the anisotropy effect. This dispersion usually lies in the megacycle range (see section 4.1.5). These effects are extremely characteristic of the length, shape and kinetic properties of polyions.

4.1.4.3. Conformational changes of DNA in the bulk of solution: The main mechanism of polarization is the transport of the positive counter-ions from one end of the DNA molecule towards the other end (Fig. 33). A consequence of the resulting depletion of the counter-ions from one end of the molecule is the reduction in the screening of the Coulomb repulsion between the phosphate groups at that end.

The enhanced repulsion between the phosphate groups may overcome the attraction due to hydrogen bonds and stacking forces and strand separation, i.e. partial denaturation, takes place [256] (Fig. 33). Strand separation starts to occur at field strengths higher than a certain threshold value E_t, which is higher with shorter DNA fragments and a higher ionic strength (Table 7). Short DNA fragments in solutions of high ionic strength are not denatured by the electric field. At field strengths just above the threshold value, the strand separation of long DNA (several thousands of base pairs) usually takes a few milliseconds, whereas higher field strengths may accelerate this up to a few microseconds [257].

4.1.4.4. Dissociation effect of the electric field: If the electric field strength exceeds the values of 10^3 V/cm other, nonlinear effects may come into play. The force now becomes strong enough to overcome the electrostatic attrac-

Table 7. Threshold field strengths E_t [kV · cm^{-1}], [257]

Number of DNA base pairs	Ionic strengths		
	10^{-4}	$10^{-3.5}$	10^{-3}
880	30	38	–
30000	3	8	13

tion between the polyion and its ionic atmosphere, producing a charge separation. This effect is analogous to the first and second Wien effects of normal electrolytes [252].

4.1.4.5. Conformational changes of DNA on the electrode surface: Nucleic acid bases, as well as nucleosides and nucleotides, are adsorbed on the hydrophobic electrode surface and may associate there and form a compact surface film [258–263]. With increasing electric field of the electrode the surface film ruptures. At potentials around −1.2 V disturbance of the double-helical conformation of native DNA on the electrode surface was observed [264–268]. At this potential some segments of the DNA molecule are adsorbed onto the electrode surface via the bases, while other unadsorbed segments of the same molecule are repulsed from the electrode surface, mainly electrostatically due to their negatively charged phosphate groups. The DNA molecule anchored in the electrode surface at this potential may thus be subjected to considerable strain which could result in the disruption of the double helix [264–268]. Corformation changes of DNA on the electrode surface were observed for the first time by Miller [269]. The recent results were reviewed by Paleček [270].

4.1.5. Dielectric properties of DNA solutions: If a solution of polyelectrolyte is placed in an electric field E, the counter-ion atmosphere is polarized, as has been mentioned in section 4.1.4.1. The polarization of a sample is represented by a vector P. The magnitude of the polarization is equal to the resultant dipole moment per unit volume. For an isotropic substance (i.e. a substance with properties independent of direction) the vector P lies in the same direction as the vector E. The magnitude of the polarization of a sample containing a large number of molecules is proportional to the magnitude of the electric field strength (see p. 478 in ref. 168):

$$P = \varepsilon_0 \chi E \quad (16)$$

where χ is the electric susceptibility. Rather than using the electric susceptibility it is more common to use the dielectric constant or permittivity ε, which is defined for the isotropic case by

$$\varepsilon = D/E \quad (17)$$

where E is the magnitude of the electric field strength and D is the magnitude of the electric displacement D defined by

$$D = \varepsilon_0 E + P \tag{18}$$

For isotropic media these vectors all have the same direction.

Four contributions to polarization must be considered: electronic, vibrational, interfacial and orientation polarization.

The first contribution, $\alpha_e E_i$, is from the field-induced displacement of the average positions of the electrons relative to the nuclei of the molecule. This electronic contribution is practically independent of the temperature and is proportional to the internal electric field strength E_i. The proportionality constant α_e is called mean molecular electronic polarizability (see section 4.1.1.1, Eqn. 5).

The second contribution, $\alpha_v E_i$, is from the deformation of the nuclear skeleton of the molecule by the electric field. The proportionality constant α_v is called mean molecular vibrational polarizability. To the first approximation α_v is independent of temperature.

With charged polyelectrolytes such as DNA, the interfacial term $\alpha_i E_i$ may become predominant. It arises from the displacement of counter-ions along the polyion, under the influence of the external electric field (see section 4.1.4.1).

Orientation polarization is due to the partial alignment of permanent dipoles. The extent to which dipoles can be oriented by an applied field was calculated by Debye using the Boltzmann distribution law. If the energy of the dipole in the field is small compared with kT, it may be shown that the contribution per molecule in a gas to the mean moment in the direction of the field is given by $\mu^2 E_i/3\,kT$, where E_i is the magnitude of the internal field. As the temperature is increased, the thermal agitation becomes more vigorous and fewer of the permanent dipoles are oriented in the direction of the field.

The magnitude P of the total polarization, which includes all four contributions, is equal to the mean moment in the direction of the field multiplied by the number N molecules per unit volume ($N = N_A/V$) (see p. 22 in ref. 271):

$$P = \frac{N}{3\,\varepsilon_0}\left(\alpha_e + \alpha_v + \alpha_i + \frac{\mu^2}{3\,kT}\right) E_i \tag{19}$$

where α_e, α_v and α_i are the electronic, vibrational, and interfacial polarizabilities, respectively, μ is the permanent dipole moment, $\varepsilon_0 = 8.854 \times 10^{-12}\,J^{-1}\,C^2\,m^{-1}$ is the permittivity of vacuum, and N_A the Avogadro number (see p. 22 in ref. 271).

The electronic contribution is the only one acting at optical frequencies, because only the electrons can move fast enough to follow the high-frequency electric field. Therefore the optical polarizability α^0 is identical with α_e. The vibrational term α_v contributes at frequencies equal to or lower than infrared ones. Permanent moments and interfacial polarization contribute at low frequencies (radio frequencies).

If the concentration of polar molecules is low, it can be shown that $E_i = E(\varepsilon + 2)/3$, where E_i is the internal field and E the applied field. Introducing $N = N_A \rho/M$, we obtain

$$\mathcal{P} = \frac{\varepsilon - 1}{\varepsilon + 2}\frac{M}{\rho} = \frac{N_A}{3\varepsilon_0}\left(\alpha_e + \alpha_V + \alpha_i + \frac{\mu^2}{3kT}\right) \tag{20}$$

The quantity $\mathcal{P}$ is called the molar polarization, and ρ is the density. This equation is accurate for dilute gases and holds approximately for slightly polar liquids, but it does not apply to liquids of high dielectric constant because of the approximation taking $E_i = E(\varepsilon + 2)/3$. In equation 20 short-range molecular interactions are neglected; more complicated theories have been developed by Onsager and Kirkwood to take these effects into account [271, 272].

The dependence of the dielectric constant of a typical polar substance on frequency is shown in Figure 34. At low frequencies all the terms on the right-hand side of equation 19 contribute. As the frequency is increased above the radio frequency range the orientation polarization eventually becomes negligible because there is insufficient time for molecular orientation to occur before the field is reversed (relaxation effect), and the dielectric constant decreases. In the region where the dielectric constant is labelled ε_∞ the polarizability is made up of vibrational, α_v, and electronic, α_e, terms. The decrease in the dielectric constant with increasing frequency is called dispersion. The term dispersion is used in analogy with optical dispersion, which refers to the dependence of the refractive index n (the ratio of the velocity of light in vacuum to that in a medium) upon frequency, and indicates how light of various wavelengths would be dispersed by a prism of the

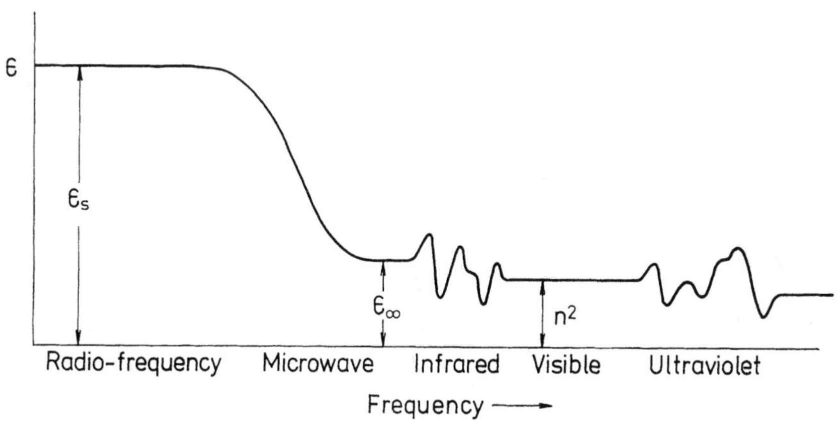

Fig. 34. Variation of dielectric permittivity with frequency; ε_s – static dielectric permittivity, ε_∞ – high frequency permittivity, n – refractive index (see p. 482 in ref. 168).

medium (see p. 516 in ref. 168). The refractive index normally increases on going to higher frequencies (shorter wavelengths): long-wavelength red light has a lower refractive index than short-wavelength violet light. In the neighbourhood of absorption lines in the infrared and ultraviolet region there is an anomalous dispersion decrease in refractive index with frequency (Cotton effect, see p. 483 in ref. 168). If the dielectric constant ε and refractive index n are measured at the same frequency, Maxwell showed that

$$\varepsilon = n^2 \tag{21}$$

The refractive index depends on the wavelength of the light and on the temperature, which are usually specified with subscripts and superscripts, respectively. Thus, n_D^{25} indicates a refractive index at 25 °C taken with monochromatic yellow light of the D line of sodium (see p. 482 in ref. 168).

To a first approximation the refractive index n is related to the optical polarizability α^o by the equation (p. 13 in ref. 271)

$$n - 1 = \frac{N\,\alpha^o}{2\,\varepsilon_0} \tag{22}$$

The relaxation effect that results in a dielectric dispersion is accompanied by an increase in dielectric losses due to the dissipation of energy during the relaxation process. Energy stored per one half of the cycle of a.c. electric field is not completely returned during the second half of the cycle and is changed to heat. (During reorientation of molecules under the influence of an a.c. electric field the hydrodynamic friction forces hinder reorientation, and heat appears due to friction.) A more complete description of the macroscopic dielectric properties of materials provides the so-called complex permittivity

$$\varepsilon^* = \varepsilon' - \mathrm{i}\,\varepsilon'' \tag{23}$$

The real part of ε^* represents the dielectric permittivity ε' as is defined by equation 17. The imaginary part ε'' represents the dielectric losses. The dielectric loss parameter ε'' is related to the dielectric conductivity σ_a [273] and dielectric loss angle δ [274]:

$$\sigma_a = \varepsilon''\,\omega \tag{24}$$

$$\mathrm{tg}\;\delta = \varepsilon''/\varepsilon' \tag{25}$$

The dielectric conductivity σ_a adds to the ionic conductivity of the solution [273].

The frequency dependence of ε^* may be written [275] as

$$\varepsilon^*(f) = \varepsilon_\infty + (\varepsilon_s - \varepsilon_\infty) \int g(\tau)\, \mathrm{d}\tau/(1 + \mathrm{i}\,\omega\tau) \tag{26}$$

The relaxation time τ is defined as the time in which polarization decreases to e^{-1} of the original value. The relaxation time τ is simply related to the rotatory diffusion coefficient Θ (p. 370 in ref. 276):

$$\tau = \frac{\Theta}{2} \tag{27}$$

The dielectric heterogeneity of the sample (e.g. presence of more than one kind of molecule in the sample) can be characterized by the relaxation times distribution function $g(\tau)$.

The form of equation 26 may be estimated for simple polyelectrolytes:

(i) A rod-like polyanion such as short native DNA is expected to yield a simple dispersion curve with one relaxation time [277]. (The relaxation mechanism is essentially determined by the exponential decay of the counter-ion atmosphere polarization or decay of the polyanion orientation.) Equation 26 changes to a more simple form known as the Debye equation of dielectric dispersion [278]:

$$\varepsilon^* = \varepsilon_\infty + \frac{\varepsilon_s - \varepsilon_\infty}{1 + i\omega\tau} \quad (28)$$

$$\varepsilon' = \varepsilon_\infty + \frac{\varepsilon_s - \varepsilon_\infty}{1 + (\omega\tau)^2} \quad (29)$$

$$\varepsilon'' = \varepsilon_\infty + \frac{\varepsilon_s - \varepsilon_\infty}{[1 + (\omega\tau)^2]\,\omega\tau} \quad (30)$$

The graphic representation of ε'' versus ε' (the so-called Cole–Cole or Nyquist plot) is a complete semicircle with the centre on the ε' axis (Fig. 35).

(ii) More flexible polyelectrolytes (high-molecular-weight native DNA or denatured DNA) may be represented as a rigid sequence of rod-like subunits. Here, in general, more dispersion regions are expected, each of which is characterized by a different relaxation time [277]. K.S. Cole and R.H.

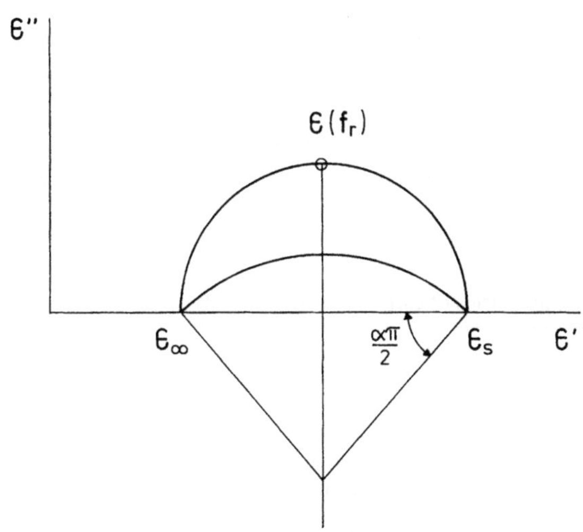

Fig. 35. Cole – Cole plot; f_r is the relaxation frequency.

Cole [279] introduced a distribution parameter α into the Debye equation (Eqn. 28) [280, 281]:

$$\varepsilon^* = \varepsilon_\infty + \frac{\varepsilon_s - \varepsilon_\infty}{1 + (i\,\omega\,\tau_0)^{1-\alpha}} \tag{31}$$

where τ_0 is the mean relaxation time.

The plot of ε' versus ε'' makes a circle with the centre below the ε' axis (Fig. 35). The angle between the abscissa and the line connecting the centre of the circle and the intersection of the circle with the ε' axis is expressed by $\alpha\,\pi/2$.

It follows from the early measurements of Takashima [280–282] that the low frequency dielectric constant of the solution of 0.1 mg/ml native herring sperm DNA, m.w. 3×10^6, is about 380 and the relaxation frequency about 400 Hz. Denaturation causes a decrease in the low frequency permittivity of up to about 110 and an increase of the relaxation frequency to about 5 kHz [280–282]. The parameter α has a value of 0.4–0.6 [283, 284].

It has been shown that DNA aqueous solutions, like other highly charged polyelectrolytes, display two low frequency dielectric dispersions [285]. The first dispersion (α dispersion), due to counter-ion fluctuations along the polyion under the external field, appears in a frequency range 10 and 10^4 Hz, depending essentially on the molecular weight of DNA; both the dielectric increment and the relaxation time are approximately proportional to the square of the length of the rod-like molecules (Table 8). The second dispersion (β dispersion), ranging from about 10 kHz to a few MHz, has been generally explained by the Maxwell-Wagner type of polarisation where the dielectric properties of the solution are given by a polymer domain and the outer medium, with different conductivities and dielectric constants [285].

At MHz frequencies the dielectric dispersion is smaller than at low frequencies and independent of molecular weight. Transverse counter-ion polarization around the helix could account for this dispersion [284].

The dielectric properties of high molecular weight DNA in the presence of monovalent and bivalent ions were extensively examined by Takashima [280–282, 286], but the experimental investigation was limited to dispersion

Table 8. Dielectric properties of calf thymus DNA solutions [284]

M.w.	c	ε_s	ε_∞	τ [s]
2.0×10^5	50 μM	26	3.5	5.5×10^{-5}
7.5×10^5	81 μM	120	3.0	5.8×10^{-4}
1.2×10^6	54 μM	105	1.2	2.6×10^{-3}
1.2×10^6	92 μM	178	2.5	2.1×10^{-3}
1.85×10^6	45 μM	180	4.3	5.7×10^{-3}
1.85×10^6	95 μM	350	4.0	3.7×10^{-3}

at lower frequencies. With low molecular weight DNA (m.w. = 3.2×10^5), both low- and high-frequency dispersions were observed.

From the measurements of the complex permittivity of aqueous Na-DNA gels at microwaves frequencies (10 GHz) in the temperature interval from −15 to 45 °C, it was possible to study the hydration of DNA. The results yield a value of 35 water molecules per one nucleotide interacting with DNA. The presence of strong and weakly bound water was considered. The modified water exhibits a mean relaxation time three times greater value than that of bulk water [287].

4.1.6. Electric properties of DNA in solid state: Samples of nucleic acid bases in solid state are usually prepared by vacuum evaporation and deposition on some substrate (e.g. quartz), where they form layers of a thickness of about 0.5–2.5 μm. Another way of preparing solid state samples is to press powders of bases, nucleosides or nucleotides into discs or to prepare sandwich-like platelets of single crystals 0.25–0.75 mm thick. Both end faces have an area of about 10 mm^2 and are then in contact with evaporated gold [288–291].

Samples of synthetic polynucleotides or natural nucleic acids can be prepared as oriented films by one-directional stroking of a homogeneous gel of nucleic acid with a glass rod on some supporting membrane (e.g. polytetrafluoroethylene, [292]).

4.1.6.1. Electric conductivity and photoconductivity: Samples of bases, nucleosides and nucleic acids in solid state are semiconductors, i.e. the temperature dependence of their electric conductivity is in the absence of light (the so-called "dark conductivity") obeys the Arrhenius law

$$\sigma = \sigma_0 \exp(-W_a/kT) = \sigma_0 \exp(-W_g/2kT) \qquad (32)$$

where W_a stands for the activating energy and $W_g = 2W_a$ for the forbidden energy gap separating a conduction band from a filled band. A similar law was found for the temperature dependence of photoconductivity ([289, 290, 292]; p. 641 in ref. 293).

The values of the forbidden energy gap W_g of the dark conductivity for nucleic acid bases [288] and DNA (p. 661 in ref. 293) lie in the range 1.8–2.6 eV; the specific conductivity σ_0 varies over the very broad range of $10^{-16} - 10^{11}\ \Omega^{-1}\ \mathrm{cm}^{-1}$ ([288–290]; p. 661 in ref. 293). With increasing humidity of the sample the specific conductivity increases and the type of conductivity probably changes from the hole and electronic type to proton conductivity [294].

D.c. conductivity of adenine sulphate at room temperature was found to be lower than $10^{-16}\ \Omega^{-1}\ \mathrm{cm}^{-1}$. The a.c. conductivity at 100 Hz is more than three orders of magnitude higher than the d.c. conductivity. At low temperatures (around 100 K) the frequency dependence of a.c. conductivity is proportional to ω^s ($\omega = 2\pi f$, where f is frequency) with $s > 1$ over the whole studied frequency range of 10 Hz–100 kHz. The exponent s decreases very weakly with increasing temperature and is equal to 1.06 above 1 kHz at room temperature. Conductivity is probably controlled by electronic impurities [291].

The activation energy of native DNA in the solid state is 1.2 eV, for denatured DNA $W_a = 1.5$ eV and the specific conductivity is 10^{-15}–10^{-10} Ω^{-1} m^{-1} [295]. Activation energy decreases and specific conductivity increases with increasing water content in the sample [295]. At low levels of hydration conductivity is of the electronic type; at hydration over 36% the proton conductivity and at hydration 70% the ionic type of conductivity participate. It has been suggested that electronic conductivity is associated with the overlap of π orbitals of nucleic acid bases [295, 296]. D.c. conductivity is predominantly of the ionic type [295]. Voltampére characteristics are linear up to the electric field strength in the order of 10^6 V m^{-1} [297].

The photoconductivity of solid state samples of DNA and synthetic polynucleotides was observed in the infrared, visible, and ultraviolet ranges of the spectrum [292, 297, 298]. The photocurrent in the excitation region below 4.5 eV arises from bolometric effects. Only above 4.5 eV does some sort of true photoconductivity coexist with the bolometric effect [297].

4.1.6.2. Dielectric, ferroelectric and piezoelectric properties: Knowledge of the low frequency dielectric permittivity of solid samples of bases and nucleic acids is essential for a rough estimate of the polarization energy of an electron and a hole in the sample [289, 290] as well as for the study of the dynamics of the DNA double helix (p. 662 in ref. 293; [295]).

The permittivity of adenine in the solid state changes by less than 2% in a frequency range form 10 to 100 kHz and increases monotonously with increasing temperature, the increase being about 5% within the temperature range of 100–320 K. At 300 K the values of relative permittivity are 4.4 ± 0.3 for adenine hemisulphate hydrate and 6.3 ± 0.5 for adenine sulphate [291].

In the frequency range 10 mHz–10 GHz four ranges of dielectric relaxation of the solid samples of DNA were observed:

(1) Low frequency dispersion (around 10 MHz) results from the electrode polarization.
(2) Relaxation at 10–100 Hz and at 500 Hz is explained by jumps of Na^+ ions in the double helix and by the movements of bases and the sugar–phosphate backbone.
(3) At 10 kHz relaxation due to the reorientation of bound water molecules and free bases at the end of the DNA double helix is observed.
(4) The high frequency region of dispersion (around 10 GHz) results from the dipolar relaxation of free water molecules [295, 299].

Polarization in the solid state samples of DNA can be sustained a long time after the removal of the electric field. Solid DNA can thus be regarded as an electret [295, 300] and shows ferroelectric properties [301–304]. An electret is a dielectric that contains a net orientation of dipoles or a space charge resulting form a trapped charge [305]. Electrets are most often formed by applying a high electric field to a dielectric at some elevated temperature and then cooling the sample with the field applied to freeze the dipole orientation. In electrets the polarization or space charge is a non-equilibirium condition. In ferroelectric materials, however, the ferroelectric state, a state in which a large net polarization exists, is an equilibrium condition. One way in which ferroelectricity might occur is by what some call the polarization catastrophe, a condition in which polarization in a dielectric becomes very large. As long as the polarization, P, is very small it is directly proportional to the applied field, E, with a proportionality constant χ, called susceptibility, $P = \chi E$. However, when polarization becomes large, it begins to contribute to the electric field at the site of the dipoles so that the effective electric field becomes $E + \lambda P$, where λ is a proportionality constant. Now $P = \chi (E + \lambda P)$ and the solution for P yields $P = [\chi/(1 - \lambda\chi)]E$. Therefore, as $\lambda\chi$ approaches one, susceptibility becomes extremely large and the material becomes ferroelectric [305].

Ferroelectric materials should exhibit a hysteresis loop if the relation between P and E is measured [305]. Hysteresis loops of solid DNA samples have been studied by several authors [301–305].

With solid state samples of bases, nucleosides and oriented DNA films the piezoelectric effect was observed [306–308]. A material is piezoelectric if a stress produces a change in polarization. The polarization can either be zero before the stress is applied or the stress can change the magnitude of an already existing polarization [305]. The piezoelectric effect was observed in a number of biopolymers and living tissues [309, 310].

4.1.7. Nucleic acids in electromagnetic fields: The mechanism of the interaction of nucleic acids with electromagnetic fields of different frequencies is the subject of spectroscopy. Atomic spectra involve only transitions of electrons from one electronic energy level to another. Molecular spectra involve transitions between rotational and vibrational energy levels in addition to electronic transitions. The chemical and physical effects of various types of radiation are quite different, and these differences can be understood in terms of the differing energies of the photons. The energy E of a single photon is given by

$$E = h\nu \tag{33}$$

where h is Planck constant, 6.626×10^{-34} J s (p. 487 in ref. 168).

In the absorption process a photon of electromagnetic radiation is absorbed by a molecule that is in a particular quantum state characterized by a certain wave function and a definite energy E''. There is a transition to another quantum state, characterized by another wave function and a higher energy E'. In the emission process, a photon is emitted by a molecule which makes a transition from a higher energy state to a lower energy one. For either the absorption or emission process, the total energy must be conserved. This leads to the relation

$$h\nu = E' - E'' = \Delta E \qquad (34)$$

where E' is the energy of the higher energy state, E'' is the energy of the lower energy state, and ν is the frequency of the radiation. (p. 489 in ref. 168.)

Electronic transitions lead to absorption spectra in the visible and ultraviolet regions. For organic compounds this involves promotion of electrons in n, σ and π orbitals in the ground state to σ^* and π^* antibonding orbitals in the excited state. Since n electrons do not form bonds there are no antibonding orbitals associated with them. In the excited state the density of electrons between the nuclei is lower than in the ground state and thus the molecule is in a higher energy state (p. 514 in ref. 168). The energy for the lowest electronic transition is that required to raise an electron from the highest filled level to the lowest unfilled level.

The energy required for a $\sigma \rightarrow \sigma^*$ transition is generally so high that the absorption occurs in the vacuum ultraviolet. Thus, saturated hydrocarbons and other compounds in which all valence shell electrons are involved in single bonds do not absorb in the ordinary ultraviolet or visible regions.

Ultraviolet absorptions due to $n \rightarrow \sigma^*$ transitions are shown by compounds that contain nonbonding electrons on oxygen, nitrogen, sulphur or halogen atoms.

Groups such as C=C, C=O, –N=N– and –N=O, which cause absorption above 180 nm, are called chromophores. Unsaturated molecules containing heterocyclic atoms, such as nucleic acid bases, can exhibit $n \rightarrow \pi^*$ and $\pi \rightarrow \pi^*$ transitions. The stronger absorption band of nucleic acid bases around 260 nm is due to $\pi \rightarrow \pi^*$ transition and the weaker absorption band around 280 nm is due to $n \rightarrow \pi^*$ transition (p. 201 in ref. 312, [311]).

4.1.7.1. Opitcal activity – circular birefringence and circular dichroism: The dependence of refractive index and absorption coefficient on wavelength can also be measured with circularly polarized light. In circularly polarized light the electric vector rotates as the light beam advances. If the electric vector rotates clockwise as observed when facing the light source, the light is said to be right-handed circularly polarized light. If it rotates counterclockwise, it is termed left-handed circularly polarized light. When left and right circularly polarized beams of equal intensity are combined, they yield plane-polarized light. In plane-polarized light the electronic vector remains in a plane.

A separated beam of circularly polarized light may be obtained by passing the plane-polarized light through a quarter-wave plate oriented at 45° to the direction of the electric vector of the polarized light. Since the quarter-wave plate may be inclined to the right or to the left, both right and left circularly polarized light may be obtained in this way.

Figure 36 gives the variation of refractive index (dispersion curve) and the variation of absorption coefficient (absorption curve) with frequency for an optically active material measured with left and right circularly polarized light. The difference Δn in refractive index for the two components is referred to as circular birefringence, and the difference ΔA in absorption is referred to as circular dichroism. The plot of Δn (or of the rotation angle ϕ) versus λ is referred to as optical rotatory dispersion (ORD) spectrum, and the plot of ΔA versus λ is referred to as the circular dichroism (CD) spectrum (p. 517 in ref. 168; [313, 314]).

When a rotatory dispersion curve rises on going to shorter wavelengths it is referred to as normal, because this is the normal pattern for a refractive index curve. When an absorption band causes the effects shown in Figure 36 the whole phenomenon is referred to as the Cotton effect. In contrast to ordinary dispersion a strong absorption band may or may not produce a large effect on rotatory dispersion, and a weak absorption band may produce a large effect on rotatory dispersion.

Substances that show circular birefringence and circular dichroism are said to be optically active. They can be divided into two classes: one in which optical activity is found only in the crystal form, e.g. quartz, and the other in which optical activity is found in the solid, gaseous and liquid states of the pure substance or in solutions. Optical activity arises in the former group due to a right- or left-handed helical structure in the crystal and disappears when this structure is melted. Substances in the latter category are optically active because of the asymmetry of the molecule itself. For a molecule whose mirror image is not superimposable on the molecule (so-called chiral molecules), left and right circularly polarized light beams have different

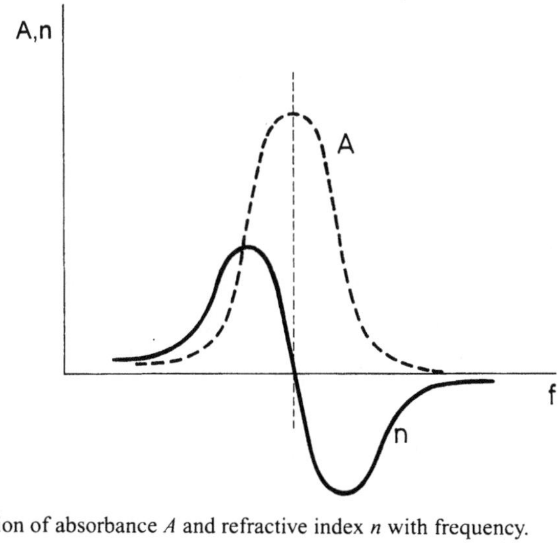

Fig. 36. Variation of absorbance A and refractive index n with frequency.

refractive indices and correspondingly different absorption coefficients (p. 519 in ref. 168).

Nucleic acids are chiral molecules. ORD and especially CD spectroscopy has been used extensively to investigate their conformation [315–319].

4.1.7.2. Optical anisotropy – linear birefringence and linear dichroism: The refractive index and absorption coefficient can also be measured with plane-polarized light. We speak about linear birefringence and linear dichroism. Substances that show linear birefringence and linear dichroism are said to be optically anisotropic. The phenomenon of optical anisotropy has been mentioned in section 4.1.1.2 and will be treated in more detail in section 4.3.1.

4.1.7.3. Anomalous resonant microwave absorption of DNA: Anomalous resonant absorption of electromagnetic energy by aqueous solutions of oligonucleotides at frequencies in the microwave region (gigahertz) has been theoretically predicted [320–322]. The experimental results are controversial [323–326]. The resonance at microwave frequencies could result from the compressional elasticity of the nucleic acid molecule [327]. The compressional elasticity constant can be determined form the sound-wave velocity [328, 329]. A mechanical resonance appears at the wavelengths corresponding to the odd multiple of $L/2$, where L is the length of DNA chain.

Hydrodynamic coupling of such vibrations to the surrounding solvent would preclude the existence of sharp resonances. When the molecular motion and the nearby solvent motion are even slightly decoupled, a well-defined resonant absorption peak appears.

As the DNA compresses along its length, the linear charge density varies and a "slip" electric field responds. The molecular compression amplitude is resonant and so is the "slip" electric field. Substantially more energy is stored in the electric fields and ion currents than in the DNA molecular motion. The fields and currents are a low dissipation system, however, and thus the combination of a slipping molecule and the coupled electric field is still a high-quality vibrating system.

The origin of resonant absorption could be connected with a dissipative term arising from the slippage of the counter-ion atmosphere over the surface of the DNA molecule. The electric field at the interface acts to shear the negatively charged DNA molecule from the positively charged counter-ion atmosphere, pushing them in opposite directions. The counter-ions respond to two sets of forces. Because they are charged they respond to the microwave electric field and because of their viscous coupling to the water, they are pushed along by any difference between their velocities and the streaming velocity of the water. As long as some slipping at the interface arises, relatively strong, narrow absorption lines are to be expected.

Measurement of the microwave absorption is very difficult and the observed phenomena are extremely weak – not much above the noise level. The apparatus may be susceptible to the generation of similar resonance absorption peaks unrelated to the sample. Consequently, the published phenomena of resonant absorption at microwave frequencies must therefore be accepted with caution.

4.2. Magnetic properties

4.2.1. Diamagnetic anisotropy: Paramagnetic molecules (molecules with a permanent magnetic dipole moment, magnetic susceptibility $\chi > 0$) are aligned by a magnetic field in the direction of the vector of magnetic induction B. In diamagnetic molecules (magnetic susceptibility $\chi < 0$) the magnetic field induces a dipole moment opposite to the direction of B. The orientation of diamagnetic molecules in a magnetic field depends on their diamagnetic anisotropy $\Delta\chi = \chi_{\parallel} - \chi_{\perp}$ where $\chi_{\parallel}$ and $\chi_{\perp}$ are the magnetic susceptibilities in the directions parallel and perpendicular to the molecular rotational symmetry axis. The magnetic field causes rotation of diamagnetic molecules so that the axis of the lowest numerical magnetic susceptibility comes nearer to the field direction.

Diamagnetic molecules with $|\chi_{\parallel}| < |\chi_{\perp}|$ (i.e. $\Delta\chi > 0$) orient their rotational symmetry axis parallel to B (Fig. 37a). Such orientation is shown by bacteriophages *Pf1* and *fd* [330], fibrinogen and fibrin [331], skeletal muscle actin [332] and α-helical synthetic polypeptides such as poly (γ-benzyl glutamate) and poly(γ-ethyl glutamate) [333]. The origin of the magnetic anisotropy of phages and polypeptides is the α-helical peptide bonds arranged roughly parallel to the phage and/or polypeptide long axis.

Diamagnetic molecules with $|\chi_{\parallel}| > |\chi_{\perp}|$ (i.e. $\Delta\chi < 0$) orient their rotational symmetry axis in a plane perpendicular to B (Fig. 37b). Such orientation is shown by bases of DNA, native DNA [334], double-helical synthetic polynucleotides such as poly(A) · poly(U) and poly(A) · 2poly(U) [333], and collagen [335].

Orientation in a magnetic field was observed with tRNA aggregates [336] and bacterial cell walls [337], as well as with several spherical plant and animal viruses [338]. Magnetic anisotropy was not observed with red blood cells [339].

The origin of the magnetic anisotropy of the polynucleotides is the diamagnetic anisotropy of bases [334, 340–343]. Diamagnetism results form the induction of orbital currents in the electron clouds of molecules by an external magnetic field [344]. The delocalized π electrons of bases (Fig. 23) can be induced to circulate about the whole ring when the magnetic field is perpendicular to the plane of the ring, but not when the field is directed along the plane of the ring. Therefore $|\chi_{\parallel}| > |\chi_{\perp}|$.

4.2.2. Degree of orientation of nucleic acids in a magnetic field: The degree of orientation of a molecule with a magnetic anisotropy $\Delta\chi$ in a

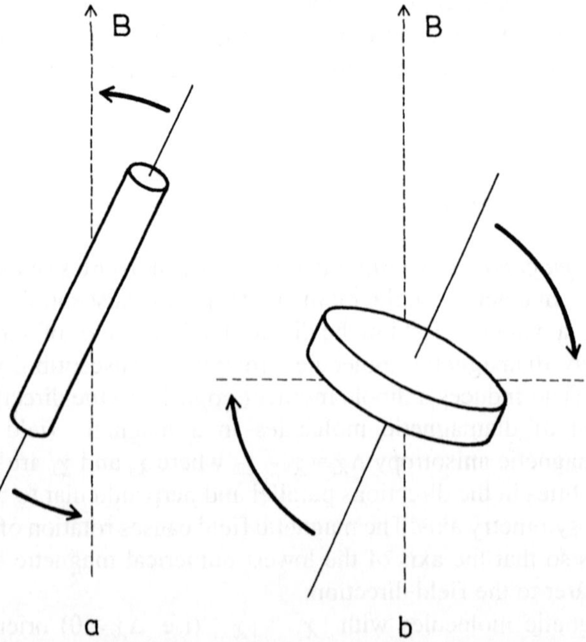

Fig. 37. Orientation of diamagnetic molecules in a magnetic field. (a) diamagnetic anisotropy; $\Delta\chi$ is positive ($|\chi|_{\parallel} < |\chi|_{\perp}$); (b) diamagnetic anisotropy; $\Delta\chi$ is negative ($|\chi|_{\parallel} > |\chi|_{\perp}$).

magnetic field depends on the ratio β between the magnetic and thermal energies:

$$\beta = \frac{\Delta\chi B^2}{kT} \tag{35}$$

Because the diamagnetic anisotropy $\Delta\chi$ is usually very low, $\beta \ll 1$, and from equation 3.10 it can be seen that $\beta = \Phi$, i.e. the degree of orientation is given by β.

Diamagnetic anisotropy of DNA base pairs exceeds that of benzene ($\Delta\chi_{\text{benzene}} = -9.48 \times 10^{-28}$ J T^{-2} [334]) by a factor of 1.098 for an adenine · thymine pair, and of 0.827 for a guanine · cytosine pair [334]. At room temperature and conventionally available magnetic fields (B around 1 T) the magnetic orientation of single, even though strongly anisotropic, molecules such as benzene or DNA bases is thus very small. According to equation 35 the degree of orientation is $\beta = 10^{-17}$ [345]. However, β can be increased when a great number of such molecules, N, are rigidly fixed together parallel to one another, since the effective diamagnetic anisotropy of such a molecular cluster is proportional to N [345]. The degree of orientation β is then given by

$$\beta = \frac{N\Delta\chi B^2}{kT} \tag{36}$$

for a rigid rod. In the case of a flexible polymer chain

$$\beta = \frac{2P}{l_0} \frac{\Delta\chi B^2}{kT} \tag{37}$$

where P is the persistence length and l_0 the length of a monomer unit [334].

For native calf thymus DNA (m.w. 4.1×10^6) in 0.15 M NaCl with 0.015 M sodium citrate buffer at pH 7.0, β was found to be equal to 0.88% at $B = 12$ T [334]. Taking $l_0 = 0.34$ nm it was calculated from equation 37 that $N = 262$ and $P = 44.5$ nm. The method of magnetic orientation thus makes it possible to measure persistence lengths of polymer samples. Conformational changes of DNA in a magnetic field were not observed.

4.3. Determination of the degree of DNA orientation in solution in electric and magnetic fields

4.3.1. Methods based on optical anisotropy: If the optical polarizabilities α^o of a substance have different values in different directions, the substance is called optically anisotropic. The measure of the anisotropy is the difference $\Delta\alpha$ between polarizabilities $\alpha_{\parallel}$ and $\alpha_{\perp}$ in two mutually perpendicular directions (see p. 18 in ref. 271), as has already been mentioned in section 4.1.1.2 (see p. 18 in ref. 271):

$$\Delta\alpha = \alpha_{\parallel} - \alpha_{\perp} \tag{38}$$

The polarizability $\alpha_{\parallel}$ is usually defined as the polarizability in the direction of the long axis of a rod-like or ellipsoidal molecule or in the direction of some external field that can orient the molecules.

The polarizabilities of molecules at wavelengths in the ultraviolet and visible regions are predominantly the polarizabilities of σ and π electrons. The σ electrons are almost isotropic and do not contribute to anisotropy. In the DNA molecule optical anisotropy is due almost completely to π electrons of the rings of heterocyclic bases. Delocalized π electrons are polarized mainly in the plane of the bases and very weakly in the perpendicular direction. (In other words, the transition dipole moment μ associated with the $\pi \rightarrow \pi^*$ absorption band is oriented in the plane of the bases.) Therefore, the optical polarizability in the direction parallel to the plane of the bases (i.e. perpendicular to the long axis of DNA), $\alpha_{\perp}$, is very much larger than the polarizability $\alpha_{\parallel}$ in the direction parallel to the long axis of DNA, $\alpha_{\perp} >> \alpha_{\parallel}$ (p. 29 in ref. 271). (In section 4.1.1.2 the symbols $\alpha_{\parallel}$ and $\alpha_{\perp}$ were used for the polarizabilities in the directions parallel and perpendicular to the plane of the bases, respectively.) Consequently, because absorbance A and refractive index n depend on α^o; see Eqn. 23, p. 13 in ref. 271, the linearly polarized light is most absorbed when the electric vector of the light is oriented in the plane of the bases, i.e. perpendicular to the long axis of DNA. If we denote the absorbances of the light linearly polarized in the directions parallel and perpendicular to the long axis of DNA as $A_{\parallel}$ and $A_{\perp}$ respectively, then $A_{\perp} > A_{\parallel}$ and the difference ΔA, called linear dichroism

$$\Delta A = A_{\parallel} - A_{\perp} \tag{39}$$

is negative for DNA [346, 347].

The same is true for the refractive index n. The difference

$$\Delta n = n_{\parallel} - n_{\perp} \tag{40}$$

is called linear birefringence and is negative for DNA (p. 30 in ref. 271).

In a solution of anisotropic molecules, because of random orientations, no anisotropy is observable. If the anisotropic molecules are oriented, the optical anisotropy of the solution can be observed and will increase with the degree of orientation. The classical methods for the determination of the degree of orientation are thus measurements of the linear dichroism (Eqn. 39) and linear birefringence (Eqn. 40), where $n_{\parallel}$, $n_{\perp}$, $A_{\parallel}$ and $A_{\perp}$ are the refractive indices and absorbances of the solution for the light linearly polarized parallel and perpendicular to the applied orienting (usually hydrodynamic, electric or magnetic) fields.

The birefringence, if present, may be observed at any wavelength while the dichroism (if present) will necessarily appear in the absorption bands only. The main advantage of the measurement of birefringence compared with dichroism is the possibility of attaining much higher sensitivities, since birefringence is usually measured in the visible range out of the absorption bands, and without the need for monochromatic light (see p. 4 in ref. 271). Thus the birefringence measurement is used for determination of a very low degree of orientation in the magnetic field. However, dichroism measurements give more information than birefringence (p. 4 in ref. 271). Analysis of the dependence of dichroism on the wavelength is easier than the analysis of the birefringence spectrum (an S-shaped Cotton effect).

The birefringence Δn as well as the dichroism ΔA are proportional to the square of the applied orienting field strength (pp. 27, 35 in ref. 271; [345]):

$$\Delta n = K \lambda E^2 \tag{41}$$

$$\Delta A = K_D \lambda E^2 \tag{42}$$

$$\Delta n = CM \lambda H^2 \tag{43}$$

where K is the Kerr constant, K_D the dichroism Kerr constant, and CM the Cotton-Mouton constant. Equation 41 is known as the Kerr law.

The absorption of plane-polarized light depends generally on the angular displacement γ of the molecular transition dipole moment μ with respect to the electric vector E of the polarized light. The absorption is proportional to the square of the scalar product μE of these two vectors and hence to $\cos^2\gamma$:

$$A_\gamma = A_0 \cos^2 \gamma \tag{44}$$

Thus A_γ is the absorbance of light, whose electric vector forms the angle γ with the transition moment μ, and A_0 is the absorbance of light polarized parallel to the transition moment μ. (A_γ is the absorbance of a polarized light by a molecule with a transition dipole μ oriented at an angle γ to the plane of polarized light) [348]. Since in the absence of any orientation all values of γ are equally probable, $\cos^2\gamma$ will take on average a value of 1/3 when a large number of randomly oriented molecules are considered. Consequently, the average value of A_γ will be

$$\overline{A_\gamma} = \overline{A_0}\ \overline{\cos^2\gamma} = \frac{A_0}{3} = A \tag{45}$$

where A is the absorbance of the non-oriented sample or absorption of conventional nonpolarized light (p. 33 in ref. 271; [349]). In an orthogonal coordinate system the x-, y-, and z-components of the vector A_γ follow the relation [348]

$$A_x + A_y + A_z = A_0 = 3A \tag{46}$$

When an electric field inducing orientation of the molecules is applied along the z-axis, the absorbance A of the light polarized parallel to the z-axis is equal to the z-component of A_γ:

$$A_\parallel = A_z \tag{47}$$

The molecules are oriented only in the z-direction, and are not forced to orient in either the x- or in y-directions, i.e. they are free to rotate around the z-axis and, consequently, $A_x = A_y$. Thus, the absorbance $A_\perp$ for light polarized perpendicular to the z-axis is [348]

$$A_\perp = \frac{A_x + A_y}{2} \tag{48}$$

A rearrangement using equations 46 and 47 yields (p. 35 in ref. 271, [349])

$$A = \frac{1}{3} A_\parallel + \frac{2}{3} A_\perp \tag{49}$$

It is useful to define reduced dichroism as the ratio $\Delta A/A$, where $\Delta A = A_\parallel - A_\perp$ (Eqn. 38). For reduced dichroism $\Delta A/A$ it follows from equations 46–49 (p. 36 in ref. 271)

$$\frac{\Delta A}{A} = \frac{3}{2}(3\cos^2\gamma - 1) \tag{50}$$

The angle γ between the electric vector of light and the transition dipole moment μ of the molecule depends on the orientation of the molecule with respect to the polarization plane of the light. When an external electric field induces complete orientation of the molecules with their main axis along the direction of the electric field, $\gamma = \alpha$, where α is the angle between the transition dipole moment μ of the molecule and the main axis of the molecule.

Generally the degree of orientation of molecules Φ varies between zero at zero electric field to unity at infinite field and the reduced dichroism is given by the equation [350]

$$\frac{\Delta A}{A} = \frac{3}{2}(3\cos^2\alpha - 1)\,\Phi \tag{51}$$

The reduced dichroism can be either positive or negative, depending upon whether α lies close to 0° or 90°. The limits for complete orientation ($\Phi \rightarrow 1$) are [350]

$$\Delta A/A \rightarrow 3 \qquad \text{for } \alpha = 0°$$

$$\Delta A/A \rightarrow -3/2 \qquad \text{for } \alpha = 90°$$

DNA in the B-conformation has $\alpha = 90°$ and $\Delta A/A \rightarrow -3/2$.

From the experimentally determined value of the reduced dichroism the average angle α of the transition dipole moment μ with respect to the long axis of the molecule can be evaluated with the aid of equation 51 [350–353]. It follows from equation 51 that the dichroism is zero when $\cos^2\alpha = 1/3$, i.e. when $\alpha = 54°45'$ (see p. 37 in ref. 721; [354–356]. This is the so-called "magic angle". If the angle between the polarization plane of the polarized light and the vector of the external electric field is equal to this magic angle, the absorbance of the polarized light is not influenced by orientation of the molecules, i.e. the absorbance is independent of the electric field strength [357]. Absorbance changes found after the application of the electric field at magic angle measurements cannot be attributed to orientation phenomena, but result from field-induced reactions or conformational changes [350, 355, 356, 358–360].

From electric dichroism measurements the kinetic parameters of the orientation in the electric field (rotation time constant), polarizability, curvature, bending amplitude, and bending time of curved DNA fragments [360], as well as the rate constant of the helix–coil transition [361], can be determined experimentally and compared with theoretical calculations [362, 363].

4.3.2. NMR methods: In high-resolution NMR experiments performed with gases and liquids it is usually supposed that the solute molecules move randomly in solution, the direction-dependent nuclear interactions are averaged by the motion of the solute molecules, and only mean values can be recorded. Thus chemical shifts and indirect spin–spin couplings behave as scalar quantities and magnetic dipolar and electric quadrupolar interactions are averaged to zero [341].

If the solute molecules do not have a random orientation but are oriented in some way (e.g. by electric or magnetic fields), much more information is available from the NMR spectra (the NMR analogue of the Kerr and Cotton-Mouton effects). The orientation of molecules that possess magnetic anisotropy can be achieved by the magnetic field of an NMR spectrometer. Due to the orientation of solute molecules the anisotropic intramolecular interactions, such as dipolar and quadrupolar spin couplings, are partly preserved, leading to splitting of the resonance lines in the NMR spectrum [341]. These NMR alignment experiments make it possible to determine some molecular parameters and to study intermolecular interactions.

In this kind of experiment the main problem is the small size of the measured effect. A very strong magnetic field of sufficient homogeneity is therefore required. The first magnetic field alignment NMR spectra were detected in 1978 by Lohman and MacLean [364, 365]. At the fields available nowadays of up to 14.57 T (^{1}H frequency of 620 MHz), the effect of the partial orientation of solute molecules possessing a magnetic anisotropy can be well observed [341]. A 900 MHz ($B = 21.2$ T) NMR spectrometer has been developed, which will extend the range of molecules to be studied, including short DNA sequences [366].

NMR magnetic field alignment experiments provide a simple and efficient method for measuring the magnetic susceptibility anisotropy of the solute molecules at a low solute concentration, and at higher concentrations the intermolecular interactions of solute molecules (cluster formation, stacking, etc.) [341].

The optimum situation for obtaining accurate splitting consists of the simultaneous occurrence of small line widths, a strong magnetic field, and a large magnetic susceptibility anisotropy $\Delta\chi$. In general, both the line width and $\Delta\chi$ increase with the size of the molecule studied. The line widths determine whether splitting can be observed [341].

If cluster formation (stacking) appears, reliable information can be obtained from the concentration dependence of the observed line splitting by fitting the curve to trail complex formation constants [341]. The measured splitting, $\Delta\gamma_i$, will be a weighted average over all species (monomers, dimers, etc.) present. When rapid exchange between the clusters occurs, assuming a single complex formation constant for the attachment of an additional molecule, the splitting $\Delta\gamma_i$ is given by [341]

$$\Delta\gamma_i = x^{(1)}\Delta\gamma_i^{(1)} + x^{(2)}\Delta\gamma_i^{(2)} + \ldots + x^{(n)}\Delta\gamma_i^{(n)} \tag{52}$$

$$x^{(n)} = \frac{n\,C_{(n)}}{C_{(1)} + 2C_{(2)} + \ldots\, n\,C_{(n)}} \tag{53}$$

where $C_{(n)}$ is the concentration of n-mer and $x^{(n)}$ its mole fraction [367].

Using equations 52 and 53 one can calculate

$$\Delta\chi^{(2)} = (1 + x^{(2)})\,\Delta\chi^{(1)} \tag{54}$$

where $\Delta\chi^{(1)}$ and $\Delta\chi^{(2)}$ are the magnetic susceptibility anisotropies of monomer and dimer, respectively. As the splitting is proportional to $\Delta\chi$, NMR provides a simple way of studying the stacking if the monomer value $\Delta\chi^{(1)}$

is known. At any concentration the amount of stacking can easily be calculated, as was demonstrated for a series of porphyrins [368].

4.3.3. Electrochemical methods: Bases, nucleosides and nucleotides are surface-active substances. They lower the value of the differential capacitance C of the electrode double layer in a certain region of potentials ψ at which they are adsorbed on the electrode surface [258–262]. In 1965 it was discovered that nucleic acid bases are able to associate at the electrode surface and to form a condensed layer there, a compact film [259]. In the region of potentials at which the surface of the electrode is covered with the compact film of bases, a sort of pit appears on the $C-\psi$ curves or alternating current (a.c.) polarograms [259]. Shortly after the discovery of the two-dimensional condensation of bases in our laboratory this interesting physical property of nucleic acid components began to be studied intensely in laboratories in Berlin [260], Jülich (Germany) [261], Georgetown [258], Norman (USA) [262] and Brussels (Belgium) [263]. The mechanism of the film formation (two-dimensional condensation) was quantitatively analyzed by Canac [369], Avrami [370], Fleischmann and Thirsk [371], de Levie [258, 372, 373], Retter *et al.* [374–376], Buess-Herman *et al.* [377, 378], Pospíšil *et al.* [380, 381], Bhattacharjee and Rangarajan [382], and others.

The ability of nucleic acid components to form a condensed layer on the electrode surface seems to be connected with the reorientation of adsorbed bases from a flat to a perpendicular position. From measurements of the differential capacitance of the electrode double layer or from electrocapillary measurements the surface concentration Γ of the adsorbed molecules and the area A required for an adsorbed molecule on the electrode surface can be determined [258, 383–390]. Using data obtained for the crystal structure of bases, the area which would be occupied by one adsorbed molecule in different surface orientations can be evaluated and compared with the experimentally determined area A. From these calculations it has been concluded that at low surface concentrations (the so-called dilute adsorption region) the adsorbed bases lie flat on the electrode surface. In a compact layer the adsorbed bases seem to adopt a perpendicular surface orientation [383, 384]. Similar reorientation from a flat to a perpendicular stance has also been observed with several nucleosides and nucleotides [378, 379, 383, 385, 386]. Some nucleosides can probably adopt two different perpendicular orientations on the electrode surface and thus form two pits (i.e. pits in two different potential regions) on the $C-\psi$ curves, for example, adenosine [383, 391] and 6-methyladenosine [383]. Surface optical measurements such as SERS (**s**urface-**e**nhanced **R**aman **s**pectroscopy) [392–396], ellipsometry [397–399] and other optical techniques [400–403] complement the polarography, voltammetry and electrocapillary experiments in order to determine more exactly the orientation of adsorbed molecules in the electric field of the electrode. Recently scanning tunneling microscopy (STM) and atomic force microscopy (AFM) have also appeared to be extremely useful in this respect [404–410].

References

1. JD Watson, NH Hopkins, JW Roberts, JA Steitz and AM Weiner, *Molecular Biology of the Gene*, The Benjamin/Cummings Publishing Co., Menlo Park (1987).
2. J Lewin, *Genes*, J Wiley and Sons, New York (1987).
3. DB Davies, W Saenger and SS Danyluk (eds), *Structural Molecular Biology: Methods and Applications*, NATO-ASI Ser. A, Plenum Press, New York (1982) Vol. 45.
4. A Kornberg, *DNA Synthesis,* Freeman Publishing, San Francisco (1974).
5. POP Ts'o in *Basic Principles of Nucleic Acid Chemistry*, POP Ts'o (ed), Academic Press, New York (1974) Vol. I, p. 453.
6. P Tollin, HR Wilson and DW Young, Nature 217 (1968) 1148.
7. AEV Haschemeyer and A Rich, J. Mol. Biol. 27 (1967) 369.
8. A Saran, D Perahia and B Pullman, Theor. Chim. Acta 30 (1973) 31.
9. RE Dickerson, J. Mol. Biol. 166 (1983) 419.
10. RE Dickerson, Sci. Amer. 249 (1983) 3.
11. RE Dickerson and HR Drew, J. Mol. Biol. 149 (1981) 761.
12. HR Drew and RE Dickerson, J. Mol. Biol. 151 (1981) 535.
13. HR Drew and RE Dickerson, J. Mol. Biol. 152 (1981) 723.
14. W Fuller, MHF Wilkins, HR Hamilton and S Arnott, J. Mol. Biol. 12 (1965) 60.
15. FM Pohl and TM Jovin, J. Mol. Biol. 67 (1972) 375.
16. AH-J Wang, GJ Quigley, FJ Kolpak, GA van der Marel, JH van Boom and A Rich, Science 211 (1980) 171.
17. J Feigon, AH-J Wang, GA van der Marel, JH van Boom and A Rich, Science 230 (1985) 82.
18. AH-J Wang, RV Gessner, GA van der Marel, JH van Boom and A Rich, Proc. Natl. Acad. Sci. USA 82 (1985) 3611.
19. AT Ansevin and AH Wang, Nucl. Acids Res. 18 (1990) 6119.
20. J Marmur, R Rownd and CL Schildkraut, Progr. Nucl. Acid Res. Mol. Biol. 1 (1963) 231.
21. W Guschlbauer, *Nucleic Acid Structure*, Springer-Verlag, Heidelberg (1976).
22. J Marmur and P Doty, J Mol. Biol. 5 (1962) 109.
23. CL Schildkraut and S Lifson, Biopolymers 3 (1965) 195.
24. WS Yen and RD Balke, Biopolymers 19 (1980) 681.
25. AA Travers and A Klug in *DNA Topology and Its Biological Effects*, NR Cozzarelli and JC Wang (eds), Cold Spring Harbor Lab. Press (1990) p. 57.
26. AA Travers, Ann. Rev. Biochem. 58 (1989) 427.
27. JA Rice, DM Crothers, AL Pinto and J Lippard, Proc. Natl Acad. Sci. USA 85 (1988) 4158.
28. L Marrot and M Leng, Biochemistry (USA) 28 (1989) 1454.
29. NC Stellwagen, Biochemistry (USA) 22 (1983) 6186.
30. JC Marini, R Weisberg and A Landy, Virology 83 (1977) 254.
31. JC Marini, SD Levene, DM Crothers and PT Englund, Proc. Natl Acad. Sci. USA 79 (1982) 7664.
32. JC Marini, PN Effron, TC Goodman, CK Singleton, RD Wells, RM Wartell and PT Englund, J. Biol. Chem. 259 (1984) 8974.
33. HM Wu and DM Crothers, Nature 308 (1984) 509.
34. S Diekmann and JC Wang, J. Mol. Biol. 186 (1985) 1.
35. EN Trifonov, CRC Crit. Rev. Biochem. 19 (1985) 89.
36. H-S Koo, H-M Wu and DM Crothers, Nature 320 (1986) 501.
37. DM Crothers, TE Haran and JG Nadeu, J. Biol. Chem. 265 (1990) 7093.
38. J Vinograd, J Lebowitz, R Radloff, R Watson and P Laipis, Proc. Natl Acad. Sci. USA 53 (1965) 1104.
39. DMJ Lilley, Chem. Soc. Rev. 18 (1989) 53.
40. W Saenger, *Principles of Nucleic Acid Structure*, Springer-Verlag, New York (1983).
41. WR Bauer, Ann. Rev. Biophys. Bioenerg. 7 (1978) 287.
42. DMJ Lilley, Topics Nucl. Acid Struct. 2 (1982) 173.

43. JC Wang, J Cell Sci. Suppl. 1 (1984) 21.
44. RD Wells, J. Biol. Chem. 263 (1988) 1095.
45. MD Frank-Kamenetskii in *DNA Topology and Its Biological Effects*, JC Wang and N Cozzarelli (eds), Cold Spring Harbor Lab. Press (1990) p. 185.
46. DMJ Lilley, Biochem. Soc. Trans. 14 (1984) 211.
47. DMJ Lilley, KM Sullivan and AIH Murchie in *Nucleic Acids and Molecular Biology*, F Eckstein and DMJ Lilley (eds), Springer-Verlag, Berlin (1987) Vol. 1, p. 126.
48. S Diekmann and DMJ Lilley, Nucl. Acids Res. 15 (1987) 5765.
49. K Nejedlý, J Klysik and E Paleček, FEBS Lett. 243 (1989) 313.
50. TM Jovin, DM Soumpasis and LP McIntosh, Ann. Rev. Phys. Chem. 38 (1987) 521.
51. MJ McLean and RD Wells, Biochim. Biophys. Acta 950 (1988) 243.
52. VI Lyamichev, SM Mirkin and MD Frank-Kamenetskii, J. Biomol. Struct. Dyn. 3 (1986) 667.
53. BH Johnston, Science 241 (1988) 1800.
54. M Frank-Kamenetskii, Nature 342 (1989) 737.
55. TR Cech, Nature 332 (1989) 777.
56. VI Lyamichev, SM Mirkin, ON Danilevskaya, ON Voloshin, SV Balatskaya, VN Dobrynin, S Filippov and MD Frank-Kamenetskii, Nature 339 (1989) 634.
57. WI Sundquist and A Klug, Nature 342 (1989) 825.
58. IG Panyutin, OI Kovalsky, EI Budowsky, RE Dickerson, ME Rikhirev and AA Lipanov, Proc. Natl Acad. Sci. USA 87 (1990) 867.
59. JH van de Sande, NB Ramsing, MW Germann, W Elhorst, BW Kalisch, E von Kitzing, RT Pon, RM Clegg and TM Jovin, Science 241 (1988) 551.
60. NB Ramsing and TM Jovin, Nucl. Acids Res. 16 (1988) 6659.
61. TH Eickbush and EN Mondrianakis, Cell 13 (1978) 295.
62. YuM Yevdokimov, SG Skuridin and VI Salyanov, Liquid Crystals 3 (1988) 1443.
63. C Robinson, Tetrahedron 13 (1961) 219.
64. E Iizuka, Polymer J. 9 (1977) 173.
65. VN Potaman, DC Alexeev, IYa Skuratovsky, AZ Rabinovich and LS Shlyakhtenko, Nucl. Acids Res. 9 (1981) 55.
66. RL Rill, PR Hillard and GC Levy, J. Biol. Chem. 258 (1983) 250.
67. F Livolant, Eur. J. Cell Biol. 33 (1984) 300.
68. R Huey and SC Mohr, Biopolymers 20 (1981) 2533.
69. LS Lerman, Proc. Natl. Acad. Sci. USA 68 (1971) 1886.
70. YuM Evdokimov, Al Platonov, AS Tikhonenko and YaM Vashavsky, FEBS Lett. 23 (1972) 180.
71. YuM Evdokimov, TL Pyatigorskaya, VA Kadykov, OF Polyvtsev, J Doskočil, J Koudelka and YaM Varshavsky, Nucl. Acids Res. 3 (1976) 1533.
72. YuM Evdokimov, SG Skuridin, VI Salyanov, G Damaschun, H Damaschun, R Missselwitz and V Kleinwächter, Biophys. Chem. 35 (1990) 143.
73. LS Lerman, LS Wilkerson, JH Venable, Jr and BH Robinson, J. Mol. Biol. 108 (1976) 271.
74. F Livolant, AM Levelut, J Doucet and JP Benoit, Nature 339 (1989) 724.
75. DE Pettijohn in *Nucleic Acids and Molecular Biology*, F Eckstein and DMJ Lilley (eds), Springer-Verlag, Berlin Heidelberg (1990), Vol. 4, p. 152.
76. R Sinden and DE Pettijohn, Proc. Natl Acad. Sci USA 78 (1981) 224.
77. K Drlica and J Rouviere-Janiv, Microbiol. Rev. 51 (1987) 301.
78. DE Pettijohn, J. Biol. Chem. 263 (1988) 12793.
79. RD Kornberg, Science 184 (1974) 868.
80. RD Kornberg, Ann. Rev. Biochem. 46 (1977) 931.
81. JT Finch, LC Lutter, D Rhodes, RS Brown, B Bushton, M Lewitt and A Klug, Nature 269 (1977) 29.
82. AD Mirzabekov and A Rich, Proc. Natl Acad. Sci. USA 76 (1979) 1118.
83. A Worcel, Cold Spring Harb. Symp. Quant. Biol. 42 (1978) 313.
84. JT Finch and A Klug, Proc. Natl Acad. Sci. USA 73 (1976) 1897.
85. AL Olins, Cold Spring Harb. Symp. Quant. Biol. 42 (1978) 325.
86. A Worcel, S Strogatz and D Riley, Proc. Natl Acad. Sci. USA 78 (1981) 1461.
87. J Sedat and L Manuelidis, Cold Spring Harb. Symp. Quant. Biol. 42 (1978) 331.

88. TR Cech and BL Bass, Ann. Rev. Biochem. 55 (1987) 599.
89. TR Cech, Science 236 (1987) 1532.
90. CC Sheldon, AC Jeffries, C Davies and RH Symons in *Nucleic Acids and Molecular Biology*, F Eckstein and DMJ Lilley (eds), Springer-Verlag, Berlin Heidelberg (1990) Vol. 4, p. 227.
91. K Hall, P Cruz and MJ Chamberlin, Arch. Biochem. Biophys. 236 (1984) 47.
92. HH Klump and TM Jovin, Biochemistry (USA) 26 (1987) 5186.
93. CC Hardin, DA Zarling, JD Puglisi, MO Trulson, PW Davis and I Tinoco Jr., Biochemistry (USA) 26 (1987) 5191.
94. CWA Pleij, K Rietveld and L Bosch, Nucl. Acids Res. 13 (1985) 1717.
95. P Dumas, D Moras, C Florentz, R Giege, P Verlaan, A Van Belkum and CWA Pleij, J. Biomol. Struct. Dyn. 4 (1987) 707.
96. JD Puglisi, JR Wyatt and I Tinoco Jr, Nature 321 (1988) 283.
97. A Bhattacharyya, AIH Murchie and DMJ Lilley, Nature 343 (1990) 484.
98. F Cramer, Progr. Nucl. Acid Res. Mol. Biol. 11 (1971) 391.
99. S-H Kim, GJ Quigley, FL Suddath, A McPherson, D Sneden, JJ Kim, J Weinzierl and A Rich, Science 179 (1973) 285.
100. A Rich, Acc. Chem. Res. 10 (1977) 388.
101 A. Rich and S-H Kim, Sci. Amer. 238 (1978) 52.
102. PR Schimmel and AG Redfield, Ann. Rev. Biophys. Bioenerg. 9 (1980) 181.
103. S-H Kim in *Topics in Molecular and Structural Biology*, S Neidle and W Fuller (eds), Vol. 1, p. 83.
104. J Lake, Ann. Rev. Biochem. 54 (1985) 507.
105. PD Lawley, Progr. Nucl. Acid Res. Mol. Biol. 5 (1966) 89.
106. C. Hélène and NT Thuong in *Nucleic Acids and Molecular Biology*, F Eckstein and DMJ Lilley (eds), Springer-Verlag, Berlin (1988), Vol. 2, p. 105.
107. DM Brown in *Basic Principles in Nucleic Acid Chemistry*, POP. Ts'O (ed), Academic Press, New York (1974).
108. PB Dervan in *Structure and Methods*, RH Sarma and MH Sarma (eds) Adenine Press, Schenectady (1990) Vol. 1, p. 37.
109. VV Vlassov, SA Gaidamakov, VF Zarytova, DG Knorre, AS Levina, AA Nikonova, LM Podust and OS Fedorova, Gene 72 (1988) 313.
110. PE Nielsen, J. Mol. Recog. 3 (1990) 1.
111. L Ehrenberg, I Fedorcsak and F Solymosy, Progr. Nucl. Acid Res. Mol. Biol. 16 (1976) 198.
112. PM Scholten and A Nordheim, Nucl. Acids Res. 14 (1986) 3981.
113. JC Furlong and DM Lilley, Nucl. Acids Res. 14 (1986) 3995.
114. N Vogt, N Rousseau, M Leng and B Malfoy, J. Biol. Chem. 263 (1988) 11826.
115. C Jeppesen and PE Nielsen, FEBS Lett. 231 (1988) 172.
116. MJ McLean and MJ Waring, J. Mol. Recog. 1 (1988) 138.
117. DE Gilbert, GA van der Marel, JH van Boom and J Feigon, Proc. Natl Acad. Sci. USA 86 (1989) 3006.
118. A Vincze, REL Hinderson, FF McDonald and NJ Leonard, J Amer. Chem. Soc. 95 (1973) 2677.
119. DP Ringer, S Burchett and DE Kizer, Biochemistry (USA), 17 (1978) 4818.
120. DS Gross and H Simpkins, J. Biol. Chem. 256 (1981) 9593.
121. D Gersanovski, P Colson, C Houssier and E Fredericq, Biochim. Biophys. Acta 824 (1985) 313.
122. M Arquilla, LM Thompson, LF Pearlman and H Simpkins, Cancer Res. 43 (1983) 1211.
123. C Houssier, MC DePauw-Gillet, R Hacha and E Fredericq, Biochim, Biophys. Acta 739 (1983) 317.
124. Z Balcarová and V Brabec, Biophys. Chem. 33 (1989) 55.
125. E Paleček, P Boubliková, F Jelen, A Krejčová, E. Makaturová, K. Nejedlý, P Pečinka and M Vojtíšková in *Structure and Methods*, RH Sarma and MH Sarma (eds), Adenine Press, Schenectady, 1990, Vol. 3: *DNA and RNA*, p. 237.
126. LR Subbaraman, J Subbaraman and EJ Behrman, Bioinorg. Chem. 1 (1971) 35.
127. E Lukášová, M. Vojtíšková, F. Jelen, T. Sticzay and E Paleček, Gen. Physiol. Biophys. 3 (1984) 175.
128. E Paleček and MA Hung, Anal. Biochem. 132 (1983) 236.

129. H Htun and JE Dahlberg, Science 241 (1988) 1791.
130. K Kayasuga, T Hahimoto, T Negishi, K Negishi and H Hayatsu, Chem. Pharm. Bull. (Tokyo) 28 (1980) 932.
131. DS Kang and RD Wels, J. Biol. Chem. 260 (1985) 7783.
132. DMJ Lilley, Nucl. Acids Res. 11 (1983) 3097.
133. Y Kohwi and T Kohwi-Shigematsu, Proc Natl Acad. Sci. USA 85 (1988) 3781.
134. G Rodighiero and G Dall'Acqua, Photochem. Photobiol. 24 (1976) 647.
135. JE Hearst, ST Isaacs, D Kanne, H Rapoport and K Straub, Quart. Rev. Biochem. 17 (1984) 1.
136. GD Cimino, HB Gamper, ST Isaacs and JE Hearst, Ann. Rev. Biochem. 54 (1985) 1151.
137. D Kanne, K Straub, JE Hearst and H Rapoport, J. Amer. Chem. Soc. 104 (1982) 6754.
138. L Kittler and G Löber, Studia Biophys. 97 (1983) 61.
139. DC Youvan and JE Hearst, Anal. Biochem. 119 (1982) 86.
140. D Rabin and DM Crothers, Nucl. Acids Res. 7 (1979) 689.
141. JM Sogo, PJ Ness, RM Widmer, RW Parish and T Koller, J Mol. Biol. 178 (1984) 897.
142. JE Hearst, Ann. Rev. Biophys. Bioenerg. 10 (1981) 69.
143. TR Cech and ML Pardue, Proc. Natl Acad. Sci. USA 73 (1976) 2644.
144. RR Sinden and DE Pettijohn, Proc. Natl Acad. Sci. USA 78 (1981) 224.
145. PD Lawley and P Brookes, Biochem. J. 89 (1963) 127.
146. ON Voloshin, SM Mirkin, VI Lyamichev, BP Belotserkovskii and MD Frank-Kamenetskii, Nature 333 (1988) 475.
147. U Siebenlist and W Gilbert, Proc. Natl Acad. Sci. USA 77 (1980) 122.
148. C Heuer and W Hillen, J. Mol. Biol. 202 (1988) 407.
149. TD Tullius, Nature 332 (1988) 663.
150. RP Hertzberg and PB Dervan, Biochemistry (USA) 23 (1984) 3934.
151. M Sawadogo and RG Roeder, Cell 43 (1985) 165.
152. TE Goyne and DS Sigman, J. Amer. Chem. Soc. 109 (1987) 2846.
153. LE Pope, KA Reich, DR Graham and DS Sigman, J. Biol. Chem. 257 (1982) 12121.
154. J Stube and JW Kozarich, Chem. Rev. 87 (1987) 1107.
155. MD Kuwabara, C Yoon, T Goyne, T Thederahn and DS Sigman, Biochemistry (USA) 25 (1986) 7401.
156. DS Sigman, Acc. Chem. Res. 19 (1986) 180.
157. LE Marshall, DR Graham, KA Reich and DS Sigman, Biochemistry (USA) 20 (1981) 244.
158. LE Pope and DS Sigman, Proc. Natl Acad. Sci. USA 81 (1984) 3.
159. PE Nielsen, C Jeppesen and O Buchardt, FEBS Lett. 235 (1988) 122.
160. TD Tullius, Ann. Rev. Biophys. Biophys. Chem. 18 (1989) 213.
161. J Portugal, Chem.-Biol. Interactions 71 (1989) 311.
162. C-HB Chen and DS Sigman, J. Amer. Chem. Soc. 110 (1988) 6570.
163. A Joachimiak, RL Kelley, RP Gunsalus, C Yanofsky and PB Sigler, Proc. Natl Acad. Sci. USA 80 (1983) 668.
164. GB Dreyer and PB Dervan, Proc. Natl Acad. Sci. USA 82 (1985) 968.
165. PG Schultz, JS Taylor and PB Dervan, J. Amer. Chem. Soc. 104 (1982) 6861.
166. PE Pjura, K Grezeskowiak and RE Dickerson, J. Mol. Biol. 197 (1987) 257.
167. ME Davis and JA McCammon, Chem. Rev. 90 (1990) 509.
168. F Daniels and RA Alberty, *Physical Chemistry*, John Wiley and Sons, New York (1975).
169. B Pullman (ed), *Intermolecular Interactions: From Diatomic to Biopolymers;* John Wiley and Sons, Chichester (1978).
170. A Beiser, *Perspectives of Modern Physics*, McGraw-Hill-Kogahusha Ltd, Tokyo (1969).
171. B Pullman and A Pullman, *Quantum Biochemistry*, Interscience Publishers, New York (1963).
172. G Del Re in *Electronic Aspects of Biochemistry*, B Pullman (ed), Academic Press, New York (1964) p. 221.
173. PW Atkins, Physical Chemistry, Oxford University Press, Oxford (1990).
174. F Jordan and B Pullman, Theor. Chim. Acta (Berlin) 9 (1968) 242.
175. G Naray-Szabo (ed), *Theoretical Chemistry of Biological Systems,* Elsevier, Amsterdam (1986).
176. M Orozco and FJ Luque, J. Computer-Aided Molec. Des. 4 (1990) 411.

177. B Pullman, *Molecular Biophysics*, Academic Press, New York (1965), pp. 137, 156.
178. H Devoe and I Tinoco Jr., J. Mol. Biol. 4 (1962) 500.
179. JS Kwiatkowski and WB Person in *Theoretical Biochemistry and Molecular Biophysics*, DL Beveridge and R Lavery (eds), Adenine Press, Schenectady (1991), Vol. 1, p. 153.
180. I Kulalowska, M Geller, B Lesyng and KL Wierzchowski, Biochim. Biophys. Acta 361 (1974) 119.
181. I Kulalowska, M Geller, B Lesyng, K Bolewska and KL Wierzchowski, Biochim. Biophys. Acta 407 (1975) 420.
182. H Berthold and B Pullmann, Biochem. Biophys. Res. Commun. 46 (1972) 125.
183. POP Ts'o in *Fine Structure of Proteins and Nucleic Acids, Biol. Macromol. Series*, GF Fasman and SN Timasheff (eds), Marcel Dekker, New York (1969), Vol. 4, p. 164.
184. MG Papadopoulos and J Waite, J. Mol. Struct. (Theochem) 171 (1988) 189.
185. R Bonaccorsi, A Pullman, E Scrocco and J Tomasi, Theor. Chim. Acta (Berlin) 24 (1972) 51.
186. A Pullman and B Pullman in *Chemical Applications of Atomic and Molecular Electrostatic Potentials*, P Politzer and DG Truhlar (eds), Plenum Press, New York (1981), p. 381.
187. J Tomasi in *Molecular Interactions*, H. Ratajczak and WJ Orville-Thomas (eds), John Wiley and Sons, New York (1982), Vol. 3, p. 119.
188. S Miertus and M Trebaticka, J. Theor. Biol. 108 (1984) 509.
189. A Pullman and B Pullman, J. Sci. Industr. Res. 39 (1980) 778.
190. GA Walker, SC Bhatia and JH Hall Jr., J. Amer. Chem. Soc. 109 (1987) 7629.
191. JE Del Bene, J. Phys. Chem. 87 (1983) 367.
192. R Wagner and W von Philipsborn, Helv. Chim. Acta 54 (1971) 1543.
193. MW Kilday, J. Res. Natl Bur. Stand. (USA) 83 (1978) 347.
194. RL Benoit and M Frechette, Can. J. Chem. 62 (1984) 995.
195. RM Izatt, JJ Christensen and JH Rytting, Chem. Rev. 71 (1971) 439.
196. JJ Christensen, JH Rytting and RM Izatt, Biochemistry 9 (1970) 4907.
197. W Guschlbauer in *Encyclopedia of Polymer Science and Engineering*, Mark-Bikales-Overberger-Menges (eds), John Wiley and Sons, New York (1988), p. 699.
198. RF Steiner and RF Beers Jr., *Polynucleotides*, Elsevier, Amsterdam (1961)
199. GD Fasman (ed), *Handbook of Biochemistry and Molecular Biology*, Chem. Rubber Co., Cleveland, Ohio (1975) Vol. 1, *Nucleic Acids*, p. 76.
200. F Jordan, Biopolymers 13 (1974) 289.
201. A Goldblum and B Pullman, Theor. Chim. Acta (Berlin) 47 (1978) 345.
202. B Pullman, A Goldblum and H Berthold, Biochem. Biophys. Res. Commun. 77 (1977) 1166.
203. A Katchalsky, Biophys. J. 4 (1964) 9.
204. B Jayaram, KA Sharp and B Honig, Biopolymers 28 (1989) 975.
205. GR Pack, G Lamm, L Wong and D Clifton in *Theoretical Biochemistry and Molecular Biophysics*, DL Beveridge and R Lavery (eds), Adenine Press, Schenectady (1991), Vol. 1, p. 237.
206. FH Hausheer, UCh Singh, TC Palmer and JD Saxe, J. Amer. Chem. Soc. 112 (1990) 9468.
207. ChA Hunter, J. Mol. Biol. 230 (1993) 1025.
208. H DeVoe in *Structure and Stability of Biological Macromolecules*, SN Timasheff and GD Fasman (eds), Marcel Dekker, Inc., New York (1969) p. 1.
209. PH von Hippel and T Schleich in *Structure and Stability of Biological Macromolecules*, SN Timasheff and GT Fasman (eds), Marcel Dekker, New York (1969), p. 417.
210. VI Poltev in *Theoretical Biochemistry and Molecular Biophysics*, DL Beveridge and R Lavery (eds), Adenine Press, Schenectady (1991), Vol. 1, p. 247.
211. POP Ts'o, IS Melvin and AC Olson, J. Amer. Chem. Soc. 85 (1963) 1289.
212. SI Chain, MP Schweizer, POP Ts'o and GK Helmkamp, J. Amer. Chem. Soc. 86 (1964) 4182.
213. C.R. Calladine and H.R. Drew, Understanding DNA, Academic Press, London (1992).
214. POP Ts'o, NS Kondo, RK Robins and AD Broom, J. Amer. Chem. Soc. 91 (1961) 5625.
215. FE Evans, DW Miller and RA Levine, J. Biomol. Struct. Dyn. 3 (1986) 935.
216. R Tribolet and H Sigel, Eur. J. Biochem. 170 (1988) 617.

217. W Guschlbauer, J-F Chanton and D Thiele, J. Biomol. Struct. Dyn. 8 (1990) 491.
218. ECM Chen and ESD Chen, Biochem. Biophys. Res. Commun. 171 (1990) 97.
219. FE Evans and RH Sarma, Biopolymers 13 (1974) 2117.
220. E Plisiewicz, E Stepien and KL Wierzchowski, Nucl. Acids Res. 3 (1976) 1295.
221. T Imoto, Biochim. Biophys. Acta 475 (1977) 409.
222. EL Farquhar, M Downing and SL Gill, Biochemistry 7 (1968) 1224.
223. KE Van Holde and GP Rossetti, Biochemistry 6 (1967) 2189.
224. Y Kyogoku, RC Lord and A Rich, J. Amer. Chem. Soc. 89 (1967) 496.
225. H Iwahashi and Y Kyogoku, J. Amer. Chem. Soc. 99 (1977) 7761.
226. A D'Albis, MP Wickens and WB Gratzler, Biopolymers 14 (1975) 1423.
227. J Langlet, C Giessner-Prettre, B Pullman, P Claverie and D Piazzola, Int. J. Quant. Chem. 18 (1980) 421.
228. O Sinanoglu in *Molecular Associations in Biology*, B Pullman (ed), Academic Press, New York (1968), p. 427.
229. G Nemethy and HA Scheraga, J. Chem. Phys. 36 (1962) 3401.
230. D Porschke and F Eggers, Eur. J. Biochem. 26 (1972) 490.
231. KJ Neurohr and HH Mantsch, Can. J. Chem. 57 (1979) 1986.
232. B Pullman, S Miertus and D Perahia, J. Theor. Biol. 50 (1979) 317.
233. R Lavey, A Pullman and B Pullman, Biophys. Chem. 17 (1983) 75.
234. A Goldblum, D Perahia and A Pullman, FEBS Lett. 91 (1978) 213.
235. I Yamaguchi and C Nagata, J. Theor. Biol. 69 (1977) 693.
236. GHF Diercksen, WP Kraemer, and BO Roos, Theor. Chim. Acta 36 (1975) 249.
237. W Kolos, Theor. Chim. Acta 51 (1979) 219.
238. WL Jorgensen, J. Amer. Chem. Soc. 101 (1979) 2001.
239. S Neidle in *Theoretical Biochemistry and Molecular Biophysics*, DL Beveridge and R Lavery (eds), Adenine Press, Schenectady (1991), Vol. 1, p. 207.
240. J Bertran, J Theor. Biol. 34 (1972) 353.
241 VI Poltev and SM Shulga, Studia Biophys. 70 (1978) 51.
242. ZG Kudritskaya and KI Danilov, J. Theor. Biol. 59 (1976) 303.
243. J Šponer, R Burcl and P Hobza, J. Biomol. Struct. Dyn. 11 (1994) 1375.
244. RL Ornstein, R Rein, DL Breen and R MacElroy, Biopolymers 17 (1978) 2341.
245. IK Yanson, AB Teplitsky and LF Sukhodub, Biopolymers 18 (1979) 1149.
246. C Salet, R Bensasson and RS Beiker, Photochem. Photobiol. 30 (1979) 325.
247. DM Soumpasis, A Garcia, R Klement and T Jovin in *Theoretical Biochemistry and Molecular Biophysics*, DL Beveridge and R Lavery (eds), Adenine Press, Schenectady (1991) Vol. 1, p. 343.
248. V Kleinwächter, Z Balcarová and J Boháček, Biochim. Biophys. Acta 174 (1969) 188.
249. DR Robinson and ME Grant, J. Biol. Chem. 241 (1966) 4030.
250. L Constantino and V Vitagliano, Biochim. Biophys. Acta 134 (1967) 204.
251. S Diekmann, W Hillen, B Morgeneyer, M Jung, RD Wells and D Porschke, Biophys. Chem. 15 (1982) 157.
252. M Eigen and G Schwarz, in *Electrolytes*, B Pesce (ed), Pergamon Press, Oxford (1962), p. 309.
253. D Porschke, Biophys. Chem. 22 (1985) 237.
254. D Porschke, Ann. Rev. Phys. Chem. 36 (1985) 159.
255. M Hogan, N Dattagupta and DM Crothers, Proc. Natl Acad. Sci. USA 75 (1978) 195.
256. M Pollak and HA Glick, Biopolymers 16 (1977) 1007.
257. S Diekmann and D Porschke, Biophys. Chem. 16 (1982) 261.
258. R de Levie, Chem. Rev. 88 (1988) 599.
259. V Vetterl, Experientia 21 (1965) 9.
260. H Jehring, *Elektrosorptionsanalyse mit der Wechselstrompolarographie*, Akademie-Verlag, Berlin (1974), p. 159.
261. YM Temerk, P Valenta and HW Nurnberg, J. Electroanal. Chem. 109 (1980) 289.
262. H Kinoshita, SD Christian, MH Kim, JG Baker and G Dryhurst in *Electrochemical Studies of Biological Systems, ACS Symposium Series*, T Sawyer (ed), American Chemical Society (1977), No. 38, p. 113.
263. G Quarin, Electrochim. Acta 29 (1984) 1707.
264. E Paleček, Collect. Czech. Chem. Commun. 39 (1974) 3449.

265. V Brabec and E Paleček, J. Electroanal. Chem. 88 (1978) 373.
266. P Valenta, HW Nürnberg and P Klahre, Bioelectrochem. Bioenerg. 1 (1974) 487.
267. F Jelen and E Paleček, Gen. Physiol. Biophys. 4 (1985) 219.
268. E Paleček in *Topics in Bioelectrochemistry and Bioenergetics*, G Milazzo (ed), J Wiley, London (1983), Vol. 5, p. 65.
269. I.R. Miller, J. Mol Biol. 3 (1961) 357.
270. E Paleček, Bioelectrochem. Bioenerg. 20 (1988) 179.
271. E Fredericq and C Houssier, *Electric Dichroism and Electric Birefringence,* Clarendon Press, Oxford (1973).
272. Ch Kittel, *Úvod do fyziky pevných látek (Introduction in the Physics of Solid Substances)*, Academia, Praha (1985).
273. M Hanss and JC Bernengo, Biopolymers 12 (1973) 2151.
274. A Surowiec, Biopolymers 17 (1978) 2279.
275. KR Foster and HP Schwan in *CRC Handbook of Biological Effects of Electromagnetic Fields*, Ch Polk and E Postow (eds), Boca Raton, Florida (1987) p. 27.
276. AG Walton and J Blackwell, *Biopolymers*, Academic Press, New York (1973).
277. F van der Touw and M Mandel, Biophys. Chem. 2 (1974) 231.
278. M Merabet and TK Bose, J. Phys. Chem. 92 (1988) 6149.
279. KS Cole and RH Cole, J. Chem. Phys. 9 (1941) 341.
280. S Takashima, J. Phys. Chem. 70 (1966) 1372.
281. S Takashima, Biopolymers 4 (1966) 663.
282. S Takashima, Biopolymers 5 (1967) 899.
283. MS Tung, RJ Molinari, RH Cole and JH Gibbs, Biopolymers 16 (1977) 2653.
284. RJ Molinari, RH Cole and JH Gibbs, Biopolymers 20 (1981) 977.
285. A Bonincontro, C Cametti, A Di Basio and F Pedone, Biophys. J. 45 (1984) 495.
286. S Takashima, Biopolymers 12 (1973) 145.
287. A Bonincontro, C Cametti, A Di Basio and F Pedone, Biopolymers 25 (1986) 241.
288. D Vasilescu in *Physico-chemical Properties of Nucleic Acids*, J. Duchesne (ed), Academic Press, London (1973) Vol. 1, p. 31.
289. E Šubertová, J Bok, P Říhák, V Prosser and E Silinsh, Phys. Stat. Sol. (a) 18 (1973) 741.
290. E Šubertová, E Silinsh and V Prosser, Czech. J. Phys. B 23 (1973) 356.
291. V Baumruk, J Kočka, J Krištofík and E Šubertová, Czech. J. Phys. B 35 (1985) 670.
292. E Šubertová, V Prosser and J Drobník, Biopolymers 8 (1969) 421.
293. V Prosser *et al., Experientální metody biofyziky (Experimental Methods of Biophysics)*, Academia, Praha (1989).
294. DD Eley, NC Lockhart and NC Richardson, J. Chem. Trans. Farad. Soc. 1, 75 (1979) 232.
295. E Jelínková, J Laudát and P Hána, Čs. Čas. Fyz. A 3 (1986) 382.
296. DD Eley and RB Leslie, Nature 197 (1963) 898.
297. E Šubertová, J Laudát and J Plášek, Studia Biophys. 96 (1983) 43.
298. RS Snart, Biopolymers 6 (1968) 293.
299. CT O'Konski, P Moser and M Shirai, Biopolymers 1 (1964) 479.
300. J Laudát, E Jelínková, M Neubert, R Bakule, J Nedbal, V Prosser, Materials Science 10 (1984) 176.
301. J Polonski, P Douzou and Ch Sadron, Biophys. Molecul. 250 (1960) 3414.
302. P Douzou, J-C Franco, J Polonski and Ch Sadron, Biophys. Molecul. 251 (1960) 976.
303. CT O'Konski and M Shirai, Biopolymers 1 (1963) 557.
304. AL Stanford, Jr., and RA Lorey, Nature 219 (1968) 1250.
305. RG Kepler, Ann. Rev. Chem. 29 (1978) 497.
306. D Vasilescu and G Mesnard, C. R. Acad. Sci. Paris 260 (1965) 4843.
307. S Toulsky and M Read, C. R. Acad. Sci. Paris 260 (1965) 7030.
308. E Fukada and Y Ando, J. Pol. Sci. 10 (1972) 565.
309. MH Shamos and LS Lavine, Nature 213 (1967) 267.
310. E Fukada, Ultrasonics (1968) 229.
311. MM Long and DW Urry in *Molecular Biology, Biochemistry and Biophysics,* E Grell (ed), Springer-Verlag, Berlin (1981), Vol. 31, p. 143.
312. H Fritzsche, L Kittler, G Löber, KE Reinert, D Tresselt, H Triebel und C Zimmer, *Strukturuntersuchungen an Bipolymeren mit spektroskopischen und hydrodynamischen Methoden*, Akademie-Verlag, Berlin (1976).

313. E Charney, *The Molecular Basis of Optical Activity, Optical Rotary Dispersion and Circular Dichroism*, John Wiley, Interscience, New York (1979).
314. F Snatzke and G Snatzke in *Analytiker-Taschenbuch*, H Kienitz, R Bock, W Fresenius, W Huber and G Tolg (eds), Springer-Verlag, Berlin (1980), Vol. 1, p. 217.
315. J Brahms and S Brahms in *Fine Structure of Proteins and Nucleic Acids, Biological Macromolecular Series*, GD Fasman and SN Timasheff (eds) Marcel Dekker, New York (1970), p. 191.
316. K Imahori and K Watanabe, J. Polymer Sci. Part C 30 (1970) 633.
317. WC Johnson Jr., Ann. Rev. Phys. Chem. 29 (1978) 93.
318. W Guschlbauer, Ch Marck and D Thiele in *Structural Aspects of Biomolecules*, R Srivasan and V Pattabhi (eds), Madras (1978) p. 228.
319. J Kypr and M Vorlíčková, J. Biomol. Struct. Dynam. 3 (1990) 67.
320. H Frohlich in *The Fluctuating Enzyme*, GR Welch (ed), John Wiley, New York (1986), p. 421.
321. AS Davydov, Physica Scripta 20 (1979) 1387.
322. AC Scott, Phys. Sci. 25 (1982) 651.
323. F Kremer, A Poglitsch, L Genzel in *Nonlinear Electrodynamics in Biological Systems*, WR Adey and AF Lawrence (eds), Plenum Press, New York (1984), p. 177.
324. F Keilmann and W Grundler in *Nonlinear Electrodynamics in Biological Systems*, WR Adey and AF Lawrence (eds), Plenum Press, New York (1984) p. 59.
325. E Postow and ML Swicord in *CRC Handbook of Biological Effects of Electromagnetic Fields*, C Polk and E Postow (eds), Boca Raton, Florida (1987) p. 425.
326. GS Edwards, CC Davis, JD Saffer and ML Swicord, Phys. Rev. Lett. 53 (1984) 1284.
327. GS Manning in *Theoretical Biochemistry and Molecular Biophysics*, DL Beveridge and R Lavery (eds), Adenine Press, Schenectady (1991) Vol. 1, p. 191.
328. LL Van Zandt and ME Davis, J. Biomol. Struct. Dynam. 3 (1986) 1045.
329. MB Hakim, SM Linsay and J Powell, Biopolymers 23 (1984) 1185.
330. J Torbet and G Maret, Biopolymers 20 (1981) 2657.
331. J-M Freyssinet, J Torbet, G Hudry-Clergeon and G Maret, Proc. Natl Acad. Sci. USA 80 (1983) 1616.
332. J Torbet and MJ Dickens, FEBS Lett. 173 (1984) 403.
333. E Iizuka, J. Appl. Polymer Sci., Appl. Polymer Symp. 41 (1985) 131.
334. GV Maret, M. Schickfus, A Mayer and K Dransfeld, Phys. Rev. Lett. 35 (1975) 397.
335. J Torbet and M-C Ronzuiere, Biochem. J. 219 (1984) 1057.
336. A Dobek and A Patkowski, J. Polymer Sci., Polymer Symp. 61 (1977) 111.
337. J Torbet and MY Norton, FEBS Lett. 147 (1982) 201.
338. J Torbet, EMBO J. 2 (1983) 63.
339. MJ Azanza, BH Blott, A del Moral and MT Peg, Biochem. Bioenerg. 30 (1993) 43.
340 A Veillard, B Pullman and G Berthier, C. R. Acad. Sci. 252 (1961) 2321.
341. EW Bastiaan, C MacLean, PCM Van Zijl and AA Bothner-By, Ann. Rev. NMR Spectr. 19 (1987) 35.
342. M Schindler, J. Amer. Chem. Soc. 110 (1988) 6623.
343. M Schindler, Magn. Res. Chem. 26 (1988) 394.
344. FA Bovey, *High Resolution NMR of Macromolecules*, Academic Press, New York (1972) p. 26.
345. G Maret and K Dransfeld, Physica 86–88B (1977) 1077.
346. E Neumann, Prog. Biophys. Molec. Biol. 47 (1986) 197.
347. D Porschke in *Structure and Dynamics of Biopolymers*, C. Nicolini (ed), Martinus Nijhoff Publishers (1987) p. 66.
348. D Porschke, J. Biomol. Struct. Dynam. 4 (1986) 373.
349. FS Allen and KE Van Holde, Biopolymers 10 (1971) 865.
350. Da-Wen Ding, R Rill and KE Van Holde, Biopolymers 11 (1972) 2109.
351. E Charney and JB Milstien, Biopolymers 17 (1978) 1629.
352. E Charney, J. Biosci. 8 (1985) 517.
353. S Diekmann and D Porschke, Biophys. Chem. 26 (1987) 207.
354. D Porschke in *Structural Molecular Biology*, DB Davies, W Saenger and S Danyluk (eds), Plenum Press, New York (1982) p. 222.
355. D Porschke and M Jung, J. Biomol. Struct. Dynam. 2 (1985) 1173.

356. D Porschke, Biopolymers 15 (1976) 1917.
357. K Yoshioka, K Kikuchi and M Fujimori, Biophys. Chem. 11 (1980) 369.
358. D Porschke in *Molecular Electro-Optics*, S Krause (ed), Plenum Press, New York (1981) p. 269.
359. D Porschke, JH Meier and J Ronnenberg, Biophys. Chem. 20 (1984) 225.
360. S Diekmann and D Porschke, Biophys. Chem. 26 (1987) 207.
361. J Antosiewicz, MV German, JH van de Sande and D Porschke, Biopolymers 27 (1988) 1319.
362. J Antosiewicz and D Porschke, J. Biomol. Struct. Dynam. 5 (1988) 819.
363. K. Yamaoka and K Fukudome, J. Phys. Chem. 92 (1988) 4994.
364. JAB Lohman and C MacLean, Chem. Phys. 35 (1978) 269.
365. JAB Lohman and C MacLean, Chem. Phys. 43 (1979) 144.
366. J Dadok and AA Bothner-By, private communication (1990).
367. PCM van Zijl, C MacLean and AA Bothner-By, J. Chem. Phys. 83 (1985) 4410.
368. AA Bothner-By, C Gayathri, PCM van Zijl, C MacLean, JJ Lai and KM Smith, Magn. Reson. Chem. 23 (1985) 935.
369. F Canac, C. R. Acad. Sci. 196 (1933) 51.
370. M Avrami, J. Chem. Phys. 8 (1940) 212.
371. M Fleischmann and HR Thirsk, Adv. Electrochem. Eng. 3 (1963) 123.
372. R Sridharan and R de Levie, J. Phys. Chem. 86 (1982) 4489.
373. R Sridharan and R de Levie, J. Electroanal. Chem. 218 (1987) 287.
374. U Retter, J. Electroanal. Chem. 106 (1980) 371.
375. U Retter, J. Electroanal. Chem. 136 (1982) 167.
376. R Philipp, J Dittrich, U Retter and E Muller, J. Electroanal. Chem. 250 (1988) 159.
377. C Buess-Herman, J. Electroanal. Chem. 186 (1985) 41.
378. H Francois, M Scharfe and C Buess-Herman, J. Electroanal. Chem. 296 (1990) 415.
379. CH Mousty and G Quarin, Electrochim. Acta 35 (1990) 1291.
380. L Pospíšil, J. Phys. Chem. 92 (1988) 2501.
381. T Wandlowski and L Pospíšil, J. Electroanal. Chem. 270 (1989) 319.
382. B Bhattacharjee and SK Rangarajan, J. Electroanal. Chem. 302 (1991) 207.
383. V Brabec, MH Kim, SD Christian and G Dryhurst, J. Electroanal. Chem. 100 (1979) 111.
384. V Brabec, SD Christian and G Dryhurst J. Electroanal. Chem. 85 (1977) 389.
385. D Krznaric, P Valenta, HW Nürnberg and M Brancia, J. Electroanal. Chem. 93 (1978) 41.
386. YM Temerk, MM Kamal, ME Ahmed and ZA Ahmed, Biochem. Bioenerg. 16 (1986) 497.
387. EG Thomas, C Buess-Herman and L Gierst, J. Electroanal. Chem. 214 (1986) 597.
388. C Buess-Herman, L Gierst, M Gonze and F Silva, J. Electroanal. Chem. 226 (1987) 267.
389. U Retter, V Vetterl and J Jursa, J. Electroanal. Chem. 274 (1989) 1.
390. T Wandlowski, J. Electroanal. Chem. 302 (1991) 233.
391. V Vetterl, Bioelectrochem. Bioenerg. 3 (1976) 338.
392. E Koglin, JM Sequaris and P Valenta, J. Mol. Struct. 60 (1980) 421.
393. JM Sequaris, J Fritz, H Lewinsky and E Koglin, J. Colloid Interfac. Sci. 105 (1985) 417.
394. V Brabec and K Niki, Biophys. Chem. 23 (1985) 63.
395. V Brabec and K Niki, Collect. Czechosl. Chem. Commun. 51 (1986) 167.
396. C Otto, TJJ van den Tweel, FFM de Mul and J Greve, J. Raman Spectr. 17 (1986) 289.
397. MW Humpreys and R. Parsons, J. Electroanal. Chem. 75 (1977) 427.
398. MW Humpreys and R. Parsons, J. Electroanal. Chem. 82 (1977) 369.
399. C-G Golander, H Arwin, JC Eriksson, I Lundstrom and R Larsson, Colloids and Surfaces 5 (1982) 1.
400. HW Nürnberg in *Bioelectrochemistry 1*, G Milazzo and M Blank (eds) Plenum Press (1983) p. 183.
401. J Heesemann and HP Zingsheim in *Membrane Spectroscopy*, E Grell (ed), Springer-Verlag, Berlin (1981) p. 172.
402. YR Shen, Nature 337 (1989) 519.
403. DJ Blackwood and S Pons, J Electroanal. Chem. 247 (1988) 277.
404. H Neddermeyer, Crit. Rev. Solid State and Material Sci. 16 (1990) 309.
405. LA Nagahara, T Thundat, PI Oden and SM Lindsay, Ultramicroscopy 33 (1990) 107.
406. RJ Driscoll, MG Youngquist and JD Baldeschwieler, Nature 346 (1990) 294.

407. WM Heckl and DPE Smith, J. Vac. Sci. Technol. B 9 (1991) 1159.
408. WM Heckl, DPE Smith, G Binnig, H Klagges, TW Hansch and J Maddocks, Proc. Natl Acad. Sci. USA 88 (1991) 8003.
409. WM Heckl, Thin Solid Films 210/211 (1992) 640.
410. A Schaper, Lia T Pietrasanta and TJ Jovin, Nucleic Acids Res. 21 (1993) 6004.

Appendix A. Glossary

A	absorbance; area of adsorbed molecule on the electrode surface
a.c.	alternating current
B	magnetic induction
BAA	bromoacetaldehyde
c	concentration
C	solubility, Coulomb
CAA	chloroacetaldehyde
CD	circular dichroism
CM	Cotton-Mouton constant
D	electric displacement
DEPC	diehtylpyrocarbonate
DMS	dimethylsulphate
DNA	deoxyribonucleic acid
e	electron charge
E	electric field strength; energy
E_i	internal electric field strength
ENU	N-ethyl-N-nitrosourea
$g(t)$	distribution function of relaxation times
G	free energy
h	Planck constant
I	ionization potential
k	Boltzmann constant
K	Kerr constant
K_D	dichroism Kerr constant
K_S	salting-out constant
l_0	length of a monomer unit of a polymer
L	linking number
MOP	methoxypsoralen
MPE	methidium propyl-EDTA-Fe^{2+}
m.w.	molecular weight
n	refractive index
N	ionic concentration
N_A	Avogadro number
NMR	nuclear magnetic resonance
P	polarization; persistence length

Ph-Cu^{+} 1,10-phenanthroline-copper complex
RNA ribonucleic acid
mRNA messenger RNA
rRNA ribosomal RNA
tRNA transfer RNA
S sedimentation coefficient;
entropy
t, T temperature
T_m melting temperature
T twisting number
V electrostatic potential
W writhing number
W_a activation energy
W_g forbidden energy gap
W_L London dispersion energy
α angle between transition dipole moment and molecule main axis;
distribution parameter
α^o optical polarizability
α_e electronic polarizability
α_i interfacial polarizability
α_v vibrational polarizability
α' volume polarisability
$\Delta\alpha$ polarizability anisotropy
β ratio between electrostatic (magnetic) potential energy
and thermal energy
γ ratio between electrostatic potential energy and thermal energy
for molecules with an induced dipole moment;
activity coefficient
ε_0 permittivity of vacuum
ε_s low-frequency (static) dielectric permittivity
ε_∞ high-frequency dielectric permittivity
ε^* complex permittivity
ε'' dielectric loss parameter
δ loss angle;
a molecular orbital
θ direction of permanent electric dipole moment;
rotatory diffusion coefficient;
surface concentration
λ molecular angular momentum quantum number;
wavelength
μ electric dipole moment
ν frequency;
number of charges
$\Delta\nu$ splitting of NMR lines
π a molecular orbital

ρ	probability density; charge density
σ	electric conductivity; a molecular orbital
σ_a	dielectric conductivity
τ_0	relaxation time
ϕ	degree of orientation
ϕ_{CN}	glycosidic torsion angle
χ	electric susceptibility; magnetic susceptibility
ψ	wave function; electrostatic potential
ω	angular frequency

Bioelectrochemistry of Biomacromolecules
ed. by G. Lenaz and G. Milazzo†

CHAPTER 2
Proteins: Structure, folding and function

Dan W. Urry[#*] and Chi-Hao Luan[#†]

[#] *The Laboratory of Molecular Biophysics, School of Medicine, University of Alabama at Birmingham, University Station, Birmingham, Alabama, USA*
[†] *Bioelastics Research, Ltd., Birmingham, Alabama, USA*

This chapter is dedicated to the memory of Professor Lorenzo Gotte whose work provided critical stepping stones toward elucidation of the structure and function of the extracellular matrix protein, elastin.

* Author to whom correspondence should be addressed.

1. Introduction

Living organisms evolved by developing the capacity to convert an available form of energy into a useful form of energy; how well they thrive depends on how effectively a needed conversion is performed.

An easily recognized form of energy is motion. We are familiar with the motion of the winds, with the expansion of heated, compressed gases moving a car, propelling a plane or lifting a rocket, and with the turning on of an electric current to drive the rotary motion of an electric motor. More enigmatic is the motion of living organisms, coming seemingly spontaneously from within. Whether seen as the pseudopodial migration of a single cell, or as the rearrangements of components within a cell, or the beating of cilia or wagging of flagella attached to the surface of a cell, or as the movements of humans, it is changes in the structure of proteins that give rise to motion. While nucleic acids provide the blueprint as to what proteins are produced, it is proteins that are the key players in the moving, transforming, functioning, living entity.

2. Protein structure

In the production of protein in living organisms the information for the sequence of amino acid residues, that is the protein, is encoded in the sequence of bases of deoxyribonucleic acid (DNA). The megamolecular weight DNA is transcribed into a megamolecular weight messenger ribonucleic acid (mRNA) which becomes edited into a shortened mature mRNA and assembled into ribosomes which are the protein producing bodies. Lower molecular weight RNA called transfer RNA, tRNA, can recognize and base pair with a triplet of bases in the mature mRNA. With a triplet codon and four different bases there are 64 different triplets possible; three triplets signal "stop translation", a fourth triplet either signals "start translation" or codes for the amino acid methionine, and the remaining 60 triplet codons code for the other nineteen naturally occurring amino acids [1]. The amino acids leucine, arginine and serine can each be coded for by any of six triplets (i. e. have a degeneracy of six) whereas only one codon can code for tryptophan and one for methionine. Each of the some sixty tRNAs, charged with its appropriate amino acid, base pairs by the hydrogen bonding of its three-base codon to the mRNA to result in a base sequence-specified alignment of amino acids into a growing protein chain. In this biological synthesis it is the α-carboxyl group of the amino acid that is activated for coupling with the α-amino group to form the peptide bond, as is done in the chemical synthesis of peptides. The resulting chain molecule of regularly recurring peptide bonds is called a polypeptide, or when of sufficient size, a protein. As first proposed by Linderström-Lang [2], protein structure can be analyzed in terms of the sequence of amino acids called the primary structure, in terms of the hydrogen bonding between peptide moieties referred to as the secondary structure, and in terms of the folding and interacting of the side chains that define the amino acid residue as the tertiary structure. Protein molecules can also associate as multimers, and this is termed quaternary structure. It is in these terms of primary, secondary, tertiary and quaternary structure that protein structure will be briefly discussed below.

2.1. Primary structure

The amino acid building blocks of protein are given by the general chemical formula for amino acids, H_2N–CHR–COOH, where R stands for the side chain which chemically distinguishes the amino acid. On removal of a water molecule, HOH, two amino acid molecules are combined with the formation of the peptide bond, –CONH–. The repeating unit of a polypeptide or protein is represented as (–NH–CHR–CO–), which is referred to as the amino acid residue. The twenty different amino acid residues are listed in Table 1 with their complete name, the chemical structure of the R group, a three letter abbreviation and a single letter code.

Table 1. List of amino acid residues

Residue name	R group	Abbreviation	Letter code
Glycine	–H	Gly	G
Alanine	$-CH_3$	Ala	A
Valine	$-CH(CH_3)_2$	Val	V
Leucine	$-CH_2CH(CH_3)_2$	Leu	L
Isoleucine	$-CH(CH_3)CH_2CH_3$	Ile	I
Serine	$-CH_2OH$	Ser	S
Threonine	$-CH(OH)CH_3$	Thr	T
Cysteine	$-CH_2SH$	Cys	C
Methionine	$-CH_2CH_2SCH_3$	Met	M
Proline	$-CH_2CH_2CH_2-$	Pro	P
Aspartic acid	$-CH_2COOH$	Asp	D
Asparagine	$-CH_2CONH_2$	Asn	N
Glutamic acid	$-CH_2CH_2COOH$	Glu	E
Glutamine	$-CH_2CH_2CONH_2$	Gln	Q
Lysine	$-CH_2CH_2CH_2CH_2NH_2$	Lys	K
Arginine	$-CH_2CH_2CH_2NHC(NH)NH_2$	Arg	R
Histdine	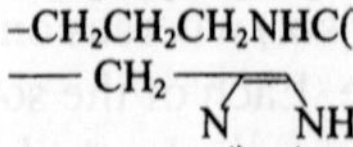	His	H
Phenylalanine	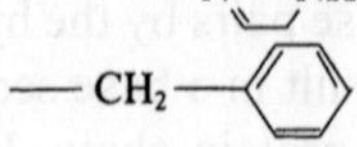	Phe	F
Tyrosine	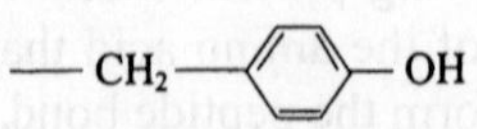	Tyr	Y
Tryptophan	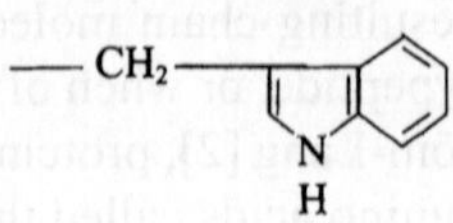	Trp	W

The primary structure can be represented by the sequence of residues indicated either by a sequence of three letter abbreviations or by a sequence of single letter codes. The sequence of bovine tropoelastin is given in Figure 1, where the sequence has been listed as the series of polypeptide sequences which end in potential cross-linking sequences running from the amino terminus at top left to the carboxyl terminus at bottom right. By the post-translational modification of lysyl oxidation to convert the lysyl ε-amino side chain terminus to an aldehyde, the protein chains become cross-linked into an insoluble fibrous elastin. Other post-translational modifications of amino acid residues within proteins include hydroxylation of prolyl residues (important in the formation of strong collagen fibres), phosphorylation of Ser, Thr or Tyr residues (important in the structure/function relationships of many proteins), glycosylation of Ser, Thr or Lys residues, methylation/demethylation of Glu residues (involved in the flagellar motor of bacteria), and so on.

The primary structure can also be represented in terms of a single residue transition temperature plot (see Fig. 2), where a measure of the hydrophobicity of a residue is represented by a transition temperature, T_t,

	GGLGPGVKPAK
W13	PGVGGLVGPGLGAGLGALPGAFPGALVPGGPAGAAAAYK**AAAK**
W8b	AGAAGLGVGGIGGVGGLGVSTGAVVPQLGAGVGAGVKPGK
W11b,c	VPGVGLPGVYPGGVLPGAGARFPGIGVLPGVPTGAGVKPK
W10	APGGGGAFAGIPGVGPFGGQQPGVPLGYPIKAPK
W14b	LPAGYGLPYKTGK
W11a	LPYGFGPGGVAGSAGKAGYPTGTGVGPQ**AAAAAAKAAAK**
W7	LGAGGAGVLPGVGVGGAGIPGAPGAIPGIGGIAGVGAPD**AAAAAAAAAKAAK**
W4	FGAAGGFP GVGVP GVGVP GVGVP GVGVP GVGVP GVGVP GVGVP GVGVP GVGVP GVGVP GVGVP GAVSP**AAAAKAAAKAAK**
W8a	FGARGGVGIGGIPTFGVGPGGFPGIGDAAAAQ**AAAAAKAAK**
W3	IGAGGVGALGGLVPGAPGAIPGVPGVGGVPGVGIP**AAAAAKAAAK**
W6	AAQFGLGPGVG VAPGVG VVPGVG VVPGVG VAPGIGLGPGGVIGAGVP**AAAKSAAKAAAK**
W-14	AQFRAAAGLPAG VPGLGVGVG VPGLGVGVG VPGLGVGAG VPGFGAVPGTL**AAAKAAK**
W5	FGPGGVGALGGVGDLGGAGIPGGVAGVGP**AAAAAAKAAAK**
W9	AAQFGLGGVGGLGVGGLGAVPGAVGLGGVSP**AAAAKAAK**
exons 4,3	FGAAGLGGVLGAGQPFPIGGVAARPGFGLSPIFPGGAGGLGVGGKPPK
exons 2,1	PFGGALGALGFPGGACLGKSCGRKRK

Fig. 1. Primary structure of bovine ligamentum nuchae elastin. The sequence data, reported by Yeh *et al.* (1987), are arranged according to sequences between cross-link-forming lysines and labelled according to Sandberg *et al.* (1985) in analogy with porcine aortic elastin. The recognized desmosine and isodesmosine (cross-link) forming sequences are boldfaced and are KAAK and KAAAK, generally associated with strings of alanines. At the heart of the primary structure is the longest sequence, W4; it contains the polypentapeptide of elastin, which was extensively characterized by Urry (1988). The next longest sequence, W6, contains a repeating hexapeptide followed in length by RP-71 (W14) which contains a repeating nonapeptide. Also notable are serines, S, threonines, T, and tyrosines, Y, in sequential proximity to charged lysine, K, or arginine, R, residues. These are potential phosphorylation sites. (Reproduced with permission from [114].

the temperature of the inverse temperature transition. The derivation and significance of these values will be given in some detail below. This plot allows ready visualization of both primary structural and certain functional elements. The negative deflections near 120 on the ordinate indicate the lysine residues which are the potential cross-links. Thus the periodic locations of these residues can be seen. The upward deflections are for the more hydrophobic or oil-like residues, and it is possible to note the increased grouping of hydrophobic residues between residues 160 and 210. What is perhaps most striking is the repeating sequence indicated as W4. To a lesser extent repeating sequences are clear in the W6 and W14 sequences, and in general a Pro-Gly repetition is seen throughout the primary structure. As will be discussed below, these Pro-Gly sequences introduce a secondary structural feature called a β-turn. Thus, little more than a glance at the single residue transition temperature plots provides ready insight into specific features of the primary structure. Since the work of Anfinsen [3], it has been appreciated that the primary structure of a protein is sufficient to specify what the three-dimensional functional protein can be. The challenge of

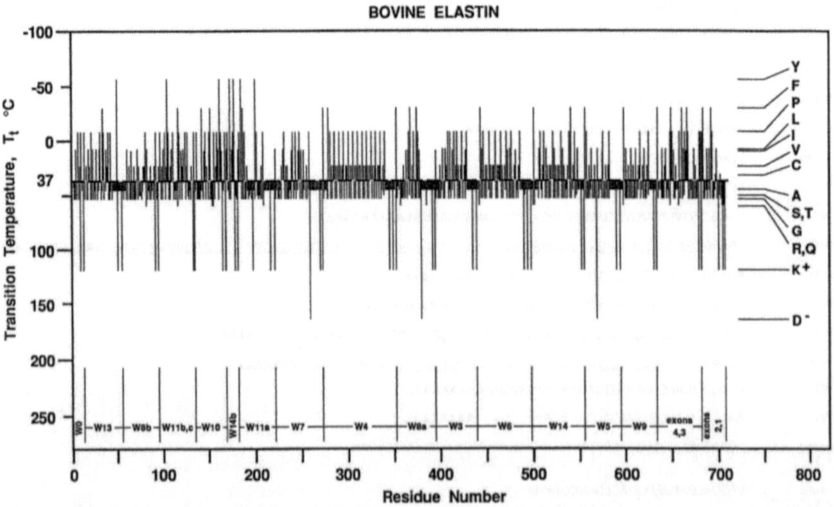

Fig. 2. Single residue transition temperature profile for bovine tropoelastin, the precursor protein of fibrous bovine elastin. The ordinate is the temperature, T_t, of the inverse temperature transition for poly [f_V(VPGVG), f_X(VPGXG)] with T_t extrapolated to $f_X = 1$. At the right are indicated the more common residues of elastin, i. e. Y, F, P, L, etc. Note the repetitive nature of the primary structure which is most evident in the W4 and W6 sequences but which is also apparent to different degrees in most of the sequences that occur between the Lys(K^+) residues which become involved in the cross-links. As the value of T_t is inversely proportional to the hydrophobicity, this is actually a single residue hydrophobicity plot with the more hydrophobic residues being displaced upward and the polar residues downward.

understanding protein structure/function is to understand protein folding and changes in protein folding that result in function. This may be done in terms of the analysis of secondary, tertiary and quaternary structures.

2.2. *Secondary structure*

Secondary structural features are defined by the hydrogen bonding involving the backbone CO of one residue and the backbone NH of another residue. These may initially be delineated in terms of the number of atoms required to achieve the hydrogen-bonded ring.

2.2.1. The α-helix: Following the work of Pauling, Corey and Branson [4], the systematic consideration of helices was performed by Donohue [5]. These helices are specified by two numbers: one gives the number of amino acid residues per turn of helix and the second, placed as a subscript to the first number, gives the number of atoms in the hydrogen-bonded ring. The helical structures listed in the Donohue nomenclature are 2.2_7, 3_{10}, 3.6_{13}, 4.3_{14}, 4.4_{16} and 5.1_{17}. Three of these have been assigned Greek letters: $\alpha(3.6_{13})$, $\pi(4.4_{16})$ and $\gamma(5.1_{17})$. Out of this family of helices, the α-helix (3.6_{13}) originally described by Pauling *et al.* [4] and the much less common 3_{10}-helix of Donohue have been observed in the crystal structures of proteins and polypeptides. The helical sense for L·amino acid residues is right-handed. A stereo pair drawing of the right-handed α-helix of L·amino acids is shown in Fig. 3 A.

2.2.2. β-pleated sheet structures: The β-pleated sheet structures are formed from an extended chain which may be viewed as a degenerate helix with a two-fold screw axis but with no hydrogen bonding within the extended chain. The hydrogen bonding in the β-pleated sheet structure is between chains. There are two types of β-pleated sheet structures, parallel and anti-parallel, based on whether adjacent chains run in parallel directions or alternately in opposite directions. These structures, described by Pauling and Corey [6], are shown in stereo pairs in Figs. 3 B and C. It is the hydrogen-bonding pattern between chains that immediately distinguishes the two β-pleated sheet structures. In this case hydrogen-bonded rings formed by a pair of interchain hydrogen-bonds can be described. For the anti-parallel β-pleated sheet, Fig. 3 B, there are alternately 10 and 14 atoms in the hydrogen-bonded rings, whereas for the parallel β-pleated sheet, Fig. 3 C, there is a single 12-atom hydrogen-bonded ring with five atoms from one chain and seven atoms from the adjacent chain. As may be seen in the stereo view or by looking at right angles to a chain direction from the plane of the sheet, there is a pleated appearance, hence its name.

2.2.3. The β-turns: While the right-handed α-helix is the most common secondary structural feature based on the analysis of 29 globular protein structures [7], even more common than the β-pleated sheet is the β-turn. The β-turn as a conformational feature was first appreciated by Venkatachalam [8] who referred to it as a $4 \rightarrow 1$ hydrogen-bond in which the NH of residue 4 hydrogen-bonded back to the CO of residue 1. It was independently described by Geddes *et al.* [9] based on X-ray studies of Chysopa silk where it was the conformational feature which allowed a single extended chain to fold back on itself to form the hydrogen-bonding pattern of the anti-parallel β-pleated sheet. This conformational feature of a single chain folding back and hydrogen-bonding, as in the anti-parallel β-pleated sheet, is called a cross-β-structure.

The most commonly occurring β-turn involves the Pro^2-Gly^3 residues as shown in Fig. 4 A for the repeat pentamer sequence of elastin [10]. The β-turn is characterized by a 10-atom hydrogen-bonded ring, but instead of functioning to generate a helix as in the 3_{10}-helix, the β-turn commonly functions to reverse the chain direction as is most apparent in a cross-β-structure, as shown schematically in Fig. 4 D. β-turns are also referred to as reverse turns, hairpin turns, 3_{10} bends or β-bends. β-turns are of two main types depending on the orientation of the end peptide moiety formed from

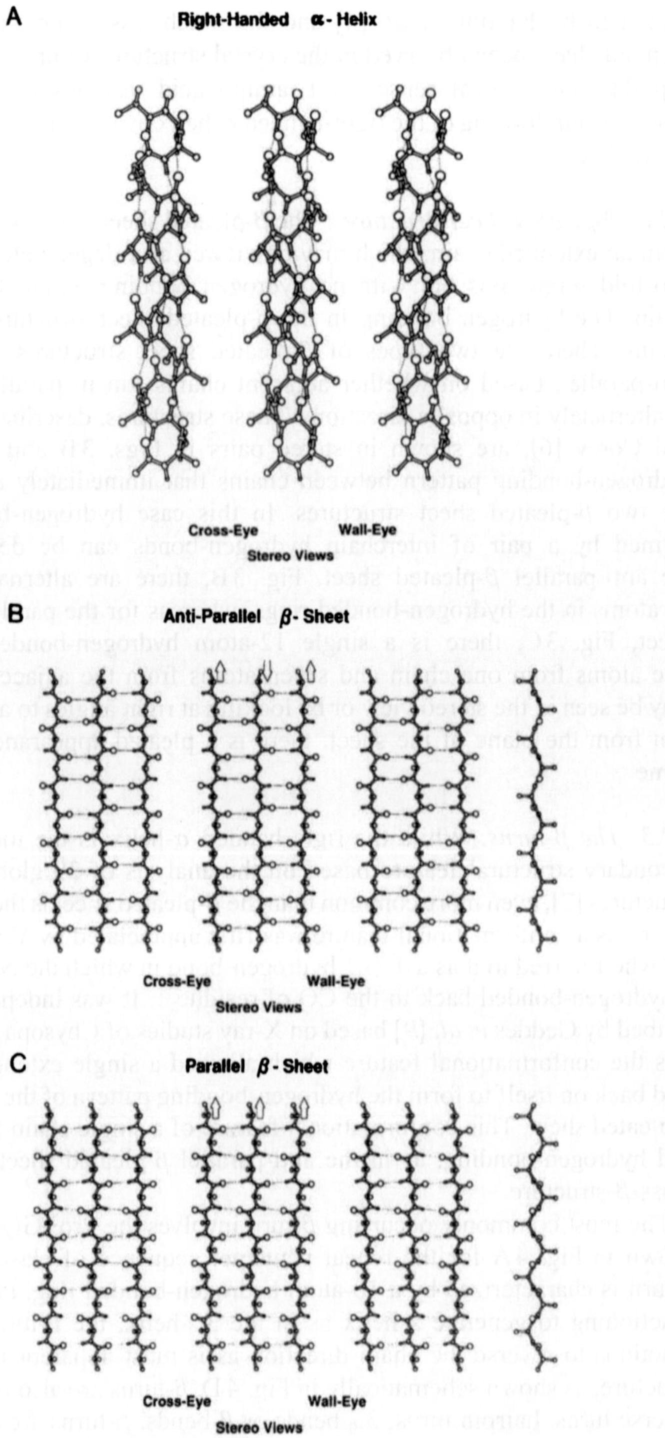
A
Right-Handed α- Helix
Cross-Eye
Wall-Eye
Stereo Views
B
Anti-Parallel β- Sheet
Cross-Eye
Wall-Eye
Stereo Views
C
Parallel β- Sheet
Cross-Eye
Wall-Eye
Stereo Views

Fig. 3.

residues $i + 1$ and $i + 2$. The Pro[2]-Gly[3] β-turn of Fig. 4A is Type II as is that shown in Fig. 4B, whereas if the end peptide moiety is rotated by approximately 180° such that the CO points into the plane rather than out it is the Type I β-turn as in Fig. 4C. The role of the β-turn in the cross-β structure is shown in Fig. 4D.

β-turns are common at the surface of globular proteins where they serve to turn the chain back into the globule. They are common in polypeptide antibiotics and hormones. The term β-turn was first used to describe the solution conformation of the valinomycin-K^+ complex [11] and immediately thereafter of oxytocin and deamino oxytocin [12] (see Figs. 5A and B). Both structures were subsequently verified by crystal structure, the former five years [13] and the latter sixteen years later [14, 15]. The conformation of deamino oxytocin may be seen as a short cross-β structure closed by a disulfide bridge.

2.2.4. The collagen triple stranded helix: Another regular secondary structure involving hydrogen-bonding between chains is that of the triple stranded helix of collagen. The complex step of developing the hydrogen-bonding pattern for three associating chains was performed by Ramachandran and Kartha [16] and by Rich and Crick [17]. Collagen is essentially a polytripeptide in which Gly must occur every third residue. The structure is given in stereo view in Fig. 6 using the coordinates of Miller and Scheraga [18] where the three chains are seen to twist around each other with a right-handed helical sense. This structure derives significance from being the structure of the most common mammalian protein, collagen, which constitutes some 30% of the protein of the human body. The triple stranded helix also occurs as part of other proteins as in the immune protein complement C1q [19, 20] and in the Type I macrophage scavenger receptor [21].

Fig. 3. Secondary structural features of polypeptides. (A) Stereo pair drawings of a right-handed α-helix of poly-L-alanine. The α-helix consists of a chain of repeating amino acids wound into a spiral chain. The spiral is held together by hydrogen bonds between each carbonyl group and the amide group of the fourth peptide residue down the chain, i.e. between the CO of i and the NH of $i + 4$. In each case the pair formed by the left and centre structures are arranged for cross-eye viewing while that of the center and right are arranged for wall-eye viewing. (B) Stereo pair plots of poly-L-alanine in the antiparallel β-sheet configuration. The arrows on the central set of chains indicate the alternating chain directions. (C) Stereo pair plot of poly-L-alanine in parallel β-sheet configuration. The arrows indicate the parallel chain direction. In the β-sheet configuration two or more peptide chains are held together laterally by hydrogen bonding. In the antiparallel β-sheet the peptide chains are aligned alternately in opposite directions while in the parallel β-sheet the peptide chains are aligned in the same direction. At the right-hand side a single chain is shown on edge to demonstrate the pleated nature of the chains in both (B) and (C).

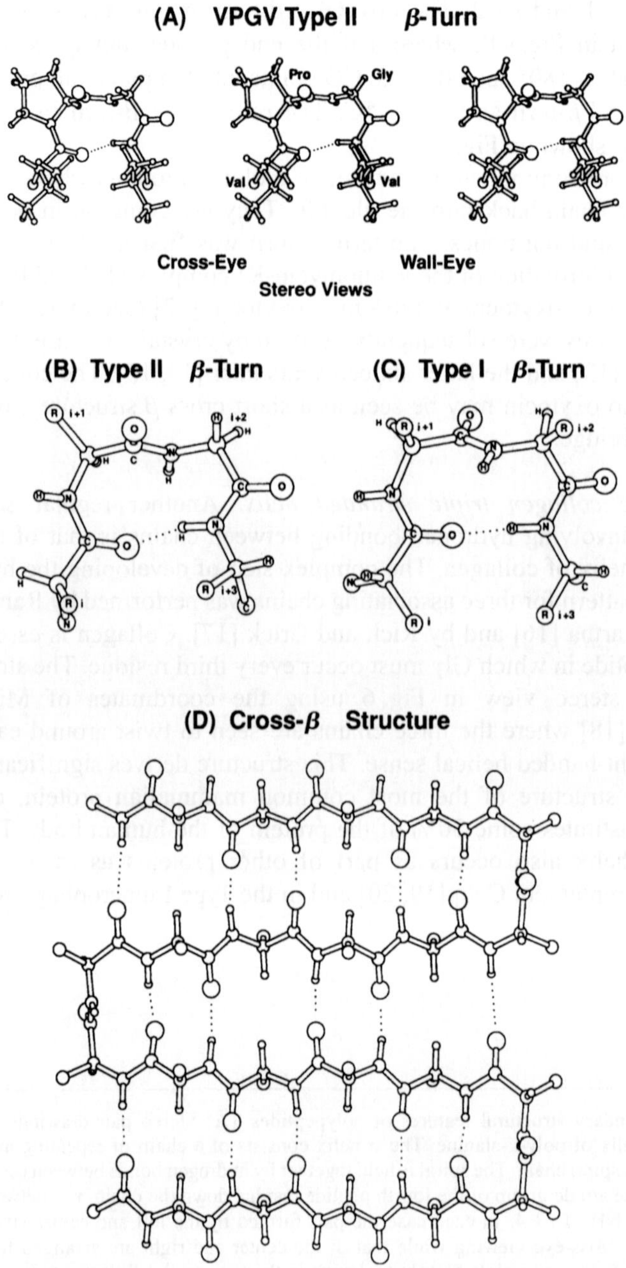

Fig. 4. β-turn configuration. A. Stereo pair drawings of Type II β-turn, involving the Pro²-Gly³ residues of VPGV, which occurs in the repeat pentamer sequence VPGVG. B. Type II β-turn in which the CO of the end peptide moiety projects out of the plane. C. Type I β-turn in which the end peptide moiety formed from residues $i+1$ and $i=2$ is rotated by approximately 180° such that the CO projects into the plane. D. Cross-β-structure in which β-turns function to fold a chain back on itself to form an antiparallel β-pleated sheet hydrogen bonding pattern resulting in a ribbon-like structure.

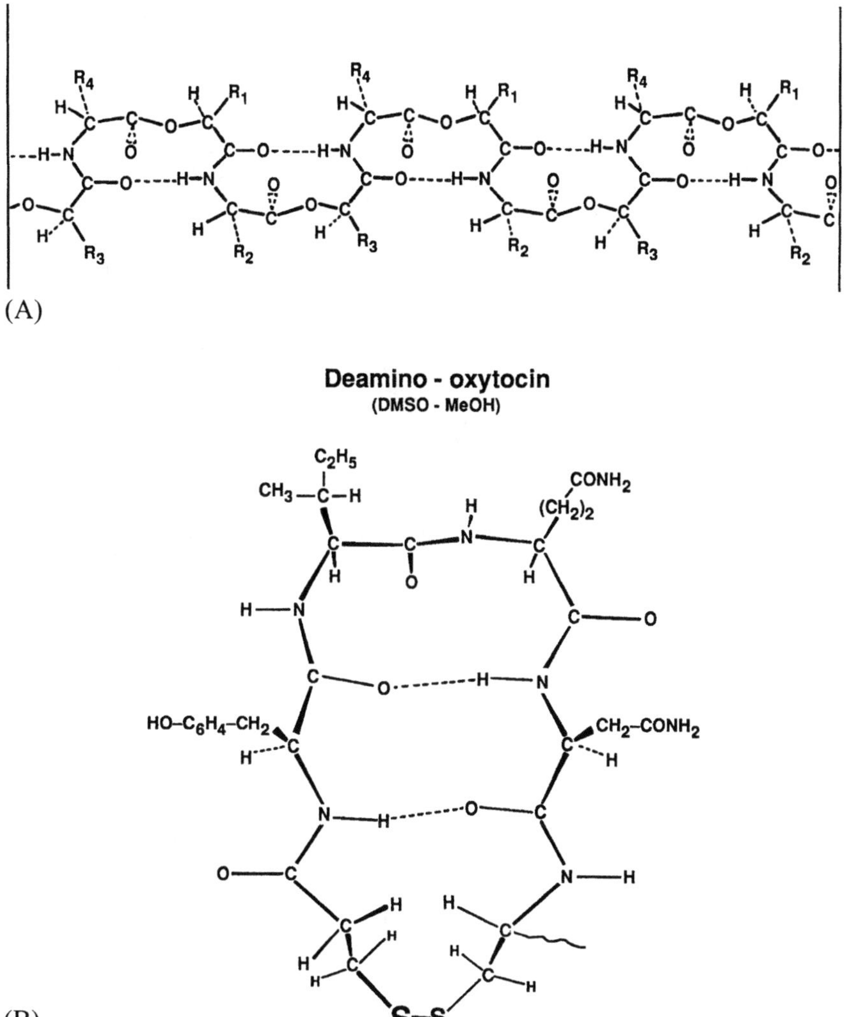

Fig. 5. A. β-turn of the valinomycin-potassium complex. Valinomycin, a cyclododecadepsipeptide, is comprised of a repeating tetramer with the absolute configuration [L·Lac–L·Val–D·HyV–D·Val–L·Val]$_3$. This diagram gives an unrolled prospective of the molecule in which all the amide protons are hydrogen bonded. It is essentially a series of β-turns in which the end peptide moiety is replaced with an ester group. (Reproduced with permission from [11]. B. β-turn structure of the ring component of deamino-oxytocin. The β-turn involves the sequence –L·Tyr–L·Ile–L·Glu–L·Asp. (Reproduced with permission from [12]).

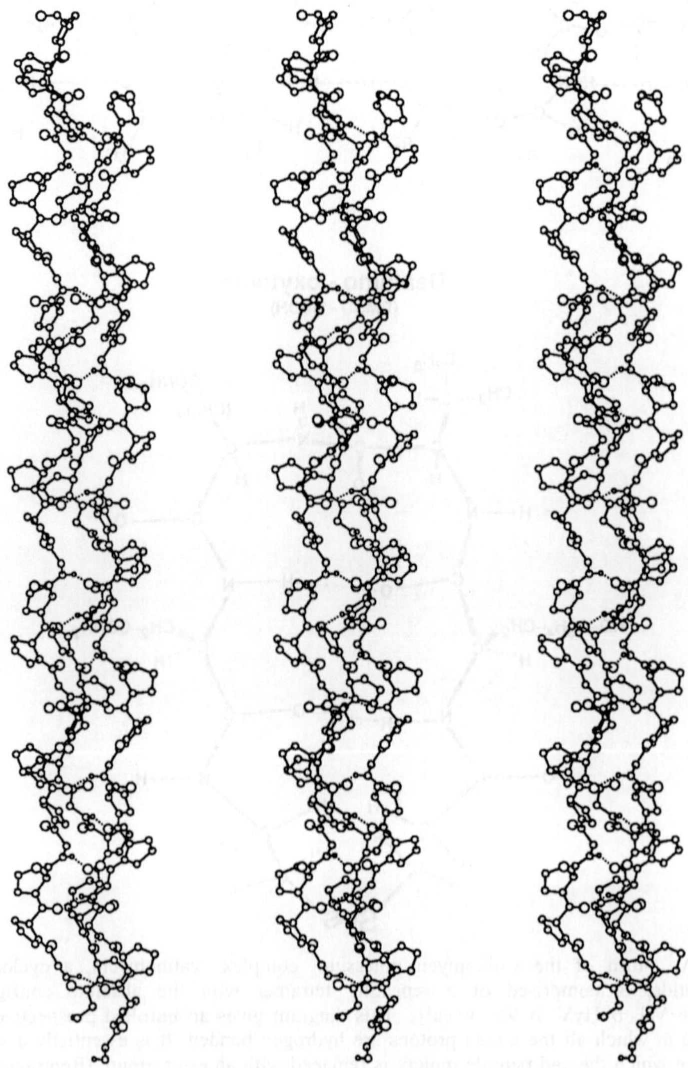

Cross-Eye **Wall-Eye**

Stereo Views

Fig. 6. Stereo pair plots of the triple-stranded collagen helix generated from the coordinates of the repeating GPP unit of Miller and Scheraga [18].

Fig. 7. Space filling model of the valinomycin-potassium ion complex. In this three-dimensional "core" structure the potassium ion resides in the central polar core. Elucidation of the molecular structure provides insight into the mechanism of potassium ion transport and into the basis for ion selectivity. The polar core is too small to accommodate a sodium ion, thus accounting for the potassium ion selectivity of the carrier. (Reproduced with permission from [22].)

2.2.5. The β-helix: While the ribosomal synthesis of polypeptides and proteins utilizes only L-amino acid residues, there are cytoplasmic enzymes in lower organisms that synthesize peptides containing D-amino acids. These peptides commonly exhibit antibiotic activity. The valinomycin structure of Figure 5 A, shown with a space filling model as the K^+ complex in Fig. 7 [22], is a cyclododecadepsipeptide comprising a repeating tetramer with the absolute configurations being (L-L-D-D) [23]. Valinomycin is a transmembrane ion carrier which is very selective for K^+ over Na^+ [24]. Another interesting transmembrane ion transport system is the pentadecapeptide with the optical configurations $(LD)_7L$, i.e. a repeating LD dipeptide. The most common sequence is referred to as Gramicidin A, HCO–L-Val^1–Gly^2–L-Ala^3–D-Leu^4–L-Ala^5–D-Val^6–L-Val^7–D-Val^8–L-Trp^9–D-Leu^{10}–L-Trp^{11}–D-Leu^{12}–L-Trp^{13}–D-Leu^{14}–L-Trp^{15}–$NHCH_2CH_2OH$ [25]. This was the first characterized ion-selective transmembrane channel [26, 27]. The Gramicidin A transmembrane channel derives from the head-to-head (amino end to amino end) dimerization of two molecules in a single-stranded β^6-helical conformation [28–30]. As shown in Fig. 8 [31], the hydrogen-bonding pattern between turns of each monomer is that of the parallel β-pleated sheet (see Fig. 3C above), whereas the hydrogen-bonding pattern at the head-to-head junction is that of the anti-parallel β-pleated sheet (see Fig. 3B above). For this reason the name was changed from the $\pi_{\text{L, D}}$-helix [28] to the β-helix. While the structure shown in Figs. 8 and 9 is given as left-handed as originally proposed [28], there is good evidence that the right-handed structure forms and could be the functional channel [32; 33; Jing and Urry, unpublished results]. It should also be appreciated that Gramicidin A forms many interesting double stranded β-helices as described by Veatch and Blout

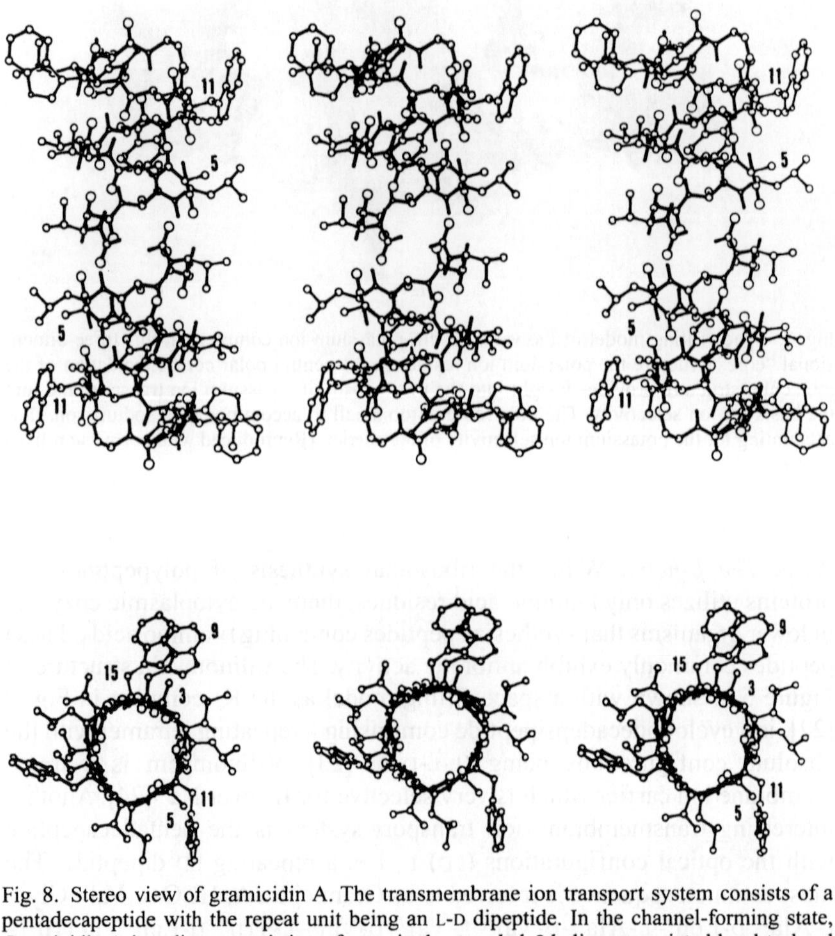

Fig. 8. Stereo view of gramicidin A. The transmembrane ion transport system consists of a pentadecapeptide with the repeat unit being an L-D dipeptide. In the channel-forming state, Gramicidin A is a dimer consisting of two single-stranded β-helices associated head-to-head (amino end to amino end). Top diagrams represent a side view while the bottom represents a channel view. See text for further discussion. (Adapted with permission from [31].)

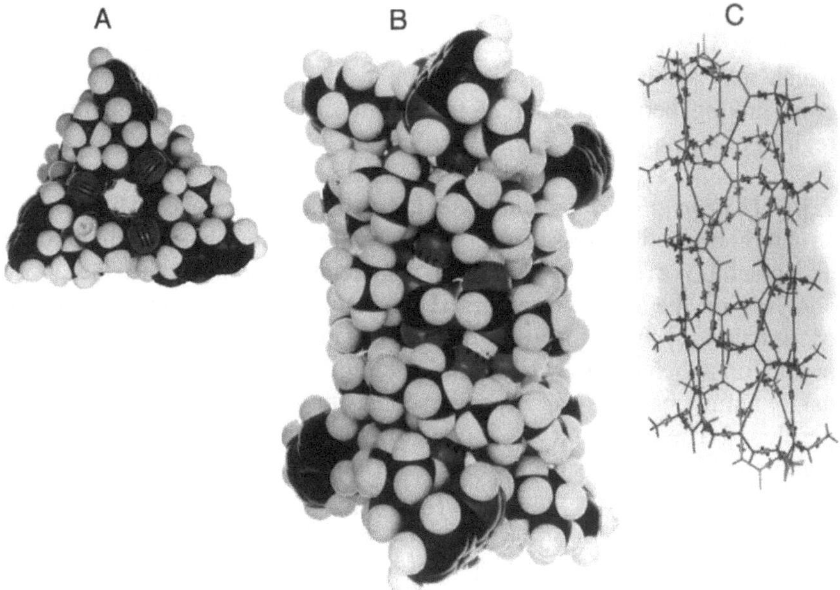

Fig. 9. Proposed structure of the gramicidin A transmembrane channel. This is the head-to-head diamerized, left-handed, single-stranded $\beta^{6.3}_{3.3}$-helix (Urry *et al.*, 1971). (A) Channel view with a 4 Å maximal diameter. (B) Side view showing the two formyl hydrogens at the center of the structure. (C) Wire model of side view oriented as in B showing the interturn hydrogen bonding pattern of the parallel-β-pleated sheet within each monomer and of the antiparallel β-pleated sheet between monomers. (Adapted with permission from D.W. Urry, M.M. Long, M. Jacobs and R.D. Harris, Ann. New York Acad. Sci, 264 [1975] 203–220).

[34]. A very attractive aspect of the structure of the β-helix of Gramicidin A is apparent in Fig. 9; the helix is cloaked in hydrophobic side chains for spanning a lipid membrane, and it has a channel along the helix axis suitable for passage of monovalent cations [28, 30]. Similarly, the valinomycin K^+ complex of Fig. 7 places the potassium ion in an ion-selective polar core in a small structure cloaked with hydrophobic residues suitable for the carrier mechanism of ion transport across a lipid layer. Thus the molecular structures provide an immediate qualitative idea of the cation transport functions, which may be of particular interest in a treatise on Biolelectrochemistry.

2.3. *Tertiary structure*

Interactions between side chains of amino acid residues in polypeptides and proteins will be considered tertiary structure. These interactions include: 1. hydrophobic side chain interactions which are important in protein folding, macromolecular assembly and, as will be discussed at some length below, in free energy transduction processes; 2. ion-pairing interactions which may stabilize specific structural features; and 3. disulfide bridge cross-linking between different polypeptide chains or between different regions of a single chain.

2.3.1. Hydrophobic side chain association: The importance of the tertiary structural effect of the association of hydrophobic side chains in the folding and assembly of proteins has been appreciated for many decades. Key analyses of the effect were given by Kauzmann [35] and Edsall and McKenzie [36]. In the pursuit of an understanding of this process and of a prediction scheme for hydrophobic folding, many hydrophobicity scales using numerous physical properties and statistical analyses have been developed. The first scale was derived by Nozaki and Tanford [37]; the most commonly used scale is that of Kyte and Doolittle [38], and more than forty other scales have been compared [39, 40]. The primary emphasis in the use of these scales has been the prediction of transmembrane helices in membrane proteins. The development of yet another hydrophobicity scale will be presented below. The features which justify yet another hydrophobicity scale are that it is the first scale which is based directly on the hydrophobic folding process of interest, that in its further development certain more complex interactional effects can be evaluated, and that it is required to design polypeptides capable of free energy transduction as well as to understand better the effects of amino acid substitutions on protein folding and stability.

With regard to globular protein structure it is appreciated that hydrophobic residues tend to be buried away from the aqueous interface. An example of hydrophobic side chain association leading to folding of a specific structure is seen in the temperature-induced folding of the polypentapeptide of elastin; on raising the temperature a series of Pro^2-Gly^3 Type II β-turns are described as folding into the β-spiral structure as depicted in Fig. 10 [41, 42] in a process of optimizing intramolecular hydrophobic contacts. In this case the dominant intramolecular interactions leading to thermally-induced hydrophobic folding are the interactions of the Pro^2-β-CH_2 moieties across the turn of a β-spiral to the Val^1-γCH_3 moieties [43].

2.3.2. Ion pairing side chain interactions: There is much that can be written about the charge-charge interactions of side chains and their relevance, for example, to protein folding [44]. Here one specific structural effect will be noted. When sequences contain, for example, a Glu^i-X-Y-Z-Lys^{i+4}, the ion pairing between the side chains stabilizes the α-helical secondary structure which has 3.6 residues per turn [45]. The $i \ldots i+4$ interaction stabilizes the α-helix more than the $i \ldots i+3$ interaction. Reversal of the Glu and Lys residues is not as effective because the dipole moment of the right-handed α-helix would not be favorably oriented with respect to the ion pair location.

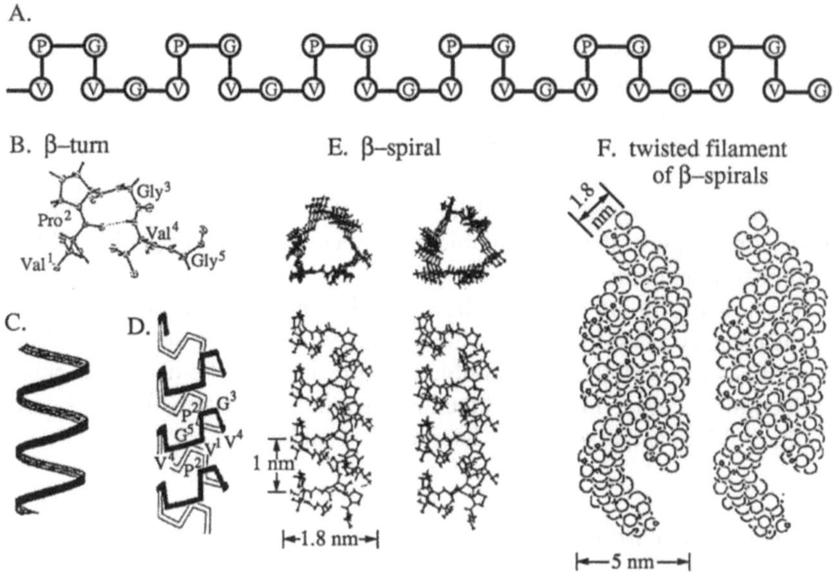

Fig. 10. (A) Molecular structure of the repeating polypentapeptide (VPGVG) of elastin. (B) The dominant secondary structural feature is the β-turn consisting of a 10-membered hydrogen bonded ring utilizing the Val^1 CO and the Val^4 NH. (Reproduced with permission from Cook *et al.*, 1980). (C) On raising the temperature the polypentapeptide wraps up into a helical structure in the process of optimizing intramolecular hydrophobic contacts. (D) The Pro^2-Gly^3 Type II β-turns function as spacers between the turns of the helical structure utilizing interturn hydrophobic contacts. C and D are schematic representations, reproduced with permission from Urry (1983). (E) Detailed representation of helical structure, called a β-spiral, axis view (above) and side view (below) and both as stereo pair sets for cross-eye viewing. (F) Spherical united residue representation of several β-spirals supercoiling to form the twisted filament much as visualized by electron microscopy with negative staining of initial aggregates formed on raising the temperature above 30 °C. (Reproduced with permission from [42]).

2.3.3. Disulphide bridging: Disulphide bridge formation is a common way of having reversible cross-linking, particularly within the cell where reduction and freeing of the cross-link can more readily occur. This, of course, utilizes the side chains of two cysteine residues. The disulphide bridge is an important structural element of many globular proteins but interestingly it is less frequently used in extracellular structural proteins.

2.3.4. A common folding hypothesis: In the hierarchical folding of a protein, the common initial step is considered to be the forming of secondary structural features along the sequence (dictated most dominantly by the sequence and less so by the solvent and temperature), with highest probability of α-helices and β-turns and a somewhat less probable association of β-chains. These would be the more stable secondary structural features. In the case of poly(VPGVG) this is the formation of the repeating Type II Pro^2-Gly^3 β-turns. Next there would occur the tertiary folding of these secondary structural features and additional chain segments with the formation of hydrophobic domains or pockets of the more hydrophobic residues; formation of this dominant hydrophobic domain would then be accompanied by the formation of opportunistic secondary structures. This would be followed by the association of hydrophobic side chains to form a hydrophobic pocket utilizing the less hydrophobic clusters of side chains, which would again be accompanied by hydrogen-bond formation that is opportunistic within the loose constraints of the hydrophobic fold. Accordingly, hydrophobic folding could itself be hierarchical.

It is now beginning to appear, however, that hydrophobic association can be the primary event followed by the sorting out of the best secondary structure consistent with the hydrophobic domain. This will be discussed below in Section 7 after having treated the inverse temperature transition of hydrophobic folding and assembly in Sections 3, 4 and 6.

2.4. *Quaternary structure*

In general, quaternary structure may be considered the association of separate polypeptide chains of either identical or related primary, secondary and tertiary structures. When large multiple domains occur within a very large protein and the interconnecting polypeptide chains do not appear essential to the interactions of the domains, the interactions between such domains are here considered to be more quaternary than tertiary in nature. Quaternary structural interactions are prominent in the multi-subunit proteins, e.g. in the interesting cooperatively effects exhibited by multi-subunit enzymes [46–49], in the association of subunits to form ion-selective transmembrane channels [50–54] and pumps [55–58], in the assembly and disassembly of actin and tubulin filaments [59], in the association of two and three stranded α-helical fibrous proteins [60, 61] both intracellularly (e.g., myosin) and extracellularly (e.g., fibrin), in the association of triple stranded collagen helices to form banded collagen fibres (see below), and in the association of tropoelastin β-spirals to form the twisted filaments of fibrous elastin (see below). Here will be briefly noted the supercoiling of α-helices and of β-spirals, and covalent cross-links that form between protein molecules utilizing amino acid side chains.

2.4.1. Supercoiling of α-helices: The left-handed supercoiling of right-handed α-helices to form multiple stranded components of fibrous proteins was early described by Pauling and Corey [62] and by Crick [63]. The current perspective is that α-fibrous proteins intracellularly appear to be double stranded [64] whereas extracellularly they appear to be triple stranded [60]. Intracellular examples of double stranded α-helical rods are myosin, tropomyosin, paramyosin, M-protein and desmoplakin [60]. Extracellular examples of triple stranded α-helical rods are fibrinogen, laminin, tenascin and the macrophage scavenger receptor protein [21, 61]. The primary structural clue to the occurrence of such structures is the finding of the heptad repeat, as early emphasized for myosin [65]. The heptameric repeat is approximately twice the 3.6 residues per turn of the α-helix such that every seventh residue defines a ridge along the α-helix. When each seventh residue is hydrophobic, this allows for association of the helices by hydrophobic interactions.

The single residue transition temperature profiles for the primary structure of the myosin rod [65] is given in Fig. 11. The negative deflections from the 37 °C line are due to the polar residues. Most apparent are the deflections near 120 °C for Lys^+, near 170 °C for $AspCOO^-$ and near 250 °C for $GluCOO^-$. These show a periodic placement along the α-helix. The positive deflections show the positions of the hydrophobic residues where the periodic spacing is also seen. Rarely are two adjacent hydrophobic residues seen; only once is there a string of three hydrophobic residues, and never a string of more than three hydrophobic residues. Generally there are two or three residues between hydrophobic residues as would be required to give a sidedness to a structure with 3.6 residues per turn; commonly there are seven residues between hydrophobic residues.

An interesting, more diverse structure is that arising form the primary structure of the Type I macrophage scavenger receptor [21]. It contains, in addition to a proposed triple stranded helical rod, a transmembrane segment and a triple stranded collagen-like region. An eleven residue mean transition temperature plot is given in Fig. 12 A where there is only one hydrophobic

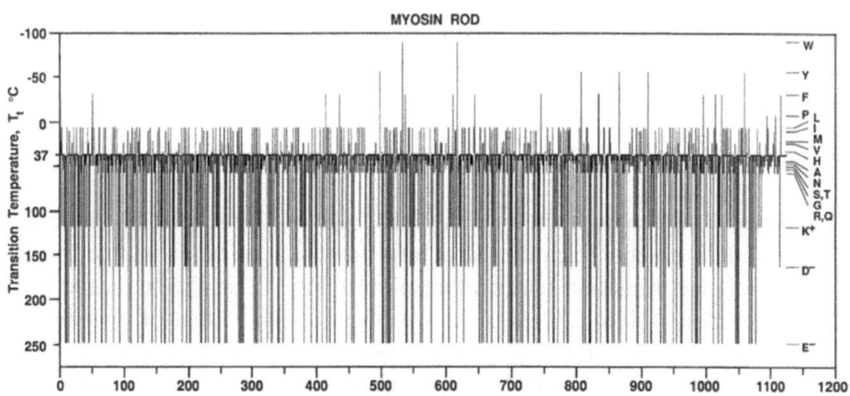

Fig. 11. Single residue transition temperature profile for the primary structure of myosin. At the right is indicated the identity of the residues corresponding to the various deflections. Deflections in a positive direction from 37 °C indicate hydrophobic residues while those in a negative direction from 37 °C indicate polar residues.

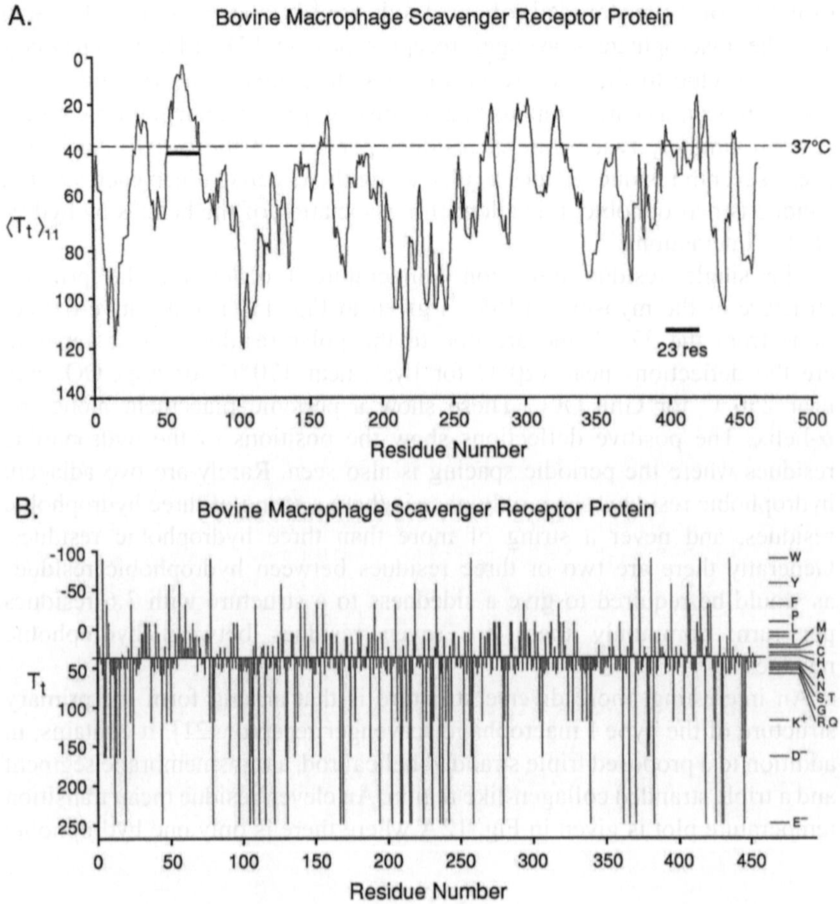

Fig. 12. Bovine macrophage scavenger receptor protein. (A) Eleven residue mean transition temperature plot. The bar extending from residue 50 to 73 indicates the position of the hydrophobic residues responsible for the transmembrane segment. (B) Single residue transition temperature plot. (See text for discussion).

sequence long enough to span a lipid bilayer. This defines the transmembrane segment. The other structural features can be considered in the single residue transition temperature plot of Fig. 12B. The sequence from approximately 110 to 270 exhibits a pattern similar to that of myosin (see Fig. 11) except that there are more pairs (nine) of hydrophobic residues and even two sequences of three hydrophobic residues in a sequence, and consequently some reduction in polar residues. These differences could be expected to be more conducive to the association of three α-helical strands. The triple stranded collagen-like sequence, which would also dictate the association of three chains, is apparent on noting the recurrence of glycine every third residue for 24 tripeptide repeats starting at residue 272 and ending at residue 343.

2.4.2. Supercoiling of β-spirals: As we have seen, periodicities in the primary structures of myosin (represented in Fig. 11) and in the macrophage scavenger receptor (represented in Fig. 12) dictate interesting secondary and quaternary structural effects. Returning now to examine the single residue transition temperature profile for bovine tropoelastin in Fig. 2, the periodicities in this protein become even more striking. Consistent with this, it was shown early that fibrous elastin is comprised of parallel aligned twisted filaments as in the supercoiling of structural entities [66, 67]. A structural entity within tropoelastin that has been shown to form twisted filaments is the β-spiral of poly (VPGVG). A depiction of the supercoiling of the β-spiral of poly(VPGVG) to form a twisted filament is given in stereo view in Fig. 10E.

2.4.3. Covalent cross-linking of structural proteins: Three notable sites which use different modes for cross-linking are the extracellular matrix, the serum, and within the cell. In the extracellular matrix a principal player is the enzyme lysyl oxidase which oxidizes the ε-NH_2 moiety of peptidyl lysine to form the aldehyde, α-amino adipic δ-semi-aldehyde [68]. The aldehyde can then combine with a lysine side chain to form the Schiff base, anhydrolysinonorleucine [69], or with another aldehyde to form the aldol condensation product. These cross-links are common to collagen and elastin. A more complex cross-link found in elastin requires four lysine residues. Three are oxidized to the aldheyde and these combine with a lysine to form a tetrasubstituted pyridinium cross-link called desmosine and its isomer isodesmosine [70]. An example of cross-linking in the serum using amino acid side chains is found in the formation of the fibrin clot, where factor XIII, a transglutaminase, reacts a glutamine carboxamide side chain with a lysine side chain to form an amide cross-link, N^{ε}-(γ-glutamyl)lysine [71].

3. Inverse temperature transitions (cold denaturation) in polypeptides and proteins

On heating to elevated temperatures, proteins denature; they undergo a transition to a state of decreased order. In general this occurs on heating proteins above physiological temperature, and it is consistent with our understanding of the second law of thermodynamics wherein the order of a system must decrease on raising the temperature. Nonetheless, there is a substantial literature going back for half a century that also speaks of "cold denaturation" of enzymes (proteins) [72]. This has been seen, for example, as a loss of enzymatic activity on lowering the temperature, a loss greater than expected from the temperature dependence of the enzyme catalyzed reaction whether treated by the Arrhenius equation,

$$k' = Ae^{-E/RT} \tag{1}$$

where k′ is the reaction rate constant, A is the collision frequency, E is the activation energy, R is the gas constant and T is the absolute temperature in degrees Kelvin, or whether treated by Eyring rate theory:

$$k' = \frac{kT}{h} e^{-\Delta H^\dagger/RT} e^{\Delta S^\dagger/R} \tag{2}$$

where k is the Boltzmann constant, h is Planck's constant, $\Delta H^\dagger$ is the enthalpy (or heat) of activation and $\Delta S^\dagger$ is entropy of activation. Insight into "cold denaturation" can be gained by studying the relatively hydrophobic sequences of elastin.

3.1. Crystallization on raising the temperature of cyclic analogs

One prominent repeating sequence in elastin is VAPGVG which repeats eight times in a continuous sequence in human elastin with only two substitutions [73–75]. The high polymers of the repeating hexapeptide and lower

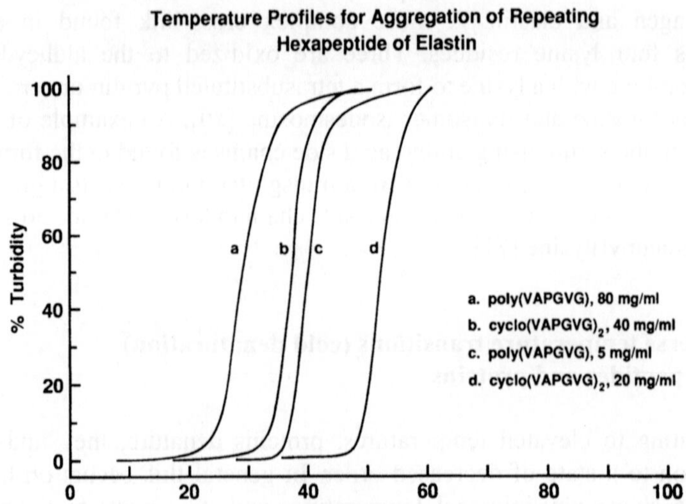

Fig. 13. Temperature profiles of turbidity formation, TPτ, of repeating hexapeptide of elastin-concentration dependence. The aggregation of the polymers upon raising the temperature was monitored by following turbidity formation spectrophotometrically at 300 nm. The aggregation temperatures of high molecular weight poly(VAPGVG) (curves a and c) and low molecular weight cyclo(VAPGVG)$_2$ (curves b and d) of the repeat sequence were compared at two concentrations. In the low molecular weight oligomers, the molecules were cyclized to eliminate the effects of the polar end groups on the association.

(A)

(B)

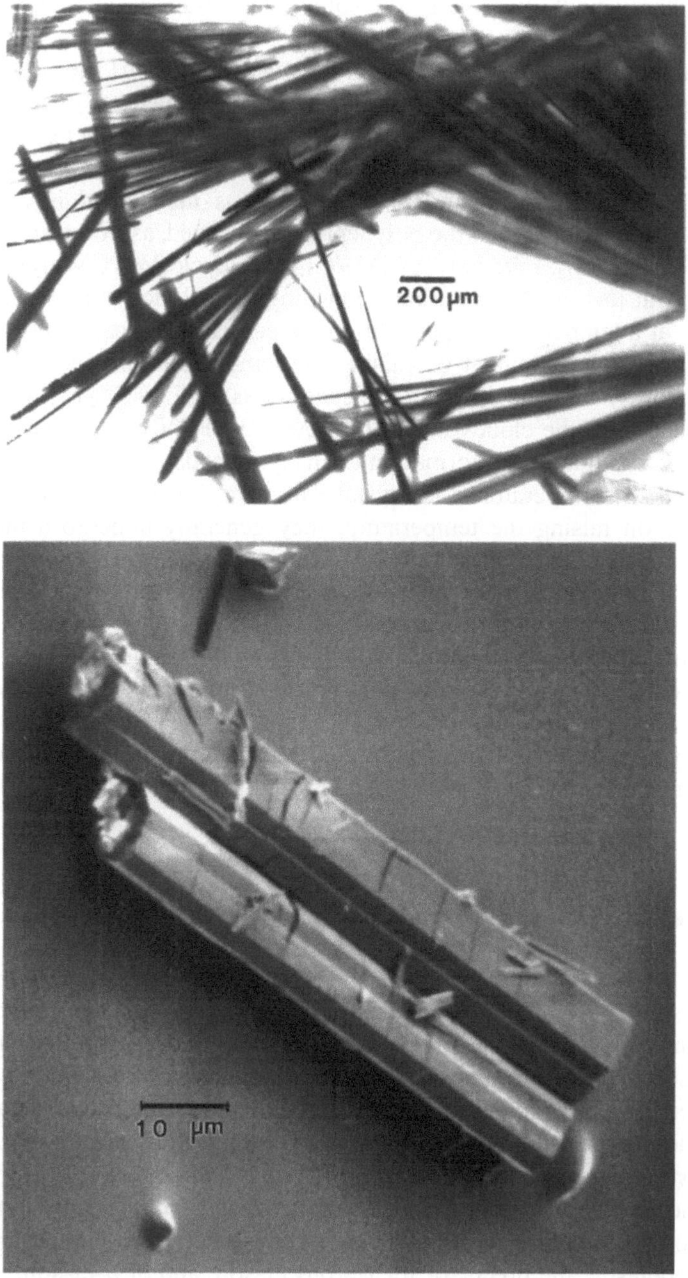

Fig. 14. Cyclododecapeptide crystal formation upon raising the temperature. (A) Light micrograph of crystals of the cyclododecapeptide in water. The fine needle-like crystals obtained on raising the temperature are birefringent, as observed with cross-polarizers. (B) Scanning electron micrograph of Au-Pd-coated cyclododecapeptide crystals. (Reproduced with permission from [77].)

molecular weight oligomers are soluble in water at low temperature, but they aggregate on raising the temperature [76]; this is because of the balance between the polar peptide moieties, CONH, and the hydrophobic (apolar) side chains. In studying shorter sequences, this balance is complicated by the polar amino and carboxyl moieties at the termini of the short sequence or due to the contributions of blocking groups. The exact same balance of apolar and polar moieties is maintained in a simpler molecular structure when the polypeptide is cyclized, as, for example, with cyclo$(VAPGVG)_2$. Like poly(VAPGVG) itself, cyclo$(VAPGVG)_2$ is soluble in water at low temperature, but on heating the solution abruptly becomes cloudy at a temperature dependent on concentration. Temperature profiles for turbidity development on raising the temperature are shown for both molecular systems at different concentrations in Fig. 13 [77]. On closer examination of the cloudy solution of cyclo$(VAPGVG)_2$ it is seen that the cloudiness is due to the formation of many crystals (see Fig. 14) [78]. On lowering the temperature the crystals dissolve. Most commonly, crystals dissolve on raising the temperature; they generally undergo a thermally induced transition of dissolution. Cyclo$(VAPGVG)_2$ does exactly the inverse; it undergoes a thermally induced transition of crystallization on raising the temperature. Accordingly, the phenomenon is referred to as an *inverse temperature transition.* While lowering the temperature and thereby causing the crystals to dissolve does not bring to mind "cold denaturation" (indeed the term denaturation would be inappropriate to describe dissolution of a crystal), the underlying driving forces involved in heat induced crystallization and heat induced increase in enzymatic activity (naturation) are the same.

3.2. Self-assembly of polypeptides and proteins on raising the temperature

The self-assembly of polypeptides and proteins into more ordered structures on raising the temperature is readily observed in the connective tissue proteins collagen and elastin. The banded nature of collagen fibers has long been seen in electron micrographs, and the early work of Gotte and colleagues [66, 78, 79] and of Clearly and colleagues [67] demonstrated the presence of parallel-aligned filaments within fibrous elastin and within reconstituted elastin [80]. The work of Gross and colleagues [81] showed that the warming of triple stranded collagen dispersed in an appropriate salt solution resulted in reconstituted banded collagen fibres. The self-assembly of tropoelastin, the precursor for fibrous elastin, and of α-elastin, a 70 kD chemical fragmentation product of fibrous elastin, on raising the temperature of molecularly dispersed solutions is demonstrated in Figs. 15A and B [81, 83]. In each case the early small aggregates were caught on a carbon coated grid and negatively stained with uranyl acetate, oxalic acid at pH 6.3, and in both cases filamentous aggregates were observed with fila-

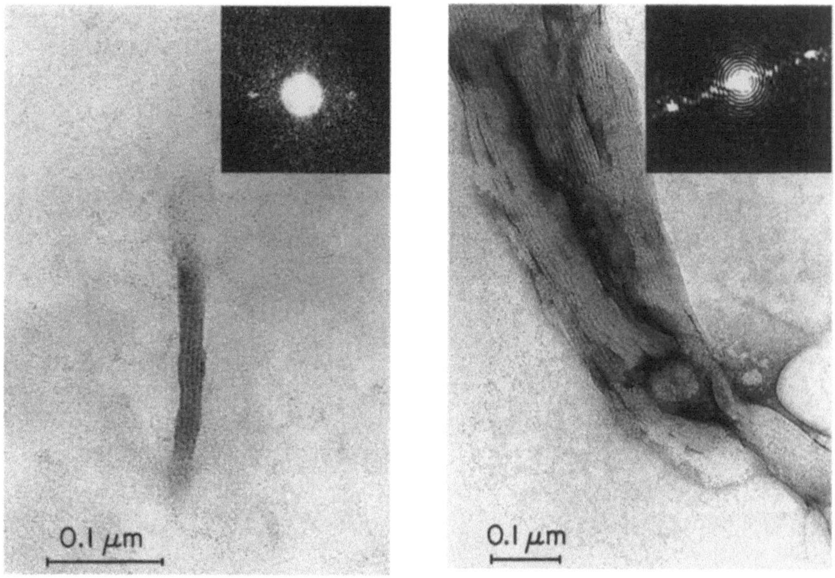

(A) (B)

Fig. 15. Electron micrographs of aggregates of tropoelastin and α-elastin. Negative staining electron microscopy clearly demonstrates the ordered arrangement into fibrous structures which develop upon raising the temperature of solutions of components of fibrous elastin. (A) Tropoelastin, the precursor molecule of fibrous elastin. (Reproduced with permission from [82].) (B) α-Elastin, a chemical fragmentation product of fibrous elastin. (Reproduced with permission from [81]). The inserts are optical diffraction patterns obtained from the electron micrographs and showing major diffraction spots similar to those found for fibrous elastin by Gotte *et al.* [66] (Reproduced with permission from [83]).

ments of approximately, 5 nm diameter in analogy to the natural fibrous elastin [66, 79].

Self-assembly on raising the temperature has also been observed with poly(VPGVG) [84] and with Lys and Glu-containing analogs in which amide cross-linking during the dynamic aggregational process fixed the aggregates such that the formation of anisotropic fibres could be observed [85] (see Fig. 16). The anisotropic fibres could be seen to be comprised of fibrils which in turn were comprised of parallel aligned 5 nm diameter filaments with a twisted filament substructure.

3.3. *Cold denaturation in proteins*

There is a large body of literature, most extensively reviewed by Privalov [72], which describes or infers reversible loss of globular protein activity or function on lowering the temperature. This has been called "cold denaturation", and it is, of course, the same reversible molecular process as the

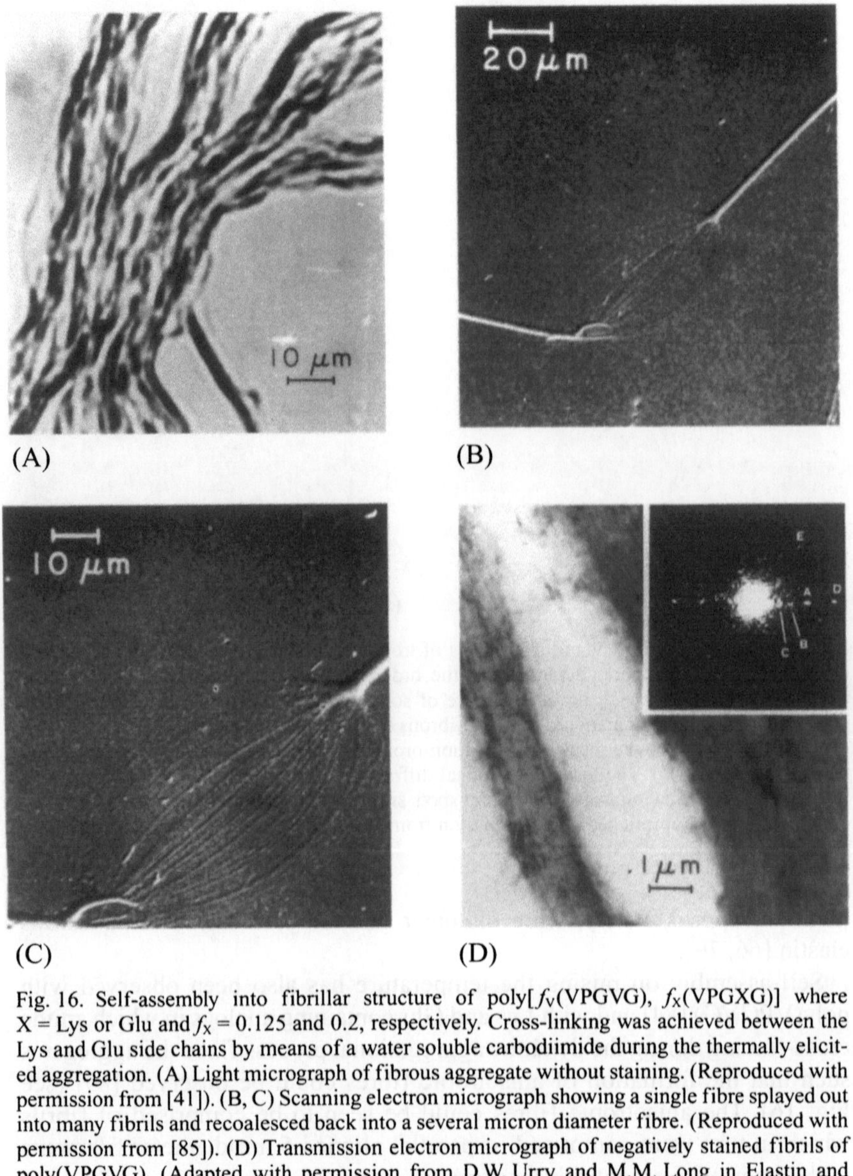

Fig. 16. Self-assembly into fibrillar structure of poly[f_V(VPGVG), f_X(VPGXG)] where X = Lys or Glu and $f_X = 0.125$ and 0.2, respectively. Cross-linking was achieved between the Lys and Glu side chains by means of a water soluble carbodiimide during the thermally elicited aggregation. (A) Light micrograph of fibrous aggregate without staining. (Reproduced with permission from [41]). (B, C) Scanning electron micrograph showing a single fibre splayed out into many fibrils and recoalesced back into a several micron diameter fibre. (Reproduced with permission from [85]). (D) Transmission electron micrograph of negatively stained fibrils of poly(VPGVG). (Adapted with permission from D.W. Urry and M.M. Long in Elastin and Elastic Tissues, L.B. Sandberg, W.R. Gray and C. Franzblau (editors), Plenum Press, New York, Adv. Exp. Med. Biol., 79 (1977) [685–714]). Insert: Optical diffraction of electron micrograph showing numerous reflections with a dominant reflection (A) near 5 nm as well as an off-equatorial reflection of E (4.5 nm). (Reproduced with permission from [84].)

observed reconstitution of protein structure and function by raising the temperature through the relevant transition temperature range. Reconstitution (otherwise called folding and self-assembly) of extracellular fibrous proteins due to heating to physiological temperatures was demonstrated by electron microscopy for collagen in 1955 [86], for a degradation product of elastin in 1968 [67 and references therein], and for the precursor protein of elastin in 1974 [82]. On the other hand, cold denaturation of globular proteins was early inferred from the temperature dependence of enzyme activity [87–90], predicted on the basis of scanning calorimetry studies [91], or observed spectroscopically under mild denaturing conditions [92, 93]. The latter methods have generally been required to deduce cold denaturation of globular proteins as the cold denaturation temperatures have been below 0 °C. Most recently, cold denaturation has been reported for chymotrypsin and lactate dehydrogenase [94, 95] and these reports have been considered to be the "first successful direct observation of a major cold-induced transition", presumably referring to enzymes or other functional globular proteins such as myoglobin. That this is a more general property of proteins much earlier characterized in components of fibrous proteins is unambiguously documented earlier in this chapter. Most importantly for our concerns, the modulation of this type of folding and assembly transition (an inverse temperature transition) is proposed as a major mechanism whereby proteins function in general and specifically whereby free energy may be converted by a protein from one form into another. But first it is useful to consider the dependencies of the temperature of the inverse temperature (cold denaturation) transition T_t, in order that it can be used as a tool with which to analyze specific free energy transductions.

4. Changing the temperature, T_t, of the inverse temperature transition for folding and assembly

What determines the temperature, T_t, of an inverse temperature transition in polypeptides and proteins and what is a good model system for studying the various dependencies of T_t? It was seen in Fig. 13 that the value of T_t is concentration-dependent for poly(VAPGVG) and for cyclo(VAPGVG)$_2$. For poly(VPGVG), however, it is possible to find an upper concentration limit for high molecular weight polymers (> 50 kDa) at which further increases in concentration cause no further lowering of the value of T_t. As seen in Fig. 17, the temperature range over which aggregation occurs on raising the temperature becomes lower and narrower on increasing the concentration and reaches a low temperature limit at a high concentration near 40 mg/ml. Accordingly, this concentration of poly(VPGVG) can be used and other variables considered for their effect on the value of T_t. Additional advantages of the poly(VPGVG) system are: (1) that it is optimally poised for the transition to occur between room temperature and physiological tem-

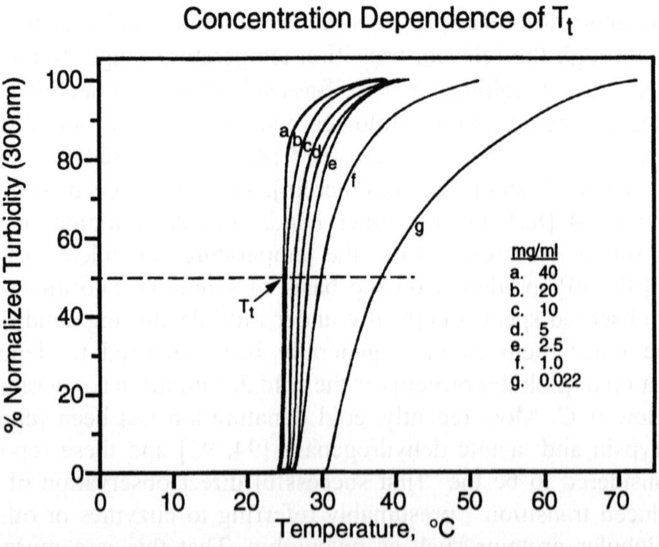

Fig. 17. Concentration dependence of the transition temperature, T_t, for high molecular weight poly(VPGVG). The temperature range over which aggregation occurs becomes lower and narrower with increasing concentrations of polypentapeptide reaching a low temperature limit at concentrations near 40 mg/ml. (Reproduced with permission from [120].)

peratures; (2) that the folding and assembly transition which occurs as part of the aggregation process is entirely reversible in water; and (3) that the model system allows for chemical modification when done with proper knowledge of conformation. Of the five residue positions in (Val^1-Pro^2-Gly^3-Val^4-Gly^5)$_n$ only position four can be extensively varied [96]. Position one utilizes β-branching which is present only in Val and Ile residues [97]; Ala in position one leads to irreversible precipitation on raising the temperature [98]. The Pro residue in position two is essential for the β-turn formation and for the $Val^1\,\gamma CH_3$-$Pro^2\,\delta CH_2$ hydrophobic contacts within the pentamer and for the $Val^1_i\,\gamma CH_3$-$Pro^2_{i+3}\,\beta CH_2$ hydrophobic contacts between turns of the β-spiral [99, 100]. Even an Ala in position 3 alters the β-turn and leads to a different but interesting polymer that, on γ-irradiation to form a cross-linked matrix, reversibly forms a relatively stiff plastic-like state on raising the temperature [101]. Even the addition of a CH_2 moiety in position 5, as in Ala^5, results in a granular precipitate which when γ-irradiation cross-linked is inelastic and fragments when extended [102]. Fortunately, position four can be varied extensively. In general, this position will be used.

4.1. Dependence of T_t on amino acid composition and a new hydrophobicity scale

For determining the effect of amino acid composition on the temperature, T_t, of the inverse temperature transition, the general formula can be written poly[f_V(VPGVG), f_X(VPGXG)] where f_V and f_X are mole fractions of each pentamer in the polypentapeptide and $f_V + f_X = 1$. In practice each pentamer is synthesized and then added to the polymerization flask in the desired ratios (mole fractions). The actual incorporation is verified by nuclear magnetic resonance spectra and amino acid analyses. When using the temperature profiles of aggregation (turbidity formation), the solution is initially clear until some onset temperature for turbidity, and the turbidity then reaches a maximum (100%). T_t is defined as the temperature at which the turbidity is half maximal (i.e. 50%). The dependence of T_t for a series of amino acids with values of $f_X = 0.2$ is given in Fig. 18 [103] for a medium of phosphate buffered saline (0.15 N NaCl, 0.01 M phosphate).

A related measure of the transition temperature is obtained from differential scanning calorimetry. The inverse temperature transition is an endothermic transition, as seen in Fig. 19, and the derivative of the thermogram with respect to temperature provides a well-defined temperature near the onset of the transition which is indicated as T_b (T_b is very close in value to T_t). For the series of non-polar amino acid residues in water, X = Trp, Tyr, Phe, Leu, Ile, Met, Ala and Gly, the value of T_b is plotted as a function of f_X [104]. The plots of T_b vs. f_X are essentially linear and there is a general distribution with the more hydrophobic residues exhibiting transi-

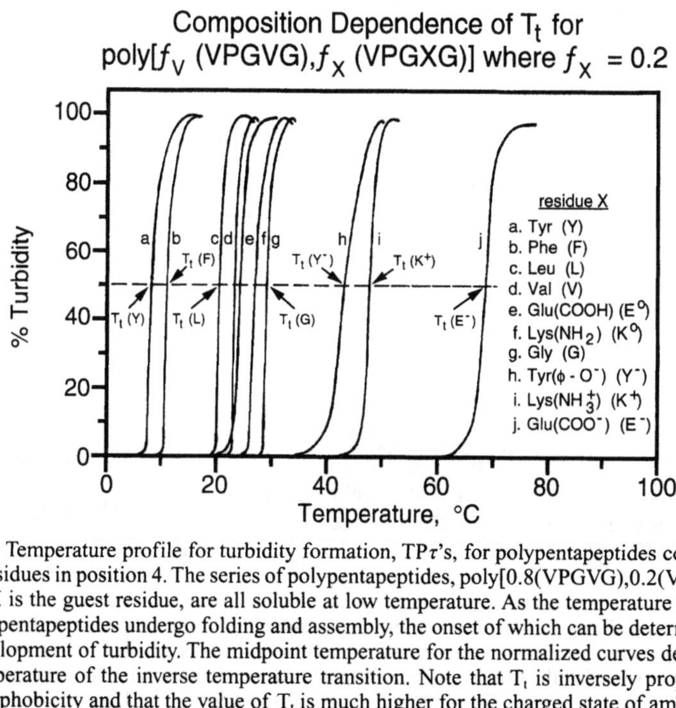

Fig. 18. Temperature profile for turbidity formation, TPτ's, for polypentapeptides containing guest residues in position 4. The series of polypentapeptides, poly[0.8(VPGVG),0.2(VPGXG)] where X is the guest residue, are all soluble at low temperature. As the temperature is raised, the polypentapeptides undergo folding and assembly, the onset of which can be determined by the development of turbidity. The midpoint temperature for the normalized curves defines T_t, the temperature of the inverse temperature transition. Note that T_t is inversely proportional to hydrophobicity and that the value of T_t is much higher for the charged state of amino acids with functional side chains. (Reproduced with permission from [124].)

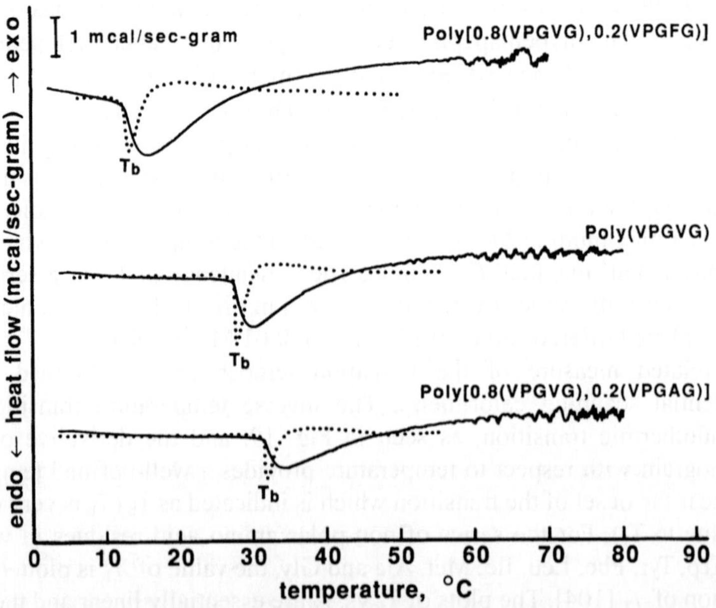

Fig. 19. Differential scanning calorimetry of three polypentapeptides in water. The solid line represents the thermogram and the dotted line the derivative of the thermogram with respect to temperature. The transition temperature is defined as T_b and is obtained from the extremum of the derivative curve. The downward deflection of the curves at the transition temperatures indicates the endothermic nature of the transition. (Reproduced with permission from [104].)

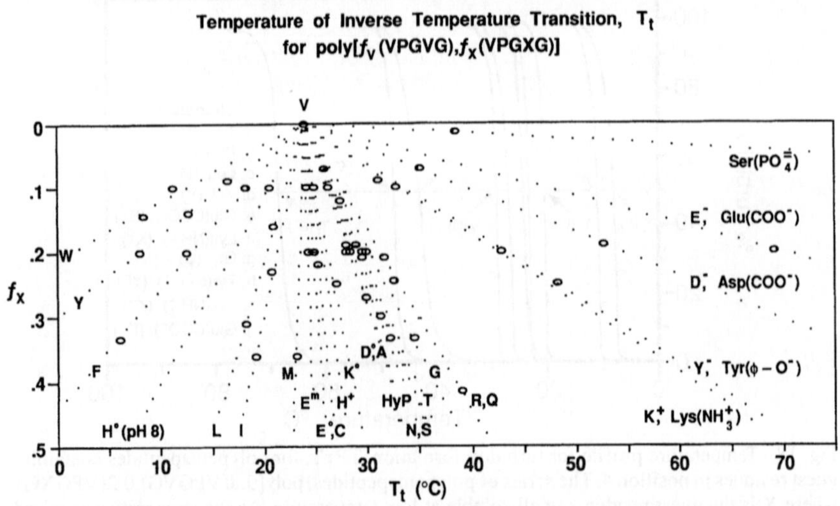

Fig. 20. Dependence on mole fraction of guest residue, f_X, of the temperature (T_t) of the inverse temperature transition. The value of T_t, plotted as a function of f_X, is seen to be very nearly linear such that the values may be extrapolated to a reference value for f_X to give a comparison of hydrophobicities. The reference value is determined at $f_X = 1$ and this value is used in Table 2. (Reproduced with permission from [103].)

Table 2. T_t-based hydrophobicity scale for proteins. T_t = temperature of inverse temperature transition for poly[f_V(VPGVG), f_X(VPGXG)] in 0.15N NaCl

Residue X		T_t, linearly extrapolated to $f_X = 1$	Correlation coefficient
Lys (dihydro NMeN)[a]		−130 °C	1.000
Trp	(W)	−90 °C	0.993
Tyr	(Y)	−55 °C	0.999
Phe	(F)	−30 °C	0.999
His (imidazole)	(H^O)	−10 °C	1.000
Pro	(P)[b]	(−8 °C)	calculated
Leu	(L)	5 °C	0.999
Ile	(I)	10 °C	0.999
Lys (6-OH tetrahydro NMeN)[a]		15 °C	1.000
Met	(M)	20 °C	0.996
Val	(V)	24 °C	reference
Glu($COOCH_3$)	(E^m)	25 °C	1.000
Glu(COOH)	(E^O)	30 °C	1.000
Cys	(C)	30 °C	1.000
His (imidazolium)	(H^+)	30 °C	1.000
Lys(NH_2)	(K^O)	35 °C	0.936
Pro	(P)[c]	40 °C	0.950
Asp(COOH)	(D^O)	45 °C	0.994
Ala	(A)	45 °C	0.997
HyP		50 °C	0.998
Asn	(N)	50 °C	0.997
Ser	(S)	50 °C	0.997
Thr	(T)	50 °C	0.999
Gly	(G)	55 °C	0.999
Arg	(R)	60 °C	1.000
Gln	(Q)	60 °C	0.999
Lys(NH_3^+)	(K^+)	120 °C	0.999
Tyr(ϕ-O^-)	(Y^-)	120 °C	0.996
Lys(NMeN, oxidized)[a]		120 °C	1.000
Asp(COO^-)	(D^-)	170 °C	0.999
Glu(COO^-)	(E^-)	250 °C	1.000
Ser($PO_4^{=}$)		1000 °C	1.000

[a] NMeN is for N-methyl nicotinamide pendant on a lysyl side chain, i.e. N-Methyl nicotinate attached by amide linkage to the ε-NH_2 of Lys. The most hydrophobic reduced state is N-methyl-1,6-dihydronicotinamide (dihydro NMeN), and the second reduced state is N-methyl-6-OH, 1,4,5,6-tetrahydronicotinamide (6-OH tetrahydro NMeN).

[b] The calculated T_t value for Pro comes from poly(VPGVG) when the experimental values of Val and Gly are used. This hydrophobicity value of −8 °C is unique to the β-spiral structure where there is hydrophobic contact between the $Val_i^1 \gamma CH_3$ and the adjacent $Pro_i^2 \delta CH_2$ and the interturn $Pro_{i+3}^2 \beta CH_2$ moieties.

[c] The experimental value determined from poly[f_V(VPGVG), f_P(PPGVG)].

Adapted with permission from references [17].

tions at lower temperatures and the less hydrophobic residues exhibiting transitions at higher temperatures.

Using data such as those of Fig. 18 with additional values of f_X and for each of the amino acid residues, the plots of f_X vs. T_t of Fig. 20 are obtained which provide a set of relative hydrophobicities based on the temperature of the inverse temperature transition as listed in Table 2 [103]. There is an ongoing refinement of the values and their use and there are important issues such as distance dependence between apolar and polar (e.g. charged) moieties to be clarified. Nonetheless, for the low substitution range, which in general is the relevant situation, the dependence of T_t on amino acid composition has shown that it is appropriate to use this scale in the design of polypeptides with desired transition temperatures.

The data of Table 2 were used in the transition temperature plots of Figs. 2, 11 and 12. In Fig. 12A when using a sliding window average for eleven residues, a single hydrophobic sequence is found to be long enough to span a lipid membrane; this is the transmembrane α-helix of bovine

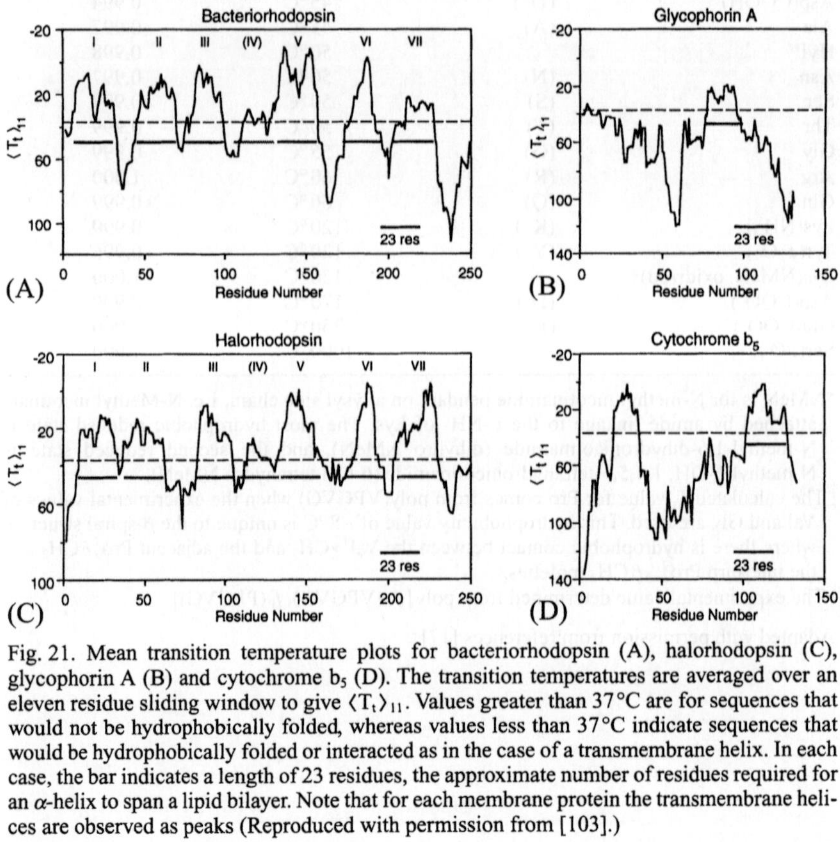

Fig. 21. Mean transition temperature plots for bacteriorhodopsin (A), halorhodopsin (C), glycophorin A (B) and cytochrome b_5 (D). The transition temperatures are averaged over an eleven residue sliding window to give $\langle T_t \rangle_{11}$. Values greater than 37 °C are for sequences that would not be hydrophobically folded, whereas values less than 37 °C indicate sequences that would be hydrophobically folded or interacted as in the case of a transmembrane helix. In each case, the bar indicates a length of 23 residues, the approximate number of residues required for an α-helix to span a lipid bilayer. Note that for each membrane protein the transmembrane helices are observed as peaks (Reproduced with permission from [103].)

Type I macrophage scavenger receptor protein. In Fig. 21 [103], the single transmembrane sequences of glycophorin A and human cytochrome b_5 are apparent as are six of the seven transmembrane helices of bacteriorhodopsin and halorhodopsin. The seventh transmembrane helix (labelled IV) is not as apparent because it contains 3 Ala and 5 Gly residues, but this sequence would also be capable of forming an α-helix that would span a lipid layer. While there is much refinement yet to be done in this and other applications of the hydrophobicity scale based on the temperature, T_t, of an inverse temperature transition, such as deducing when a Glu, Asp or Lys side-chain would be charged or uncharged, it should be apparent that the scale provides an index of hydrophobicity. Significantly, it is the first hydrophobicity scale based directly on the hydrophobic folding process.

4.2. Dependence of T_t on degree of ionization of functional side chains

As is apparent in Table 2, the value of T_t for the Glu, Asp, Lys and Tyr residues changes dramatically with a change in the state of ionization of the side chain. This is most pronounced for the Glu residue where on extrapolation to $f_E = 1$ the change in value of T_t is over 200 °C. Figure 22 A is a graph of the pH dependence of the temperature profiles for aggregation of poly[0.8(VPGVG), 0.2(VPGEG)] where the change in T_t for $f_E = 0.2$ is 45 °C as is also shown in Fig. 18 (i.e. 24 °C for COOH and 69 °C for COO^- [105]). Using the data of Fig. 22 A, a plot of T_t versus pH is seen in Fig. 22 B and in Fig. 22 C is a plot of T_t versus α where α is the degree of ionization obtained from an acid-base titration curve. It is remarkable that in water less than two carboxylate moieties in 100 residues can raise the transition temperature so much as to not be measurable either by means of temperature profiles of aggregation or by differential scanning calorimetry. This effect is greatly damped in the presence of 0.15 N NaCl such that for $f_E = 0.2$ the change in T_t is 20 to 25 °C for two carboxylates per 100 residues.

4.3. Dependence of T_t on phosphorylation/dephosphorylation of the Ser side chain

The reversible process of phosphorylation/dephosphorylation of a Ser or Thr residue is not readily achieved chemically, as dephosphorylation results in a residue with an α-β saturation as in the formation of a dehydroalanine on dephosphorylation of serine-phosphate. Fortunately, the enzymes (kinases) which phosphorylate Ser, Thr or Tyr residues require only a short sequence of residues to specify an active site whereas phosphatases are less specific. For the test of the effect of phosphorylation/dephosphorylation the kinase site of lysozyme, RGYSLG, was chosen and polymerized with IPGVG to give poly[30(IPGVG), (RGYSLG)]. In the appropriate medium

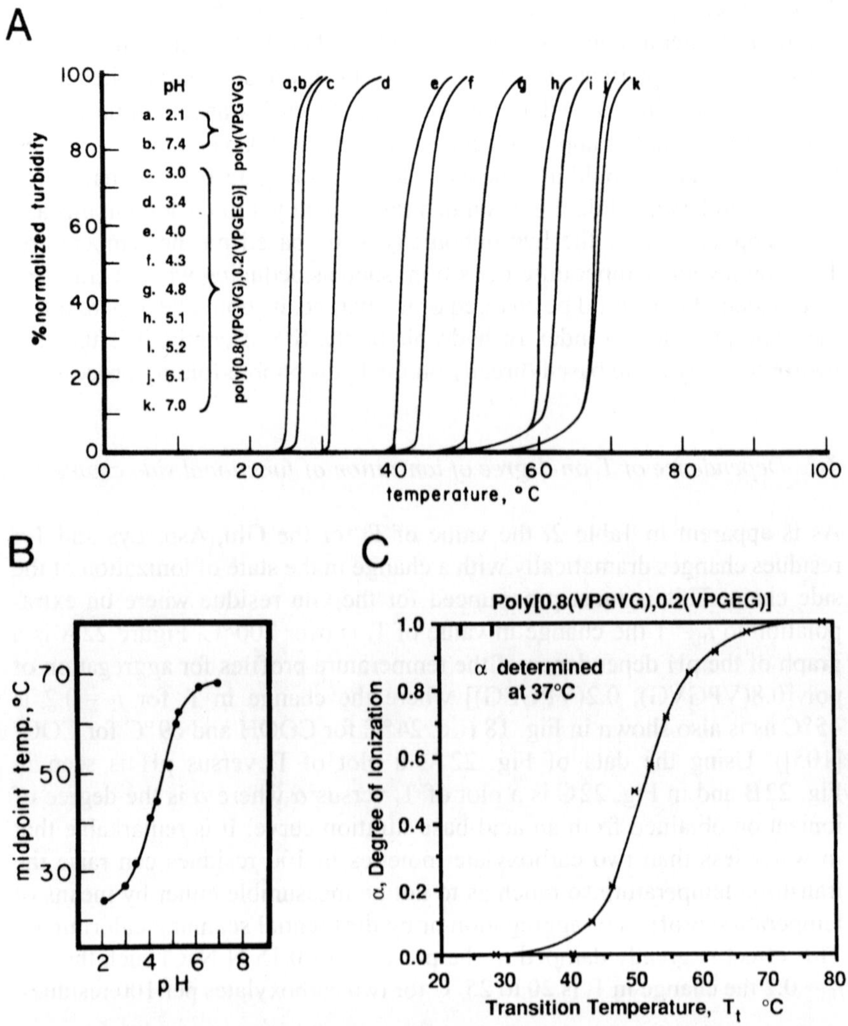

Fig. 22. The pH dependence of the transition temperature, T_t, for polypentapeptide containing a glutamic acid substitution. (A) The pH dependence of the temperature of the inverse temperature transition, T_t, for poly[0.8(VPGVG),0.2(VPGEG)] where there are 4 Glu residues per 100 residues. For the Glu-containing polymer there is a dramatic increase in the temperature of transition on increasing the pH (curves c through k) whereas the transition temperature of poly (VPGVB) over the pH range of 2.1 to 7.4 is independent of pH. (B) Plot of T_t versus pH using midpoint temperatures from data of Fig. 22A gives a pKa value of 4.4 for poly[0.8(VPGVG),0.2(VPGEG)]. (A and B adapted with permission from [105].) (C) Plot of T_t versus α, the degree of ionization, obtained using a 37 °C acid-based titration curve for poly[0.8(VPGVG),0.2(VPGEG)]. These data were obtained in 0.15 N NaCl.

this polymer exhibited a value of T_t of approximately 18 °C. On using porcine heart 3′-5′-cyclic AMP-dependent protein kinase to phosphorylate to a level of about 50%, the value of T_t was raised to 32 °C [106]. The increase in value of T_t due to phosphorylation was reversed on dephosphorylation using calf intestine alkaline phosphatase [106]. When calculated on a per pentamer basis and extrapolating to $f_{sp} = 1$, i.e. to an equivalent Ser-phosphate in each pentamer, the value of T_t for Ser-phosphate would be 1050 °C, rounded to 1000 °C in Table 2. This represents the largest shift in T_t due to a chemical perturbation that has yet been observed, being some fourfold greater than the effect of Glu(COOH → COO^-).

Bayer and Krebs [107] stated in the preface to their book *The Enzymes* that "the modulation of enzymic or other protein-dependent processes by protein phosphorylation or dephosphorylation has emerged as the most widespread and important control achieved by covalent modification." It is not unreasonable in view of the findings reviewed here to anticipate that this is substantially due to the effect of phosphorylation on hydrophobic folding and assembly, perhaps even more so than the effect of phosphorylation on charge-charge interactions.

4.4. Dependence of T_t on solutes

Addition of salts as well as organic solutes and solvents to the aqueous solutions shift the value of T_t. In particular, the lyotropic (Hofmeister) series [108, 109] is apparent in the systematic progression in the effect of sodium salts on the value of T_t. The relative effects of several sodium salts are shown in Fig. 23A for poly(VPGVG). As with degree of ionization or with phosphorylation, the more intense the ionic nature, the more effective the species is in shifting the temperature of the transition. The direction of the shifts, however, is reversed. When a serine of the polypentapeptide becomes phosphorylated or a Glu carboxyl becomes a carboxylate, that is, in a polymer-based change, the temperature, T_t, of the inverse temperature transition for folding and assembly increases. When similar species are added to the solution as solutes, e. g. the phosphate and carbonate anions, the relative magnitudes of the shifts are similar, but the value T_t decreases for the solvent-based effect instead of increasing as occurs for the polymer-based effect. On a per mole pentamer basis the magnitude of the shift is also much greater for the polymer-based effect. The solvent-based effect is presumably due more to solute effects on the activity of water when there are no charges on the polypentapeptide than to interactions of the ion directly with the polypeptide.

The effect of organic solutes and solvents on T_t is again different depending on the solute. For example, addition of 15% or more dimethyl sulfoxide or ethylene glycol by volume lowers T_t [110] whereas addition of urea raises T_t [111]. Interestingly, increasing amounts of ethylene glycol and or urea

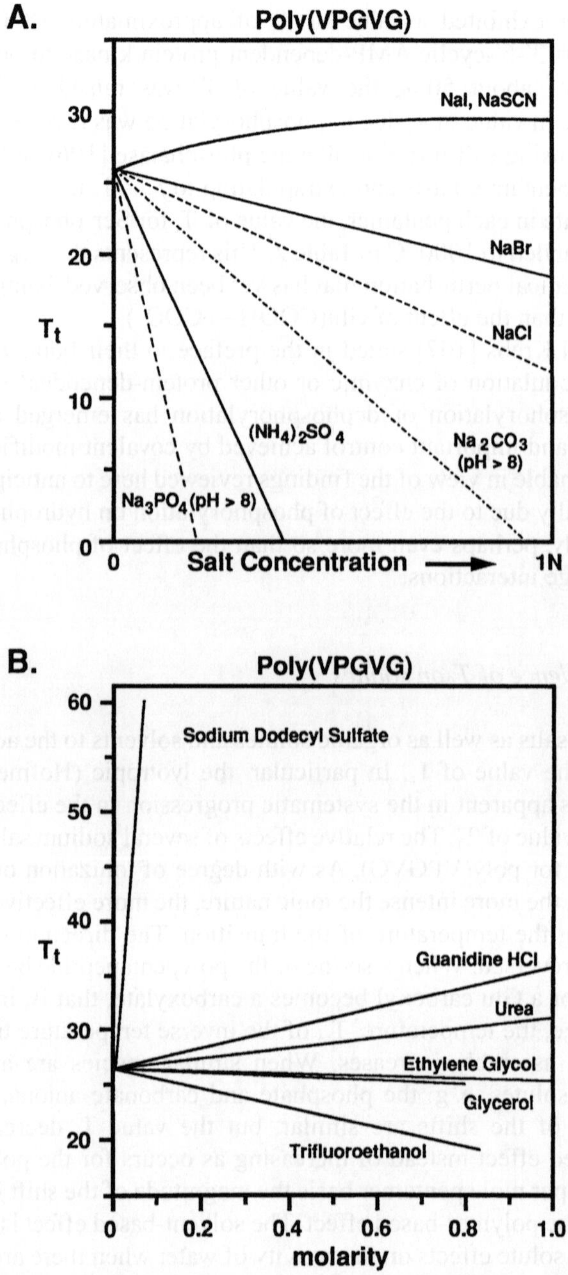

Fig. 23. (A) Effects of sodium salts on the value of the transition temperature, T_t, of poly(VPGVG). The greater the ionic nature of the salt the greater the effect of increasing concentrations on decreasing the value of T_t, that is, increasing concentrations of more ionic species (Na_2CO_3, Na_3PO_4) significantly decrease T_t. (B) Effects of organic solutes and solvents on the value of T_t for poly(VPGVG). Note the extraordinary effect of sodium dodecyl sulfate on raising the value of T_t, that is, on effecting hydrophobic unfolding and disassembly. (Reproduced with permission from [124].).

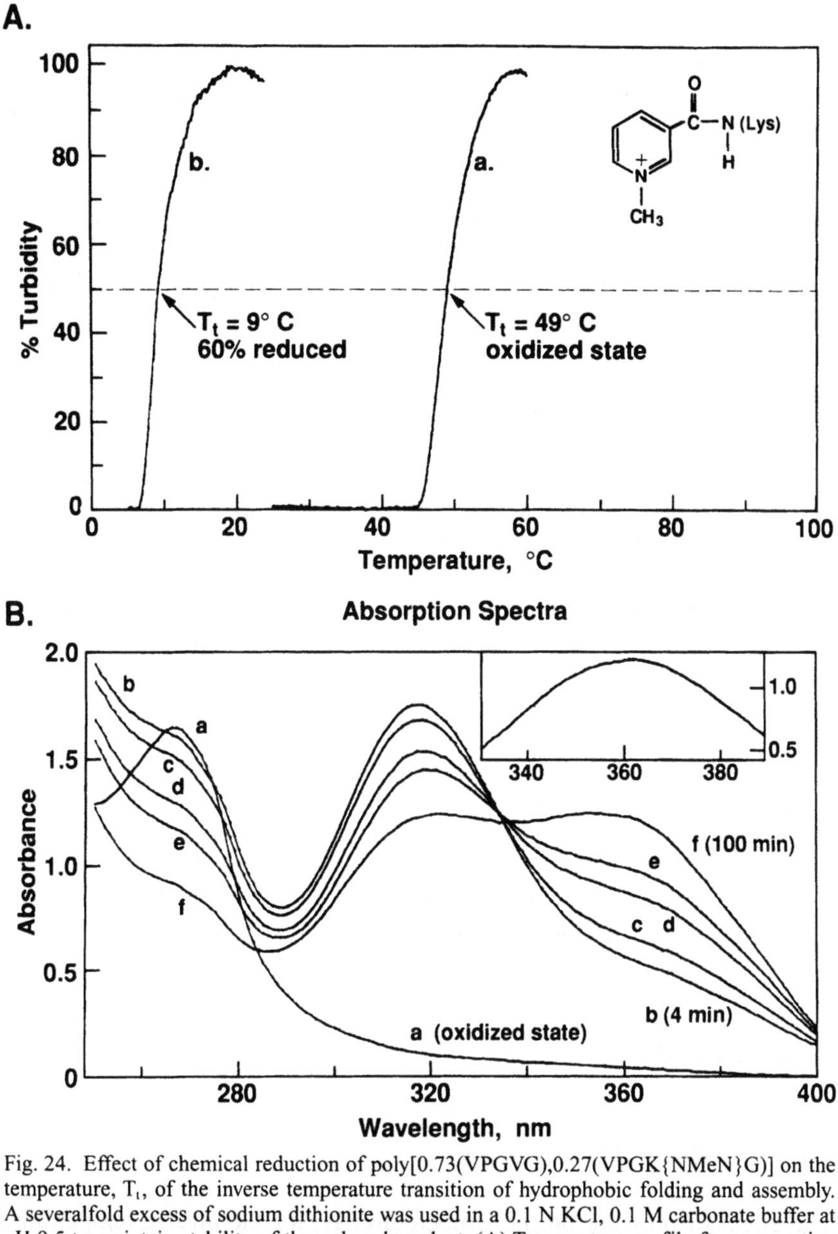

Fig. 24. Effect of chemical reduction of poly[0.73(VPGVG),0.27(VPGK{NMeN}G)] on the temperature, T_t, of the inverse temperature transition of hydrophobic folding and assembly. A severalfold excess of sodium dithionite was used in a 0.1 N KCl, 0.1 M carbonate buffer at pH 9.5 to maintain stability of the reduced product. (A) Temperature profile for aggregation showing T_t to change from 49 °C for the oxidized state to 9 °C for partial reduction. (B) Absorption data used to verify the reduced species as the N-methyl-1,6 dihydronicotinamide and to calculate the percent of reduction using an $\varepsilon^{360} \cong 7060$ and the resolved 100 min curve shown as the inset. (Reproduced with permission from [112].)

decrease the magnitude of the endothermic heat of the transitions [110, 111]. Replacement of H_2O by D_2O lowers the value of T_t a few degrees and increases the heat of the transition [111]. As seen in Fig. 23B, trifluoroethanol (known to be a helix inducer) lowers T_t, favouring hydrophobic folding; guanidinium chloride is more effective in raising T_t than urea, that is, it is more effective in achieving hydrophobic unfolding and disassemble; and the organic solute with the most dramatic effect on raising T_t, that is, on causing hydrophobic unfolding and disassembly, is sodium dodecyl sulfate. It would appear that there can be no hydrophobic folding even at relatively low molar concentrations of sodium dodecylsulfate.

4.5. Dependence of T_t on the redox state of a prosthetic group

When N-methyl nicotinamide, NMeN, is attached to the Lys(K) side chain as in poly[0.73(GVGVP), 0.27(GK{NMeN}GVP)] and the 0.1 N KCl, 0.1 M potassium carbonate buffer system at pH 9.5 is used to stabilize the reaction products, the value of T_t is 49°C for the oxidized state (see Fig. 24). When the nicotinamide is chemically reduced and estimated to be 60% reduced by absorption data, the value of T_t is lowered by 40°C to 9°C [112]. On a molar basis, changing from one state to another of the N-methyl nicotinamide redox couple appears to be as effective in changing the value of T_t as changing between protonated states of the $COOH/COO^-$ chemical couple of the Glu residue. Electrochemical reduction has also been achieved by means of a three-electrode potentiostat, and again reduction lowers the value of T_t [112]. This brings inverse temperature transitions of hydrophobic folding and assembly directly into the realm of bioelectrochemistry. The reduction product characterized in Fig. 24 is the N-methyl-1,-6-dihydronicotinamide which is relatively unstable and converts after several hours to N-methyl-6-hydroxy-1, 4, 5, 6 tetrahydronicotinamide, which is less hydrophobic with a reference T_t value of 15°C (see Table 2) as opposed to 120°C for the oxidized state and –130°C for the dihydro reduced state.

4.6. Dependence of T_t on pressure

The effect of pressure on the temperature, T_t, for the inverse temperature transition for folding and assembly depends on the amino acid composition [113]. As seen in Fig. 25, increasing the pressure has only a small effect on T_t for poly(VPGVG). When aromatic residues are included, however, as in poly[f_V(VPGVG), f_X(VPGXG)] where X = Trp, Phe or Tyr, the effect of pressure is quite significant in changing the value of T_t. The volume changes (ΔV) can be obtained by combining differential scanning calorimetry (DSC) data with the pressure data. From the DSC data the heat of the transitions,

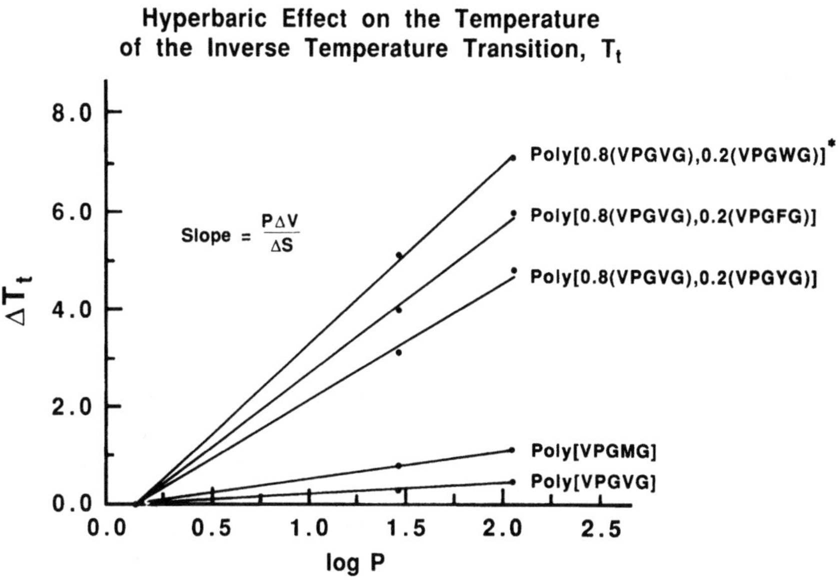

* Corrected for linear adjustment of f_w from 0.143 to 0.2 for proper comparison to the other aromatic residue containing polymers. The low pressure dependence of ΔT_t for the M and V containing polypentapeptides should be reduced by one-fifth for comparison in this figure.

Fig. 25. The effects of pressure on the temperature, T_t, of the inverse temperature transition. Increasing pressure increases the value of T_t with the most significant changes occurring with those polypentapeptides containing aromatic residues. (Reproduced with permission from [113].)

ΔH_t, and the mean temperature of the transition (T_m) can be obtained. The ratio, i.e. $\Delta H_t/T_m$, provides an experimental estimate of the entropy change for the transition, ΔS_t. As the slope of the curves in Fig. 25 is $P\Delta V/\Delta S$, the combined DSC and pressure data provide an estimate of ΔV. The volume change in cm^3 per mole of pentamer, VPGXG, is calculated to be -430 cm^3/mole for VPGWG, -420 cm^3/mole for VPGFG and -260 cm^3/mole for VPGYG, but only -3 cm^3/mole for VPGVG. One interpretation is that pressure leads to an unfolding when in the proper temperature range because the volume occupied by a water molecule hydrating the hydrophobic (primarily aromatic) moiety is less than the volume occupied by a water molecule of bulk water. Clearly the unfolded state occupies less volume than the folded state.

The above provides a molecular perspective of what is happening. Of particular interest here is the magnitude and direction of the change in T_t for a given change in pressure. In particular, a change in pressure from 1.3 to 115 atm extrapolated to $f_X = 1$ results in values of ΔT_t of 36 °C for VPGWG, 29 °C for VPGFG, 23 °C for VPGYG, about 1 °C for VPGMG and about 0.5 °C for VPGVG. As will be discussed below, the process of changing T_t, i.e. achieving ΔT_t, can, when at the correct temperature, lead to a change in folding or a change in assembly of the elastic protein-based polymer and this can be used in the cross-liked matrix to perform mechanical work. First, however, two additional means of varying T_t will be discussed.

4.7. Dependence of T_t on light with chromophore attached to a side chain

When chromophores such as azobenzene ({AB}, $\phi - N = N - \phi - NH_2$) or cinnamic acid ({CA}, $\phi - CH = CH - COOH$), where ϕ stands for the Phenyl moiety, are attached to a functional side chain by an amide linkage, it has been found that the absorption of light resulting in a structural transition, for example from *trans* to *cis*, in the chromophore causes an increase in the value of T_t. In particular, when azobenzene is attached by an amide linkage to a Glu side chain, as in poly[0.68(VPGVG),0.32(VPGE{AB}G)], irradiation by 350 nm light causes a marked decrease in absorbance as the chromophore converts from *trans* to *cis*. For a 50% substitution and at a buffered pH of 4.1, on illumination the absorption is seen to decrease as shown in Fig. 26A and the value of T_t shifts from 32° to 42 °C as seen in Fig. 26B [114]. The *trans* isomer is more hydrophobic than the *cis* isomer, such that the value of T_t increases when exposed to 350 nm light. The effect with azobenzene is reversible for the polymer described above.

A similar light-effected increase in T_t has been found for cinnamic acid attached to a lysine side chain, as in poly[f_V(VPGVG),f_K(VPGK{CA}G)].

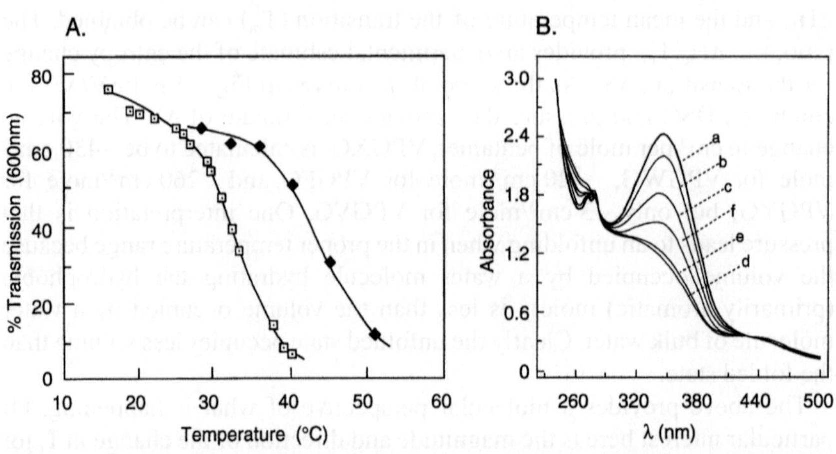

Fig. 26. The effects of light on the temperature, T_t, of the inverse temperature transition for azobenzene {AB} attached by amide linkage to a Glu side chain in poly[0.68(VPGVG), 0.32(VPGE{AB}G)]. (A) Temperature profile for percentage transmission (rather than turbidity), showing the T_t near 32 °C before intermittent irradiation with 350 nm light and shifting to near 42 °C after irradiation. (B) Absorption spectra, curve a, before flashes of 350 nm light progressing to curve d after 45 s of 350 nm light.

The photochemical reaction again causes a decrease in hydrophobicity. This light reaction, however, is irreversible in the systems studied so far. Studies on the effect of light on model proteins containing both Asp and substituted Lys residues, as well as some replacement of Val by Phe, now become possible (Heimbach, Peng, Gowda and Urry, in preparation). In this more complex model protein, the cinnamic acid is attached to the Lys side chain and light again increases the value of T_t; but of even greater significance, light causes a measurable change in the pKa of the Asp residue coupled by hydrophobic folding, as will be discussed below when treating photochemical transduction in 6.7.

4.8. Relaxation of ΔT_t due to ionization by salts

Because of its potential significance to protein mechanisms, the effect of adding salts on the large increase in T_t due to ionization is worth attention. In particular, ionization of four carboxyls to carboxylates per 100 residues, as in poly[0.8(VPGVG),0.2(VPGEG)], increases the value of T_t from 25 °C to an extrapolated 200 °C in distilled water. The addition of salts, e.g. 0.15 N NaCl and 0.01 M phosphate as in Fig. 18, sufficiently lowers the increase in T_t such that T_t can be demonstrated near 70 °C for the ionized state, as shown in Figs. 18 and 22 and in Table 2.

Addition of $CaCl_2$ instead of NaCl has an even greater effect. Accordingly, the interaction of Ca^{+2} at a site containing one or more carboxylates could be

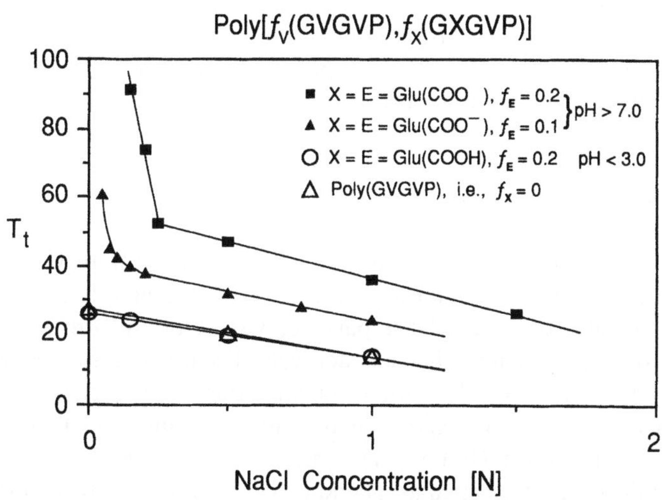

Fig. 27. Effect of NaCl on the T_t of Glu-containing polypentapeptides. Note that when the side chain is ionized small amounts of NaCl cause dramatic decreases in T_t. This is an important charge neutralization effect on lowering T_t that is expected to be relevant whenever cation interaction occurs with carboxylate anions and that will drive hydrophobic folding (unpublished data).

expected to lower T_t dramatically. If the calcium ion interaction should lower T_t from above 37°C to substantially below 37°C, the binding would drive hydrophobic folding. Some experimental results are shown in Fig. 27.

5. Development of a macroscopic assay system for observing the folding and assembly transition and its equivalent of energy conversion

Above, a number of ways of varying T_t were described. Each of these ways was expected to change the temperature at which the inverse temperature transition of polypeptide hydrophobic folding and assembly would occur. While many sophisticated physical methods have been employed to characterize and demonstrate the increase in order of folding and assembly on raising the temperature [115, 116], none is more convincing than the formation of crystals on raising the temperature of aqueous solutions of cyclo$(VAPGVG)_2$ [77] or the self-assembly into filaments and fibres of Glu and Lys containing poly(VPGVG) [85]. This is because the assembly into crystals and fibres could be seen with the unaided eye, then in better detail with the light microscope and optical diffraction of the electron micrographs, and finally at finer resolution with the electron microscope. Similarly, if one could convert the molecular process of folding and assembly into a macroscopic process easily observed by the unaided eye then it becomes possible, just as convincingly, to see the conversion, for example, of thermal or chemical energy into the macroscopic process of lifting a weight, i.e. the conversion of thermal, chemical or other energies into mechanical work. This will be demonstrated in the next section, but first the development of the elastic macroscopic system will be described and the nature of the elasticity will be briefly considered.

5.1. Formation of elastomeric matrices

In Figs. 17, 18 and 22, a clear solution at low temperature was found to turn cloudy due to heat-induced aggregation. If the cloudy suspension is allowed to stand, the more dense particles will settle to form a new, more dense, viscoelastic phase called a coacervate. The process is entirely reversible for poly(VPGVG) and position four analogues. The viscoelastic coacervate can be cast in any desired shape and cross-linked in that shape by gamma (γ-) or electron (E-) beam radiation. In particular, the coacervate of poly(VPGVG) can be shaped as a band and cross-linked to form a synthetic elastomeric strip [116]. Both carbon-13 and nitrogen-15 nuclear magnetic resonance spectra of samples enriched by more than 90% [117, 118] and amino acid analyses of poly(VPGVG) before and after cross-linking [118] demonstrate the breakdown of the amino acids due to 20 Mrad γ-irradiation

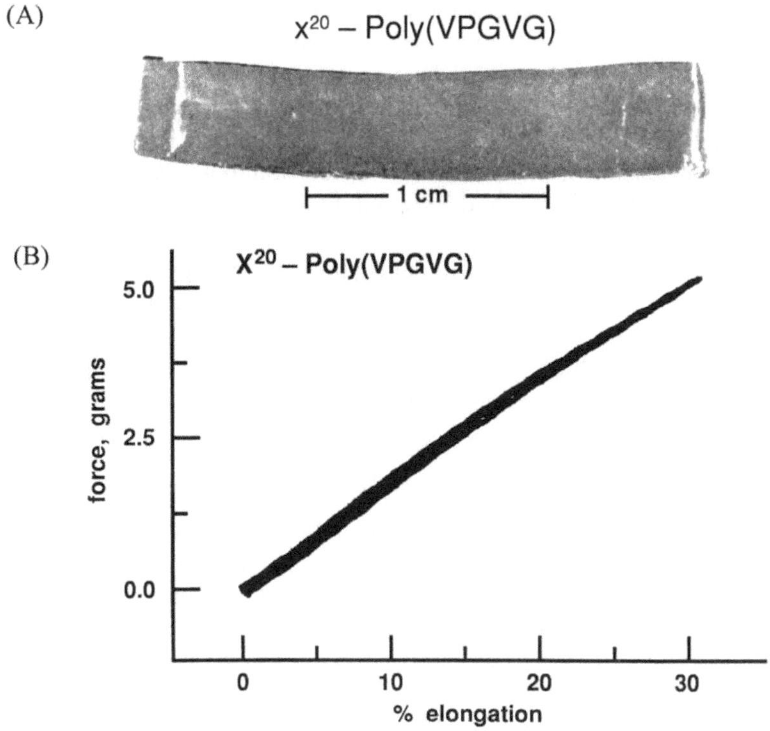

Fig. 28. (A) An elastomeric strip formed by 20 Mrad γ-irradiation of the aggregated or coacervated phase of poly(VPGVG). This material is designated X^{20}-poly(VPGVG). (B) Stress-strain curve for the above material in water at 37 °C. The experiment is carried out by clamping a strip of material in a pair of groups one of which is fixed and the other of which is connected to a force transducer. The force generated versus percent extension is determined at a fixed temperature. The initial elastic modulus is near 1×10^6 dynes/cm^2 (unpublished data).

to be below detectable limits. All of the naturally occurring amino acids thus far tested tolerate the radiation as well as position four analogues of poly(VPGVG), but there have been other sequences observed which are less tolerant of γ-irradiation. A strip of 20 Mrad γ-irradiation cross-linked poly(VPGVG), designated X^{20}-poly(VPGVG), is shown in Fig. 28A. This strip is elastomeric and exhibits stress/strain curves with little or no hysteresis as shown by numerous repetitions in Fig. 28B.

5.2. Nature of elasticity of the elastomeric matrices

Having developed the molecular system (which is capable of exhibiting perturbable inverse temperature transitions of hydrophobic folding and assembly on raising the temperature) into a state of elastomeric bands, it is useful to characterize briefly the nature of the elasticity before demonstrating the capacity of the elastomeric band to function in energy conversion by changing the temperature of the inverse temperature transition.

A thermoelasticity study of the development of elastomeric force as a function of temperature provides the traditional basis with which to deduce the nature of elasticity [120, 121]. The total elastomeric force, f, which develops on extension is made up of two contributions, one due to the increase in internal energy, f_e, and a second due to the entropic component of the force, f_s. The former arises due to the straining of bonds on extension and the latter is due to a decrease in configurational entropy on extension, that is, due to the decrease in the number of states of the same low energy that differ by existing with a different distribution of backbone rotational (torsion) angles. A useful relationship in discerning the relative magnitude of the two components of elastomeric force is

$$\frac{f_e}{f} = -T\left(\frac{\partial \ln f/T}{\partial T}\right)_{V, L, n} \tag{3}$$

where the partial is taken for conditions of constant volume, V, length, L, and composition, n, and T is in °K. That the conditions of constant, V, L and n are satisfied by poly(VPGVG) in water in the 40° to 60 °C temperature range has been demonstrated elsewhere [122].

The thermoelasticity data for X^{20}-poly(VPGVG) in water (curve B) and in ethylene glycol/water (30:70 by volume) (curve A) are plotted as $\ln(f/T)$

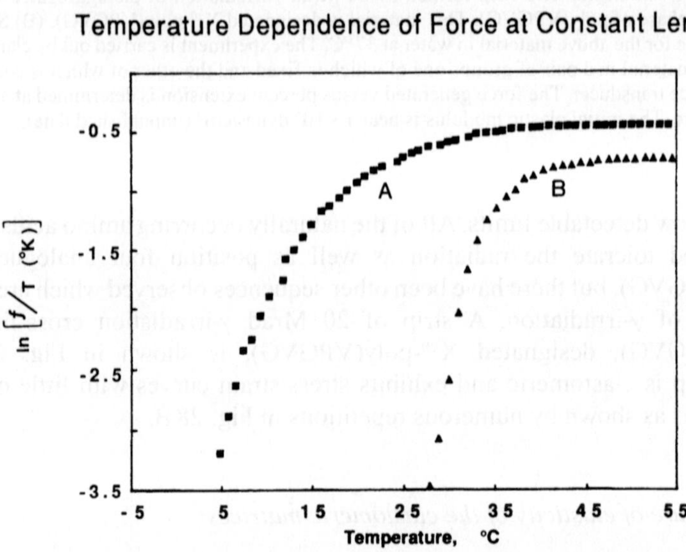

Fig. 29. Thermoelasticity studies of 20 Mrad cross-linked poly(VPGVG) A in ethylene glycol: water (3:7 by volume) and B in water. The experiments were carried out on a sample that had been extended by 50% at 37 °C. The force generated with increasing temperature is measured and the data plotted as ln(f/T) vs. temperature. (Reproduced with permission from [110].)

vs. T in Fig. 29 for the same band. In the higher temperature range, that is, once the inverse temperature transition is complete, the slope is such that the f_e/f ratio is 0.1 or less [110]. Thus, based on this classical analysis in both solvents, the poly(VPGVG)-based elastomeric band functions as a dominantly entropic elastomer.

The data in ethylene glycol/water are informative as, on adding 30% ethylene glycol, the temperature of the transition decreases and the heat ΔH_t also decreases such that the entropy of the transition ($\Delta H_t/T_m$) decreases to a small number. The mean solvent entropy change, therefore, must be very small, yet the elastomeric force increases. This finding argues that mean solvent entropy changes do not contribute a large component to the entropic elastomeric force [110]. Commonly, a dominantly entropic elastomer is considered to arise from a random chain network in which the chains tend to align on extension and thereby become more ordered. This perspective runs contrary to all that we have seen above where there is an increase in order on increasing the temperature through the range of the inverse temperature transition. Accordingly, it is difficult to accept the perspective that the result of the inverse temperature transition is a random chain network. Instead, a mechanism of damping of internal chain dynamics on extension called the librational entropy mechanism has been developed using the poly(VPGVG) system [115, 116]; it is based on many physical characterizations and molecular mechanics and dynamics calculations using relaxed and stretched states of the β-spiral of poly(VPGVG) [115, 116, 123].

An important advantage in using this elastomeric system to carry out energy conversion is that the energy barriers between different configurational states are small [117, 118]. Accordingly, throughout the application of a perturbation, the system remains in equilibrium and does not become trapped in metastable states for which the free energy changes could not be determined.

6. The ΔT_t-mechanism for energy conversion in polypeptides and proteins

To transduce is to convert energy (or power, energy per unit time) from one form into another. Free energy transduction is the process of converting one form of free energy into another. Free energy is given by the product of two variables, i.e. the product of an intensive variable times an extensive variable. The extensive variable is a capacity factor whereas the intensive variables relevant to protein function include mechanical force, temperature, pressure, chemical potential, electrochemical potential, and the frequency of electromagnetic radiation. Proteins are transducers capable of performing the pairwise free energy transductions of Fig. 30 [124]. The elastic

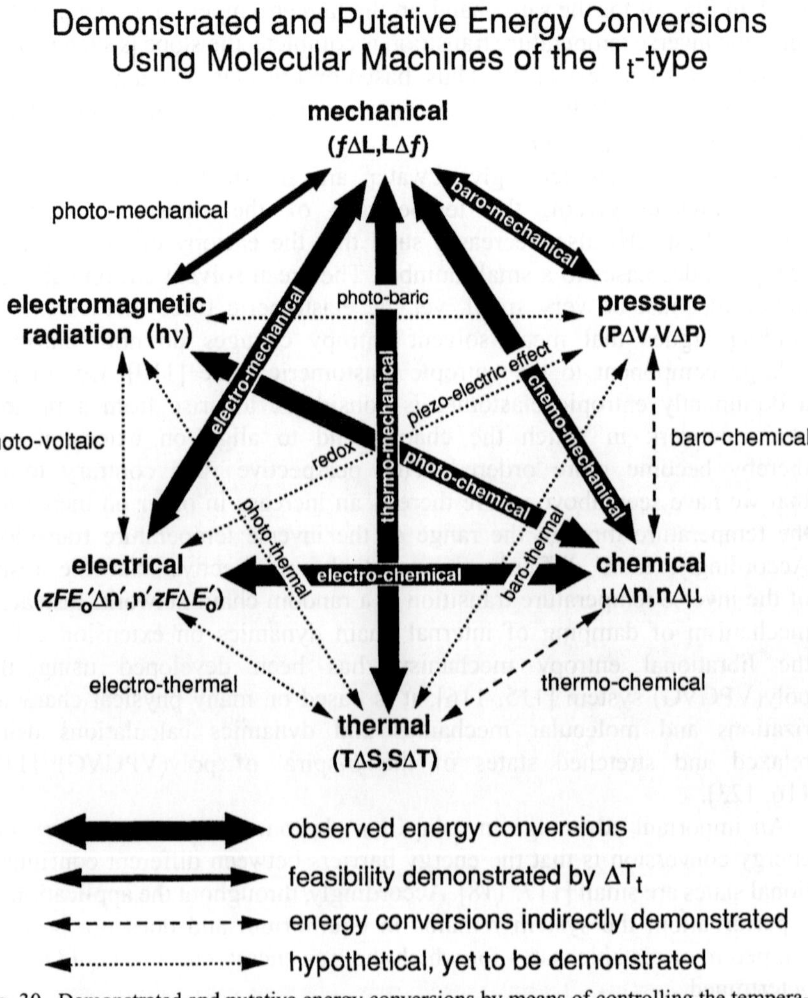

Fig. 30. Demonstrated and putative energy conversions by means of controlling the temperature at which inverse temperature transitions of hydrophobic folding and assembly occur. Adapted with permission from [124]. See text for discussion.

protein-based polymers can be designed to perform a number of the free energy transductions depicted in Fig. 30, and it is proposed that, with proper use of the inverse temperature transition mechanism, all these free energy transductions would become possible. To our knowledge these include all the free energy conversions of which living organisms are capable with the possible exception to sense the earth's magnetic field by migrating animals.

If a pair of fine elastic bands fixed at their distant ends were attached at correctly chosen points on each side a partially folded protein and the protein was then induced to complete its folding, the elastic bands would

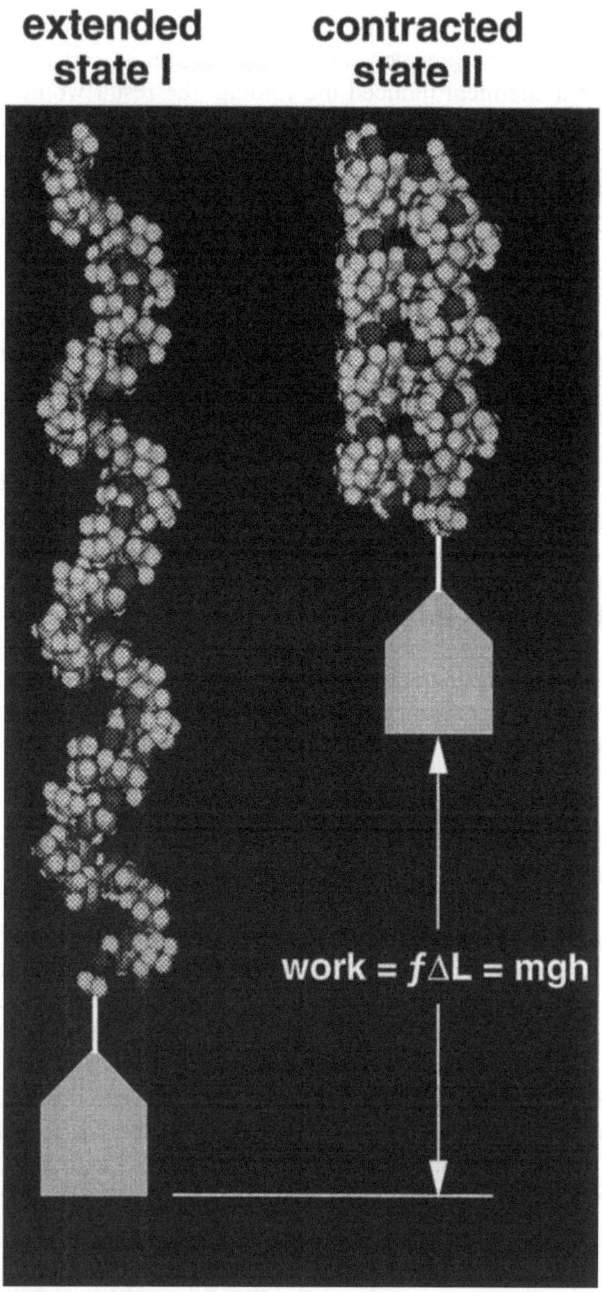

Fig. 31. Schematic representation of a single molecule of poly(VPGVG) extended and attached to a weight at left. The β-spiral structure, hyrophobically folded (contracted) state, at right, shown as having lifted the weight.

become stretched, with the development of elastomeric force which could be used to lift a weight. If an increase in temperature caused the folding, the result would be thermo-mechanical transduction. If a change in the concentration of a chemical induced the folding, the result would be chemo-mechanical transduction. If a largely unfolded poly(VPGVG) molecule, for example, at a temperature below T_t, could be fixed at one end and a weight could be attached to the other end, then raising the temperature through the transition temperature range should induce the folding of the polypeptide chain and the lifting of a weight as depicted in Fig. 31. This would be thermo-mechanical transduction.

Instead of raising the temperature to drive the folding depicted in Fig. 31, the concept of the ΔT_t-mechanism is to shift the temperature, T_t, for the hydrophobic folding and assembly transition form above the operating temperature to below. With T_t above the operating temperature the polypeptide would be unfolded; on lowering T_t to below the fixed operating temperature, the polypeptide should become folded and lift the weight. When the lowering of T_t is carried out by chemical means as described above, then the result would be chemo-mechanical transduction. While it is not feasible to work with a single polypeptide chain as represented in Fig. 31, it is possible to cross-link millions of such chains to produce the elastomeric band described above (see Fig. 28 A) and to see if a properly designed elastomeric band can perform the thermo-mechanical, chemo-mechanical, electro-mechanical, baro-mechanical and photomechanical transductions which have been deduced by studies on folding and assembly.

6.1. Driving hydrophobic folding to perform mechanical work (to produce motion)

6.1.1. Thermo-mechanical transduction (thermally-driven contraction): In a schematic representation, thermo-mechanical transduction is seen in Fig. 32 A as the folding of poly(VPGVG) from an extended state to the hydrophobically folded state on raising the temperature from 10° to 30 °C as in curve a.

In Fig. 32 B is shown a plot of the thermally driven contraction/relaxation of X^{20}-poly(VPGVG) as a function of time; it is achieved by switching the temperature between 20° and 40 °C (approach *i*). In this example, the weight being lifted can be one thousand times the dry weight of the elastomeric band doing the lifting. It is particularly interesting to note that in each subsequent cycle the same extent of contraction is achieved in the same period of time; this thermo-mechanical transduction is clearly reversible. Thus, it has been possible to demonstrate on a visual, macroscopic scale, the thermally-driven hydrophobic folding and assembly deduced at the molecular level using more sophisticated physical characterizations.

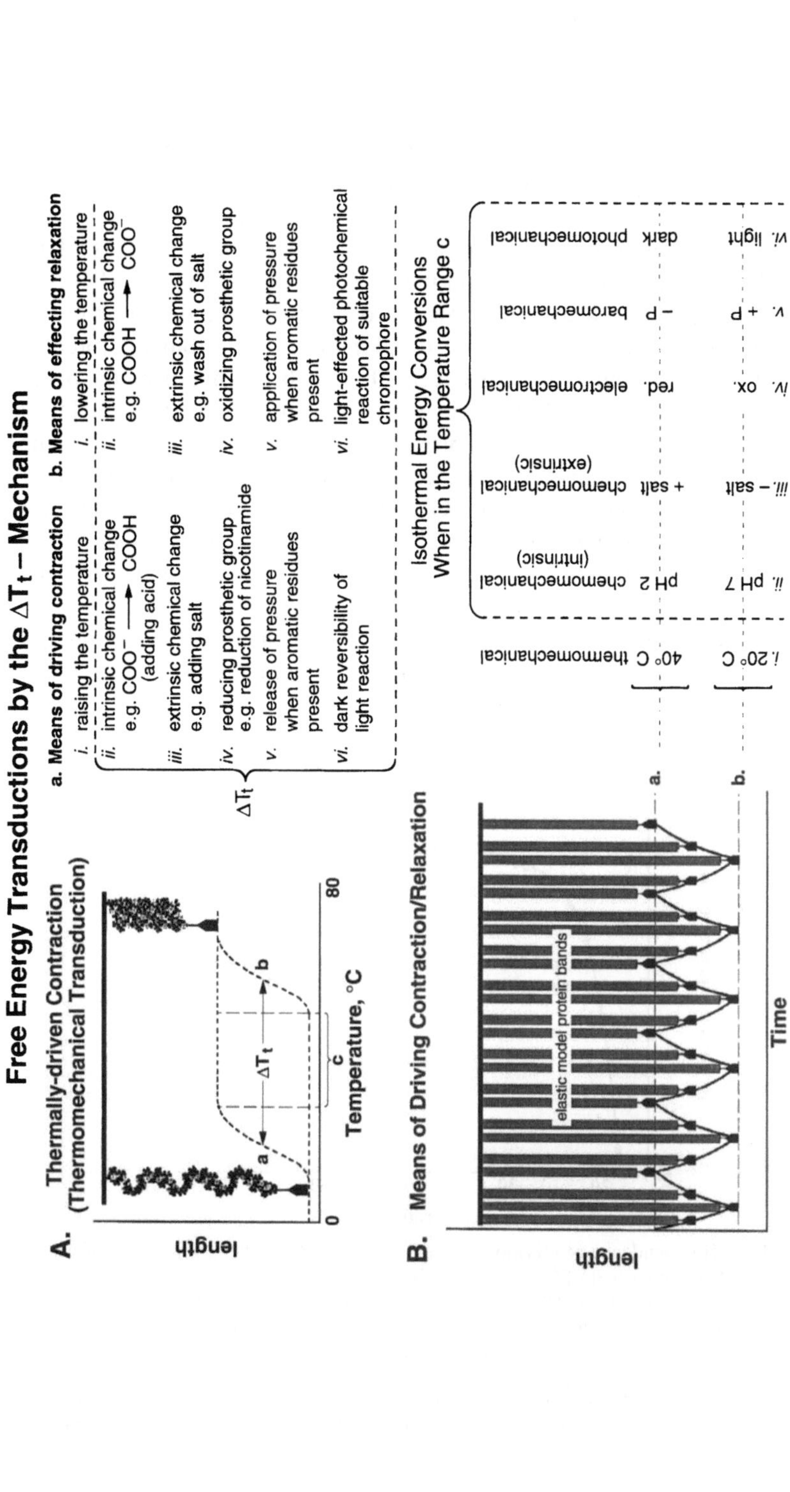

Fig. 32. Summary of energy conversions (free energy transductions) resulting in the performance of mechanical work; these are molecular engines. This is a representation of postulates I and II of Table 3 and of the experimental studies on which they are based. (Adapted with permission from [124].)

6.2.1. Chemo-mechanical transduction (chemically-driven contraction):

6.1.2.1 By changing the degree of ionization (an intrinsic chemical change). It was shown in Figs. 18 and 22 that a change in degree of ionization of poly[0.8(VPGVG),0.2(VPGEG)] markedly shifted the temperature, T_t, of the inverse temperature transition for hydrophobic folding and assembly. In phosphate buffered saline, when the degree of ionization, α, of the Glu residue carboxyl moiety is essentially zero (i.e. pH 2.1), the value of T_t is 24 °C. On 20 Mrad γ-irradiation cross-linking, i.e. for X^{20} -poly[0.8(VPGVG), 0.2(VPGEG)], and with the mechanical process being the development of elastomeric force as in Fig. 32, the development of elastomeric force in Fig. 33 at pH 2.1 is seen to occur between 20° and 30 °C. On raising the pH to 4.5, the development of elastomeric force now occurs near 50 °C [125] which, as shown in Fig. 22B, approximates the temperature, T_t, of the inverse temperature transition for folding and assembly of non-cross-linked poly[0.8(VPGVG),0.2(VPGEG)]. Accordingly, the temperature for the hydrophobic folding and assembly transition for poly[0.8(VPGVG), 0.2(VPGEG)] as a function of degree of ionization corresponds with the temperature range for the development of elastomeric force.

Now it should be possible, based on the ΔT_t-mechanism for free energy (chemo-mechanical) transduction, to hold the temperature at an inter-

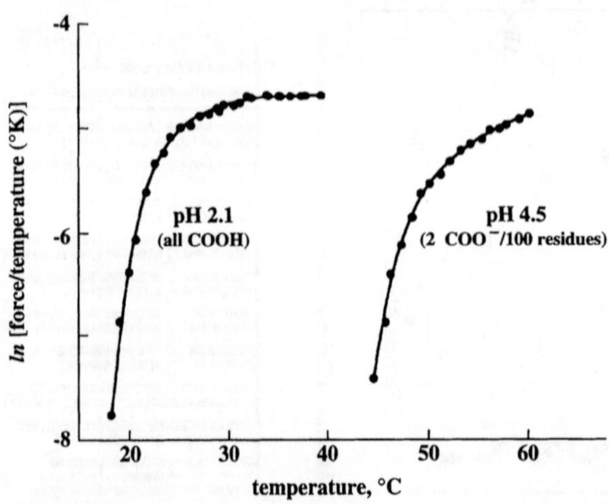

Fig. 33. The pH dependence of thermo-mechanical transduction, the development of elastomeric force of X^{20}-poly[0.8(VPGVG),0.2(VPGEG)] as a result of temperature change. At pH 2.1 where the degree of ionization of the carboxyl moiety of the Glu residue is essentially zero, the development of elastomeric force occurs between 20 °C and 30 °C. Upon raising the pH to 4.5 the development of elastomeric force now occurs between 45 °C and 60 °C. These results demonstrate that the temperature for folding of the Glu-containing polypentapeptide, as a function of the degree of ionization of the Glu carboxyl, corresponds to the temperature range for the development of elastomeric force in the cross-linked matrix (see Fig. 28A). (Reproduced with permission from [125].)

mediate 37 °C and to drive the contraction/relaxation chemically by changing the degree of ionization, i.e. by changing the pH. For a specific experimental case, at pH 2 ($\alpha \cong 0$) with a 3 gm weight, the band of X^{20}-poly[0.8(VPGVG),0.2(VPGEG)] is 4.2 mm long. On raising the pH to 7 ($\alpha \cong 1$), the band extends to 8.2 mm. Returning to pH 2 causes the band to contract and lift the weight. Many cycles can be repeated with retention of the same percent contraction. Thus, it is shown that the ΔT_t mechanism is capable of achieving chemo-mechanical transduction when a change in the degree of ionization of a moiety which is an integral part of the protein-based polymer changes the temperature of the inverse temperature transition [125]. This is referred to as polymer-based chemo-mechanical transduction and arises from an intrinsic chemical change, i.e. a change intrinsic to the polymer (see approach *ii* of Fig. 32).

It should also be noted that a comparably characterized polyelectrolyte-based chemo-mechanical transduction system which uses the carboxyl moiety in a charge-charge repulsion mechanism [126, 127] appears to require more than ten times the amount of chemical energy to carry out a similar amount of mechanical work [96, 124]. Thus, with the greater efficiency of the ΔT_t mechanism, it is anticipated that it will be utilized by living organisms whenever possible. As the source of chemical energy is the intake of food, those organisms which could carry out the required chemo-mechanical transductions with a more efficient mechanism would better survive periods of famine.

6.1.2.2. By changing the salt concentration (an extrinsic chemical change). It was seen in Fig. 23 A that addition of sodium salts changed the value of T_t. In particular the addition of NaCl leads to a decrease of T_t such that at 2 N NaCl the value of T_t would be essentially 0 °C. Even small changes in NaCl concentration can significantly change the value of T_t. Going from pure water to 0.15 N NaCl lowers the value of T_t by several degrees, and this change has been shown to lead to contraction at constant load or to cause an increase in force at fixed length [128]. In Fig. 32 approach *iii,* the lowering of the temperature range over which thermo-mechanical transduction occurs for X^{20}-poly(VPGVG) as the result of adding NaCl is also depicted. As with the thermally-driven contraction, the contraction/relaxation cycle is entirely reversible. This may be called solvent-based chemo-mechanical transduction, that is, it is due to an extrinsic chemical change.

6.1.2.3. Reversible mechanochemical engines. The McGraw-Hill Concise Encyclopedia of Science and Technology [129] defines an engine as "a machine designed for the conversion of energy into useful mechanical motion." The lifting of a weight or the moving of an object can be considered useful mechanical motion. Therefore, described above were molecular mechanochemical engines, that is, molecular machines designed for the conversion of chemical energy into useful mechanical motion. The contraction/relaxation cycles are reversible, thus the term reversible molecular mechanochemical engines [127, 130] is used. While it may seem somewhat strange as a terminology, proteins which are induced to fold or unfold by a change in the concentration of a chemical may be considered to be reversible molecular mechanochemical engines. While chemo-mechanical transduction of muscle contraction [131] is the most commonly recognized free energy transduction carried out in living organisms, other pairwise conversions of Fig. 30 do occur, as do the very fundamental photo-chemical and electro-chemical transductions of photosynthesis and respiration (specifically, oxidative phosphorylation).

6.1.3. Electro-mechanical transduction (reduction-driven contraction): It was shown in Fig. 24 that reduction of a nicotinamide moiety markedly lowers the temperature, T_t, of the inverse temperature transition of hydrophobic folding and assembly. In approach *iv* of Fig. 32, it is noted that reduction lowers the temperature range over which thermally-driven contraction (thermo-mechanical transduction) occurs. At an intermediate temperature reduction is seen to drive contraction with the performance of the work of lifting a weight [132]. Accordingly, electro-mechanical transduction has now been observed, as indicated in Fig. 30.

Interestingly, the three principal chemical aspects of oxidative phosphorylation have been demonstrated using the ΔT_t mechanism: a change in the redox state of a prosthetic group, a change in the concentration of protons and a change in the phosphorylation state. Changing the oxidative state of a prosthetic group of a protein in the inner mitochondrial membrane by the ΔT_t mechanism could alter the hydrophobically folded state of the protein resulting in the vectorial release of a proton, and several protons would have the capacity to drive phosphorylation through the ΔT_t mechanism [112].

6.1.4. Baro-mechanical transduction (pressure-driven relaxation): As discussed above and shown in Fig. 25, the application of pressure raises T_t. From the analysis of the data, it was deduced that increasing pressure favoured the unfolded state because the volume occupied by water molecules which would hydrate hydrophobic residues exposed to water in the unfolded state was less than the volume occupied by water molecules in bulk solution. Accordingly, when in the proper temperature range (see Fig. 32) and with an elastomeric band such as X[20]-poly[0.8(VPGVG), 02(VPGFG)] in a contracted state with a weight lifted, the application of pressure should result in a lowering of the weight. This pressure-driven relaxation has been observed as schematically shown in approach *v* of Fig. 32 [133]. It is not unreasonable, therefore, to propose that the ΔT_t mechanism of free energy transduction could be involved in the sense of touch as well as in the functional properties of protein within articular cartilage.

6.1.5. Photo-mechanical transduction (light-driven relaxation): In each of the preceding cases (thermo-, chemo-, electro- and baro-mechanical transduction) an elastomeric band comprised of model protein has been fashioned, a weight attached and the relevant energy input was seen to drive the reversible lifting and lowering of a weight. The matrices were seen reversibly to pump iron. This has yet to be done for photo-mechanical transduction using these model proteins, but light-driven relaxation (swelling) is under study with an elastic strip of X^{20}-poly[0.9(GVGVP),0.1(GK{CA}GVP)] where {CA} is cinnamic acid attached by an amide linkage to the side chain of a Lys(K) residue [Heimbach, Gowda and Urry, unpublished results]. In studies employing modified poly(N-isopropyl acrylamide) gels, such light-driven swelling has been observed and considered to be an example of photo-mechanical transduction [133]. Indeed, when Szent-Gyorgyi first observed the motion of living organisms outside the living cell by fashioning an actomyosin preparation into a fibre, adding the appropriate chemicals and observing it to contract (to deswell) [135], this was the same kind of energy conversion, in that case chemo-mechanical transduction, as considered here for photo-mechanical transduction.

6.2. Physical basis for ΔT_t mechanism

Before proceeding with more complex energy conversions, it is helpful to gain insight into the physical basis for the ΔT_t mechanisms. The basic experimental results are the finding of a stretch-induced pK shift, the demonstration that increasing hydrophobicity can result in marked pKa shifts, and the finding that the introduction of charges decreases the endothermic heat required to drive thermally hydrophobic folding and assembly.

6.2.1. An apolar-polar repulsive free energy of hydration:

6.2.1.1. Stretch-induced pKa shifts. When an acid-base titration was carried out on an elastomeric strip of X^{20}-poly[0.8(VPGVG),0.2(VPGEG)] in the absence of a load, that is without stretching, the pKa was found to be 3.99 [136]. When the elastomeric band was stretched from its hydrophobically folded and assembled state at 37 °C in a particular experiment with the load of one gram, the pKa increased to 4.84 [137]. Since the force increased, Δf was positive, but since the concentration of protons for the pKa decreased from $10^{-3.99}$ M to $10^{-4.84}$ M the change in chemical potential, $\Delta\mu$ was negative (i.e. $\mu = \text{RT}\, ln\, a$ where a indicates the activity of protons which at these proton concentrations can be written as $\mu = \text{RT}\, ln[\text{H}^+]$, but since pH = −log [H$^+$] then $\mu = -2.3$ RT pH and the change in chemical potential for 50% ionization of the polymer can be written $\Delta\mu = -2.3$ RT ΔpKa $= -2.3$ RT (4.84−3.99) or −1.2 kcal/mole at 37 °C). This means that $(\partial\mu/\partial f)_n = \alpha^{<0}$

where n is at constant composition for the polymer which is at a constant degree of ionization, $\alpha = 0.5$, i.e. at the pKa. With the $(\partial\mu/\partial f)_n$ being negative in this case, it is of exactly the opposite sign as occurs for the charge-charge repulsion mechanism exhibited by polymethacrylic acid, [–CCH_3COOH–CH_2–], as shown by Katchalsky *et al.* [127]. In the latter case, extension is due to charge-charge repulsion between some 50 to 60 carboxylates per 200 backbone atoms, and contraction occurs as the degree of ionization decreases to 0 to 10 carboxylates per 200 backbone atoms when the repulsion causing chain extension is relaxed. A superficially analogous situation occurs for X^{20}-poly[0.8(VPGVG),0.2(VPGEG)] with only four carboxylates per 300 backbone atoms, as this polymer is also extended when ionized and contracted when protonated. While the many fewer carboxylates should signal differences the opposite sign for $(\partial\mu/\partial f)_n$ clearly indicates a different mechanism.

This experiment may also be considered from a different perspective. As the elastomer is stretched from its hydrophobically folded state at 37 °C, hydrophobic side chains become exposed to water and necessarily become hydrated in an exothermic reaction. The experimental result is an increase in pKa even though there is an increase in water in the elastic matrix. Surprisingly, even though there is an increase in water in the matrix, conditions are less favorable for formation of the carboxylate anion. One is brought to the conclusion that the particular water that does form, i.e. water of hydrophobic hydration, is not suitable for hydration of polar species such as carboxylates.

6.2.1.2. Hydrophobic-induced pKa shifts. If it is correct that increased exposure to hydrophobic groups, as occurs on stretching of hydrophobically folded matrices, causes an increase in pKa of carboxyl moieties, then it should be possible to change the pKa by changing the hydrophobicity of the model protein. Specifically, an increase in the number of more hydrophobic residues would be expected to increase the pKa as well. Indeed, the pKa for the Glu residue in poly[0.8(IPGVG),0.2(IPGEG)] is higher than in poly[0.8(VPGVG),0.2(VPGEG)] by more than one pH unit at 37°C [137]. The addition of one CH_2 moiety pentamer for a composition containing four Glu residues per 100 residues increases the pKa.

More recently a dozen polymers of poly[f_V(IPGVG)f_X(IPGXG)] have been synthesizes each for X = Glu and X = Asp with f_X varying from 1 to 0.06 [138, 139]. In both of these cases a hydrophobically dominated domain could readily be delineated in which the largest pKa shifts (increases in pKa) occurred as f_X approached zero, i.e. as the polymer became more hydrophobic and there were fewer carboxyl moieties. The same study has also been carried out for X = Lys in which case the largest pKa shifts (decreases in pKa) occurred again as f_X approached zero [140]. Detailed discussions comparing electrostatic- and hydrophobic-induced pKa shifts are available in the original publications [138–140].

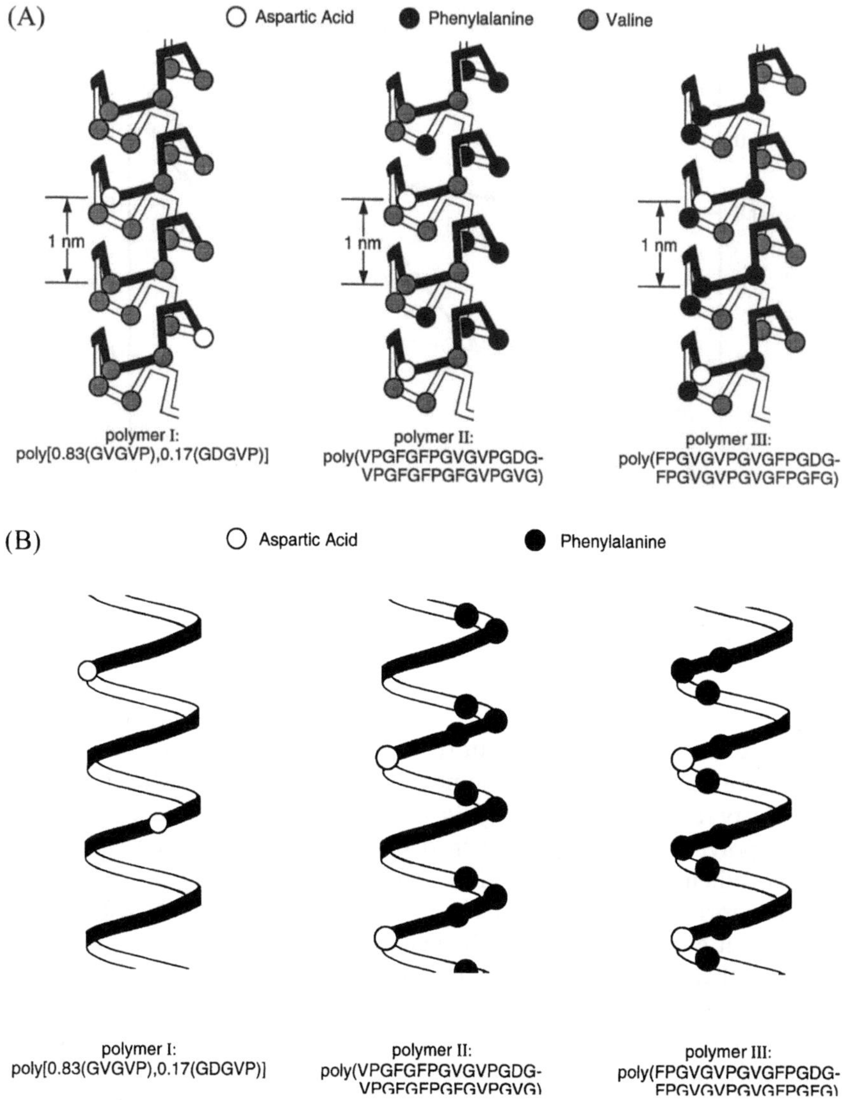

Fig. 34. Representations ((A) including β-turns; (B) as a simplified helix) of polytricosapeptides (polymers II and III) in which there are a single Asp ○ residue and five Phe ● residues per 30mer. In polymer II the Phe residues are as far removed as possible from the Asp residue; in polymer III the Phe residues are placed as proximal to the Asp residue as possible with retention of the β-spiral structure. (C) Acid-base titrations of the Asp residue showing the large pKa shifts on going form no Phe residues in polymer I, to five removed Phe residues in polymer II with a ΔpKa of 2.8, to five proximal Phe residues with a remarkable ΔpKa of 6.2. (Reproduced with permission from [141].)

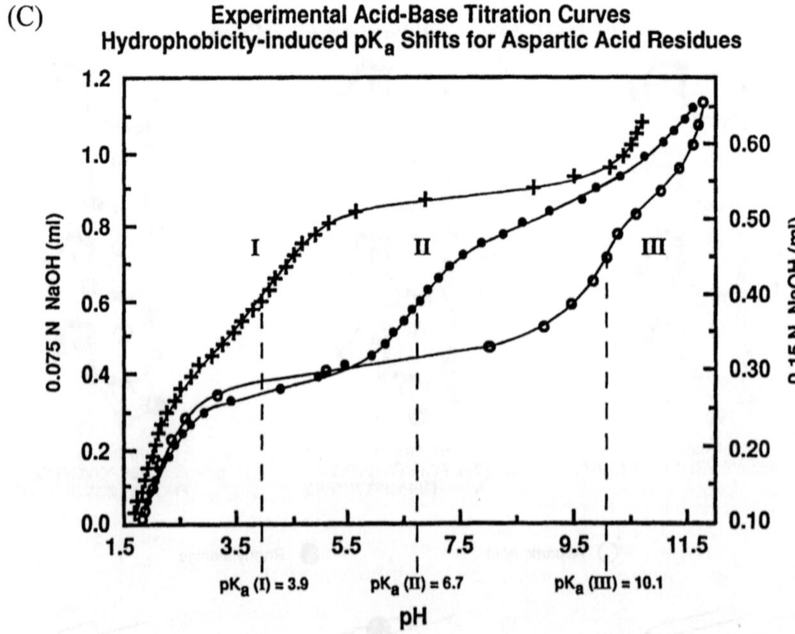

Fig. 34. (C)

By preparing model proteins of fixed sequence, the importance of hydrophobic-induced pKa shifts becomes even more evident. Model proteins with thirty residue repeats in Fig. 34 were prepared and structurally verified. These polytricosapeptides contained one Asp and five Phe residues per 30mer. When the five Phe residues were placed structurally as far away as possible form the Asp residue, as in polymer II, the pKa was shifted form 3.9 to 6.7, but even more remarkably when the five Phe residues were placed structurally as close as possible to the Asp residue consistent with retention of the β-spiral structure, as in polymer III, the pKa was found to be shifted to 10.1 [141]. Thus, it is possible to design model proteins with extraordinarily large hydrophobic-induced pKa shifts. Again, please refer to the original publications of the polytricosapeptide studies for more detailed discussion [141, 142].

6.2.1.3. Charge-induced destruction of waters of hydrophobic hydration. In the immediately preceding studies, the variable was hydrophobicity and the observed response was change in pKa, i. e. a change in the charged state that resulted. The following study is the reverse: the number of charged residues is changed and the effect on the underlying basis of hydrophobicity, that is, on the amount of water of hydrophobic hydration, is observed by differential scanning calorimetry. Recall that the formation of the water of hydrophobic hydration is exothermic. With the soluble model proteins in water below T_t, the temperature is raised above T_t drive hydrophobic folding and assembly by an endothermic reaction which destructures the water of hydrophobic hydration. Thus, when the temperature is scanned from below T_t to above T_t for poly[0.8(VPGVG),0.2(VPGEG)] in solution in a differential scanning calorimeter, the endothermic heat of the inverse temperature transition is a measure of the amount of water of hydrophobic hydration that is destructured. At low pH (2.5) where all of the Glu residues are COOH, the ΔH is 0.97 kcal/mole of pentamers, but at pH 4.1 where there are less than 2 COO^- moieties per 100 residues, the heat of the endothermic transition has been reduced to 0.27 kcal/mole pentamers [143]. We interpret this to mean that the formation of two carboxylates per 100 residues has destroyed almost three quarters of the thermodynamically measurable water of hydrophobic hydration, that is, the requirement of solvating two carboxylates per 100 residues necessarily resulted in the destruction of three-fourths of the water of hydrophobic hydration.

6.2.1.4. Competition between apolar and polar residues for hydration. The above experiments are collectively interpreted to indicate that there exists a competition for hydration between hydrophobic (apolar) and charged (polar) species constrained to exist in model proteins capable of hydrophobic folding or of hydrophobic assemblage. This is referred to as an apolar-polar repulsive free energy of hydration [96, 124]. When the hydrophobic moiety and the charged moiety are at sufficient distance, both receive their full free energy of hydration by the organization of water around them. But as the two moieties become more proximal, the intervening water molecules can no longer simultaneously be in the appropriate structure for each and there is an increase in the free energy designated as ΔG_{ap}. The hydrophobic-induced pKa shifts are a measure of ΔG_{ap} such that $\Delta G_{ap} = \Delta\mu = -2.3$ RT ΔpKa.

To attempt a better visualization of what is happening, an example of a clam-shaped globular protein is depicted in Fig. 35. The clam-shaped globular protein has two states, closed and open. It is closed by hydrophobic folding because T_t is below the operating temperature. On opening, water of hydrophobic hydration that does form would be thermodynamically unfavorable, maintaining the equilibrium shifted toward the close state. Should a phosphorylation occur, this most polar moiety would destructure the proximal water of hydrophobic hydration thereby decreasing the thermodynamic driving force for closure. In part B of Fig. 35, it is noted that there is remaining water of hydrophobic hydration such that if a complementary surface could be found, by virtue of the structure and the limitation to two states of the globular protein, phosphorylation which destroys hydrophobic folding could be used to drive hydrophobic assembly.

6.2.2. Non-linear hydrophobic-induced pKa shifts: The next piece of information that is essential to the design of model proteins capable of using hydrophobic folding to interconvert energies, not including mechanical

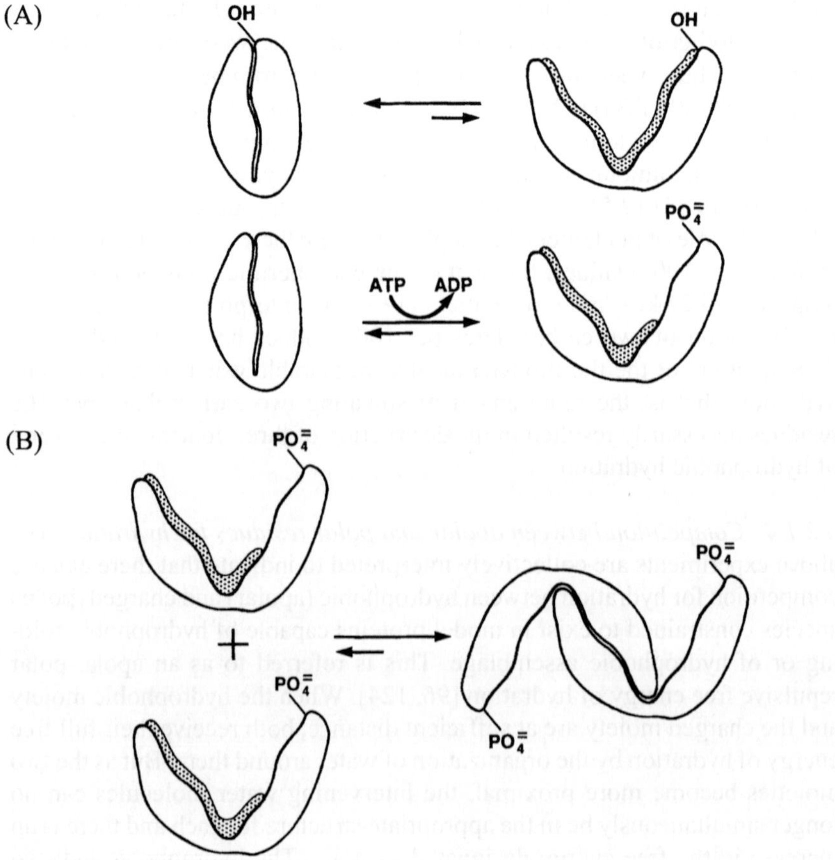

Fig. 35. (A) Phosphorylation driven unfolding and (B) Phosphorylation driven dimerization. Hypothetical clam-shaped globular protein capable of two states, open and closed (hydrophobically folded). The shaded area indicates water of hydrophobic hydration at a temperature above the value of T_t. Phosphorylation is seen to destructure more proximal water of hydrophobic hydration. See text for discussion. (Reproduced with permission from [124].)

work, is the non-linearity of hydrophobic-induced pKa shifts. This becomes apparent using the set of structures in Fig. 36 where for the Asp residue the presence of two Phe residues per 30mer which are most proximal by primary structure causes a 0.4 pH shift in pKa; the presence of the three interturn Phe residues causes a pKa shift 0.7; several combinations of the two sequence-proximal Phe residues with two interturn Phe residues causes pKa shift up to 1.4 whereas the combined set of five Phe residues, as seen in Fig. 34, causes the very large pKa shift of 6.1. Similar but somewhat lower shifts are exhibited by the Glu residue (see Fig. 36B).

Now if a given energy input to one of a pair of functional groups (the other of which is ionizable) results in a given change in hydrophobicity, ΔHPB, then a greater shift in pKa occurs with a greater hydrophobic background.

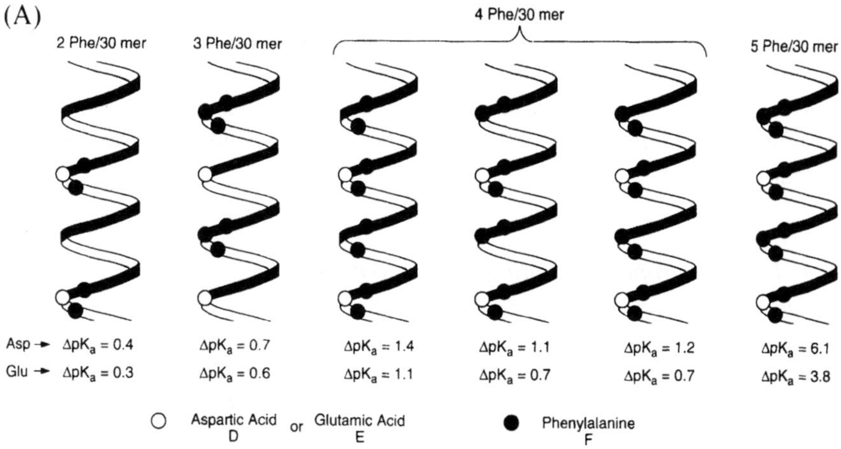

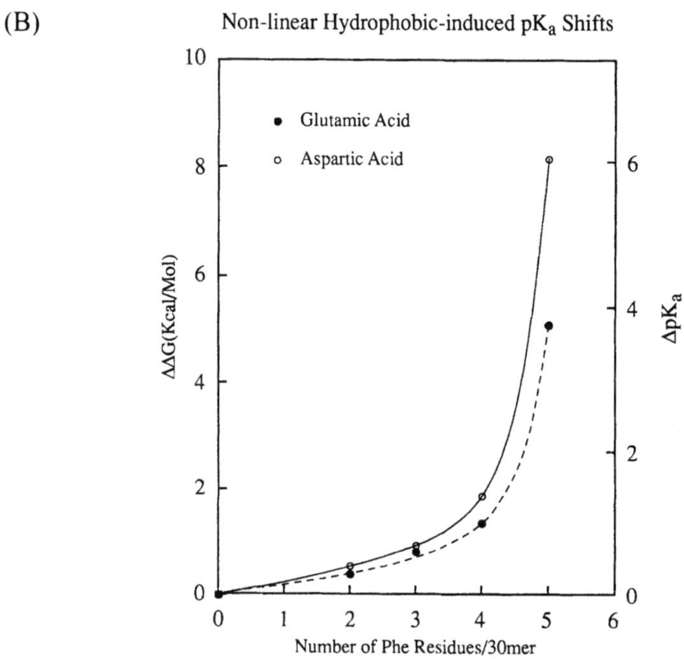

Fig. 36. (A) Representation of a series of polytricosapeptides containing one Asp or one Glu per 30mer and with varying numbers and locations of Phe residues. Below are listed the ΔpKa values from 3.9 for Asp and 4.3 for Glu. (B) Plots of ΔpKa or $\Delta\Delta G(=\Delta\mu)$ versus number of Phe residues per 30mer. For the 4 Phe/30mer structures, the permutation with the largest ΔpKa was used. The pKa shifts are seen to be very non-linear with increases in hydrophobicity (Reproduced with permission [144].)

Since the magnitude of the shift in pKa can be a measure of the chemical work possible, more chemical energy is obtained from the fixed energy input as the background hydrophobicity is increased. This concept is called *poising*, and it becomes possible to design model proteins with an increased efficiency of energy conversions by replacing Val residues by Phe residues as discussed below.

6.3. Using hydrophobic folding to interconvert energies not involving mechanical work

Using the model protein constructs of Fig. 37, it becomes possible to attach an N-methyl-nicotinamide moiety to the Lys(K) side chain and to input electrical energy to shift the pKa of the Asp residue to achieve electro-chemical transduction, or to attach a cinnamic acid moiety to the Lys(K) side chain such that the absorption of 350 nm light can effect a pKa shift to achieve photo-chemical transduction. These free energy transductions are briefly reported below.

6.3.1. Electro-chemical transduction (reduction-driven proton uptake): The model protein designed to observe electro-chemical transduction is the polytricosapeptide, poly[K{NMeN}PGDGFP GVGVP GVGVP GFGVP GVGVP GVG]; it contains two Phe(F) residues and an N-methyl nicotinamide attached to a Lys(K) side chain situated across the β-turn from the Asp(D) residue as shown in Fig. 37, in analogy to polymer III written for the {CA} adduct. As seen in Fig. 38, when an acid-base titration is carried out on the oxidized state a pKa is observed at 8.5. This is taken to be that of the Asp residue, shifted from its usual value near 3.9 by the hydrophobicity of the two Phe residues and the Lys-NMeN moiety. After reduction of the nicotinamide with dithionite in carbonate buffer at a pH near 9 and dialysis removing all solute but the polytricosapeptide, the reduction product N-methyl-6-hydroxy-1,4,5,6-tetrahydronicotinamide is obtained. On titration of the reduced product the pKa is found to have increased to 11.0 (see Fig. 38) [145]. This is one pH unit higher than observed with five Phe residues, as shown in Figs. 34 and 36. Presumably, the very hydrophobic Lys-N-methyl-6-hydroxy-1,4,5,6-nicotinamide side chain so proximal to the Asp residue, plus the two Phe residues, constitutes a slightly greater hydrophobicity than the five Phe residues positioned as in Figs. 34 and 36. Thus, the results are interpreted to be a reduction-effected increase in hydrophobicity which results in a hydrophobic-induced pKa shift and as such the result would represent the first example of electro-chemical transduction in a synthetic model protein. Such a finding provides an example of a reduction-driven proton uptake which is relevant to the first free energy transductional step of oxidative phosphorylation which results in the development of a proton gradient across the inner mitochondrial membrane. This will be briefly discussed below.

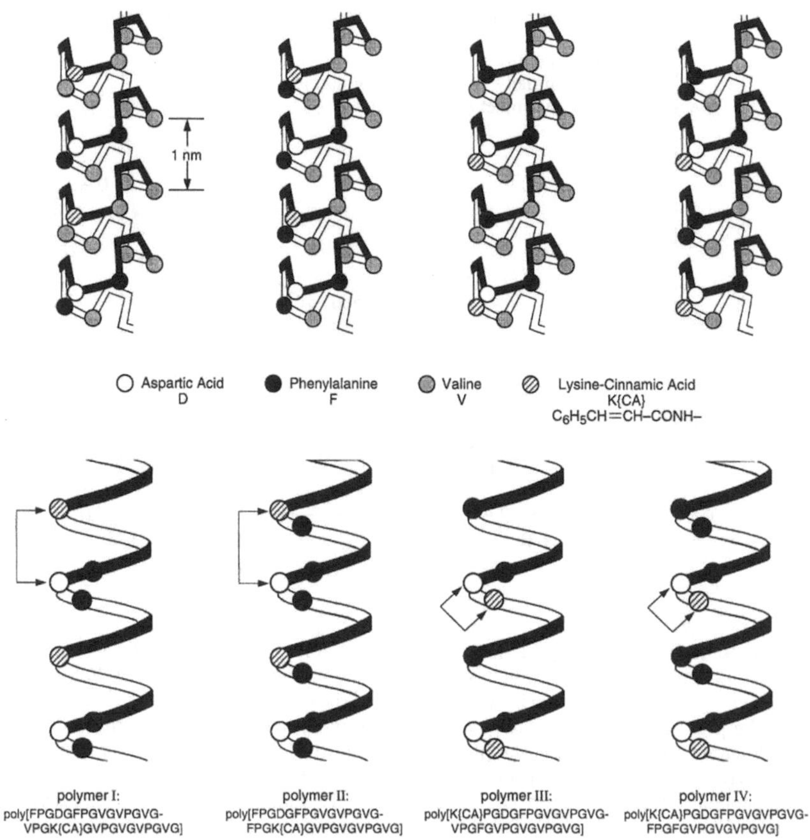

Fig. 37. Polytricosapeptides designed to demonstrate electro-chemical and photo-chemical transduciton. There are two functional groups, an Asp residue and a Lys residue, at interturn positions (displaced by fifteen residues) in polymers I and II and at positions i and $i+3$ of the same β-turn (i.e. cross-β-turn) in polymers III and IV. Also, the interturn positioning can be with two (polymer I) or with three (polymer II) Phe residues, and the cross-β-turn position can be with two (polymer III) or with three (polymer IV) Phe residues. To the Lys residue a cinnamic acid {CA} chromophore is attached by amide linkage for the photo-chemical transduction studies. Entirely equivalently, N-methyl nicotinamide {NMeN} replaces the cinnamic acid adduct for the electro-chemical studies (unpublished representation).

6.3.2. Photo-chemical transduction (light-driven proton release): A model protein designed to observe photo-chemical transduction, poly[FPGDGFP GVGVPGVGFPGK{CA}GVP GVGVP GVG], contains three Phe(F) residues and a Lys(K) with an attached cinnamic acid chromophore across the turn of the β-spiral from the Asp(D) residue, as shown in Fig. 37 as polymer II. When an acid-base titration is carried out on this model protein, it is anticipated that irradiation to drive the chromophore from trans to cis would lower the pKa of the Asp residue. Such a result is expected as the absorption of light decreases the hydrophobicity of the chromophore attached to the Lys(K) residue and as such would result in a relaxation of the

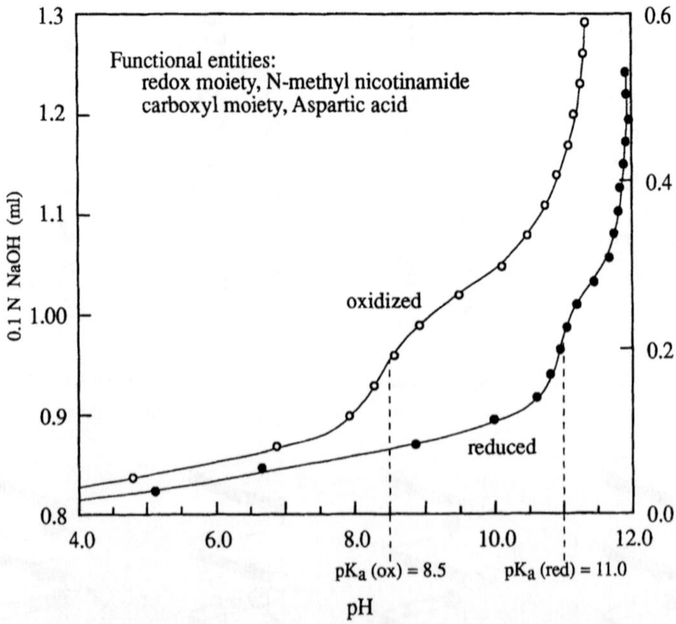

Fig. 38. Reduction-induced pKa shift in a designed polytricosapeptide with two functional entities. Electro-chemical transduction demonstrated using polymer III of Fig. 37 with N-methyl nicotinamide attached by amide linkage to the lysyl side chain. When in the oxidized state, the pKa is 8.5; when the reduction is to the more hydrophobic N-methyl-6-hydroxy-1,4,5,6-tetrahydronicotinamide, the pKa is increased to 11.0. Thus, when at an intermediate pH of, say 10, reduction will result in the uptake of a proton. By lowering the background hydrophobicity, the pKa values can span a lower pH range (Reproduced with permission from [145]).

hydrophobic-induced pKa shift. At the appropriate pH, this would be a photon-driven release of a proton. The two functional groups are coupled by virtue of being part of the same hydrophobic folding and assembly domain.

6.4. Postulates for energy conversion by means of hydrophobic folding and assembly (inverse temperature transitions)

The family of energy conversions that now appear possible by controlling hydrophobic folding and assembly in model proteins may be formalized by organization into a set of three postulates and an associated total of fifteen corollaries, one for each of the pairwise energy conversions of Fig. 30. The postulates and corollaries are detailed in Table 3 and as such will not be restated here. To elaborate briefly, however, postulates I and II are concerned with what may be called molecular engines, as they are machines that result in the performance of useful work on input of the appropriate energy. The molecular systems capable of these transductions have been called first-

Table 3. Postulates for energy conversion by means of hydrophobic folding and assembly (inverse temperature transitions).

Postulate I: The input of thermal energy to a protein capable of *hydrophobic folding and assembly* on raising the temperature from below to above the temperature, T_t, of an *inverse temperature transition* can result in motion and the performance of mechanical work.

Corollary:	Thermo-mechanical transduction

Postulate II: Any energy input that changes the temperature, T_t, at which an inverse temperature transition occurs can be used to produce motion and perform mechanical work.

Corollary 1:	Chemo-mechanical transduction
Corollary 2:	Electro-mechanical transduction
Corollary 3:	Baro-mechanical transduction
Corollary 4:	Photo-mechanical transduction

Postulate II: Different energy inputs, each of which can individually drive hydrophobic folding to produce motion and perform mechanical work, can be converted one into the other (transduced) by means of the inverse temperature transition with the correctly designed coupling and T_t value.

Corollary 1:	Electro-chemical transduction
Corollary 2:	Electro-thermal transduction
Corollary 3:	Baro-electrical transduction
Corollary 4:	Photo-voltaic transduction
Corollary 5:	Thermo-chemical transduction
Corollary 6:	Photo-thermal transduction
Corollary 7:	Baro-thermal transduction
Corollary 8:	Baro-chemical transduction
Corollary 9:	Photo-baric transduction
Corollary 10:	Photo-chemical transduction
Corollary 11:	Chemo-chemical transduction

order molecular machines of the T_t-type, as the hydrophobic folding and assembly process is directly utilized in the performance of mechanical work. Postulates II and III utilize the ΔT_t mechanism to achieve the energy conversion. The energy conversions of postulate III are of particular interest as they use the hydrophobic folding and assembly process to couple functionalities with the result of interconverting energies not involving mechanical work. The molecular systems capable of doing so are called second-order molecular machines of the T_t-type.

7. The ΔT_t hydrophobic paradigm for protein folding and function

In considering protein structure it has been natural to progress conceptually, as in Section 2 of this chapter, from primary to secondary to tertiary and finally to quaternary structure. A similar hierarchical thought process is naturally considered when addressing protein folding. When considering function, it has also been alluring to consider primary and secondary structural elements before addressing tertiary structural contributions. The ΔT_t hydrophobic paradigm for protein folding and function alters those perspectives; it suggests a primacy for the hydrophobic folding and assembly process, i.e. for inverse temperature transitions, with changes in hydrogen-bonding having less prominence. In particular, in the ΔT_t hydrophobic paradigm there is a new aspect of achieving function by changing the temperature, T_t, at which the inverse temperature transition of hydrophobic folding and assembly occurs. In fact, the majority of the energy conversions which have enabled living organisms to exist, to evolve and to thrive have been demonstrated in Section 6 simply by utilizing the new perspectives in designing model proteins which perform energy conversion functions by means of controlling the temperature at which the hydrophobic folding and assembly transition occurs. This section makes a brief effort to provide a glimpse of the world of protein structure/function as seen through control of inverse temperature transitions.

7.1. Experimental and theoretical basis of hydrophobic collapse as an initial protein folding event

Progress on both experimental and theoretical fronts in the drive towards an understanding of protein folding is bringing the role of hydrophobic folding to the foreground.

7.1.1. The hen lysozome folding studies of Dobson and coworkers: Dobson and coworkers have most effectively utilized a set of three experimental approaches to make clear the initial folding events of hen lysozyme [146, 147]. Using far ultraviolet stopped-flow circular dichroism (CD) studies, it was found that the CD pattern of essentially native hen lysozyme was obtained within four milliseconds. This would generally be interpreted to mean that the α-helix and β-chain secondary structures have already formed, but it should be appreciated that circular dichroism provides a measure of the sum of the relative orientations of peptide moieties and the relative orientations of a residue's R group with respect to its carbonyl moiety [148, 149]. In the second experimental approach, pulsed hydrogen exchange labeling studies demonstrated that, after the first several milliseconds, all of the peptide NH hydrogens were still rapidly exchanging. This result argues that no stable secondary structure of α-helix and β-chains, etc. had yet developed. The question now becomes, what is holding the structure together such that a near native CD pattern could be observed? The answer is given by the third experimental approach, that of determining the fluorescence intensity of the half-dozen tryptophan residues in hen lysozome. By means of the fluorescence intensity it can be determined whether a tryptophan residue is exposed to water or buried within a hydrophobic milieu. The telling result was that the tryptophans were buried within a hydrophobic domain of the protein within the first several milliseconds. Arguably, the initial event in the folding of hen lysozyme is hydrophobic association, perhaps a hydrophobic collapse, which is then followed by the sorting out of the most favourable secondary structure that would fit with the hydrophobic domain(s).

From the perspective of the T_t-based hydrophobicity scale (Table 2), tryptophan (Trp, W) is by far the most hydrophobic of all residues. What is remarkable about the hen lysozyme amino acid composition is that there is a ratio of approximately one Trp for twenty residues. This is a very high ratio. For the repeating Type III domain of fibronectin, which recurs in many other proteins, it is one Trp for every 90 or more residues which is the principal feature for this common repeating sequence (see Fig. 41 below). It is therefore perhaps understandable that hydrophobic collapse would be a dominant feature of the folding of hen lysozyme. The T_t for the inverse temperature transition of hen lysozyme would be very low – well below 0 °C.

7.1.2. Hydrophobic nucleation, the hydrophobic zipper and hydrophobic collapse: The importance of hydrophobic nucleation in protein folding was early appreciated by Matheson and Scheraga using side chains of residues i and $i+4$ of the β-turn [150]. More recently, Dill and coworkers presented a theoretical argument for cooperative hydrophobic folding [151] using a binary lattice model of a polymer chain comprised of hydrophobic and polar units. Briefly, if the association of a pair of hydrophobic subunits placed a second pair and a third pair, etc., of hydrophobic units in proximity, then the hydrophobic folding would be cooperative. This was called the hydrophobic zipper model [151], and it can be related to the association of β-chains or the wrap-up to form an amphiphilic α-helix as considered below. With the appropriate positioning in the primary structure of a substantial number of hydrophobic residues, a hydrophobic collapse such as that of hen lysozyme could be expected. The inverse temperature transition for the poly(GVGVP)-type of β-spiral structure formation has long been discussed as an optimization of interturn hydrophobic contacts and could be called a helical or β-spiral hydrophobic zipper.

7.2. T_t-based hydrophobicity plots and sequence editing to identify directing of secondary structure by hydrophobic residues

The above emerging emphasis on hydrophobic association as an initial event in protein folding allows for ready utilization of studies of the last decade on the rules governing hydrophobic folding and assembly as developed from the elastic model proteins. In particular, it provides an immediate use for the T_t-based hydrophobicity scale and suggests that it is an appropriate scale with which to analyze hydrophobic periodicities in the primary structure that would result from the primacy of hydrophobic association.

7.2.1. The cross-β, β-pleated sheet and β-barrel structures: There are two natural extensions of the theoretical hydrophobic zipper model of Dill that would allow it to become practical: one is to introduce the peptide residue as the repeating unit with its particular stereochemistry, and the second is to recognize that there are more and less hydrophobic residues as shown by the T_t-based hydrophobicity scale of Table 2. A significant stereochemical aspect can be shown to be associated with cross-β, β-sheet and β-barrel types of structures as demonstrated by the progressive zippering of a cross-β-structure in Fig. 39.

When a β-turn appears in a structure, the residues i and $i+3$ side chains are in position to associate hydrophobically as appreciated by Matheson and Scheraga [150]. Then as the cross-β-structure continues folding in a zipper-like manner, alternating residues in one chain that pair with alternating residues in the other chain result in a hydrophobic sideness. Thus, if β-sheets or β-barrels with a hydrophobic sideness are required to form larger hydrophobic domains, then it should become apparent by the occurrence of alternating hydrophobic residues in the primary structure. The propensity for such to occur would depend on just how hydrophobic the residues were. Accordingly, by Table 2 in order of decreasing significance there would be the Trp(W), Tyr(Y), Phe(F), Leu(L) and Ile(I) residues. In what follows, specific structural domains of proteins will be considered in this light.

7.2.2. A hydrophobically collapsed β-barrel: The human fibronectin type III domain as an example: Using the primary structure of human fibronectin [153] and the hydrophobicity scale of Table 2, the hydrophobicity plot of

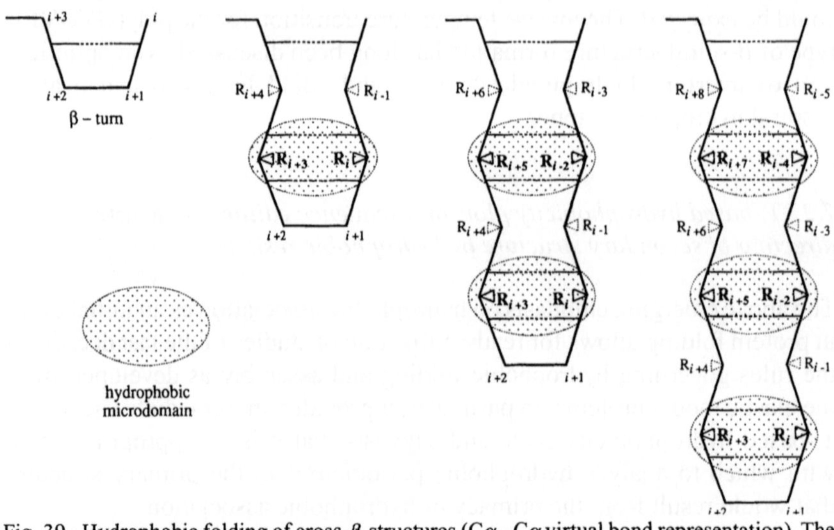

Fig. 39. Hydrophobic folding of cross-β-structures ($C\alpha-C\alpha$ virtual bond representation). The stepwise, zipper-like, formation of a cross-β-structure due to the pairing of alternating hydrophobic residues to result in antiparallel aligned β-chains with one side being hydrophobic. See text for discussion. (Reproduced with permission from [152].)

Fig. 40A is obtained. Even though there are over 2,000 residues, a glance at the plot shows the repeating sequences signaled by the recurring most hydrophobic Trp(W) residues. The repeats with Trp recurring about every 90 residues are called the Type III domains of fibronectin which, identified principally by the set of one Trp, two Tyr and one Leu residue, is known to occur in immunoglobulins, in titin, in tenascin, in cytokine receptors, and others [154]. The structure of this domain has been determined for tenascin [154] as shown in Fig. 41. It is basically a seven-stranded β-barrel substantially collapsed such that it may also be described as a four chain β-sheet

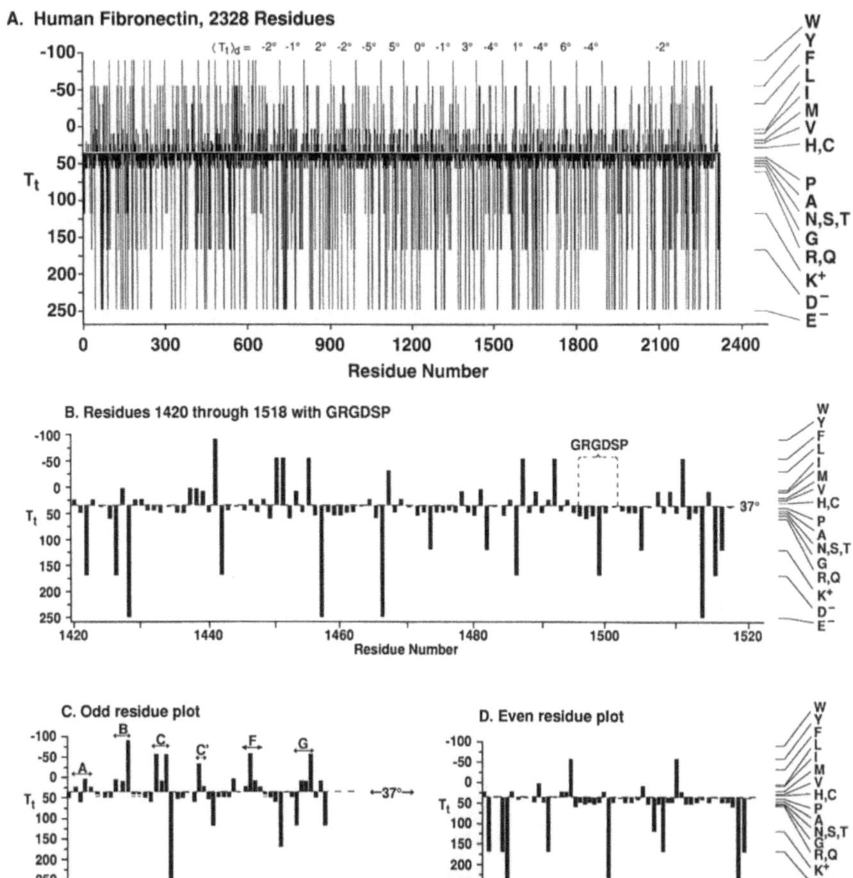

Fig. 40. Hydrophobicity plots of human fibronectin using the T_t-based hydrophobicity scale of Table 2. (A) The entire protein showing the repeating units and with mean residue hydrophobicities indicated above for the Type III domains. (B) Expanded plot of the tenth Type III domain where there are numerous examples of alternating hydrophobic residues which would define a hydrophobic side of a β-chain. (C) Plot of only the odd residues of the tenth Type III domain showing stretches of hydrophobic residues defining the hydrophobic sides of β-chains. (D) Even residue plot. (Reproduced with permission from [155].)

with a hydrophobic side associating with the hydrophobic side of a three-chain β-sheet, with the chain directions of the two β-sheets running at an angle of about 30°. The hydrophobicity plot for the tenth Type III domain of fibronectin is shown in Fig. 40B.

If the analysis of Section 7.2.1 and the perspective of Fig. 39 obtains then on sequence editing of the hydrophobicity plot, that is, on plotting the odd and even residues only, as in Figs. 40C and D, it should be possible to identify the β-chains providing the hydrophobic sideness. Indeed, β-chains are demonstrable in Fig. 40C and they can even be ranked as to which provide the greater stability to the hydrophobic domain. A comparison with the teascin Type III domain in Fig. 41 is helpful in seeing the individual chains.

7.2.3. Amphiphilic α-helices: Two apolipoprotein examples: T_t-based hydrophobicity plots for apolipoprotein III using the sequence for *Locusta migratoria* [156] and that for *Manduca sexta* [157] are given in Figs. 42A and B, respectively, and a representation of the structure of apolipoprotein III for *Locusta migratoria* due to Breiter *et al.* [158] is shown in Fig. 43. Immediately apparent in Fig. 42A and perhaps even more so in Fig. 42B is the periodicity of the hydrophobic residues which results in the formation of amphiphilic α-helices with the hydrophobic or non-polar residue occurring at every fourth or sometimes every third position. In Fig. 42A the

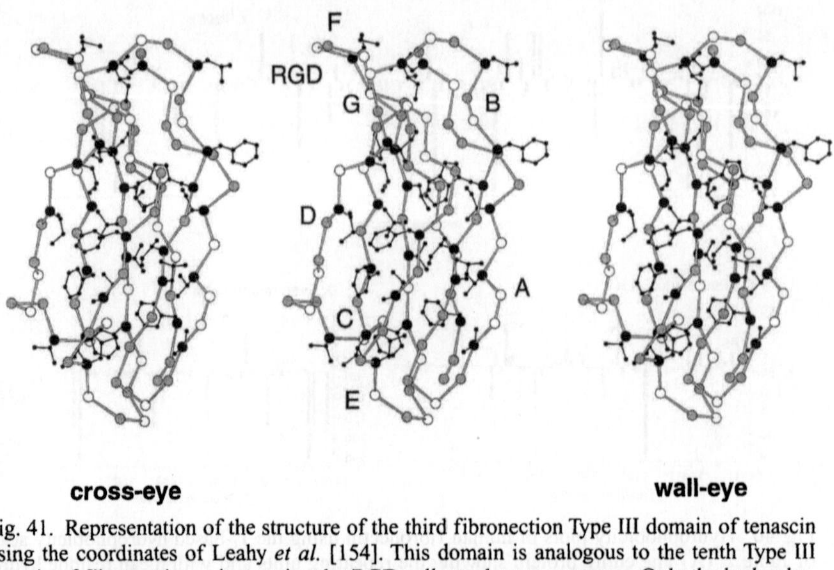

Fig. 41. Representation of the structure of the third fibronection Type III domain of tenascin using the coordinates of Leahy *et al.* [154]. This domain is analogous to the tenth Type III domain of fibronectin, as it contains the RGD cell attachment sequence. Only the hydrophobic side chains are given (Key-see figure). (Reproduced with permission from [155].)

● W, Y, F, L, I, M, V
○ all other residues
○ D, E, K, R

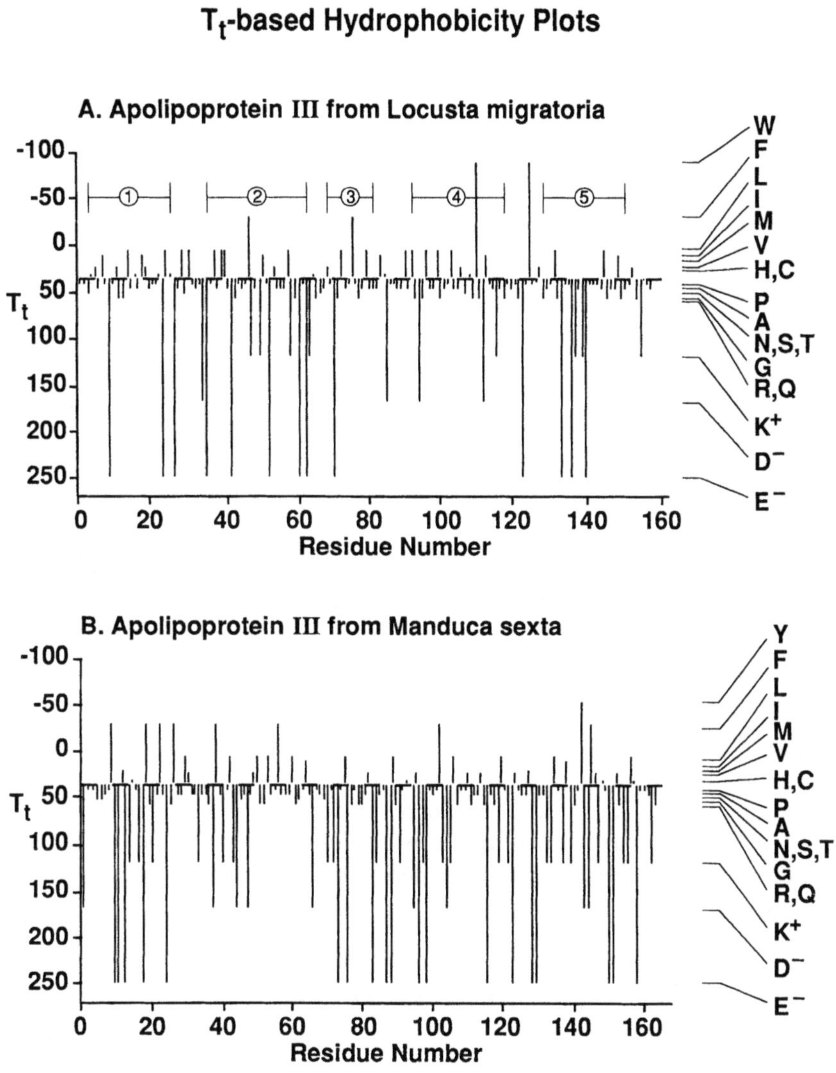

Fig. 42. Hydrophobicity plots using the T_t-based hydrophobicity scale of Table 2. (A) Apolipoprotein III from *Locusta migratoria* with the five segments of amphiphilic α-helices indicated above. (B) Apolipoprotein III from *Manduca sexta*. In both cases there is a regular recurrence of hydrophobic or non-polar residues at the fourth position and sometimes in the third position, as required for an amphiphilic α-helix. The structure for A is given in Fig. 43. (Reproduced with permission from [155].)

actual α-helical sequences are noted based on the structure in Fig. 43 with the hydrophobic side chains added such that it is possible to see where each of the tenascin hydrophobic residues resides. An analysis of the folding points has been carried out elsewhere and will not be included here for sake of brevity [155]. What is observed in these examples of a hydrophobic β-barrel and packed amphiphilic α-helices is an argument

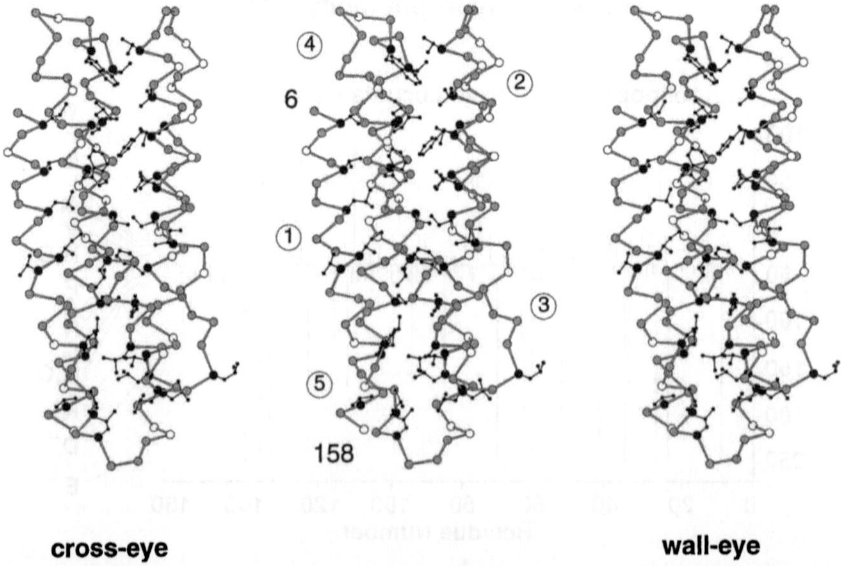

Fig. 43. Representation of the crystal structure of Apolipoprotein III from *Locusta migratoria* determined by Breiter *et al.* [158]. Only the hydrophobic side chains have been included to demonstrate the amphiphilic helices and their hydrophobic association (Key-see figure). (Reproduced with permission from [155].)
● W, Y, F, L, I, M, V
○ all other residues
○ D, E, K, R

for protein folding being dictated by periodic occurrence of hydrophobic residues in the primary structure. In what follows it will be noted that the efficiency of energy conversion depends on just how hydrophobic the folding domain is.

7.3. Efficiency if energy conversion based on non-linear hydrophobic-induced pKa shifts

The non-linear relationship between hydrophobicity and pKa shifts was shown above in Fig. 36. The efficiency of conversion of an energy input such as electrochemical reduction or the absorption of light to chemical energy differs depending on whether the hydrophobic domain associated with the conversion is more or less hydrophobic. This is depicted in Fig. 44. The reduction of nicotinamide to a specific reduction product or the absorption of light by a particular chromophore resulting in a specific photochemical result each represent a specific change in hydrophobicity, ΔHPB. The magnitude of the ΔHPB can be largely independent of the environment in which it occurred, yet the change in pKa that it brings about can be small if the domain hydrophobicity is small or it can be larger if the domain hydrophobicity is larger, as indicated in Fig. 44.

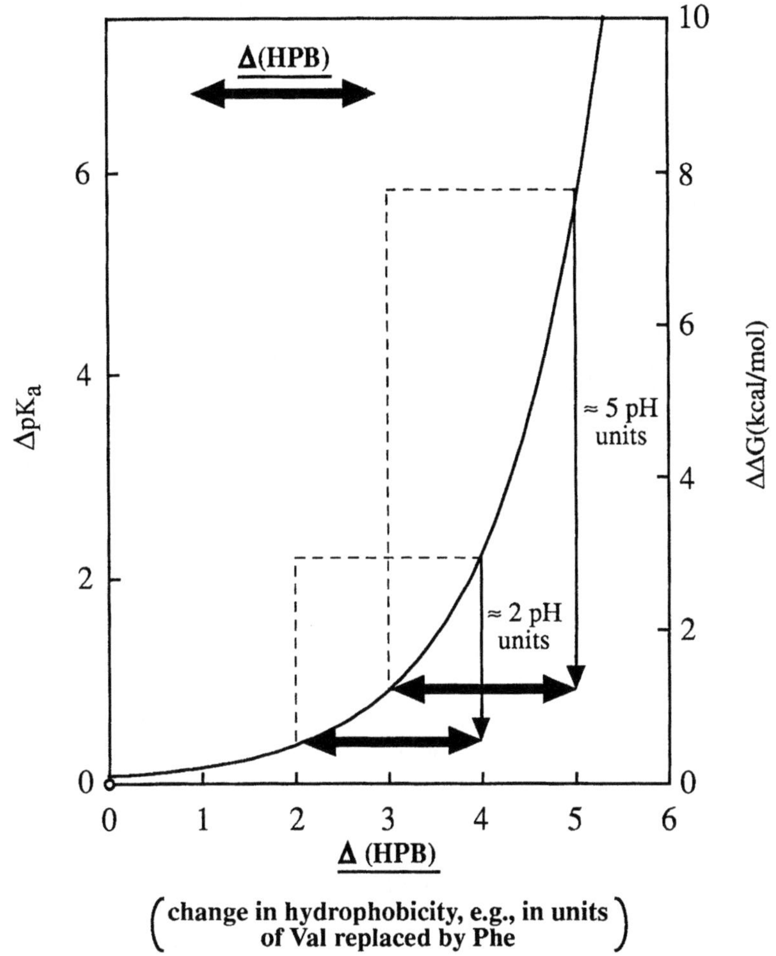

Fig. 44. Schematic representation of non-linear hydrophobic induced pKa shifts. When a change in hydrophobicity represented by the Δ(HPB) arrow is imposed on the molecular system, the magnitude of the resulting pKa change depends on the background hydrophobicity. As a ΔpKa is proportional to the chemical work resulting from the ΔHPB, increased background hydrophobicity means increased efficiency of energy conversion. See text for discussion. (Reproduced with permission from [152].)

It was to test this perspective that the structures of Fig. 37 were prepared with two and with three Phe residues per 30mer, and the data to date support Fig. 44; the pKa shift has been larger when there are three Phe residues than when there are two Phe residues. Chemical energy is the product of the change in chemical potential, $\Delta\mu$, times the change in number, Δn, for example of carboxylate moieties, i.e. chemical energy = $\Delta\mu \cdot \Delta n$. Accordingly, for a given Δn, the change in pKa, i.e. $\Delta\mu = -2.3\ RT\ \Delta pKa$, is a measure of the chemical energy resulting from a given energy input such as the energy of a photon of 350 nm light. In short, the larger the ΔpKa, the greater the efficiency of energy conversion, i.e. the more hydrophobic the domain is the more efficient is the energy conversion.

7.4. Relevance of the ΔT_t paradigm to the proton gradients and energy conversions of oxidative phosphorylation and photosynthesis

Fundamental to bioelectrochemistry and bioenergetics is the development of ion concentration gradients across membranes. It is to provide insight into this process that an attempt has been made to carry the presentation of protein structure, folding, and function to this stage in the present chapter and volume. The primary energy converting events for all such concentration gradients in animals is the reduction of prosthetic groups or electron carriers in the inner mitochondrial membrane resulting in a proton gradient. The proton gradient is then utilized, as proposed in the Mitchell chemiosmotic hypothesis, to phosporylate ADP to produce ATP or their equivalents. ATP is then used to produce other ion gradients by the specific ion ATPases known as ion pumps. An analogous situation obtains in the photosynthetic apparatus of plants. One result of the absorption of light is a proton gradient across the thylakoid membrane of the chloroplast [159]. Thus, the recent developments in the design of model proteins that are capable of electrochemical transduction (seen as a reduction-driven proton uptake in 6.3.1) and of photo-chemical transduction (seen as a light-driven release of protons in 6.3.2) become central to bioelectrochemistry and bioenergetics.

The key elements of the ΔT_t hydrophobic paradigm that enable electrochemical and photo-chemical transduction are the hydrophobic-induced pKa shifts and the change in hydrophobicity on the T_t scale due to the reduction of a prosthetic group such as a nicotinamide (see Fig. 24) or resulting from the absorption of a photon of light. Given the non-linear nature of hydrophobic induced pKa shifts seen in Fig. 36, incorporation of such an energy conversion process into a lipid bilayer not only provides a barrier across which a proton gradient could be sustained but also contributes to the hydrophobic milieu which would ensure efficient energy conversion.

7.5. Relevance of the ΔT_t hydrophobic paradigm to muscle contraction

There are two aspects of the ΔT_t hydrophobic paradigm for protein folding and function of apparent relevance to muscle contraction. One is the dramatic effect of phosphorylation on raising the value of T_t (see Table 2) and thereby causing hydrophobic unfolding as schematically depicted in Fig. 35 with dephosphorylation, almost certainly wherever possible, ensuring hydrophobic folding. The second is the effect of cation interaction at charged carboxylate sites, i.e. of charge neutralization, in dramatically lowering the value of T_t as seen in Fig. 27.

The structure of the myosin cross-bridge with the light chains attached has recently been published, including the phosphorylation and calcium ion binding sites [160]. In short, by the ΔT_t hydrophobic paradigm for protein folding and function, a hydrophobic collapse by the coordinated events of Ca^{2+} binding at charged carboxylates and dephosphorylation could very effectively provide the basis for the power stroke of muscle contraction.

8. Concluding comment

Understanding the mechanism whereby a protein functions has two critical elements: one is to know the complete molecular structure for each functional state, and the second is to understand the driving forces involved in going from one to the other. Thus, even if the complete molecular structure for the sodium pump, the Na-K ATPase, become known in each of its states, the actual mechanism will not be known until the forces that cause the transformation between states are fully understood. The most definitive way to obtain the latter knowledge is to design molecular structures that will function in a predictable way based on the perspective being tested. This has been done and is being done within the context of the ΔT_t hydrophobic paradigm of protein folding and function. Molecular structures are being designed for specific functions as tests of the level of understanding and the assay can even be visual or as simple as an acid-base titration.

Acknowledgments

This work was supported in part by contracts N00014-89-J-1970 from the Department of the Navy, Office of Naval Research; F33615-93-C-5378 from Wright-Patterson Air Force Base; and by grant HL-29578 from the National Institutes of Health. Thanks are extended to Betty A. Cox and R. Dean Harris for their contributions to the earlier draft of this manuscript.

References

1. See for example S. Prentis, *Biotechnology, A New Industrial Revolution*, George Braziller, Inc., New York (1984).
2. KU Linderstrom-Lang, *Proteins and Enzymes, Lane Medical Lectures*, Stanford University Press, Stanford, CA (1952).
3. CB Anfinsen, Science 181 (1973) 223–230.
4. L Pauling, RB Corey and HR Branson, Proc. Natl. Acad. Sci. USA 37 (1951) 205–211.
5. J Donohue, Proc Natl Acad. Sci. USA 39 (1953) 470–478.
6. L Pauling and RB Corey, Proc. Natl. Acad. Sci. USA 37 (1951) 729–740.
7. YP Chou and GD Fasman, J. Mol. Biol. 115 (1977) 135–175.
8. CM Venkatachalam, Biopolymers 6 (1968) 1425–1436.
9. AJ Geddes, KD Parker, EDT Atkins and EJ Beighton, J. Mol. Biol. 32 (1968) 343–358.

10. WJ Cook, HM Einspahr, TL Trapane, DW Urry and CE Bugg, J. Am. Chem. Soc. 102 (1980) 5502–5505.
11. M Ohnishi and DW Urry, Science 168 (1970) 1091–1092.
12. DW Urry, M Ohnishi and R Walter, Proc. Natl. Acad. Sci. USA 66 (1970) 111–116.
13. K Neupert-Laves and M Dobler, Helv. Chim. Acta 58 (1975) 432–442.
14. SP Wood, IJ Tickle, AM Treharne, JE Pitts, Y Mascarenhas, JY Li, J Husain, S Cooper, TL Blundell, VJ Hruby, A Buku, AJ Fischman and HR Wyssbrod, Science 232 (1986) 633–636.
15. DW Urry, Int. J. Pept. Protein Res. 30 (1987) 701–703.
16. GN Ramachandran and G Kartha, Nature 174 (1954) 269–270; Nature 176 (1955) 593–595.
17. A Rich and FHC Crick, Nature 176 (1955) 915–916.
18. MH Miller and HA Scheraga, J Polymer Sci Symp. No. 54 (1976) 171–200.
19. GN Ramachandran in *Proceedings of the Second SUNYA Conversation in the Discipline Biomolecular Stereodynamics,* RH Sarma (ed.), Adenine Press, New York (1981) Vol. II, pp. 1–36.
20. B Brodsky-Doyle, KR Leonard and KBM Reid, Biochem. J. 159 (1976) 279–286.
21. T Kodama, M Freeman, L Rohrer, J Zabrecky, P Matsudaira and M Krieger, Nature 343 (1990) 531–535.
22. DW Urry and M Ohnishi, in *Spectroscopic Approaches to Biomolecular Conformation*, DW Urry (ed.), Am. Med. Assoc. Press, Chicago, Illinois (1970) pp. 263–300.
23. MM Shemyakin, EI Vinogradova, MY Feigina, NA Aldanova, NF Loginova, ID Ryabova, IA Pavlenko, Experientia 21 (1965) 548–552.
24. P Läuger, Science 178 (1972) 24–30.
25. R Sarges and B Witkop, Biochemistry 4 (1965) 2491–2494.
26. SB Hladky and DA Haydon, Nature 225 (1970) 451–453.
27. E Bamberg, H-J Apell and H Alpes, Proc. Natl. Acad. Sci. USA 74 (1977) 2402–2406.
28. DW Urry, MC Goodall, JD Glickson and DF Mayers, Proc. Natl. Acad. Sci. USA 68 (1971) 1907–1911.
29. GN Ramachandran and R Chandrasekaran, Indian J Biochem. Biophys. 9 (1972) 1.
30. DW Urry in *Topics in Current Chemistry*, FL Boschke (ed.), Springer-Verlag, Heidelberg, Germany (1985) Volume 128, pp. 175–218.
31. AS Arseniev, IL Barsukov, VF Bystrov, AL Lomize and Yu A Ovchinnikov, FEBS Letters 186 (1985) 168–174.
33. LK Nicholson and TA Cross, Biochemistry 28 (1989) 9379–9385.
34. WR Veatch, ET Fossel and ER Blout, Biochemistry 13 (1974) 5249–5256.
35. W Kauzmann, Adv. Protein Chem. 14 (1959) 1–63.
36. JT Edsall and HA McKenzie, Adv. Biophys. 16 (1983) 53–183.
37. Y Nozaki and C Tanford, J. Biol. Chem. 246 (1971) 2211–2217.
38. J Kyte and RF Doolittle, J. Mol. Biol. 157 (1982) 105–132.
39. JL Cornette, KB Cease, H Margalit, TL Spouge, TA Berzofsky and C DeLisi, J. Mol. Biol. 195 (1987) 659–685.
40. M Degli Esposti, M Crimi and G Venturoli, Eur. J. Biochem. 190 (1990) 207–219.
41. DW Urry, Ultrastruc. Pathol. 4 (1983) 227–251.
42. DW Urry, CM Venkatachalam, MM Long and KU Prasad, in *Conformation in Biology,* R Srinivasan and RH Sarma (eds.), Adenine Press, USA (1982) GN Ramachandran Festschrift Volume, pp. 11–27.
43. DK Chang, CM Venkatachalam, KU Prasad and DW Urry, J. Biomol. Structure and Dynamics 6 (1989) 851–858.
44. KA Dill, Biochemistry 29 (1990) 7133–7155.
45. S Margusee and RL Baldwin, Proc. Natl. Acad. Sci. USA 84 (1987) 8898–8902.
46. J Monod, J-P Changeux and F Jacob, J. Mol. Biol. 6 (1963) 306–329.
47. J Monod, J Wyman and J-P Changeux, J. Mol. Biol. 12 (1965) 88–118.
48. D Blangy, H Buc and J Monod, J. Mol. Biol. 31 (1968) 13–35.
49. YV Griko, SY Venyaminov and PL Privalov, FEBS Letters 244 (1989) 276–278.
50. W Stuhmer, F Conti, H Suzuki, X Wang, M Noda, N Yahagi, H Kubo and S Numa, Nature 339 (1989) 597–603.
51. W Stuhmer, Annu. Rev. Biophys. Biophys. Chem. 20 (1991) 65–78.

52. WA Catterall, Science 253 (1991) 1499–1500.
53. WA Catterall, T Scheuer, W Thomsen and S Rossie, Ann. N. Y. Acad. Sci. 625 (1991) 174–180.
54. T Konno, C Busch, E von Kitzing, K Imoto, F Wang, J Nakai, M Mishina, S Numa and B Sakmann, Proc. R. Soc. Lond. B (1991) 69–79.
55. M Misra, HC Beall, KA Taylor and HP Ting-Beall, J. Struct. Biol. 105 (1990) 67–74.
56. AB Maunsbach, E Skriver and H Hebert in *The Sodium Pump: Structure, Mechanism and Regulation, The Rockefeller University Press* (1991), Soc. of Gen. Physiol. Series, pp. 159–172.
57. NM Green and DH MacLennan, Biochem. Soc. Trans. 17 (1989) 819–822.
58. J-D Horisberger, V Lemas, J-P Kraehenbühl and BC Rossier, Annu. Rev. Physiol. 53 (1991) 565–584.
59. TL Hill, Biophys. J. 49 (1986) 981–986.
60. C Cohen and DAD Parry, Proteins: Structur, Function and Genetics 1 (1990) 1–15.
61. JF Conqy and DAD Parry, Int. J. Biol. Macromol. 13 (1991) 14–16.
62. L Pauling and RB Corey, Nature 171 (1953) 59–61.
63. FHC Crick, Acta Cryst. 6 (1953) 689–697.
64. JF Conway and DAD Parry, Int. J. Biol. Macromol. 12 (1990) 328–334.
65. AD McLachlan, Annu. Rev. Biophys. Bioeng. 13 (1984) 167–189.
66. L Gotte, D Volpin, RW Horne and M Mammi, Micron 7 (1976) 95–102.
67. EG Cleary and WJ Cliff, Exp. Mol. Pathol. 28 (1978) 227–247.
68. SR Pinnell and GR Martin, Proc. Natl. Acad. Sci. USA 61 (1968) 708–716.
69. C Franzblau in *Comprehensive Biochemistry*, M Florkin and E Stotz (eds.), Elsevier, Amsterdam (1970), Vol. 26c, pp. 659–712.
70. SM Partridge, Gerontologia 15 (1969) 85–100.
71. J McDonagh in *Hemostasis and Thrombosis*, RW Coleman, J Hirsch, VJ Mardin and EW Salzman (eds.), JB Lippincott Co., Philadelphia, PA (1987) pp. 289–300.
72. PL Privalov, Critical Reviews in Biochemistry and Molecular Biology 25 (1990) 281–305.
73. L Sandberg, J Leslie, C Leach, V Torres, A Smith and D Smith, Pathol. Biol. 33 (1985) 266–274.
74. H Yeh, N Ornstein-Goldstein, Z Indik, P Sheppard, N Anderson, J Rosenbloom, G Cicila, K Yoon and J Rosenbloom, Collagen and Related Research 7 (1987) 235–247.
75. Z Indik, H Yeh, N Ornstein-Goldstein, P Sheppard, N Anderson, J Rosenbloom, L Peltonen and J Rosenbloom, Proc. Natl. Acad. Sci. USA 84 (1987) 5680–5684.
76. DW Urry in *Methods in Enzymology*, LW Cunningham and DW Frederiksen (eds.), Academic Press, Inc., New York (1982) Volume 82, 673–716.
77. DW Urry, MM Long and H Sugano, J Biol. Chem. 253 (1978) 6301–6302.
78. L Gotte and A Serafini-Frascassini, J Atheroscler. Res. 3 (1963) 247–255.
79. L Gotte, MG Giro, D Volpin and RW Horne, J. Ultrastruct. Res. 46 (1974) 23–33.
80. WJ Cliff and EG Cleary, Austr. J. Exp. Biol. Med. Sci. 46 (1968) 8.
81. J Gross, JH Highberger and FO Schmitt, Proc. Natl. Acad. Sci. USA 41 (1955) 1–7.
82. BA Cox, BC Starcher and DW Urry, Biochim. Biophys. Acta 317 (1973) 209–213.
83. BA Cox, BC Starcher and DW Urry, J. Biol. Chem. 249 (1974) 997–998.
84. D Volpin, DW Urry, BA Cox and L Gotte, Biochim. Biophys. Acta 439 (1976) 253–258.
85. D Volpin, DW Urry, I Pasquali-Ronchetti and L Gotte, Micron. 7 (1976) 193–198.
86. DW Urry, K Okamoto, RD Harris, CF Hendrix and MM Long, Biochemistry 15 (1976) 4083–4089.
87. VP Maier, AL Taptel and DH Volman, J. Amer. Chem. Soc. 77 (1955) 1278–1280.
88. GB Kistiakowsky and R Lumry, J. Amer. Chem. Soc. 71 (1949) 2006–2013.
89. JL Kavanau, J. Gen. Physiol. 34 (1950) 193–209.
90. FH Johnson, H Eyring and BJ Stover in *The Theory of Rate Processes in Biology and Medicine*, John Wiley & Sons, Inc., New York (1974) pp. 155–272.
91. JF Brandts, J. Amer. Chem. Soc. 86 (1964) 4291–4301.
92. B Chen and JA Schellman, Biochemistry 28 (1989) 685–691.
93. LC Antonino, RA Kautz, T Nakano, RO Fox and AL Fink, Proc. Natl. Acad. Sci. USA 88 (1991) 7715–7718 and references therein.

94. F Franks and RHM Hatley, Cryo-Letters 6 (1985) 171–180.
95. RHM Hatley and F Franks, Cryo-Letters 7 (1986) 226–233.
96. DW Urry, Prog. Biophys. Molec. Biol. 57 (1992) 23–57.
97. DW Urry, MM Long, RD Harris and KU Prasad, Biopolymers 25 (1986) 1939–1953.
98. RS Rapaka, K Okamoto and DW Urry, Int. J. Pept. Protein Res. 12 (1978) 81–92.
99. DW Urry, DK Chang, R Krishna, DH Huang, TL Trapane and KU Prasad, Biopolymers 28 (1989) 819–833.
100. DK Chang, CM Venkatachalam, KU Prasad and DW Urry, J. Biomol. Structure and Dynamics 6 (1989) 851–858.
101. DW Urry, J Jaggard, KU Prasad, T Parker and RD Harris in *Progress in Biomed. Polym.*, CG Gebelein (ed.), Plenum Publishing Co., New York, NY (1991) 265–274.
102. DW Urry, TL Trapane, MM Long and KU Prasad, J. Chem. Soc., Faraday Trans. 179 (1983) 853–868.
103. DW Urry, DC Gowda, TM Parker, CH Luan, MC Reid, CM Harris, A Pattanaik and RD Harris, Biopolymers 32 (1992) 1243–1250.
104. DW Urry, C-H Luan, TM Parker, DC Gowda, KU Prasad, MC Reid and A Safavy, J. Am. Chem. Soc. 113 (1991) 4346–4348.
105. DW Urry, Intl. J. Quant. Chem.: Quant. Biol. Symp. 15 (1988) 235–245.
106. A Pattanaik, DC Gowda and DW Urry, Biochem. Biophys. Res. Comm. 178 (1991) 539–545.
107. PD Boyer and EG Krebs in *The Enzymes*, PD Boyer and EG Krebs (eds.), Academic Press, Inc., Orlando, Florida (1986) (See the Preface).
108. F Hofmeister, Arch. Exp. Pathol. Pharmakol. 24 (1888) 247.
109. PH Yancey, ME Clark, SC Hand, RD Bowlus and GN Somero, Science 217 (1982) 1214–1222.
110. C-H Luan, J Jaggard, RD Harris and DW Urry, Intl. J. of Quant. Chem.: Quant. Biol. Symp. 16 (1989) 235–244.
111. C-H Luan and DW Urry, J. Phys. Chem. 95 (1991) 7896–7900.
112. DW Urry, LC Hayes, DC Gowda, CM Harris and RD Harris, Biochem. Biophys. Res. Comm. 188 (1992) 611–617.
113. DW Urry, LC Hayes, DC Gowda and TM Parker, Chem. Phys. Lett. 182 (1991) 101–106.
114. LA Strzegowski, MB Martinez, DC Gowda, DW Urry and DA Tirrell, J. Am. Chem. Soc. 116 (1994) 813–814.
115. DW Urry in *Molecular Conformation and Biological Interactions*, P Balaram and S Ramaseshan (eds.), Indian Academy of Science Bangalore, India (1991) pp. 1–29.
116. DW Urry, J. Protein Chem. 7 (1988) 1–34; 81–114.
117. DW Urry, TL Trapane, M Iqbal, CM Venkatachalam and KU Prasad, Biochemistry 24 (1985) 5182–5189.
118. DW Urry, TL Trapane, RB McMichens, M Iqbal, RD Harris and KU Prasad, Biopolymers 25 (1986) 209–228.
119. DW Urry, TM Parker, A Nicol, A Pattanaik, DS Minehan, DC Gowda, C Morrow and DT McPherson, Am. Chem. Soc., Div. Polym. Mater.: Sci. and Eng. 66 (1992) 399–402.
120. PJ Flory, a Ciferri and CAJ Hoeve, J Polymer Sci 45 (1960) 235–236.
121. KL Dorrington and NG McCrum, Biopolymers 16 (1977) 1201–1222.
122. DW Urry, TL Trapane and KU Prasad, Biopolymers 24 (1985) 2345–2356.
123. ZR Wasserman and FR Salemme, Biopolymers 29 (1990) 1613–1631.
124. DW Urry, Angew. Chem. Int. Ed. Engl. 32 (1993) 819–841.
125. DW Urry, B Haynes, H Zhang, RD Harris and KU Prasad, Proc. Natl. Acad. Sci. USA 85 (1988) 3407–3411.
126. A Katchalsky, J. Polym. Sci. 7 (1951) 393–412.
127. A Katachalsky, S Lifson, I Michaeli and M Zwick, in *Size and Shape of Contractile Polymers: Conversion of Chemical and Mechanical Energy*, A Wasserman (ed.), Pergamon Press, New York, NY (1960) pp. 1–40.
128. DW Urry, RD Harris and KU Prasad, J. Am. Chem. Soc. 110 (1988) 3303–3305.
129. McGraw-Hill Concise Encyclopedia of Science and Technology, 2nd ed. McGraw-Hill, New York (1989) 711.
130. W Kuhn, B Hargitay, A Katchalsky and H Eisenberg, Nature 164 (1950) 514–516.
131. GH Pollack, Muscles and Molecules, Ebner and Sons, Seattle, Washington (1990).

132. DW Urry, LC Hayes and DC Gowda, Biochem. Biophys. Res. Comm. 204 (1994) 230–237.
133. DW Urry, LC Hayes, TM Parker and RD Harris, Chem. Phys. Ltrs. 201 (1993) 336–340.
134. M Irie, Pure Appl. Chem. 62 (1990) 1495–1502.
135. A Szent-Györgyi, Acta Physiol. Scand. 9 (1945) 1–116.
136. DW Urry, SQ Peng, L Hayes, J Jaggard and RD Harris, Biopolymers 30 (1990) 215–218.
137. DW Urry, SQ Peng and TM Parker, Biopolymers 32 (1992) 373–379.
138. DW Urry, SQ Peng and TM Parker, J. Am. Chem. Soc. 115 (1993) 7509–7510.
139. DW Urry, SQ Peng, TM Parker, DC Gowda and RD Harris, Angew. Chem. 105 (1993) 1523–1525; Angew. Chem. Int. Ed. Engl. 32 (1993) 1440–1442.
140. DW Urry, SQ Peng, DC Gowda, TM Parker and RD Harris, Chem. Phys. Ltrs. 225 (1994) 97–103.
141. DW Urry, DC Gowda, SQ Peng, TM Parker, N Jing and RD Harris, Biopolymers 34 (1994) 889–896.
142. DW Urry, DC Gowda, SQ Peng TM Parker and RD Harris, J. Am. Chem. Soc. 114 (1992) 8716–8717.
143. DW Urry, C-H Luan, RD Harris and KU Prasad, Polym. Preprints, Am. Chem. Soc., Div. Polym. Chem., Inc. 31 (1990) 188–189.
144. DW Urry, DC Gowda, SQ Peng and TM Parker, Chem. Phys. Ltrs. 239 (1995) 67–74.
145. DW Urry, LC Hayes, DC Gowda, SQ Peng and N Jing, Biochem. Biophys. Res. Commun. 210 (1995) 1031–1039.
146. CM Dobson, PA Evans and SE Randford, Trends in Biochem. Sci. 19 (1994) 31–37.
147. PA Evans and SE Randford, Curr. Op. Struct. Biol. 4 (1994) 100–106.
148. DW Urry, Annu. Rev. Phys. Chem. 19 (1968) 477–530.
149. DW Urry, in *New Comprehensive Biochemistry, Modern Physical Methods in Biochemistry*, A Neuberger and LLM Van Deenen (eds.), Elsevier Amsterdam, The Netherlands (1985) pp. 275–356.
150. RR Matheson, Jr and HA Scheraga, Macromolecules 11 (1978) 819–829.
151. KA Dill, FM Fiebig and HS Chan, Proc. Natl. Cad. Sci. USA 90 (1993) 1942–1946.
152. DW Urry, Int. J. Quant. Chem.: Quant. Biol. Symp. 21 (1994) 3–15.
153. AR Kornblihtt, K Umezawa, K Vibe-Pedersen and FE Baralle, EMBO. J. 7 (1985) 1755–1759.
154. DJ Leahy, WA Hendrickson, I Aukhil and HP Erickson, Science 258 (1992) 987–991.
155. DW Urry and CH Luan, Am. Chem. Soc. Symp. Series "Proteins at Interfaces" (1995) 92–110.
156. MR Kanost, MS Boguski, M Freeman, JI Gordon, GR Wyatt and MA Wells, J Biol. Chem. 263 (1988) 10568–10573.
157. KD Cole, GJP Fernando-Warnakulasuriya, MS Boguski, M Freeman, JI Gordon, WA Clark, JH Law and MA Wells, J. Biol. Chem. 262 (1987) 11794–11800.
158. DR Breiter, MR Kanost, MM Benning, G Wesenberg, JH Law, MA Wells, I Rayment and HM Holden, Biochemistry 30 (1991) 603–608.
159. L Stryer (ed.), Biochemistry 3rd ed., Freeman, New York (1988).
160. I Rayment, WR Rypniewski, K Schmidt-Base, R Smith, DR Tomchick, MM Benning, DA Winkelmann, G Wesenberg and HM Holden, Science 261 (1993) 58–69.

Appendix A. Glossary

X^{20}	20 Mrad γ-ray cross-linking
T_t	Temperature for the onset of an inverse temperature transition
ΔS_t	Transition entropy
ΔH_t	Transition enthalpy
ΔT_t	Change in T_t
ΔG	Change in Gibb's free energy

$\Delta\mu$	Change in chemical potential
ΔpKa	Change in pKa
γ	Degree of ionization
ΔHPB	Change in hydrophobicity
kD	Kilo Dalton
N	Normality
M	Molarity

Bioelectrochemistry of Biomacromolecules
ed. by G. Lenaz and G. Milazzo†

CHAPTER 3
Electrochemistry of redox proteins

H. Allen O. Hill, C. Bruce Moore and David M. A. NabiRahni*

Inorganic Chemistry Laboratory, University of Oxford, Oxford, UK

1. Introduction

The use of electrochemical techniques has, for many years, been standard in fields ranging from solid state physics to organic chemistry. Their widespread use is a result of their versatility in the study of numerous thermodynamic and kinetic problems. For instance, the comparatively simple technique of D.C. cyclic voltammetry can yield a range of information such as: a) the formal reduction potentials of electroactive species; b) the number of electrons transferred in electrode reactions, and the rate constants of such reactions; c) whether or not the electrode reaction is diffusion controlled, or involves

* On sabbatical leave from: Pace University, Department of Chemistry, Pleasantville, NY 10570, USA.

adsorbed species; d) the chemical stability (or fate) of transformed species; and e) mechanistic information such as the detection of catalytic processes.

It is not surprising therefore that, for many years, electrochemical techniques have been employed in the study of redox active biological molecules. Many of the fundamental processes in nature rely upon the redox processes of constituent biomolecules. For example, cell respiration involves the stepwise oxidation of organic stubstrates *via* a cascade of redox reactions, while photosynthetic organisms participate in light-induced electron transport *via* chains of electron carriers such as cytochromes and iron-sulphur proteins [1].

The aim of this review is to survey the techniques commonly used and the results obtained in the electrochemical studies of proteins and enzymes, with emphasis on direct electron transfer, as well as the use of oxidoreductase enzymes in the development of biosensors.

2. Protein electrochemistry

Prior to 1977 it was believed that the direct electrochemistry of redox proteins was not possible [2]. Studies of this kind were rare, electrochemical work being confined to indirect electrochemistry involving the use of small electron carriers as electrochemical mediators. Three reasons were given for this failure:

a) The slow rates of electron transfer to or within proteins.
b) Adsorption at the electrode surface.
c) Slow diffusion of the protein.

2.1. Electron transfer to or within the protein

Proteins (and especially enzymes) are much larger, and more anisotropic, than most inorganic complexes. The distance between an electrode surface and the protein active centre will consequently be greater for many relative orientations of electrode and protein, and electron transfer rates may consequently be reduced.

Nevertheless, rates of electron transfer *between* many redox proteins are often comparatively fast. The origin of this effect lies in the amino acid residues which are present around the active sites of electron transfer proteins. In many cases, these result in charged "patches" on or near the surface of the protein. These are arranged in such a way as to promote favourable relative orientations of the two proteins: the redox sites within the proteins are brought as close together as possible. Dutton *et al.* [3] among others, have shown that the rates of electron transfer between protein redox sites are essentially distance dependent. Hence, due to the "recognition" of a protein by its redox partner, nearly every collision between the two proteins is a productive encounter.

This effect, however, causes electrochemistry to be critically dependent upon how the protein interacts with the electric field at the electrode-solution interface. The orientation of the protein with respect to the electrode surface, and so the ease of electron transfer, will depend upon the favourability of this interaction.

2.2. *Adsorption at electrode surfaces*

It is well known that proteins display a tendency to adsorb strongly at electrode surfaces [4, 5]. This may block any electron transfer to and from proteins in solution, whether or not the adsorbed protein retains its activity. The development of electrode surface modifiers has, however, led to the possibility of the "design" of electrode surfaces in such a way as to prevent such adsorption.

2.3. *Proteins have small diffusion coefficients*

The electrochemistry of species in solution is ultimately limited by the mass transport of reactant and product to and from the electrode. Proteins will usually be slower to diffuse in aqueous solution relative to smaller molecules. This is due both to their greater physical size and their possession, usually, of polar side chains that interact strongly with solvent molecules and ions. In the case of reversible (i. e. diffusion-controlled) electron transfer, the electrochemistry of proteins will therefore be limited by their inherently low rates of diffusion.

As will be seen in the following section, conventional "clean" metal electrodes generally provide very poor surfaces for the direct oxidation or reduction of redox proteins. This is not surprising in view of the generally accepted requirement [6], for electron transfer *between* protein molecules, of the fast and reversible formation of a precursor donor-acceptor complex. A plain metal electrode will present to the protein few of the molecular features required for the formation of such a precursor complex. Not only will this drastically reduce the rate of protein-electrode electron transfer, but may also lead to irreversible adsorption of the protein onto the electrode surface, coupled with conformational changes and subsequent loss of protein activity.

Perhaps it is not surprising, therefore, that it was not until 1977 that the first reports of direct reversible type electrochemistry for proteins were made. In these pioneering studies on horse heart *cytochrome c*, Eddowes and Hill [7] described how a gold electrode modified by 4,4′-bipyridyl enabled a well-defined one-electron Fe(III)/Fe(II)/in cytochrome *c* to be observed. Yeh and Kuwana [8] also reported direct electron transfer of the same protein at a tin-doped indium oxide surface.

3. The nature of the electrode surface

It would be desirable, therefore, to produce at the electrode surface the essential elements of macromolecular recognition, i.e. the complementary hydrogen bonding, salt-bridging and hydrophobic contacts that are important in a protein's interaction with its physiological partners. The electrode could, therefore, be tailored to the requirements of the chosen biological system. Such "tailoring" is now possible for most redox proteins *via* the use of a range of surface modifiers.

3.1. *Plain metal surfaces*

3.1.1. Mercury electrodes: The first detailed polarographic study of cytochrome *c* was reported by Betso, Klapper and Anderson [9]. Since that report, the reaction of cytochrome *c* at mercury electrodes has been extensively reviewed by Bowden *et al.* [10]. Several models for the electron transfer of cytochrome *c* at mercury electrodes have been proposed, but it is clear that the most important feature is that of fast irreversible adsorption of the protein onto the electrode surface. This would explain the concentration dependence of the electron transfer kinetics of cytochrome *c* at mercury electrodes, which ranges from polarographically reversible at low concentrations to quasi-reversible and irreversible at higher concentrations [9, 11].

3.1.2. Gold Electrodes: A "bare" gold electrode placed in aqueous solution does not provide a bare gold surface at which electron transfer can take place. Rather, depending on the applied potential, pH and other conditions, gold oxides may be formed on the electrode surface. Some electrochemical redox processes can be completely inhibited by even a small covering of such an oxide [12]. Trace impurities adsorbed from solution or from the protein, or absorption of the protein, itself can also partially block an electrode. It is now clear, for example from *in situ* STM studies [13], that roughening, annealing, gold oxide formation and certain dynamic events can strongly influence the interfacial electrochemical properties of the gold surface [14].

The most extensively studied redox protein at gold electrodes is cytochrome *c*. It is apparent that well-defined electrochemistry of reduction of cytochrome *c* at the thermodynamic potential does in fact occur at bare metal electrodes including gold [15, 16], *provided* the protein, buffer, and electrolyte are highly purified and scrupulously clean conditions are maintained. However, under many experimental conditions, a *modified* electrode is required in order to observe voltammetry at the thermodynamic potential at a gold electrode (i.e. passivation or blocking of the surface rapidly occurs). Ellipsommetric studies have shown that irreversibly adsorbed cytochrome *c* can convert to cytochrome *c* in the bulk solution, although in such a way as to produce only a reductive, not an oxidative, response [17].

3.1.3. Other plain metal electrodes: Platinum and silver, and to some extent nickel, have also been used in the study of redox proteins. Irreversible reduction of cytochrome *c* has been shown to occur at platinum [9], silver [18] and nickel [19] electrodes, although at an even slower rate than at plain gold [20]. It has been suggested that reduction of cytochrome *c* at platinum may be occurring *via* electrogenerated hydrogen [21]. Surface enhanced resonance Raman spectroscopy has been used by Cotton *et al.* [18] to demonstrate the effects of adsorption at silver surfaces on the electrochemistry of cytochrome *c*. Cytochrome *c* has also been shown to adsorb strongly onto platinum electrodes [22].

A number of metallic-type oxide conductors and semi-conductors have been shown to present suitable surfaces for redox protein electron transfer. Indeed, as mentioned earlier, Yeh and Kuwana [8] were among the first to report direct, reversible redox protein electron transfer. Here well-behaved cyclic voltammetry (and differential pulse voltammetry) of cytochrome *c* was obtained at a tin-doped indium oxide electrode. For such electrodes the amphoteric nature of the surface hydroxyl groups leads to a surface charge due to unequal adsorption of H^+ and OH^- [23–24]. Hence quasi-reversible voltammetric responses have been obtained for cytochrome *c*, azurin, ferrodoxin and plastocyanin at thin film ruthenium dioxide electrodes [25] under suitable solution conditions (suitable pH, ionic strength, etc.).

3.2. Graphite electrodes

Rapid and reversible electrochemistry is observed at certain forms of graphite more often, and for more redox proteins, than at any other unmodified electrode surface. The characteristics of graphite electrodes may be most easily understood in terms of those of pyrolytic graphite. This material is formed by the deposition of carbon from the vapour phase and has a crystalline structure most closely approximating that of the ideal graphite crystal [26]. Due to its highly ordered structure, pyrolytic graphite is anisotropic and provides two highly distinctive types of surface depending upon the plane along which it is cleaved [27]. The parallel or basal plane provides a hydrophobic surface, while the edge face (formed by cutting across the aromatic rings) displays additional surface structure (i.e. various CO functionalities). Comparison of the direct electrochemistry of cytochrome *c* (see Fig. 1), ferrodoxin and rubredoxin at the edge and basal planes of a pyrolytic graphite electrode [28] revealed the catalysis of heterogeneous electron transfer at the edge surface. The surface oxidised functional groups (negatively charged at pH 7.0) localized at the edge plane can be seen to impart considerable hydrophilicity and ionic character to the surface which aids binding of the protein at the electrode prior to electron transfer.

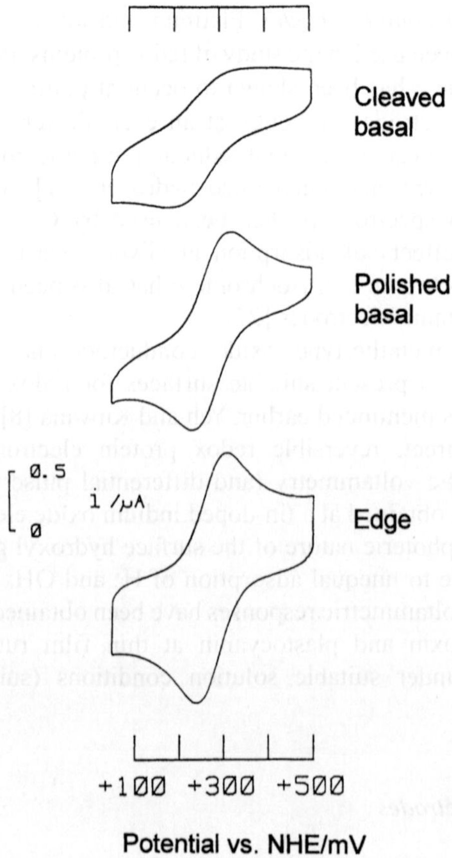

Fig. 1. Cyclic voltammograms (CVs) of a 200 μM solution of cytochrome *c* in 0.1 M phosphate buffer/0.1M KCl at EPG and BPG electrodes (scan rate 50 mVs^{-1}).

3.3. *Organic salt electrodes*

It was once believed that direct electrochemical oxidation of several flavoenzymes had been achieved [39] at electrodes made from the conducting organic salt, tetrathiafulvalinium tetracyanoquinodimethianide (or TTE; TCNQ). However, as Kulys has pointed out [30], the oxidation of the active centres of glucose oxidase and xanthine oxidase proceeds by way of mediation. Mediators are formed in the layer near the electrode surface due to a slight dissolution of the organic metal. Because the organic metal compounds oxidise the reduced flavin enzymes at a high rate, a small amount of dissolved compounds is enough to mediate the reaction with high efficiency.

3.4. *Modified electrode surfaces*

The realisation that suitably modified (i.e. functionalised) electrode surfaces could allow stable and reversible direct electrochemistry (without problems such as irreversible protein adsorption) came in the late 1970s. As stated earlier, the first such report came from Eddowes and Hill [7] in 1977, who found that essentially reversible cyclic voltammetry of horse mitochondrial cytochrome *c* could be achieved at a gold electrode onto

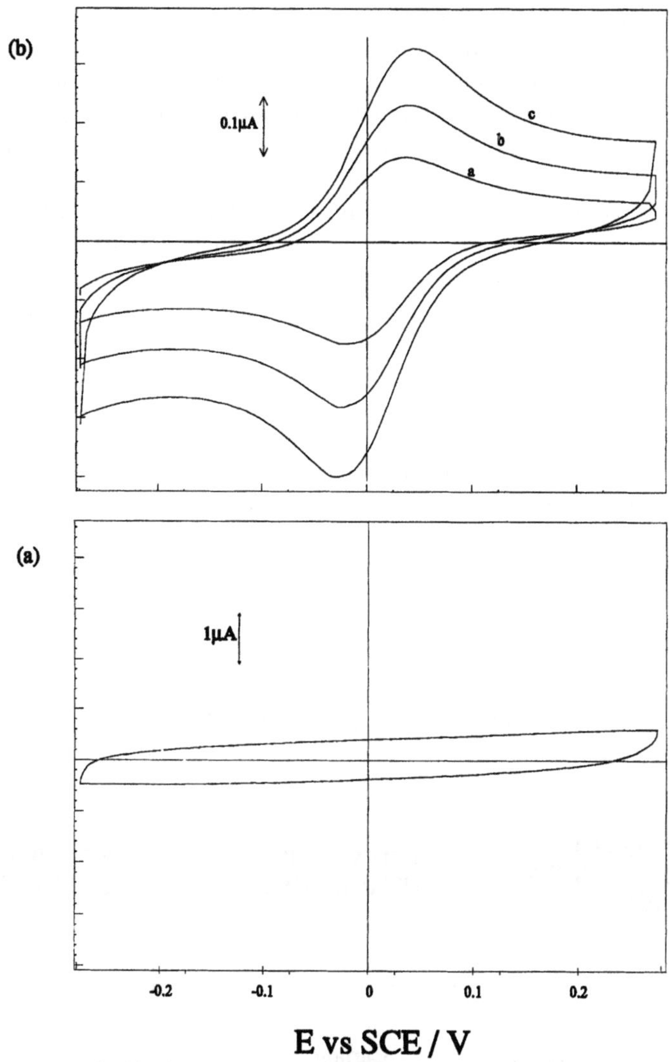

Fig. 2. (a) A typical background CV profile for a 0.1 M phosphate buffer/0.1M KCl solution at a gold electrode. Scan rate 20 mVs $^{-1}$. (b) The CV of a 100 μM cytochrome *c* in phosphate buffer at a modified gold electrode, at scan rates: (a) 5 mVs $^{-1}$, (b) 10 mVs $^{-1}$, and (c) 20 mVs $^{-1}$.

which was adsorbed the reagent 4,4′-bipyridyl. The result is shown in Fig. 2. The results indicate a one-electron process controlled by linear diffusion, the value of E^0 (+ 0.255 V), given by $(E_{pc} + E_{pa})/2$, being in good agreement with values determined by potentiommetry (+ 0.260 V). It was proposed [31] that the organic adsorbate allowed electron transfer to occur *directly* by providing functionalities at the electrode surface with which the protein could interact specifically and reversibly and thereupon donate or accept electrons rapidly. It was thus termed a *promoter* as opposed to a *mediator*, since the latter is considered to convey electrons in bulk solution.

Since these early studies, the field of modified electrodes has developed to such an extent that it is now possible to begin to design the electrode/electrolyte interface. A variety of approaches to the modification of electrode surfaces has been pursued; some of these are presented in Fig. 3. These methods have been reviewed [6, 14, 28] and so will only be summarised here.

In order to facilitate binding of the protein to the electrode surface in a manner that is conductive to electron transfer, there must be functional groups at the electrode-solution interface to which the protein can bind, either directly or indirectly. To achieve a near-reversible, diffusion-dominated electrode reaction, both the association and dissociation rates must be fast.

A more clearly defined picture of the structural requirements for successful promotion of electrochemical activity was provided by Allen *et al.* [32], who assessed a series of organic compounds in terms of their ability to promote the direct electrochemistry of cytochrome *c* at a modified gold electrode (Fig. 4). They are all bi- (or multi-)functional molecules of the type **X~Y**, such that **X** is a substituent which allows binding to the metal electrode surface (for example a pyridyl, phosphine, sulphydryl or thioether) and **Y** is functional group (for example a carboxyl or pyridyl) which interacts transiently with some part of the protein surface.

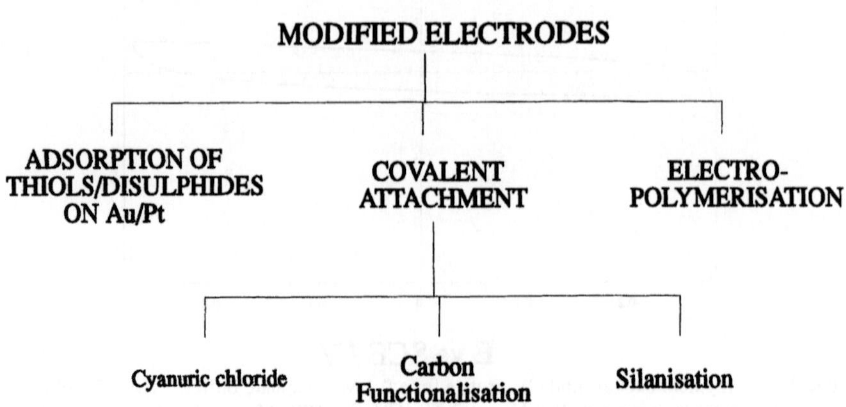

Fig. 3. Schematic representation of the various approaches to electrode modification.

With such a large selection of promoters, it is possible to achieve the electrochemistry of most redox proteins. Few small redox proteins fail to give acceptable electrochemistry, and results are often such that the half-wave potential corresponds to the thermodynamic potential of the protein. However, one must bear in mind that the experimental value reported is for protein-promoter-electrode interaction; if this is significant, for example when a metal ion or its complex is employed as an ancillary promoter, then a difference may be observed. Such redox potentials are not without interest: it may be that the redox potential of, for example, the complex [33] of magnesium with plastocyanin, may provide a more meaningful description of the complex of the protein in the photosynthetic unit than the redox potential of the "free" protein.

Modification of the electrode by adsorption of modifier from solution generally yields monolayer coverages. Similar coverages are achieved by the covalent attachment methods mentioned (i.e. cyanuric chloride [34], direct reaction with acidic and carbonyl functionalities on graphite electrodes [35],

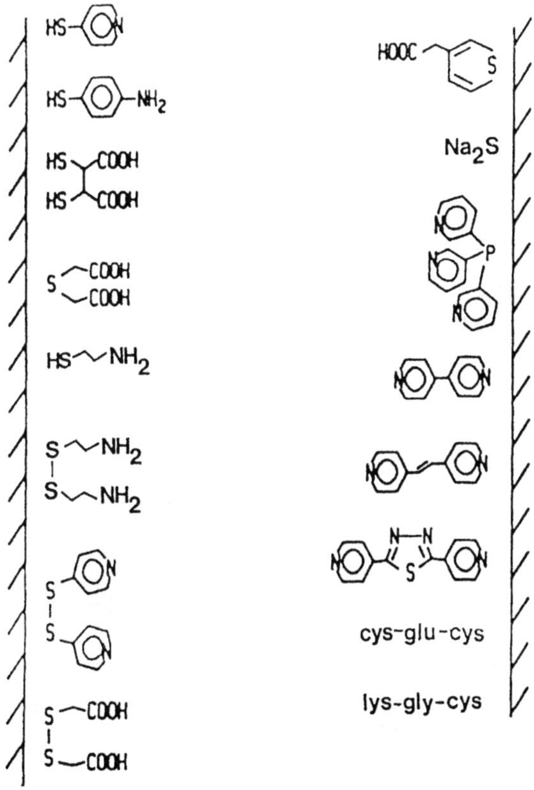

Fig. 4. Structures and proposed surface conformations of surface modifiers for the promotion of protein electrochemistry.

and silanisation [36]. The above studies depended upon surface modifiers which were irreversibly adsorbed onto the electrode surface. However, surface modification can also be achieved by species in solution which are only reversibly adsorbed. Such behaviour is exhibited principally by two types of cationic species: multivalent cations (such as Mg^{2+}) [32] and aminoglycosides such as neomycin and gentamycin [36].

Multivalent cations, such as Mg^{2+} and $Cr(NH_3)_6^{3+}$, have been shown to be capable of promoting the electrochemistry of proteins possessing negatively charged interaction domains at negatively charged surfaces. This phenomenon has been observed at both metal oxide and pyrolytic graphite electrodes, but has been most extensively studied at the latter [33]. The approach of the proteins, which carry an excess of negatively charged residues, to the nega-

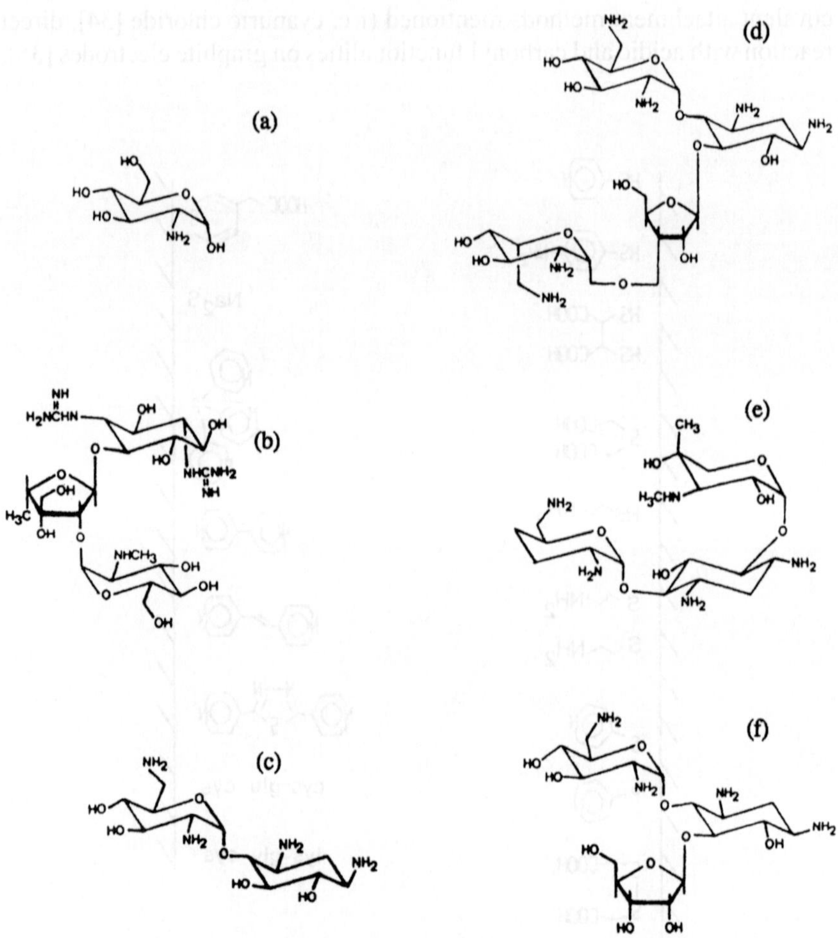

Fig. 5. Part of the aminoglycoside family: a) glucosamine, b) dihydroxystreptomycin, c) neamine, d) neomycin, e) gentamycin, f) ribostamycin.

tively charged electrode surface is assisted by the high positive charge density generated by multivalent cations adjacent to the electrode.

Such a mechanism has also been proposed to explain the action of the aminoglycosides mentioned. These molecules, as can be seen from Fig. 5, contain spatial arrangements of $-NH_3^+$ groups on a quasi-rigid framework. These act to promote the docking of, for example, cytochrome *c* peroxidase [37] at a pyrolytic graphite edge surface.

4. STM and AFM of electrode surfaces

It would be misleading to give the impression that these results were straightforward. Of course there were and are problems concerning the exact nature of the electrode surface. Attempts have been mode to examine these surfaces in detail, e.g. by ellipsometry [17], and more recently by Scanning Tunneling Microscopy (STM) [38]. An exact description in atomic terms still eludes us.

The discovery that STM could be operated under aqueous media [39–40] further opened the possibility that biological substances could be examined in a near-native environment. This, however, did not result in any dramatic breakthrough, mainly due to poor resolution and the fact that most biological substances are not sufficiently conducting. Atomic Force Microscopy (AFM) could overcome many of the shortcomings of STM, especially when employed in an non-contact mode so as not to disturb the soft proteins adsorbed on an electrode surface. Employing AFM, polymerisation of the blood protein fibrin from an aqueous solution onto a net that is the molecular fabric of blood clots has been demonstrated [41]. Physical parameters such as height, stiffness, and viscosity of live platelets that have undergone full activation before AFM imaging has also been demonstrated [42].

5. The microscopic model

Many early interpretations of the data from direct electrochemical studies of proteins were based on a macroscopic model of the electrode surface. This assumed that transport of redox active species to and from the electrode occurs *via* linear diffusion onto a planar disc surface; the entire surface is uniformly electroactive, and the heterogeneous electron transfer reaction can take place at any point on that surface.

However, under a host of experimental conditions, peak-shaped electrochemical responses could not be obtained. Often, when a promoter was used, rather "poor" electrochemistry resulted: when the electrode surface was modified in a variety of ways, the electrochemical results indicated "poor" electrochemistry when essentially no electrochemistry would have been expected. When the results were reassessed [43–46] in terms of the behav-

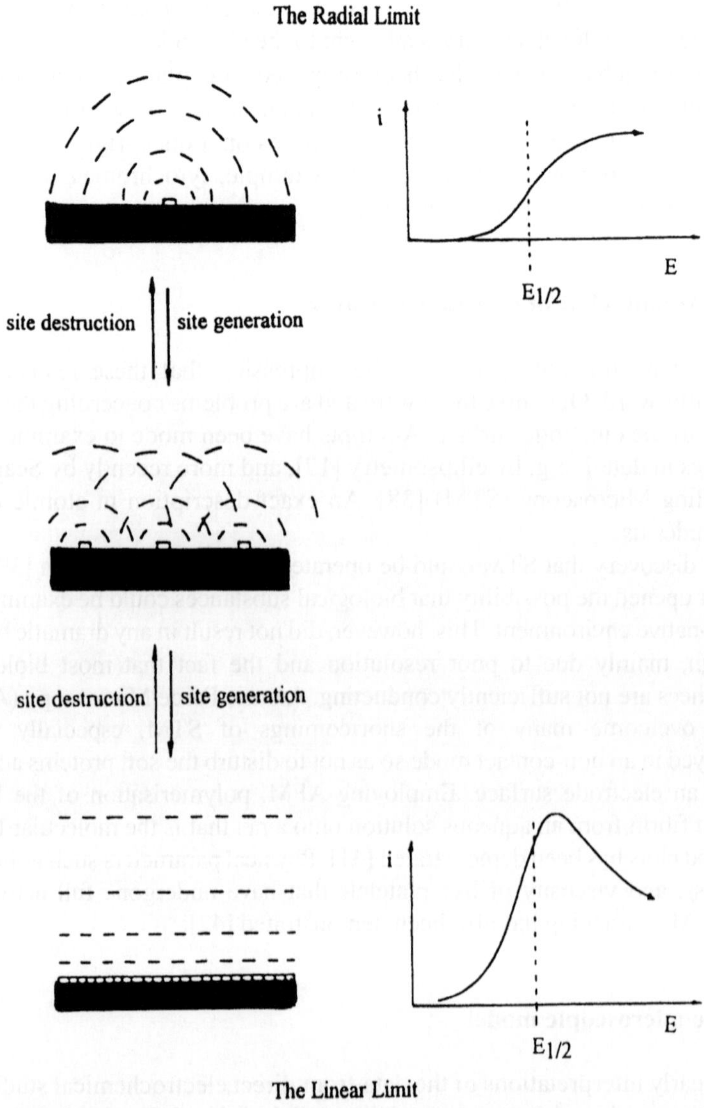

Fig. 6. Schematic representation of an electrode surface depicting the conversion of radial to linear diffusion as the density of the surface electroactive sites increase.

iour of the electroactive species at a microelectrode, i.e. one having dimensions of the order of 1 mm, they appeared consistent with rapid electron transfer rates. The difference was due to the effect of radial diffusion (Fig. 6) of the electroactive species to the microelectrode, as compared to the normally observed linear diffusion at macroelectrodes. The cyclic voltammogram of proteins whose diffusion to a microelectrode is radial in a manner resembles, albeit superficially, poor electrochemistry at a macroelectrode. Thus it was suggested that due to the non-uniformity of an electrode surface, *only an array of microscopically-sized sites are actually electroactive to protein molecules*, such that diffusion to these sites occurs in a radial manner. Detailed analysis of the experimental results showed that the electrochemistry of the proteins corresponded, under all conditions, to situations in which the heterogeneous electron transfer rate was very fast.

Another phenomenon which was elucidated by consideration of the radial diffusion was the time dependence of the electrochemistry, whether in situations corresponding to small amounts of promoter on the electrode surface, or with mixtures of promoters, or indeed of the mixture of a promoter with an inactive adsorbate. Even the gradual addition of increasing amounts of an adjunct promoter, for example $Cr(NH_3)_6^{3+}$, to a ferrodoxin solution produced a time-dependent electrochemistry. Obviously this could be associated with the rearrangement of, for example, the $Cr(NH_3)_6^{3+}$/protein complex, but many of the changes are now thought to be due either to the rearrangement of the promoters on the electrode surface, or to dissolution of the promoter followed by re-adsorption. In most cases, it is relatively easy to envisage how a mixture of promoter and inhibitor could, with time, rearrange to a set of "islands", corresponding to an array of microelectrodes on the electrode surface. There are a few examples of situations where the reverse happens [47], i.e. where two different promoters rearrange such that a protein's electrochemistry alters from a response corresponding to an electrode consisting of an array of microelectrodes to that characteristic of linear diffusion to a macroelectrode.

6. Protein – protein complexes

The majority of the so-called "redox proteins" have no biological function *alone*, but rather they are associated with other redox proteins, present in membranes in an electron transport assembly in pursuit of some metabolic process. It therefore appeared sensible to attempt to investigate protein-protein complexes electrochemically. The complexes formed between cytochrome *c* and plastocyanin, and cytochrome *c* and cytochrome b_5, have been studied [48].

The availability of electrodes which are "selective" towards the electrochemistry of the given proteins, for example, edge-plane pyrolytic graphite (EPG) for cytochrome *c*, or gold coated with the tri-peptide cys-lys-cys for plastocyanin, enabled the behaviour of the complexes to be understood. It appears that the protein, for example cytochrome *c*, which binds to the electrode surface at edge-plane pyrolytic graphite, acts as an adjunct promoter, holding plastocyanin or cytochrome b_5 in such an orientation that electron transfer to the latter can occur. This is consistent with zinc cytochrome *c* or indeed many proteins which have an overall charge opposite to that of plastocyanin, acting in the same manner.

There is one important point that emerged from these and related studies. The structures of the complexes must be considered as dynamic: they move with respect not only to each other, but to the electrode surface. The initial description of electron transfer at the electrode envisaged a static arrangement of the protein or proteins at the surface, but one must now consider a more mobile array of the protein(s) essentially moving in a lateral manner over the electrode surface until a configuration is reached where electron transfer to or from the surface is rapid. Such a dynamic view of the electrode/protein structure must involve motions within the protein. Indeed, recent studies on the electrochemistry of genetically engineered [49] variants of cytochrome *c*, and of its complexes with cytochrome b_5, indicate that subtle structural changes within cytochrome *c* affect the electrochemistry.

7. Direct enzyme electrochemistry

The procedures described so far have not, however, proved as successful in the probing of metalloenzymes. This is essentially due to the same problems encountered in studying protein electrochemistry, i.e. low diffusion coefficients of the biological "macro" molecule in aqueous solution as a result of the size of the molecules, rapid and often *irreversible* adsorption of the molecule at the electrode surface, and a redox centre "buried" within the protein and thus shielded to all but a few electrode-protein orientations. The electrochemical study of enzymes has proven more difficult, since they are generally substantially larger and more flexible than other proteins, and therefore each of the above problems is more apparent. In addition many enzymes are closely associated with membranes, and possess highly lipophilic surfaces, which further hinders aqueous studies.

Redox enzymes may be classified as being intrinsic or extrinsic in nature [50]. An intrinsic redox enzyme is one in which electron transfer associated with the catalyzed event is contained within the confines of the active site. An extrinsic redox enzyme requires an associated redox protein (cofactor) as part of the electron transfer process, and therefore there must exist one or more sites for its binding at the enzyme surface (and correspondingly one or more electron transfer pathways through the enzyme between redox centre and bonding site). It might be assumed such a binding site could allow interaction with an electrode surface and thus facilitate the heterogeneous electron transfer.

The first reports of the direct electrochemistry of enzymes involved studies with flavoenzymes, but since, in these instances, the prosthetic group is not covalently bound to the protein, and is often observed to dissociate, especially at an electrode surface, it is likely the prosthetic group was acting as a mediator in the electron transfer process. Similarly, some reports suggested that direct electron transfer of a number of enzymes had been observed at conducting organic salt electrodes, such as tetrahiafulvalinium tetracyanoquinodimethanide (TTF-TCNQ) [29]. It has now been generally accepted that the organic groups of the electrode possess sufficient solubility for slight dissolution away from the surface to occur in aqueous solution, and the resulting enzyme electrochemical reaction takes place via a mechanism involving heterogeneous catalysis by the organic salt [30, 51].

Probably the first genuine study of direct electrical communication with an enzyme involved the copper-containing laccase [52–53]. Lee *et al.* adsorbed fungal laccase A from *Polyporous versicolor* directly onto edge-oriented plane pyrolytic graphite electrodes, and observed the direct electroreduction of dioxygen catalyzed by the enzyme. A reversible electrochemical response by the adsorbed enzyme, in the absence of dioxygen, was produced in a solution saturated with 2,9-dimethylphenanthroline or 4,4′-bipyridine, which acted as promoters.

Extrinsic redox enzymes require an electron donating or accepting centre to bind to the enzyme, usually at a site not immediately near the active site. It is presumed that an electron transfer pathway exists between the site at which the ancillary protein binds and this active site region. Thus there is a part of the enzyme surface, no matter how small, across which the electron transfer can occur between electrode and enzyme. One example of such a system is *p*-cresol methylhydroxylase (PCMH). Electron transfer has been exhibited directly and reversibly at an electrode by PCMH [54]. The enzyme is a flavoxytochrome *c* of 115 kDa; it catalyses the oxidative hydroxylation of *p*-cresol to *p*-hydroxybenzyl alcohol and subsequently to *p*-hydroxybenzaldehyde. It is an $\alpha_2\beta_2$ tetramer, with two subunits containing covalently bound FADs, the remaining pair containing *c*-type heme groups. A blue copper protein is suspected [55] to be its *in vivo* electron acceptor. It has been shown that direct electron transfer between PCMH and an edge plane graphite electrode can be achieved in the presence of a range of electroinactive *cationic* species (see Table 1). Based on the structural information on PCMH [56], the most likely interaction site between enzyme and electrode is the negatively charged region of propionic acid groups on the cytochrome subunit [57].

The direct electrochemistry of methylamine dehydrogenase (MADH) from the bacterium *Methylbacterium extorquens* AM1 has recently been investigated in our laboratory [58]. MADH exhibits a pI value of 5.2, and is thus negatively charged at neutral pH. A quasi-reversible (Ep 58 mV) electrochemical response due to direct electron transfer with the enzyme was obtained at an EPG electrode, as depicted in Fig. 7. However, on addition of an aminoglycoside such as neomycin, ribostamycin, streptomycin or tobramycin, much greater faraday currents were obtained. The increase in current was further enhanced when the substrate (methylamine) was added to the working solution to a concentration of 80 μM; this was due to the catalytic turnover of the enzyme.

The direct electrochemistry of more and more enzymes is now being reported [59]. For a recent review, see Hill and Hunt [60].

Table 1. Range of compounds used as promotors for the electrochemistry of PCMH

Promotor	Catalytic current (μA) [a]	Promotor charge at pH 7.0	Concentration of promoter used (mM)
Aminoglycosides			
Glucosamine	Nil	+1	Up to 10
Dihydroxy -streptomycin	2.3	+3	2
Neamine	2.3	+4	2
Neomycin	4.2	+6	0.2
Gentamycin	3.8	+4	2
Ribostomycin	3.0	+4	2
Polyamines			
Spermidine	1.1	+3	10
Spermine	3.8	+4	2
Cyclam	Nil	+2	Up to 20
Diammac	3.5	+4	4
Inorganic cations			
K^+	Nil	+1	Up to 20
Mg^{2+}	3.4	+2	4
$Cr(NH_3)_6^{3+}$	3.5	+3	2

[a] The catalytic current was obtained using a scan rate of 10 mVs^{-1}, at pH 7.0, in the presence of *p*-cresol methylhydroxylase (1 μM) and *p*-cresol (2 mM) at an edge plane graphite electrode.

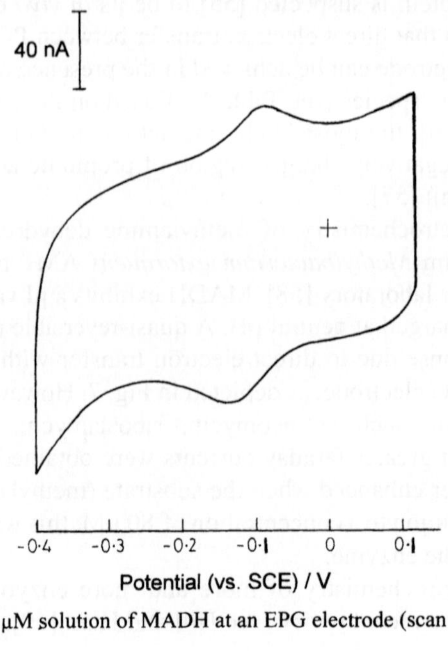

Fig. 7. A CV of a 40 μM solution of MADH at an EPG electrode (scan rate 5 mVs^{-1}).

8. Biosensors

Thirty years have elapsed since the concept of enzyme immobilisation for analytical purposes was first introduced by Clark and Lyons [61]. Biosensor research has since become a very active area. A recent review of electrochemical based biosensor research [62] cited over 75 examples of its use in clinical analysis during the years 1991 and 1992. However, the widespread acceptance and application of such devices has yet to be fully realized. This is due in part to instability, lack of adaptability, and inconsistency in the mass fabrication of such devices.

Electrochemical methods based on enzyme electrodes frequently involve monitoring either the consumption of oxygen using a Clark electrode, or the formation of hydrogen peroxide using a platinum electrode. The latter requires the application of a cathodic potential, typically +650 mV *vs.* Ag/AgCl, at which species such as ascorbic acid, acetaminophen, uric acid, and glutathione, are also electroactive [63]. The effect of such potential interferences, present in considerable concentrations in physiological fluids, renders this approach to the assay of clinical samples difficult without any pre-treatment.

The co-immobilisation of glucose oxidase and horseradish peroxidase has been useful for eliminating major interferences [64–65]. Glucose oxidase-substituted polypyrrole-coated electrodes offered a greatly improved stability [66]. The coupling of a microdialysis probe and a glucose biosensor for *in vivo* monitoring of glucose has recently been reported [67]. Implantable glucose sensors have also been evaluated *in vivo* [68], and *in vitro* [69–70], with good stability over several days. The fluctuation of dioxygen concentration as the physiological electron acceptor for oxidases can be circumvented by choosing an alternative biological (e.g. cytochrome *c*) or non-biological (e.g. ferrocene monocarboxyl acid) electron acceptors. The interest in developing micro-sensing devises for *in vivo* applications and in health care is also growing rapidly.

8.1. Glucose analysis of whole blood

The transformation of a ferrocene-based enzyme electrode for glucose from laboratory bench to a hand-held commercial device has been achieved, and the instrument marketed world wide as the ExacTech glucose meter [71] (Fig. 8). The meter comprises a pen-sized potentiostat with an LCD display for direct glucose reading. The electrode, a disposable test strip, incorporates and immobilised layer of glucose oxidase and 1,1′-dimethyl-3-(1-hydroxy-2-aminoethyl)ferrocene, coated with a hydrophilic membrane to attract a drop of blood for analysis. Each strip also contains its own reference electrode. A glucose reading is produced in 20 sec.

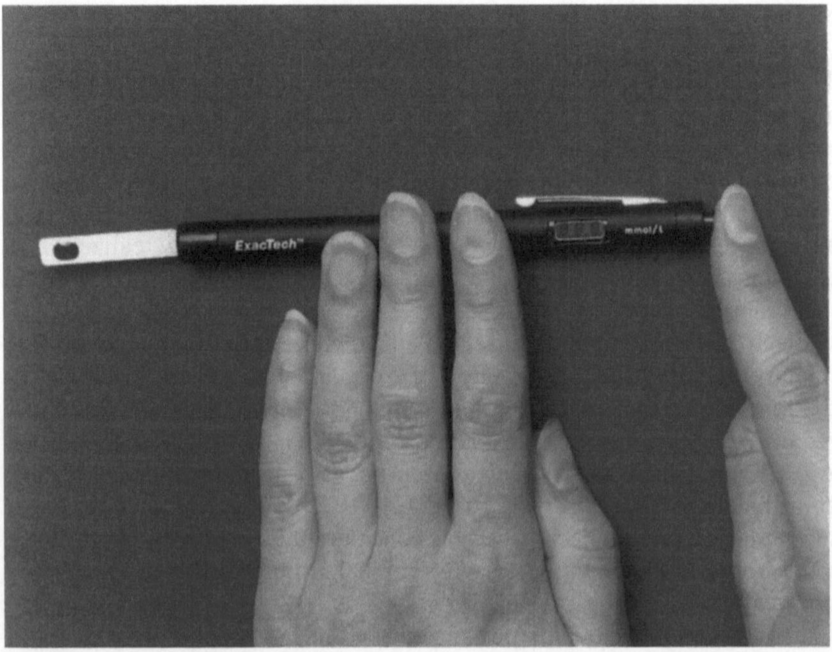

Fig. 8. The Exactech glucose pen.

8.1. Other examples

The physical principles behind the glucose biosensor may be extended to other enzymes which interact with a mediator. Two possible examples would be creatine kinase assay and cholesterol sensing.

Creatine kinase (ATP: creatine *N*-phosphotransferase, EC 2.7.3.2) catalyses the reversible transfer of a phosphate residue form adenosine-5′-triphosphate (ATP) to creatine. Elevated levels of creatine kinase (CK) in blood are linked with acute myocardial infarction. Consequently, the measurement of CK activity is of vital importance in post-myocardial infarction patients. Assays based on the strategy outlined in Fig. 9 have been developed for CK. Here, the rate of current decrease at the electrode will be proportional to the rate of creatine phosphate consumption, from which CK activity may be estimated [72]. A typical calibration curve for creatine kinase *vs.* glucose consumption is shown in Fig. 10.

Three possible ferrocene-mediated electrochemical schemes for the determination of cholesterol have been proposed [73], as outlined in Fig. 11. Each of these may be converted into a simple biosensor by immobilisation of the enzymes and mediators involved. The dehydrogenase-based assay exhibited extreme sensitivity to inhibition by surfactants, which are required to free cholesterol esters form lipoprotein complexes in whole

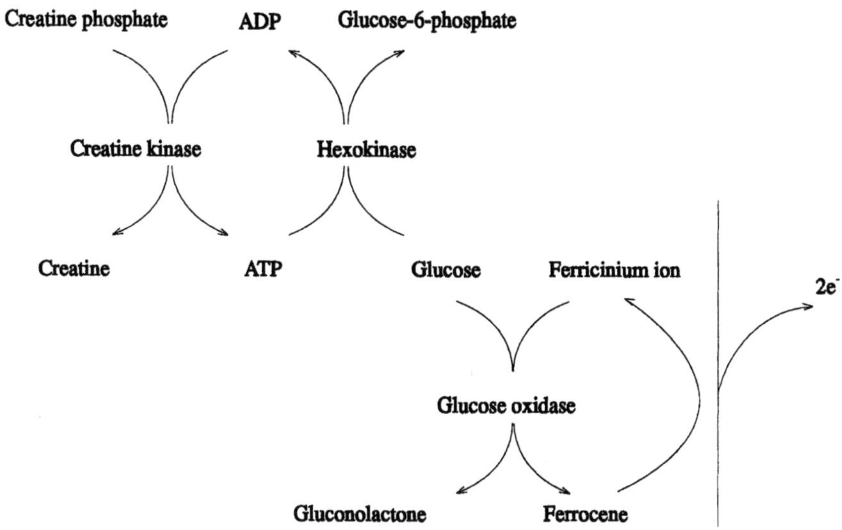

Fig. 9. Reaction scheme depicting the creatine kinase assay *via* ferrocene-mediated glucose oxidase.

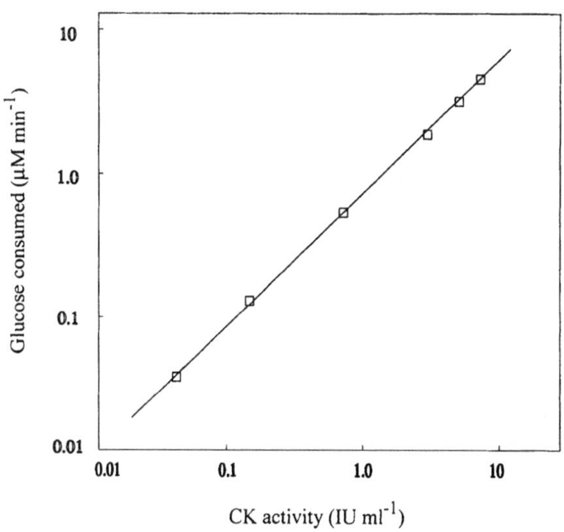

Fig. 10. Calibration curve for creatine kinase *vs* glucose consumption.

sera. The COD assay was limited by the very narrow pH range of the enzyme (pH 4.5–5.5); this is unsuitable for clinical samples (~ pH 7.4). The peroxidase-linked assay proved to be the most reliable [73], exhibiting linearity over the range 0–10 mM cholesterol. Results for cholesterol in serum as compared to a standard clinical method yielded a correlation coefficient of 0.993. However, so far neither of the creatine kinase or cholesterol systems have resulted in commercial devices.

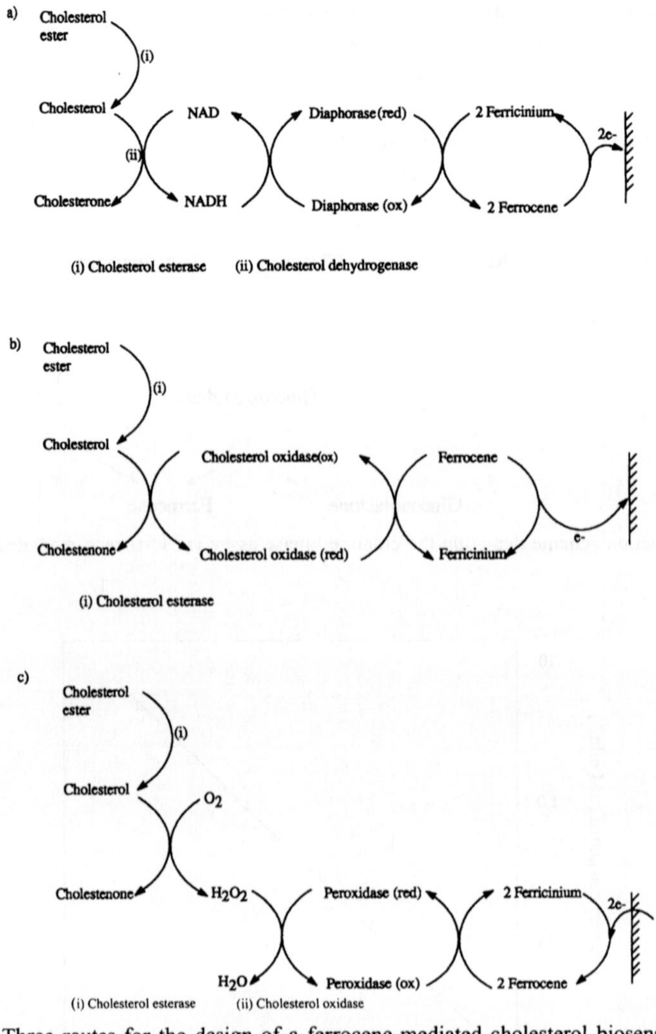

Fig. 11. Three routes for the design of a ferrocene-mediated cholesterol biosensor. (a) dehydrogenase assay; (b) oxidase assay; (c) peroxidase/oxidase assay.

9. Conclusion

The techniques which have been developed over the past fifteen years in the study of protein electrochemistry are now being applied in the study of enzyme systems. Although electrochemical communication with over twenty metalloenzymes has been achieved, most studies in this field involve the development of biosensors, where the enzyme is immobilised at the electrode surface and provides a substrate-dependent current response. The application of electrochemical techniques towards providing kinetic information on enzyme systems is only slowly developing towards the point at which it is applicable in protein systems. In the light of structural, conformational, and kinetic similarities between proteins, enzymes, and monoclonal antibodies, we expect the latter to be the new focus of electrochemical investigations with in the next decade.

Acknowledgments

CBM would like to thank the Medical Research Council and Medisense Inc. for financial support. DMANR would like to thank Pace University for the Kenan Scholarly Award.

References

1. JL Dreyer, Experientia 40 (1984) 653–675.
2. G Dryhurst, KM Kadish, F Scheller and R Renneberg, in *Biological Electrochemistry*, Academic Press New York (1982) Vol 1, chapter 7.
3. MR Gunner and PL Dutton, J. Am. Chem. Soc. 111 (1989) 3400–3412.
4. Many relevant articles are to be found in: JL Brash and TA Horbett (eds), *Proteins at Interfaces*, ACS Adv. Ser. 343 (1987).
5. FA Armstrong, HAO Hill and NJ Walton, Quart. Rev. Biophys. 18 (1986) 261–322.
6. FA Armstrong, HAO Hill and NJ Walton, Acc. Chem. Res. 21 (1988) 407–413.
7. MJ Eddowes and HAO Hill, J. Chem. Soc., Chem. Commun. 45 (1977) 771.
8. P Yeh and T Kuwana, Chem. Lett. (1977) 1145.
9. SR Betso, MH Klapper and LB Anderson, J. Am. Chem. Soc. 94 (1972) 8197–8200.
10. EF Bowden, FM Hawkridge and HN Blount, *Comprehensive Treatise on Electrochemistry*, Plenum, New York, 1985, Vol 10.
11. PA Serre, J Haladjian and P Bianco, J. Electroanal. Chem. 122 (1981) 327–330.
12. U Oesch and J Janata, Electrochim. Acta 28 (1983) 1247–1253.
13. DJ Trevor, C Chidsey and DN Loiacono, Phys. Rev. Lett. 62 (1989) 929–932.
14. AM Bond and HAO Hill, in *Metal Ions in Biological Systems*, H Sigel and A Sigel (eds) 27 (1991) 431–494.
15. DE Reed and FM Hawkridge, Anal. Chem. 59 (1987) 2334–2339
16. SC Sun, DE Reed, J Cullison, LH Rickard and FM Hawkridge, Mikrochim. Acta III (1988) 97–104.
17. D Elliott, A Hamnett, OC Lettington, HAO Hill and NJ Walton, J. Electroanal. Chem. 202 (1986) 303–314.
18. TM Cotton, SG Schultz and RP Van Duyne, J. Am. Chem. Soc. 102 (1980) 7960–7963.
19. BA Kuzentov, Bioelectrochem. Bioenerg. 8 (1981) 681–695.
20. MR Tarasevich and VA Bogdamovskaya, Bioelectrochem. Bioenerg. 3 (1976) 589–597.
21. T Kono and S Nakamura, Bull. Agr. Chem. Soc. Japan 22 (1958) 399–405.
22. J Haladjian, P Bianco and PA Serre, Bioelectrochem. Bioenerg. 6 (1979) 555–561.
23. EF Bowden, FM Hawkridge and HN Blount, J. Electroanal. Chem. 161 (1984) 355–376.
24. S-C Sun, DE Reed, JK Cullison, LH Rickard and FM Hawkridge, Mikrochim. Acta III (1988) 97–104.
25. MA Harmer and HAO Hill, J. Electroanal. Chem. 170 (1984) 369–375; MA Harmer and HAO Hill, J. Electroanal. Chem. 189 (1985) 229–246.
26. PL Walker, Am. Scientist. 50 (1962) 259–263.
27. RE Panzer and P Elving, Electrochim. Acta 20 (1975) 635–646.
28. FA Armstrong, HAO Hill and NJ Walton, Quart. Rev. Biophys. 18 (1986) 261–322.
29. WJ Albery, PN Bartlett, M Bycroft, DH Craston and BJ Driscoll, J. Electroanal. Chem. 218 (1987) 119–126.
30. JJ Kulys, J. Biosensors 2 (1986) 3–13.
31. MJ Eddowes and HAO Hill, in KM Kadish (ed), *Electrochemical and Spectroelectrochemical studies of Biological redox components*. ACS Adv. Chem. Ser. ACS, Washington; 201 (1982) 173–191.
32. PM Allen, HAO Hill and NJ Walton, J. Electroanal. Chem. 178 (1984) 69–86.

33. D Datta, HAO Hill and H Nakayama, J. Electroanal. Chem. 324 (1992) 307–323.
34. AWC Liu, P Yeh and T Kuwana, J. Electroanal. Chem. 84 (1977) 411–419.
35. PR Moses, L Weir and RW Murray, Anal. Chem. 47 (1975) 182–186.
36. RW Murray in *Electroanalytical Chemistry*, AJ Bard (ed) Marcell Dekker, New York, Vol 13, (1984).
37. FA Armstrong and AM Lannon, J. Amer. Chem. Soc. 109 (1987) 7211–7212.
38. AJ Mayne, AR Avery, J Knall, TS Jones, L Pinheiro, TRI Caraldi, HAO Hill, GAD Briggs, JB Pethica and TWH Weinberg, Farad./Disc. 94 (1992) 199–212.
39. R Sonnenfeld and PK Hansma, Science 232 (1986) 211–213.
40. HY Lin, FRF Fan, CW Lin and AJ Bard, J. Am. Chem. Soc. 108 (1986) 3838–3839.
41. B Drake, CB Prater, AL Weisenborn, SAC Gould, TR Albrecht, CF Quate, et al., Science 243 (1989) 1586–1589.
42. M Radmacher, RW Tillman, M Fritz and HE Gaub, Science 257 (1992) 1900–1905.
43. AM Bond and HAO Hill in *Metal Ions in biological Systems*, H Sigel and A Sigel (eds) 21 (1991) 431–494.
44. FA Armstrong, AM Bond, HAO Hill, ISM Psalti and CG Zoski, J. Phys. Chem. 93 (1989) 6485–6493.
45. FA Armstrong, AM Bond, HAO Hill, BN Oliver and ISM Psalti, J. Amer. Chem. Soc. 111 (1989) 9185–9189.
46. AM Bond, HAO Hill, DJ Page, NJ Walton and ISM Psalti, Eur. J. Biochem. 191 (1990) 737–742.
47. HAO Hill and GA Lawrence, J. Electroanal. Chem. 270 (1989) 309–318.
48. S Bagby, PD Barker, LH Guo and HAO Hill, Biochem. 29 (1990) 3213–3219.
49. A Burrows, LH Guo, HAO Hill, GMcClendon and F Sherman, Eur. J. Biochem. 202 (1991) 543–549.
50. FA Armstrong, HAO Hill and NJ Walton, Acc. Chem. Res. 21 (1988) 407–413.
51. JJ Kuls, Enz. Microbiol. Technol. 3 (1981) 344–352.
52. CW Lee, HB Gray, FC Anson and BG Malmstrom, J. Electroanal. Chem. 172 (1984) 289–300.
53. MR Tarasevich, AI Yaropolov, VA Bogdanovskaya and SD Varfolomev, Bioelectrochem. Bioenerg. 6 (1979) 393–397.
54. LH Guo, HAO Hill, GA Lawrence, GS Sanghera and DJ Hopper, J. Electroanal. Chem. 266 (1989) 379–390.
55. MJ Causer, DJ Hopper, WS McIntire and TP Singer, Biochem. Soc. Trans. 12 (1984) 1131–1132.
56. FS Mathews, Z. Chen and HD Bellamy, Biochemistry, in press.
57. LH Guo and HAO Hill, Adv. Inorg. Chem. 36 (1991) 341–375.
58. J Butler, HAO Hill, NI Hunt and Li Wang, in: Proc. 5th Inter. Symp. Redox Mech. Interfacial Prop. Mol. Biol. Import. Ed. FA Schultz and I. Tariguchi, 93–II (1993) 1–8.
59. A Sudeta, R Cammack, J Weiner and FA Armstrong, Biochemistry 32 (1993) 5455–5465.
60. HAO Hill and NI Hunt in *Methods in Enzymology*, Acad. Press New York 227 (1993) 501–522.
61. LC Clark and CA Lyons, Ann. NY Acad. Sci. 102 (1962) 29–45.
62. J Wang, Anal. Chem. 65 (1993) 450–453 R.
63. MA NabiRahni, G Palleschi, GJ Lubrano, JN Ngwaibi and GG Guilbault, Anal. Biochem. 159 (1986) 114–121.
64. R Maiden and A Heller, J. Am. Chem. Soc. 113 (1991) 9003–9004.
65. GJ Patterson, Electroanalysis 3 (1991) 741–742.
66. S Wolowacz, B Yon and C Lowe, Anal. Chem. 64 (1992) 1541–1543.
67. D Mascone, M Pasini and M Mascini, Talanta 8 (1992) 1039–1044.
68. D Moatti, F Capton, V Poinout, G Reach, D Bindra and Y Zhan et al., Diabetologia 35 (1992) 224.
69. G Shaw, D Claremonte and J Pickup, Biosens. Bioelectron. 6 (1991) 401.
70. *Biosensors*, AFP Turner, I Karube and GS Wilson (eds) OUP, Oxford (1989).
71. Manufactured by MediSense Inc., Abington, UK, and Cambridge, MA, USA.
72. HAO Hill and GS Sanghera, *Biosensors*, AEG Cass (ed), IRL Press, Oxford (1990).
73. JE Frew, MA Hammer, HAO Hill and SI Libor, J. Electroanal. Chem. 201 (1986) 1–12.

Bioelectrochemistry of Biomacromolecules
ed. by G. Lenaz and G. Milazzo†

CHAPTER 4
Applications of voltammetric methods for probing the chemistry of redox proteins

Fraser A. Armstrong

Inorganic Chemistry Laboratory, University of Oxford, UK

1. Introduction

Despite their recognised excellence for observing and analyzing the redox thermodynamics and kinetics of small molecule systems, voltammetric methods are relatively unfamiliar tools for investigating redox centres in proteins. Prior to breakthroughs achieved in the 1970s, it was widely believed that proteins were unable to undergo facile direct (unmediated) electron exchange with an electrode [1]. Presumed complications in applying *direct* electrochemical methods, i.e. irreversible adsorption with denaturation, and sluggish electrode kinetics due to the protein's occluded active site(s), appeared serious obstacles to progress, and interest was focused instead on *indirect* methods of electrochemical measurement. Thus potentiometry, with employment of small redox mediators to facilitate electrochemical equilibration during spectroscopically monitored titrations of redox centres, has become the established way to determine reduction potentials [2]. The first studies to demonstrate conclusively that voltammetry might be as viable for proteins (MW > 10 kDa) as for small molecules were carried out with cytochromes. Niki and co-workers found that cytochrome c_3, an active and robust electron-transport protein containing four heme groups, displays reversible diffusion-controlled electrochemistry at a mercury electrode [3], while the research groups of Hill and Kuwana independently achieved reversible, diffusion-controlled electrochemistry of mitochondrial cytochrome *c*, respectively at 4,4′-bipyridyl modified gold and at metal oxide electrodes [4, 5]. From these relatively simple experiments, formal reduction potentials and diffusion coefficients were easily measurable and observed to agree with values obtained by other methods. In identifying these experiments as having particular significance, I have taken into consideration that no potentially mediating small molecules were present (4,4-bipyridyl does not have suitable redox properties) and that the protein sample remained undamaged by contact with the electrode during the course of extensive electrolysis. For example, it was ascertained that bulk solution samples could be electrolytically cycled without mediators and with no loss of activity [6].

Extensive efforts have been made by numerous groups to determine and rationalize the various and puzzling requirements for direct electron transfer, to find other useful experimental methods and configurations, to extend voltammetric methods to other proteins including enzymes, and to apply the methods to solve problems for which they might be uniquely suited. Detailed studies of the nature of the electrode-protein interface have remained largely focused on cytochrome *c*, while the methods have been extended to other heme proteins, small Fe-S and Cu proteins, and even to multi-centred enzymes with molecular masses in excess of 100 kDa. Problem-solving has ventured into the areas of Fe-S cluster characterization (monitoring complex reactivities and labile centres) and to elucidating some subtle catalytic properties of redox enzymes.

This chapter will focus on problem solving. From a practical standpoint (and once conditions have been identified for eliciting direct electrochemistry) there are several advantages of voltammetric methods over potentiometry, even for the task of determining reduction potentials of stable and otherwise unreactive redox systems. One is that it becomes feasible to measure redox activity over a wide and continuous potential range, because measurement is no longer restricted by the properties of the mediator(s). For example the potential window available with carbon electrodes at pH 7 is typically between −1 and +0.7 V, and can be extended in either direction under appropriate conditions. This aspect is valuable for investigating systems that operate, even if only transiently, outside the normal thermodynamic range of water. Other advantages are speed, sensitivity, and removal of the need to rely on some spectroscopic characteristic of a particular oxidation state, which in many cases can only be measured indirectly, i.e. as frozen samples in EPR tubes. While I will endeavour to mention examples in which these efficiency-motivated aspects have been the foremost consideration, I will be seeking particularly to suggest and illustrate possibilities for exploiting the more unique analytical and mechanistic capabilities of voltammetric methods. We will thus be concerned mainly with applications to the study of complicated chemical behaviour associated with unstable and labile active sites, including those of enzymes, and to the mechanisms of their reactions. It will be apparent that voltammetric methods do indeed reveal properties that are not detected or appear chaotic when approached by other methods. We can also recognize easily that the voltammetric response for a particular centre is equivalent to a spectroscopic transition, and that a redox couple observed in a voltammogram is effectively a *signal* which reports on the presence of a species and its properties. We shall *not* be concerned with opinions regarding the kinetics and mechanisms of interfacial electron transfer, only that this process should be sufficiently reversible so as not to be a rate-determining factor.

The main body of the chapter divides into two parts. In Section 2, we consider the different configurations and styles of voltammetric experiments, outlining some of the advances being made to obtain reproducible, useful electrochemistry and a sounder understanding of the protein-electrode interface, and then predicting how the methods should enable us to examine complex coupled reactions. In Section 3, we will discuss some experiments that illustrate the use of the methods for solving problems (and revealing new ones) in bioinorganic chemistry and enzymology. Emphasis will be on providing a pictorial outline and thus technical or mathematical details will not be given; rather, brief assessments will be made of the various advantages and disadvantages for protein systems. For detailed descriptions of the voltammetric methods, the reader is referred to recent literature articles as well as several excellent texts [7–10]. This chapter is not intended to be a review, and thus a naturally biased proportion of the studies that I shall describe in greater detail stem from the efforts of my own colleagues.

2. Voltammetric methods for studying redox proteins

2.1. Outline and comparison of principal methods

Despite the increasing use of computer-based pulse methods for analytical applications, cyclic voltammetry currently remains the most popular tech-

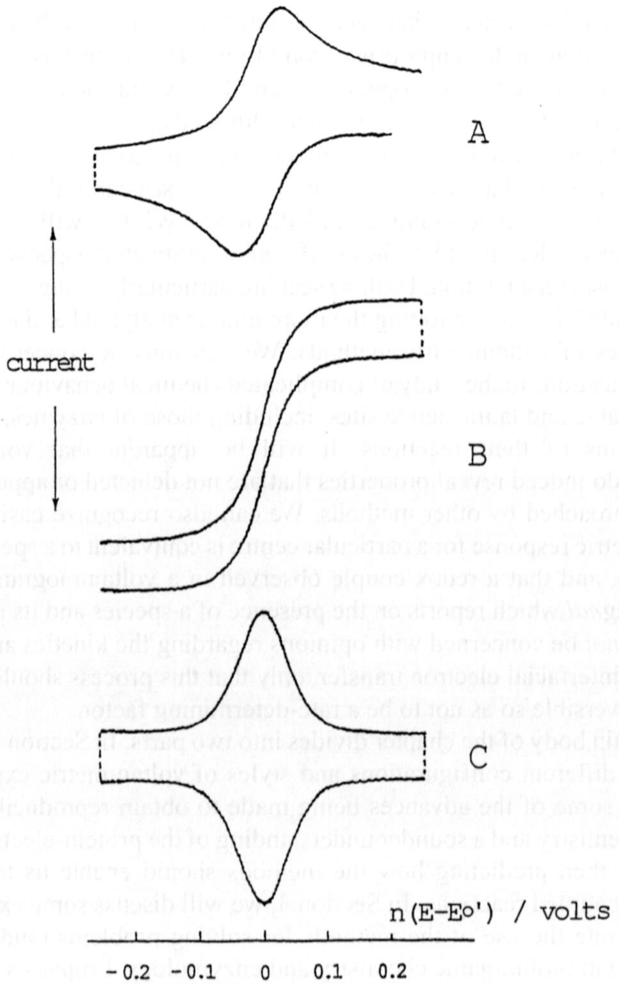

Fig. 1. Some limiting forms and analytical parameters of reversible cyclic voltammograms, as they typically appear with charging current. (A) Diffusing redox couple under conditions in which mass transport of redox-active molecules between bulk solution and a stationary planar electrode occurs by linear diffusion. (B) Diffusing redox couple under steady-state conditions, i.e. at a microelectrode or a rotating disc electrode. (C) Immobilized redox couple.

nique for making preliminary investigations of a redox system, and most of the following discussions will refer to its use. As will be emphasized throughout, the exploitation of voltammetric methods in order to visualize and quantify intrinsic properties of protein redox centres depends to a great extent upon taking into consideration the effects that arise from differing modes of mass transport of the redox active molecules. This can be illustrated very simply by reference to the representative cyclic voltammetric current responses (waveforms) shown in Fig. 1. Some relevant equations and relationships are summarized in Table 1.

Table 1. Electrochemical equations in common use

Cottrell equation; relates current to time following a potential step.

$$i(t) = \frac{nFAD_o^{1/2}C}{\pi^{1/2}t^{1/2}}$$

Randles-Sevcik Equation; in cyclic voltammetry at a planar macro-electrode, this relates peak current to scan rate (inverse time).

$$i_p = 0.4463\left(\frac{nF}{RT}\right)^{1/2} nFAD\,v^{1/2}C$$

Heyrovsky-Ilkovic Equation: relates current to potential for steady-state reversible system.

$$\log\left\{\frac{(i_{lim} - i)}{i}\right\} = \frac{nF(E - E_{1/2})}{2.3\,RT}$$

Cyclic voltammetry for immobilized redox couple. This equation relates current to scan rate and n.

$$\Psi = \frac{i}{(F^2/RT)\,n^2\,v\,\Gamma_T}$$

Butler-Volmer equation (the version shown is for immobilized redox couple); the equation relates current to applied potential and rate constant.

$$i_E = nFAk_s[\Gamma_O \exp\{\alpha nF(E - E^{0'})/RT\} - \Gamma_R \exp[(1-\alpha)\,nF(E - E^{0'})/RT\}]$$

Levich equation; relates current to electrode rotation rate

$$i_{Lev} = 0.62\,nFAD^{2/3}\omega^{1/2}\gamma^{-1/6}C$$

Michaelis-Menten equation (electrochemical version for immobilized enzyme); relates catalytic current to concentration of substrate.

$$i_{cat} = \frac{nFA\,\Gamma\,k_{cat}\,C}{C + K_M}$$

The familiar shaped voltammogram (Fig. 1 A) with oxidation and reduction currents reaching peak values at $E_p = E^0 \pm 29.5/n$ mV (thus $\Delta E_p = 59/n$ mV at 25 °C) is the limiting (ideal) result for reversible, diffusion-controlled electrochemistry at a stationary, planar electrode. The theory described in the classic paper by Nicholson and Shain also predicts that the peak current is proportional to the square root of the scan rate [11]. The same reversible, diffusion-controlled electrochemistry can also be measured under steady-state conditions [12, 13]. This typically involves either rotating the electrode to achieve hydrodynamic control over mass transport, or using a micro-electrode at which a radial diffusion field is achieved at sufficiently slow scan rates. The resulting voltammogram (Fig. 1 B) is *sigmoidal* and the current is *independent* of scan rate (i.e. time). The half-wave potential $E_{1/2}$ is essentially equal to the formal reduction potential and is independent of the direction of sweep. By contrast, the voltammogram observed for a reversible, uncoupled electrochemical reaction of a molecule *immobilized* on an electrode has the form shown in Fig. 1 C, which is independent of electrode dimensions or rotation rate. The maximum (peak) current is proportional to scan rate and occurs at the formal reduction potential [14–16]. As we will enlarge upon in Section 2.2, the *actual* electrochemical results obtained for proteins are usually less than reversible; they may involve mixtures of different mass transport characteristics 1 A, 1 B, and 1 C, and they may also involve coupling to chemical reactions. The latter feature will form the focal point of our attention later.

Aside from cyclic voltammetry, there are other methods that are capable of providing more detailed information, complementary perspectives, or greater resolution and sensitivity. Chronoamperometry and chronocoulometry are techniques in which the potential of the working electrode is stepped abruptly from one value to a new value, thus shifting the position of redox equilibrium [17]. The current (or charge, respectively) is now monitored as the system accepts or releases electrons to re-establish equilibrium. For a diffusion-controlled reaction at a stationary planar electrode, the current-time profile is given by the Cottrell equation (Table 1). The current decrease is proportional to $t^{-1/2}$, whereas for a surface-bound species, the time course of this relaxation is exponential (as is double-layer charging) [18]. These methods have been widely used in nonbiological studies and should offer interesting possibilities for analyzing "gated" electron-transfer reactions in proteins (see later), since they enable the potential-dependence of an electrode reaction to be separated from its time-dependence.

Among the digitally modulated methods we can highlight square-wave voltammetry, in which a pulsed square potential waveform is superimposed upon a step ramp and the current is sampled at stipulated times after each forward and reverse pulse [19, 20]. The individual current components and the difference are each of interest. Because the perturbing potential is modulated, the method is very sensitive; thus, it is appropriate for observing currents due to very dilute solutions, or currents obtained (as is typically the case) at partially blocked electrode surfaces. It is also more useful than linear sweep methods for resolving two or more redox processes occurring at similar electrode potentials. Indeed, with its exquisite capacity for fine, temporal control over the electrochemical driving force through variation of frequency and pulse height, square-wave voltammetry should provide a means for breaking new ground in the study and understanding of multi-step electron transfer in complex enzymes.

2.2. *Sample configurations for protein electrochemistry*

2.2.1. Definitions: Here, the term 'sample configuration' refers to the manner in which the protein molecules under observation are maintained in contact with the working electrode. We can identify two limiting cases in which the experiment addresses the protein either as: a) molecules in free exchange between the electrode and the bulk solution which is of sufficient volume to have effectively 'infinite thickness' (the free diffusion configuration); or b) as molecules immobilized on the electrode surface and undergoing no exchange with the contacting solution (the immobilized configuration). An interesting intermediate category is the situation in which molecules are free to diffuse within the confined space of a microscopically *thin layer*. Each of these will be discussed briefly, and we will then consider some applications that have arisen largely from use of the immobilized configuration.

2.2.2. The free diffusion configuration: Most protein voltammetry has been performed with the sample molecules contained in a solution of sufficient volume to allow easy handling. This requirement is typically in the range 0.25 to 0.5 ml but can be less than 0.1 ml with appropriate cell design [21, 22]. The chief aim has been to achieve a reversible electrochemical reaction at the electrode under conditions described as "semi-infinite diffusion". Communication between the electrode and species in solution is an obvious requirement for obtaining "free molecule" reduction potentials that are comparable with potentiometrically determined values, and is of course necessary for achieving bulk electrolytic redox transformations. Obviously, this necessitates a fully reversible interaction (i.e. weak binding) between the diffusing protein and sites (for this purpose loosely defined) at the electrode-solution interface at which electron transfer occurs.

The detailed mechanistic aspects of these interactions and of the electron exchange are extremely complex and remain largely unresolved even for systems that have been generally considered to exhibit ideal electrochemistry, such as cytochrome *c* at certain modified Au electrodes [23–29]. Nevertheless, the conditions for achieving electrochemistry of protein solutions have been extensively studied by many groups, and we can identify two rules, open to broad interpretation, either or both of which may apply. These are first, that the electrode should have the appropriate surface structure(s) to enable these interactions to occur, and second, these sites must not be blocked by more strongly adsorbing and effectively insulating material (particularly proteinaceous impurities). Observations leading to these rules are that: a) in general, bare metal electrodes are not successful unless modified with certain small adsorbates [28, 29], yet b) an *ultra-clean* metal surface yields an electrode response without modification [30], c) proteinaceous impurities adversely affect the electrode response [31], and d) metal oxide or polished (oxidized) carbon electrodes can yield good results [21, 32–35], although this may require establishing and optimizing attractive electrostatic interactions with the protein by inclusion of counter-charged ions [33–35]. Following the original discovery by Eddowes and Hill, there have been numerous investigations to determine how 4,4′-bipyridyl and other small adsorbates (promoters) give rise to well-behaved electrochemistry of cytochrome *c* and other proteins at a Au electrode [23–27]. Despite these efforts, which have resulted in increasingly successful and elaborate modifications, it appears that these promoters probably function in a number of different ways [27]. We note here that due to the complex inter-

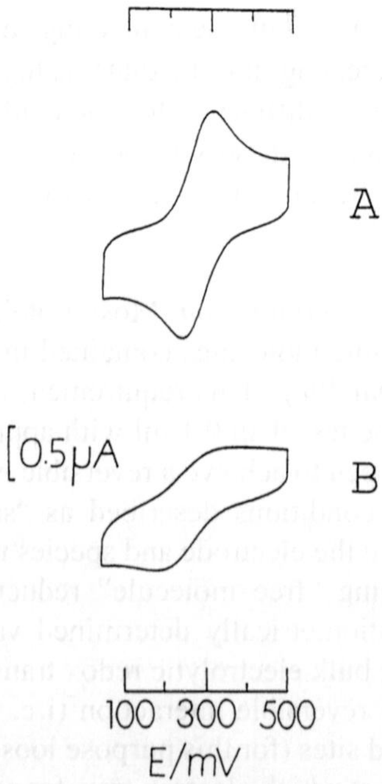

Fig. 2. Cyclic voltammograms of horse heart mitochondrial cytochrome *c*, at pH 8.0 (5 mM Tricine, 0.1 M NaCl), temperature 20°C, scan rate 20 mV s^{-1}. (A) At a freshly polished edge-oriented pyrolytic graphite electrode. (B) At a freshly cleaved basal plane pyrolytic graphite electrode.

actions involved, the term "diffusion-controlled" is not to be interpreted in its strictest sense [23].

An interesting but problematic feature of the free-diffusion configuration is the observation of weak and often ill-defined waveforms in conventional cyclic voltammetry. As an illustration, we refer to Fig. 2 which compares two voltammograms obtained for cytochrome *c*. The rounded, drawn-out current response shown in Fig. 2B does not have the sharp form of Fig. 2A. Results such as these have sometimes been interpreted as reflecting very small rate constants for heterogeneous electron transfer, i.e. an irreversible electrode reaction. To explain these observations, Bond and co-workers have proposed a "microscopic model" for protein electrochemistry which assumes that facile, i.e. reversible, interfacial electron transfer will occur *only* at those areas of the electrode surface at which the protein is able to interact in a productive manner [36–39], and that diffusional encounter at other areas of the electrode does not lead to electron exchange. On the one hand, this "all or nothing" idea parallels the much greater specificity of electron-transfer proteins for their (physiological) redox partners, when compared to a small, simple molecule such as ferrocene. Protein-protein interactions are extremely complex, requiring optimization of a multitude of electrostatic and hydrophobic contacts that orientate the partners for fast electron transfer. Hence it is not surprising that interactions with an electrode should be so critically dependent upon the nature of the electrode surface. On the other hand, proteins have a tendency to adsorb at most surfaces, and may do so in a way that prevents electron transport between the electrode and free molecules in the bulk solution, i.e. the familiar effect known as "fouling" [40]. This insulating effect can occur even if the adsorbed protein itself is electroactive.

On either of these terms, an electrode surface in contact with a protein solution may appear somewhat non-uniform at the molecular level, and a protein molecule in the solution close to the electrode will experience a mixture of diffusion fields. The outcome of this is that "active" and unfouled electrode surface sites behave as isolated microelectrodes (radial diffusion) under conditions where their separation and size are respectively greater and smaller than the time-dependent thickness of the diffusion layer. The resulting cyclic voltammogram has a steady-state waveform (Fig. 1 B). As the diffusion layer thickness increases (slower scan rates) or as active surface sites become larger or increase in number, we approach the more familiar macroscopic situation (that of a uniform planar electrode giving rise to a linear diffusion field) the result of which was shown in Fig. 1 A. In practice the resultant waveform may lie somewhere between those of Fig. 1 A and B, and may be difficult to analyze, particularly if the faradaic current is very small compared to the double-layer charging component. The electrochemistry shown in Fig. 2 B is in fact reversible [36].

This idea has been extended, by testing the effect of *deliberately* introducing electrode surface heterogeneity to create varied mixtures of diffusion fields. An example is the work of Salamon and Tollin who have studied the voltammetry of small electron-carrier proteins at metal electrodes modified by self assembly of a lipid bilayer based upon phosphatidylcholine [41, 42]. Their approach is instructive as a model for the diffusion field effects that must influence the efficiency of *physiological* electron transport between peripheral and integral membrane proteins. For cytochrome *c*, the packed bilayer is inactive but becomes active if partially substituted with hexadecylphosphate, a lipid bearing a negatively charged headgroup [41]. The resulting waveforms are sigmoidal at low substituent density, but they intensify and take on a peak-shaped form at 30 mol% hexadecylphosphate. Analysis of both the low-amplitude steady-state and the resulting peak-shaped voltammetry shows the electrochemistry to be effectively *reversible* under all conditions. The results are consistent with a transition from radial to linear diffusion. Similar results are obtained with the negatively charged protein spinach ferredoxin when, in this case, the substituent is dimethyldioctadecyl-ammonium (positively charged headgroup) [42]. Viewed at the molecular level, the necessarily weak electrostatic binding occurs only at areas containing counter-charged head groups, and the electron-transfer reaction that follows is presumably dependent upon transient reorganization of the protein/lipid complex.

We must therefore consider the likelihood that the voltammetric response of proteins may vary because of changes in diffusional behaviour rather than as a reflection of characteristic electrochemical properties. If this is so, the use of linear sweep methods at stationary planar electrodes has obvious drawbacks for the detection and analysis of complex, coupled electrochemical patterns. One way to overcome this problem is to enforce conditions of total steady-state by using a rotating disc electrode or a micro-electrode. Another way (at least to overcome the problem of weak and ill-defined current response) is to exploit the greater sensitivity of methods such as square-wave voltammetry.

2.2.3. The immobilized configuration: A different, and arguably more satisfactory, perspective is achieved by studying proteins that are immobilized on an electrode surface. The basic configuration, albeit once again the *idealized* case, is shown in Fig. 3. Protein molecules are induced to bind tightly to the electrode surface in the form of a stable, porous, mono/submonolayer film, oriented for facile electron transfer, and with retention of native structural and reactivity characteristics. The redox centres act independently and homogeneously, and remain fully accessible to ions and small reagents in the contacting electrolyte, as does all the intervening space in the interface. Ideally, the cyclic voltammetry of such systems should resemble the form shown in Fig. 1C. As described later, the voltammetric response now becomes a particularly useful signal.

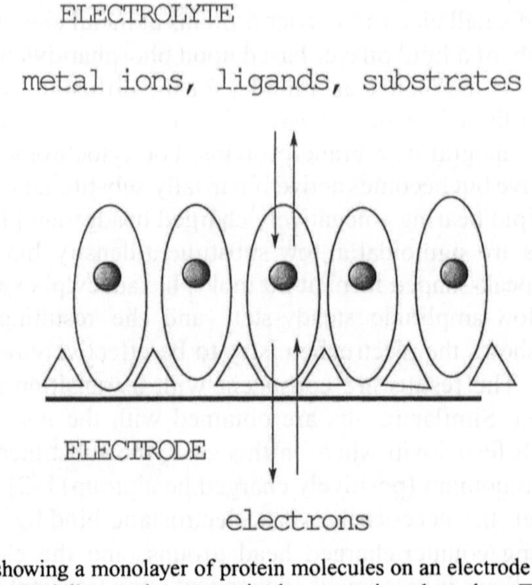

Fig. 3. Cartoon showing a monolayer of protein molecules on an electrode surface, and their interaction with potentially reactive agents in the contacting electrolyte. Triangles represent electrode surface modifications that may be required for protein attachment.

Detailed studies of the electrochemical properties of cytochrome *c* adsorbed at various surfaces have been carried out by Bowden and co-workers in their efforts to achieve and develop well-defined *protein monolayer electrochemistry* [43–50]. By analyzing both the electrochemical and protein structural (*in situ* spectroscopic) aspects, their investigations have appraised the general validity of this approach for the study of redox proteins, and provided insight into the way that a protein's properties alter between the free and the bound state. The latter question is of course relevant to the physiological function of the numerous electron-transfer proteins that shuttle continuously between bulk aqueous and membrane surface environments. As we will discuss in more detail later, the immobilized configuration has been particularly useful for studying some complicated coupled reactions of proteins, ranging from metal and ligand exchange at labile Fe-S clusters to catalytic electron transport processes in complex enzymes [51–63].

Bowden and co-workers have summarized the advantages and limitations in analyzing and interpreting the electrochemical response from immobilized proteins [50]. The theoretical aspects are well established, with the simplest model assuming a Langmuir adsorption isotherm. The electrochemical kinetics can be analyzed on the basis of the Butler-Volmer equation which is modified (see Table 1) to use surface rather than solution activities, while the electrochemical rate constant has first-order (s^{-1}) rather than the more familiar heterogeneous dimensions ($cm\ s^{-1}$) [14–16]. The formalism is readily extended to include the effects of first- and second-order coupled chemical reactions [55, 57, 58, 61–63].

To see how the immobilized configuration lends itself to particular applications, we first take a very general look at some advantages over studying freely diffusing protein molecules. For an ideally behaved system, these are as follows.

2.2.3.1. Control of the redox status of all the centres in the sample. The oxidation states of all the redox centres in the sample are directly and rapidly "locked" to the applied potential with their populations given immediately by the Nernst equation. This "direct connection" provides a much greater degree of control over the redox states of centres in the sample than is possible with the free diffusion configuration.

2.2.3.2. Waveform definition. For an ideal (homogeneous) immobilized array of reversible redox couples that are not coupled to kinetically competing chemical reactions, the cyclic voltammetry waveform is compact and finite at all scan rates because no electrons stray to or from freely diffusing molecules. This being so, the number of electroactive molecules is derived by integration.

2.2.3.3. Exploratory capabilities. The mono/submonolayer film on the electrode can be transferred sequentially between electrolyte solutions that subject the redox centres to various conditions and opportunities for reaction. This feature constitutes what is equivalent to rapid "dialysis" at the molecular level. Conditions that can be examined even include extremes of pH or other media in which the protein might not survive macrodialysis. Here, the speed of the voltammetric measurement can allow a "snapshot" to be taken before the sample deteriorates.

2.2.3.4. Economy. The amount of sample required is minuscule; a typical experiment with multiple transfers uses only about 10^{-12} to 10^{-11} mole per square centimeter of electrode surface. (As a guide, this is about 10^3 to 10^4 – fold less material than required for a typical EPR sample.)

2.2.3.5. Sensitivity and trace-level stoichiometry. Because the sample size is so small (yet is at extremely high local concentration) measurements are sensitized to trace amounts of a reactive reagent in the contacting electrolyte. Indeed, with good conditions of mass-transport (particularly as can be enforced with a rotating-disc electrode) stoichiometric reactions may be analyzed with reagent concentrations that are nanomolar or lower.

It is of course essential to ascertain that the protein remains "native" or functionally active in the adsorbed state, and in this respect both direct and indirect evidence has been used to obtain information. Direct methods are those in which the structure of the adsorbed protein or the active site are probed by *in situ* spectroscopy. Techniques include surface-enhanced Raman spectroscopy (SERS), optical electroreflectance and absorption [27, 64–68]. Indirect methods involve confirming that intrinsic properties of the protein such as catalytic activity are retained quantitatively in the adsorbed state [61, 62]. Our discussion will be limited to examples where overwhelming evidence has been obtained to show that the protein structure or function is not seriously affected by the forces at the electrode-electrolyte interface. It is also important to ascertain that the adsorbed protein sample has an acceptably low level of inhomogeneity (i.e. dispersion of reduction potentials and kinetic capabilities), and that there is minimal communication between centres in adjacent molecules. Because of their long-range separation by the intervening protein structure, interactions between centres in adjacent molecules are expected to be very weak (contrast the less ideal situation expected for small molecules [14, 18]) while inhomogeneity is very likely and should be evident from broadening of the voltammetric waves.

In some cases, no formal modification of the electrode surface (beyond normal pre-treatment) has been required to induce adsorption of electroactive, functional molecules. This has appeared to be the case for certain electron-transfer proteins and several enzymes that display catalytic activity at carbon and metal oxide electrodes [31, 32, 43–46, 59, 69–72]. To date, the most detailed examinations of the electrochemical properties of an immobilized protein have been undertaken by Bowden and co-workers [43–46].

Cytochrome *c* adsorbs at a fluorine-doped tin oxide electrode, and the resulting cyclic voltammetry has the general characteristics of an immobilized redox couple as depicted in Fig. 1C. A stable film is achieved only at low ionic strength and neutral pH, and confirmation that the adsorption process is predominantly electrostatic has been obtained by adsorption isotherm measurements at various values of pH and ionic strength [44–45]. As determined from the area under background-corrected waves, the coverage is almost equivalent to one electroactive monolayer. The half-height widths of anodic and cathodic waves are 120–140 mV at 10 mV s^{-1}, and further broadening occurs as the scan rate is increased. From the peak separation observed at different scan rates, the electrochemical rate constant was estimated to be between 1 and 10 s^{-1}, but this value also was found to depend upon the scan rate. These observations differ somewhat, but not too drastically, from the theoretical (Butler-Volmer) expectations, in that the half-height width at 25 °C should range between ca. 90 mV for a reversible reaction and $62.5/\alpha$ for an irreversible reaction [15]. The non-ideality of the electrochemistry of immobilized cytochrome *c* may reflect some dispersion in the environments of molecules in the film, i.e. the centres vary somewhat in their reduction potentials and interfacial electron θ transfer kinetics. This result is not surprising, considering the defects likely in even the most perfect of electrode surfaces.

The status of cytochrome *c* adsorbed at tin oxide (which is optically transparent) has been examined by absorption spectroscopy, exploiting the intense Soret absorption band ($\varepsilon > 100000$ M^{-1} cm^{-1}) which is sensitive to coordination and spin state [46]. It was concluded that the protein does not undergo any major structural change in either oxidation state over the normal potential range of electrochemical activity. This result is consistent with the observation that the reduction potential is similar (but not identical) to that measured for the protein in solution. In fact, $E^{0'}$ is decreased by 20 – 25 mV compared to the solution value. Willit and Bowden analyzed the thermodynamics of the adsorption of different oxidation states by measuring and comparing the temperature dependence of reduction potentials in the free and adsorbed states. The Fe(III) form adsorbs more strongly than the reduced form because it is conformationally more labile and can more readily undergo the small structural changes that are incurred [44]. Their results are of physiological relevance since it is known that cytochrome *c* reduction potentials are also lowered upon binding to natural redox partners, many of which are associated with membranes.

Similar observations have been made in studies of cytochrome *c* immobilized by embedding into a Langmuir-Bodgett monolayer of lecithin at tin oxide [73]. Within the bounds of what we consider to be retention of native characteristics, there is a delicate balance that depends on the ionic composition of the solution as well as the chemical and electrical properties of the surface. On a more general note, it certainly appears that a crucial factor for maintaining adsorbed cytochrome *c* in its native state is a clean hydrophilic surface possessing acidic functionalities (e.g. oxides). By contrast, cytochrome *c* adsorbed at bare Ag or bare Au appears to adopt, very readily, a non-native conformation distinguished by a much more negative reduction potential [27, 67, 68].

The importance of electrostatics in the protein-electrode interaction was recognized in earlier studies of protein solution voltammetry, in which it was noted that diffusion-controlled reversible electrode reactions of some negatively charged electron-transport proteins, particularly plastocyanin and ferredoxins, are promoted at pyrolytic graphite edge or metal oxide electrodes by inclusion of metal cations such as Mg^{2+} or $Cr(NH_3)_6^{3+}$ in the solution [33–35]. One way of describing how these cations function is that they create weak salt-bridge interactions between the protein surface and the weakly acidic oxide groups on the electrode. Not surprisingly, much stronger interactions are induced by complex polyamines, particularly various aminocyclitols (which possess a rigid, spatially defined array of $-NH_3^+$ groups) and polymyxin (a structurally defined "lassoo-shaped" molecule) each of which has been found to give rise to stable electroactive films of ferredoxins and other proteins at "edge"-oriented pyrolytic graphite [51–58, 61–63]. These interactions are not as well understood as is the case with cytochrome *c*, but since reduction potentials and characteristic cluster reactivities are similar to those manifested in free solution, it appears certain that the important native characteristics are retained in the adsorbed state.

The most sophisticated electrode surfaces for immobilizing proteins involve self-assembled monolayers (SAMs) of functionized alkane thiolates on an ultra-flat sputtered Au surface [47, 49, 50]. The close-packed monolayers that these molecules form have been characterized in considerable detail. In addition to the exquisite possibilities for tailoring an optimized interface, such electrodes have a low background capacitance and are thus particularly suitable for detailed examination of the faradaic response. Bowden and co-workers have studied the electrochemical behaviour of cytochrome *c* immobilized on a monolayer of various ω-alkanethiol carboxylic acids [47, 49, 50]. The idea is that the dense array of terminal carboxylate groups engage multiple salt-bridge interactions with lysines in the vicinity of the exposed heme edge of the protein, thus mimicking the expected physiological complexation of cytochrome *c* with reaction partners. A cartoon illustrating the hypothetical structure of the interface, and some results, are shown in Fig. 4. Stable films are formed only at low ionic strength. The influence of the intervening interfacial medium has been examined by varying the chain length as determined by n in the $-(CH_2)_n$-linker. As expected, the electrochemical rate constant decreases as the chain length is increased through $n = 6$ (reversible), $n = 10$ (reversible), to $n = 16$ ($k_{et} = 0.4\ s^{-1}$). Less obvious are the observations that the adsorbed protein molecules form a higher cover-

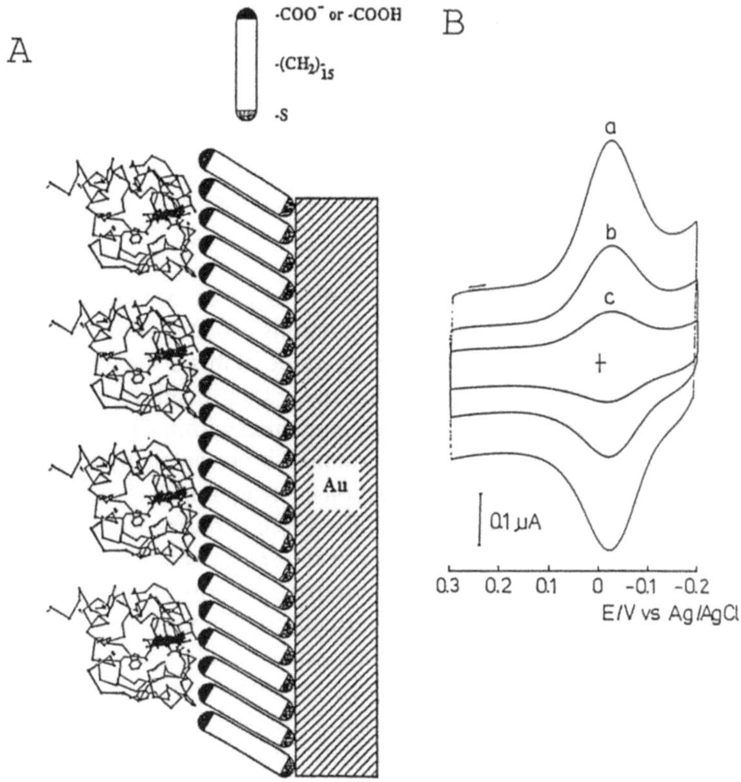

Fig. 4. (A) Cartoon showing hypothetical structure of an ideal ordered and oriented composite layer of cytochrome *c* molecules bound to a monolayer of alkanethiol carboxylic acid (16-mercaptohexadecanoic acid) at a Au electrode. (B) Cyclic voltammograms of cytochrome *c* bound to a monolayer of alkanethiol carboxylic acid (11-mercapto-undecanoic acid) at a Au electrode. Electrolyte is 4.4 mM potassium phosphate at pH 7.0. Scan rates (mV s^{-1}) are: a) 200; b) 100; c) 50. (Adapted, with permission, from Reference 49. Copyright 1993, American Chemical Society.)

age (essentially an electroactive monolayer) on the longer chain as compared to the shorter ones, and that the half-height widths are considerably larger (150–160 mV) for n = 16, vs 110–120 mV for n = 6 or 10. The long-chain system was studied in greater detail, utilizing impedance measurements as well as cyclic voltammetry, in order to explore the possibilities for investigating the more fundamental aspects of the electron-transfer mechanism.

Covalent attachment of the protein to the electrode represents the limiting case for immobilized proteins and enables some assessment to be made of the roles of orientation and rotation in interfacial electron transfer. Bowden and his group have covalently attached cytochrome *c* to monolayers of ω-alkanethiol carboxylic acids using carbodiimide coupling to protein surface lysines [48]. In this way, electroactive coverages of ca. 0.3–0.4 monolayers are produced, which by contrast with non-covalent immobilization appear to be stable indefinitely at high ionic strength. Again, as observed for electrostatically adsorbed protein, the reduction potentials are more negative than for free protein, and electrochemical rate constants decrease with increasing chain length. A similar result has been obtained by Cooper and co-workers who have immobilized cytochrome *c* at N-acetyl cysteine-modified Au [74].

2.2.4. Protein molecules confined to a thin layer: Intermediate between the free diffusion and immobilized configurations is one in which the protein is confined within a thin layer. A recent example of this is the work by Rusling and co-workers who have shown that myoglobin displays excellent direct electrochemistry characteristics when contained within a liquid crystal film of surfactant (didodecyldimethyl ammonium bromide) on pyrolytic graphite [75]. A subject of several investigations, myoglobin has shown puzzling electrochemical behaviour. Its established role is reversible dioxygen binding, not electron transfer, and studies of the sperm whale protein have provided only evidence for slow electron transfer activity at various electrodes. However, one recent report has described quasi-reversible cyclic voltammetry of horse heart myoglobin at an indium oxide electrode [76], while another study on horse skeletal muscle myoglobin showed that the cyano form is much more active than the aquo form [77]. The cyclic voltammetry obtained for horse skeletal myoglobin in the liquid crystal film represents the best electrochemical kinetics thus far reported, with electrochemical rate constants greater than 10^{-3} cm s^{-1}. The electrochemical characteristics are typical of thin-layer behaviour, the thickness of the film being ca. 20 μm. Thus, the response at high scan rates conforms to expectations for semi-infinite diffusion control, with the peak current being proportional to the square root of scan rate above 50 mV s^{-1}, while below 6 mV s^{-1} the peak current becomes proportional to scan rate. Integration of the slow scan waveforms yields the amount of sample contained in the film. The results of various spectroscopic methods (reflectance-absorbance Fourier transform infrared, UV-visible absorption and linear dichroism) suggest that the myoglobin molecules are preferentially oriented in the lamellar film, and are in the high-spin state. The protein was also shown to catalyze reduction of organohalides. This configuration may thus turn out to be extremely useful, not only in enabling study of proteins in a confined but non-bound form, but also in future technological applications. Other systems of trapping small samples of electroactive proteins within a confined film have been reported [78, 79].

2.3. *Coupled chemical reactions with particular reference to immobilized proteins*

2.3.1. Cyclic (square) schemes: Coupled chemical transformations affect the electrochemical response in different ways depending on their thermodynamic magnitude and how their rates compare with the timescale of the voltammetry. Leaving until later the special features of electrocatalytic coupling, we can immediately identify different situations by reference to the single rectangular or "square" scheme (a thermodynamic box) shown in Fig. 5. For electrochemists, this is a particularly well established concept [80, 81]. More extensive or complicated coupling requires multiple boxes "pasted together" (as will be seen in Section 3.1.1) or more dimensions, as for example in the cubic model for redox-coupled proton pumps [82]. In this scheme, two redox states of a protein, O and R, each exist in two structurally distinct forms 1 and 2. These could differ, for example, in the binding of an external small ion/molecule (such as H^+), binding of another protein, or by adopting different conformations. The square thus comprises species O_1, R_1, R_2, and O_2, which are interconnected by elementary electrochemical (E) reactions and chemical (C) reactions which may be of varied order. For simplicity we assume that these reactions occur in a stepwise manner, i.e. that concerted (diagonal) reactions do not contribute significantly. Here and for the rest of this chapter, we will be concerned mainly with results that have been obtained using the immobilized protein configuration. We consider that the voltammetric cycle commences with a population of active sites that is 100% in the state O_1, and assume all E reactions are reversible, i.e. that electrochemical rate constants k_{el} are high enough not to complicate the waveforms which will thus report on the C reactions. The following cases are represented by the cyclic voltammograms shown in Fig. 6.

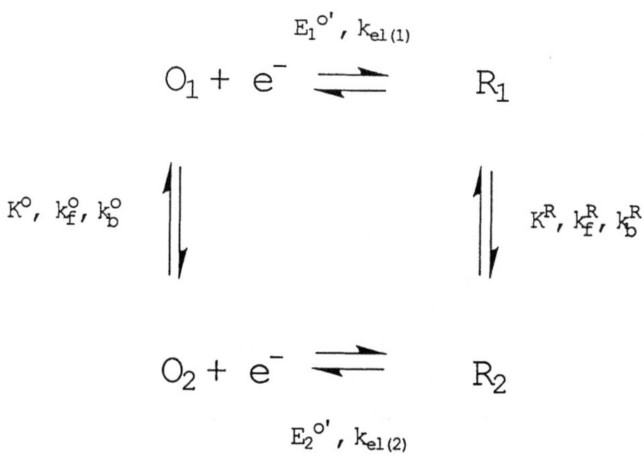

Fig. 5. Generalized square scheme for a simple, chemically coupled electrochemical reaction.

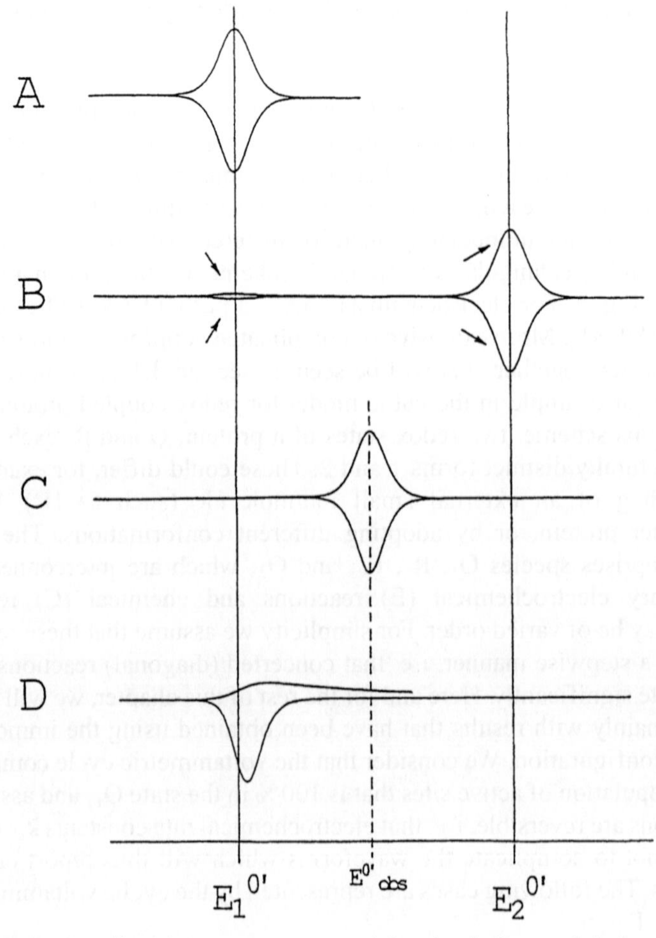

Fig. 6. Idealized cyclic voltammograms expected for chemically coupled electrochemical systems under various kinetic and thermodynamic situations as described in the text. The position of the reversible signal for a rapidly equilibrating coupled chemical system involving a second reagent X is given by the equation shown, where K^R and K^O are association constants (M^{-1}).

$$E^{0'}_{obs} = E^{0'}_1 + \frac{2.3\,RT}{F} \log \left\{ \frac{(1 + [X]\,K^R)}{(1 + [X]\,K^O)} \right\}$$

A. The simplest case. Only O_1 R_1 are thermodynamically stable (K_O and $K_R \ll 1$). Only one redox couple is observed, with a potential $E^{0'}_1$.
B. Slow, redox-induced chemical interconversion. $K_R \gg 1$ and interconversion between R_1 and R_2 is slow on the voltammetric timescale. Here, voltammograms recorded at different times reveal the gradual appearance of a new couple O_2/R_2 at a potential $E^{0'}_2$. Greater control of this interconversion is achieved when the protein is immobilized.

C. Rapid and reversible chemical interconversion. $K_R >> 1$ and interconversion between R_1 and R_2 is fast on the voltammetric timescale. The electrochemically observed reaction is thus between O_1 and rapidly equilibrating populations of R_1 and R_2. The waveform is regular and the reduction potential for the diagonal equilibrium is displaced from $E_1^{0'}$ according to standard thermodynamic relationships [54, 63].
D. An intermediate rate of reversible chemical interconversion. There are many variations but here is depicted the waveform expected if $K_O << 1$ but $K_R >> 1$ and interconversation between R_1 and R_2 occurs at rates comparable to the voltammetric timescale. Thus R_1 reacts to give R_2, and re-oxidation requires a rate-limiting C reaction, i.e. conversion of R_2 back to R_1. The re-oxidation reaction is gated, and appears as a drawn-out wave, persisting to high potentials since the C reaction has a time dependence but not a potential dependence. Quantitative information can be derived from analysis of the distorted voltammogram, ultimately using simulation procedures [55, 58].

While these arguments hold regardless of the configuration of the voltammetric experiment, the immobilized configuration affords certain simplifications and a greater range of conditions. These benefits arise from the absence of protein diffusional restrictions and the possibility for exploiting the advantages of sensitivity and trace-level stoichiometry. In Section 3.1 we outline how these ideas can be applied for detecting and characterizing active sites, and for determining the kinetics of redox-linked ligand exchange and proton transfer reactions at Fe-S clusters.

2.3.2. Catalytic electron transport: Catalytic electron transport by an immobilized enzyme can be treated in the following way, referring to the hypothetical model for a two-centre enzyme shown in Fig. 7. The simplest description of catalytic electron transport (open arrows) between the electrode and molecular substrates in solution encompasses just three steps, which can be linked to a series of resistors (R) [61, 62]. The electrode provides a tunable, variable electrochemical driving force, while control of substrate mass transport can be achieved with a rotating-disc electrode [13].

The observed current for catalytic turnover is given by the reciprocal relationship

$$1/i_{obs} = 1/i_E + 1/i_{transport} + 1/i_{cat}$$

in which the three terms correspond, respectively, to:

a. Electron exchange between the electrode and the primary electron entry/exit site on the enzyme. This stage can be described by the diffusionless form of the Butler-Volmer equation (Table 1) where the potential-dependent current i_E corresponds to the rate of this process.
b. Transport of reactants and products to and from the enzyme-loaded electrode. This can be controlled by rotating the electrode, whereupon the

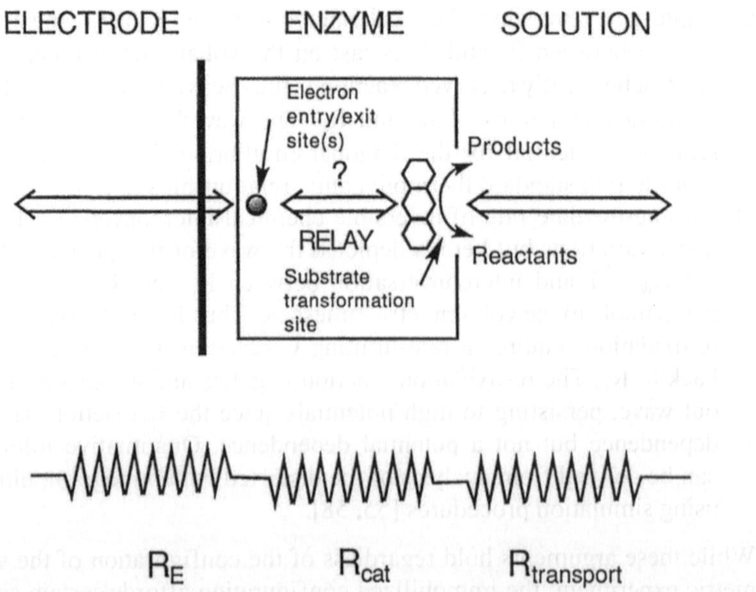

Fig. 7. Cartoon showing the sequence of steps in electron transport between an electrode and substrate molecules during catalysis by an adsorbed enzyme molecule. The steps are analogous to a series of resistors as shown.

relationship between mass-transport controlled current $i_{transport}$ (i.e. i_{Lev}) and rotation frequency ω is given by the Levich equation (Table 1).

c. Characteristic catalytic processes occurring *within* the enzyme, and which are of particular interest. These are: i) intramolecular electron transfer (perhaps a multiple site relay) between the substrate transformation site and the electron entry/exit site; ii) chemical reactions at the substrate transformation site. To an initial approximation, the enzyme's properties might be described by an electrochemical version of the Michaelis-Menten equation, as given in Table 1. The Michaelis constant K_M and turnover number k_{cat} are the generalized parameters determining i_{cat}, which becomes the dominant term if i_E and $i_{transport}$ are optimized to the degree that their reciprocals effectively vanish.

The catalytic current is a *direct* measurement of turnover *rate*. Moreover the voltammogram is a plot of rate versus electron-motive driving force.

To illustrate just the basic principles, we refer next to Fig. 8, which shows model (faradaic-only) cyclic voltammograms plausible for an immobilized enzyme having two redox centres A and B. Trace 1 is a "reversible" (non-turnover) voltammogram obtained at a stationary or rotating electrode (in this situation, both give the same result). The two pairs of peak-like signals A and B correspond to reversible redox transitions of the two centres. Trace 2 is obtained upon introducing a low concentration of reducible sub-

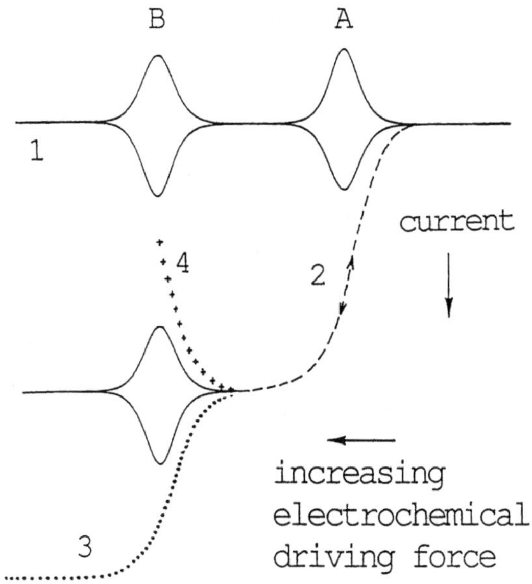

Fig. 8. Idealized cyclic voltammograms expected for catalytic electron transport by an adsorbed enzyme having two redox centres A and B. The four different traces 1–4 are explained in the text.

strate and rotating the electrode to maintain a steady state under hydrodynamic control. Signals A have converted to a sigmoidal wave, while signals B remain peak-like. The interpretation is that signals A stem from the *catalytic* site while B is not involved in turnover, at least under these conditions. Traces 3 and 4 represent two important alternative effects (each branching from trace 2) that arise from centre B. Trace 3 shows centre B providing a *second pathway* for electron transfer to A, while trace 4 shows one of the possible effects if centre B is a *regulatory* site – in this case activity is *switched off* when B is in the reduced state.

The extremely small amount of enzyme actually addressed is comparable to amounts used in conventional steady-state kinetics, yet under optimal conditions (i.e. active coverage approaching a monolayer) it may be possible to observe redox centres directly. As discussed later, this situation has been achieved for fumarate reductase, even with the relatively poor sensitivity of cyclic voltammetry. However, the amplified catalytic response remains the easiest and most reliable indication that an enzyme is interacting well with the electrode and retaining its native characteristics. With mass transport control, it is possible to achieve a long-lived steady state with substrate concentrations much lower than micromolar, while stoichiometric reactions with very tightly binding reagents may be detectable at the nanomolar level or below. Features of the catalytic wave can be correlated with redox transformations and participation of specific centres, a situation that is analogous to being able to observe steady-state and transient kinetics simultaneously on the same sample. For example, turnover can be "frozen out" by increasing the scan rate, thereby allowing extraction and analysis of the kinetics and thermodynamics of selected individual processes, including substrate and inhibitor binding. The rotating-disc electrode imposes rapid removal of free product from the vicinity of the enzyme, thus suppressing possible product inhibition.

To illustrate how these ideas are applied, we turn next to some examples in which complex chemistry has been detected and elucidated by voltammetric methods, in each case largely exploiting the special properties of the immobilized configuration. The examples touch on three aspects; the complex and still largely uncharted chemistry of Fe-S clusters, the detection and analysis of "gated" electron transfer, and the elucidation of catalytic electron transport in multi-centred enzymes.

3. Applications to studies of redox-active sites in proteins

3.1. Applications to Fe-S clusters

3.1.1. Difficulties in studying cluster reactions in proteins: It is now clear that iron-sulphide clusters have numerous biological functions, with gene expression (the so-called "iron-responsive element binding protein" (IRE-BP) which senses Fe levels to control Fe uptake and storage by eukaryotic cells) and oxygen-sensitive enzymatic regulation (as in amidotransferase), as well as acid – (aconitase, bacterial L-serine dehydratase) and redox – catalysis, (nitrogenase, hydrogenase) now eclipsing their long-established role as simple electron-transfer centres [83–87].

However, determining the details of reactions of Fe-S clusters in proteins poses many problems. They exhibit broad, featureless absorption spectra, and most analytically-useful spectroscopies require cryogenic temperatures. Consequently it is usually difficult to monitor identifiable species in the time-domain mode. Furthermore, in cases where the reduction potential is very negative (i.e. < -550 mV) it becomes impossible to generate interesting reduced species by using aqueous chemical titrants such as sodium dithionite. Clusters are thermodynamically unstable with respect to oxide formation and tend to be O_2-sensitive even in higher oxidation levels. Given this situation, our knowledge could indeed remain limited (albeit pragmatically) to those systems that enjoy the greatest thermodynamic or kinetic stability and/or can be readily isolated, a notion that applies even to the more detailed studies on synthetic cluster analogues [88]. Thus, by contrast with the past emphasis on isolation and characterization of "static" species, the systems that we need to be able to characterize are the "*vacillatory*" centres – those lacking sufficient stability in any single state to be studied readily. These indeed are the catalysts and sensors, mentioned above, that are now becoming a major focus of interest.

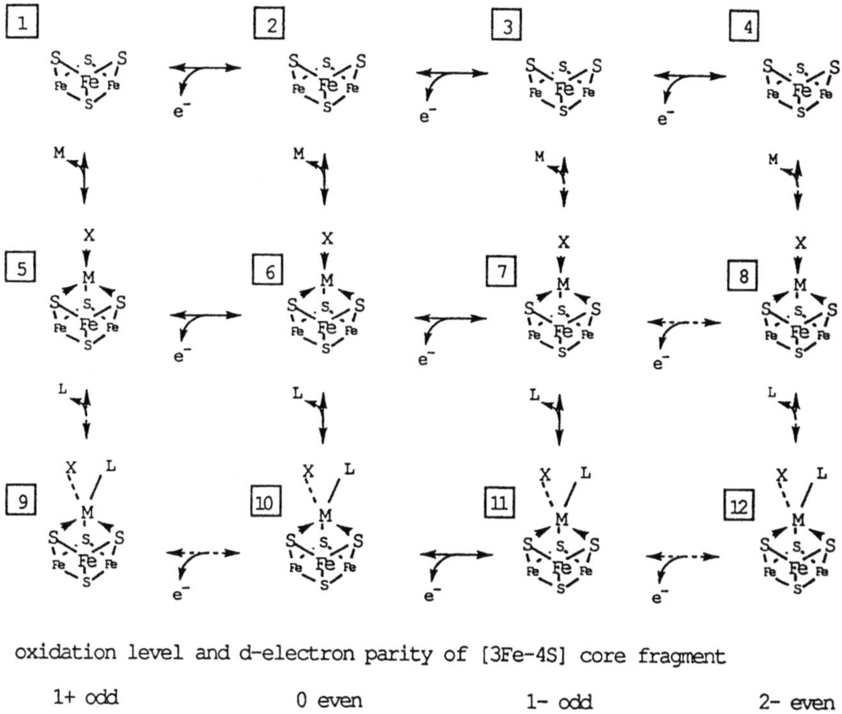

Fig. 9. Reaction "grid" depicting possible reactions of a [3Fe-4S] cluster which, in various oxidation levels, may coordinate a fourth metal atom and accompanying ligands. The grid may be regarded as a network of square schemes.

To illustrate how the complexity of the chemistry can be mastered by voltammetric experiments, we next consider a hypothetical reactivity scheme based upon the chemistry that can stem from a [3 Fe-4 S] cluster bound within a protein such that the vacant corner is open for coordination of a fourth metal ion M. Such examples are numerous, the best example being inactive 3 Fe aconitase which is activated by reduction and addition of Fe(II), but analogous reactions occur with much simpler proteins as described below [53, 54, 56, 89–92]. The possible reactions are described by network of square schemes pasted together to form a grid (Fig. 9) featuring electrochemical (*horizontal*) and chemical (*vertical*) transitions (E and C reactions respectively). The [3 Fe-4 S] "parent" may undergo successive one-electron transfers (E reactions 1↔2, 2↔3, and 3↔4). Since [3 Fe-4 S] is a ligand akin to a crown thioether but having redox-tunable donor capability, clusters 1 to 4 will have differing propensities for coordinating M (metallation according to C reactions 1↔5, 2↔6, 3↔7 and 4↔8). The resulting cubanes [M 3 Fe-4 S] (linked via M to ligand X, which may be an amino-acid residue) are also related by successive electron transfers (E reactions 5↔6, 6↔7 and 7↔8). Clusters 5 to 8 have differing propensities for substitutional or additive ligation by an alternative (typically *exo*genous) ligand L (C reactions 5↔9, 6↔10, 7↔11 and 8↔12). The *exo*-ligated clusters are again related by electron-transfer steps (reactions 9↔10, 10↔11 and 11↔12). Although Fig. 9 appears complicated, it does in fact constitute a relatively simple set of reactions compared to the more extensive chemistry of entire cluster assembly or fragmentation.

To see how known "stable" clusters fit into this scheme, we note that structures 1 and 2 are the well-characterized $[3Fe\text{-}4S]^{1+}$ cluster and its one-electron product $[3Fe\text{-}4S]^0$, while structures 6 and 7, with M = Fe(II), X = cysteine thiolate and reaction 6↔7 respectively represent the common forms and redox interconversion of typical $[4Fe\text{-}4S]^{2+/1+}$ clusters [63]. Structure 5 with M = Fe(II) is equivalent to oxidized "HiPIP" ("High-potential Iron Protein" and is unusual in stabilizing $[4Fe\text{-}4S]^{3+}$ rather than $[4Fe\text{-}4S]^{1+}$). Only species with resultant odd-electron parity are readily detectable by normal (i.e. perpendicular-mode) EPR, which is the primary technique for studying these systems [93].

Vacillatory clusters may exist in *several* of the above states depending upon the electrochemical potential and availability of M and L. In practice, the observed chemical interconversions (which can be very rapid) often correspond to *diagonal* transitions, i.e. they involve oxidation or reduction *in addition to* ligation or metallation. This complexity, along with the above-mentioned difficulties in distinguishing species, makes these systems extremely difficult to study. The immobilized configuration is particularly suited for studies of vacillatory systems since the minuscule sample of protein molecules can be subjected to various external chemical environments (e.g. the vertical transitions in Fig. 9) while remaining under strictest possible electrochemical control (horizontal transitions). The otherwise complicated scheme of Fig. 9 is now reduced to a collection of square schemes that can be isolated from each other. For chemical reactions that are relatively slow on the voltammetric timescale, the signals provide identification labels for monitoring changes in cluster type (case B of Fig. 6). Cluster instability, a severe problem in traditional studies of clusters, can become an asset in the immobilized configuration, since transformations are easy to detect but remain controllable via the applied potential.

3.1.2. Voltammery of ferredoxins immobilized on an electrode: Voltammetry has been useful for mapping out reactivities of the [3Fe-4S] cluster that are expressed in certain proteins. Two subjects that have been studied in detail are the 7Fe ferredoxin III from *Desulfovibrio africanus* (Da. Fd III) and the 7Fe ferredoxin I from *Azotobacter vinelandii* (Av. Fd I). The cluster domains in each of these proteins are depicted in a simple schematic manner in Fig. 10. The scheme shown for Fd III is based upon the amino acid sequence and spectroscopic data [94], while the structure of Fd I is known in considerable detail from X-ray crystallography [95].

Azotobacter vinelandii Fd I

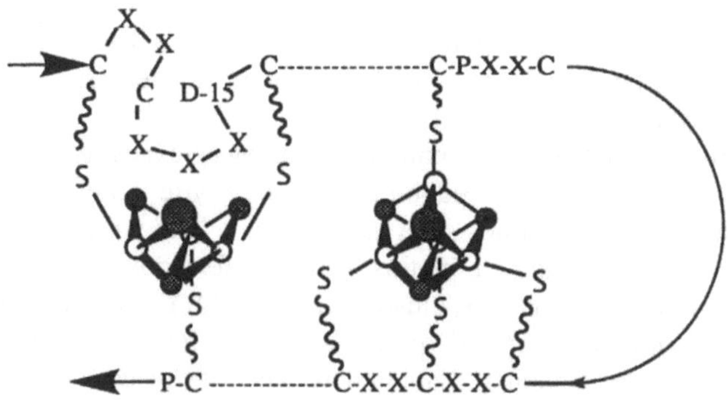

Desulfovibrio africanus Fd III

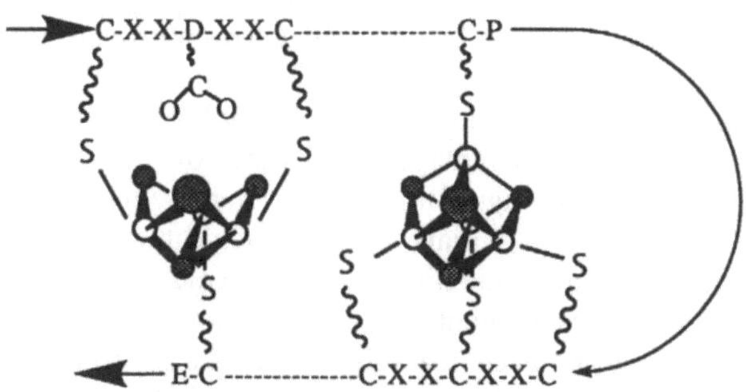

Fig. 10. Schematic structures of *Desulfovibrio africanus* (Da) Fd III and *Azotobacter vinelandii* (Av) Fd I, showing the modes of ligation of the two clusters.

The [3 Fe-4 S] cluster in Da. Fd III occupies a binding domain that resembles typical [4 Fe-4 S] clusters except that one of the four cysteines is replaced by aspartate (D) [94]. In the "0" oxidation level (see Fig. 9) the cluster is very reactive and readily takes up Fe(II) from solution to give a $[4\,Fe\text{-}4\,S]^{2+}$ cluster that exhibits the unusual spin state S = 3/2 when reduced to the (1+) oxidation level [92]. This transformation is similar to the process that occurs during activation of aconitase [83] and is observed with several other small proteins, examples being the 3 Fe ferredoxins form *Desulfovibrio gigas* and *Pyrococcus furiosus* [89–91]. The latter protein contains just one cluster, and also has an aspartate in one of the positions normally occupied by cysteine [92].

The [3Fe-4S] cluster in Fd I does not take up Fe(II) from solution, although a protein containing two [4Fe-4S] clusters can be obtained by reconstitution after unfolding [96]. This inactivity is presumably because the vacant subsite is blocked in the native protein, and neither C-11 nor D-15 are suitably positioned to act as ligands for a fourth metal. As will be discussed later, D-15 may mediate proton transfer between the cluster and the solvent.

Proteins such as ferredoxins which carry an excess of negative charge can be immobilized at an edge-oriented pyrolytic graphite electrode in the presence of polyamines such as neomycin. The voltammograms shown in Fig. 11 were obtained in this way for ferredoxins that each contain a [3Fe-4S] and a [4Fe-4S] cluster, i.e. Av. Fd I, Da. Fd III and the ferredoxin from *Sulfolobus acidocaldarius* (Sa Fd). Of the signals observed, the two at most positive potential (labelled A′ and B′) in each example have been assigned on the basis of the corresponding bulk solution voltammetry and spectroscopic examination of samples prepared by controlled potential electrolysis. The forms of the voltammetric responses are in many cases surprisingly close to theoretical expectations. For example, signal A′ of Av. Fd I, which is due to the couple $[3Fe\text{-}4S]^{1+/0}$ (reaction 1↔2 of Fig. 9) has half-height widths of between 90 and 100 mV [57], while the theoretical value at 0 °C is 83 mV [15]. Signal B′ corresponds in each case to the $[4Fe\text{-}4S]^{2+/1+}$ cluster (reaction 6↔7, when M = Fe). Signal B′ in the voltammogram of Fd I occurs at a particularly low potential, and serves to illustrate a generally advantageous feature of voltammetric methods mentioned at the beginning of this chapter, namely that they enable determination of reduction potentials that lie outside the capabilities of stable titrants or the thermodynamic limits for solvent electrolysis. The $[4Fe\text{-}4S]^{1+}$ cluster in Fd I requires a potential < −700 mV for its generation in solution, but this was readily achieved in a bulk electrolysis experiment on the closely related protein from *Azotobacter chroococcum* [97].

3.1.3. Evidence that [3Fe-4S] clusters exhibit a stable all-Fe(II) form: A general observation made with film voltammograms of proteins containing a [3Fe-4S] cluster is that they exhibit a pronounced pH-dependent redox couple (signal C′ in Fig. 11, see legend for explanation) at very negative electrode potentials (ca. −700 mV at pH 7). It thus appears that [3Fe-4S] clusters may be unique, at least among the simple clusters, in that they have the ability to exist in more than two oxidation states within a protein (c.f. the [2Fe-2S] and [4Fe-4S] clusters). This was first evident from studies on Da. Fd III, for which other potentially redox-active groups (such as non-ligating cysteines) are absent, and it was noted that signal C′ vanishes when the [3Fe-4S] cluster is transformed into a [M3Fe-4S] cubane (see next section) [52]. The number of electrons involved in the signal was determined to be *two* by measuring the charge passed within the baseline-corrected wave and comparing the values with that of signal A′ which (since assigned

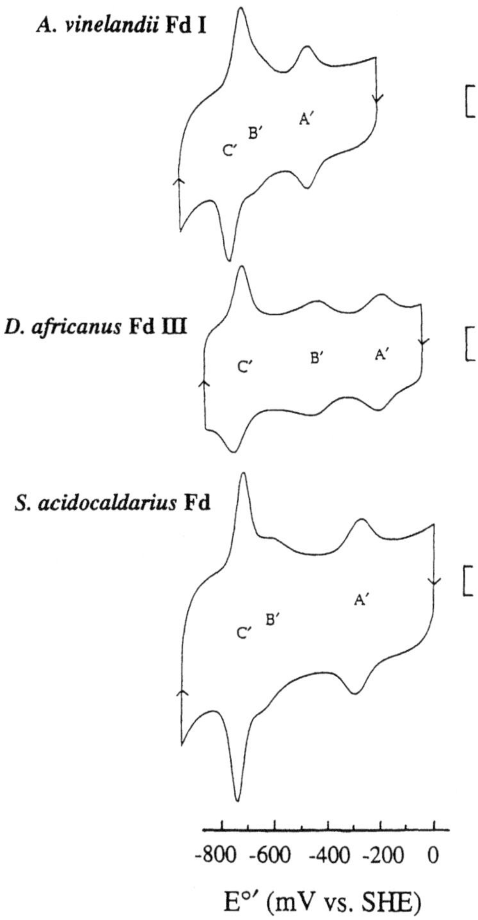

Fig. 11. Typical cyclic voltammograms of 7 Fe ferredoxins immobilized at an edge-oriented pyrolytic graphite electrode at 0 °C. Conditions (not optimized to reveal couples in clearest detail) are as follows. Top: *Azotobacter vinelandii* Fd I; co-adsorbate polymyxin, pH 7.0 scan rate 20 mV s^{-1}, current scale 0.5 μA (note overlap between signals B′ and C′ at this pH); Centre: *Desulfovibrio africanus* Fd III; co-adsorbate neomycin, pH 7.0 scan rate 191 mV s^{-1}, current scale 5 μA; Bottom: *Sulfolobus acidocaldarius* Fd; co-adsorbate polymyxin, pH 7.3, scan rate 10 mV s^{-1}, current scale 0.2 μA.

as $[3Fe\text{-}4S]^{1+/0}$) is a one-electron reaction. The narrow width of signal C′, evident under certain conditions, indicate that the two electrons may be taken up and released in an effectively concerted manner. The redox process is very sensitive to pH with both thermodynamics and kinetics being reflected in interesting ways.

The model that is emerging invokes the formation of a novel cluster species in which all Fe sites are formally Fe(II) and electron addition to $[3Fe\text{-}4S]^0$, i.e. a combination of E reactions 2↔3 and 3↔4 in Fig. 9, is accompanied by protonation, possibly at each of the three μ_2 sulphido atoms which are expected to become appreciably electron rich in such a highly reduced state. Direct proton binding of this type would be able to compensate for the otherwise unacceptable increase in negative charge. Surprisingly the $[3Fe\text{-}4S]^{2-}$ cluster appears to be remarkably inert, as evidenced both from the persistence of the reoxidation wave even after several minutes, and by the lack of any significant rate of electrocatalytic H_2 production. It remains to be seen whether this state has any direct physiological relevance.

3.1.4. Reversible metallation of [3Fe-4S] clusters: Another aspect of current interest is the occurrence and properties of heterometal sulphide clusters [88]. For example, the crystal structure determination of nitrogenase revealed a novel [Mo7Fe-9S] cluster consisting of cubane-type fragments [Mo3Fe-3S] and [4Fe-3S] linked by three μ_2 sulphides [85]. Several questions arise. The [4Fe-4S] cluster is familiar to all bioinorganic chemists, but to what degree might other (heterometal) clusters be chemically feasible? If so, what spectroscopic characteristics are observed and what chemical factors influence or indeed insure the prevalence of *all-Fe* clusters? To help answer these questions, voltammetric techniques have been applied to study reversible binding of a variety of metal ions at the [3Fe-4S] cluster in Da. Fd III [51, 53, 54, 56]. As a result, heterometal clusters with various M have been "synthesized" and their stabilities and properties determined.

A multi-pot cell, in which four sample compartments radiate from a central compartment containing the reference electrode, has proved to be a very convenient device for carrying out metallation and ligation reactions at proteins immobilized on the electrode [63]. After coating the electrode with the ferredoxin/polymyxin mixture, the electrode is first cycled in buffer solution containing EGTA in order to equilibrate the film, and is then transferred to a solution containing the metal ion to be tested. Cluster interconversions are then executed under the control of the applied potential and monitored through their voltammetric response. Verification of product formation has been carried out by performing the appropriate "bulk" solution transformations and characterizing the product by various spectroscopic methods. The film voltammetric method is thus seen here as a "trailblazing" technique, but it also yields quantitative data on reduction potentials (which are again verified by bulk solution experiments), and enables acquisition of equilibrium constants and kinetic data that would be very difficult to obtain by other methods. Table 2 lists some such information on heterometal [M3Fe-4S] cluster products generated in Da. Fd III and discovered by voltammetry. The non-cluster ligand(s) to M remain unestablished at this point, but H_2O or carboxylate from aspartate are likely candidates.

A recent example of a cluster metallation reaction is shown in Fig. 12 and it demonstrates how voltammetric methods can make the initial and most obvious detection of a new species [56]. Here, a PGE electrode on which is coated a neomycin-stabilized film of Da. Fd III was introduced to a

Table 2. Novel [M3Fe-4S] clusters detected and quantified by voltammetry of *Desulfovibrio africanus* Fd III adsorbed on pyrolytic graphite electrodes in presence of neomycin or polymyxin

Cluster/couple	K_d for M	Rate of formation	$E^{0'}$/mV
$[Fe3Fe\text{-}4S]^{2+/1+}$	30 μM	slow	−393
$[Zn3Fe\text{-}4S]^{2+/1+}$	1.6 μM	intermediate	−492
$[Cd3Fe\text{-}4S]^{2+/1+}$	0.8 μM	intermediate	−569
$[Tl3Fe\text{-}4S]^{2+/1+}$	1.5 μM (1+) 34 mM (2+)	fast	81
$[Cu3Fe\text{-}4S]^{2+/1+}$	<< μM (1+)	slow for Cu(II)	148

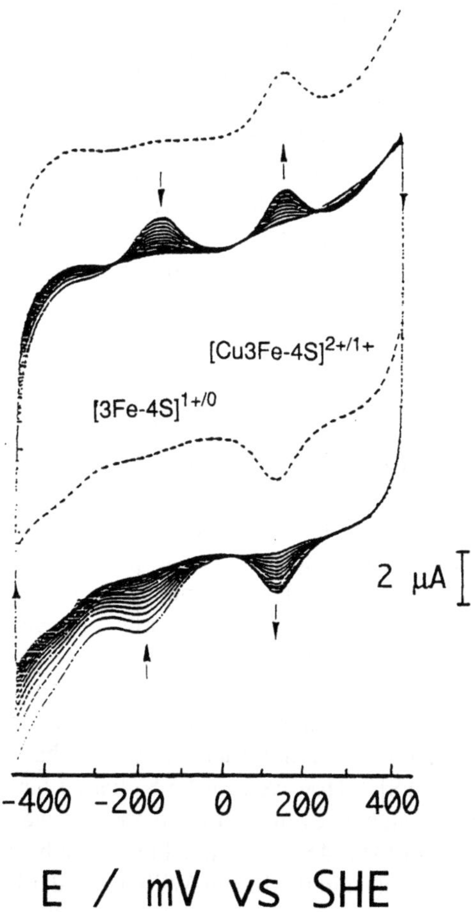

Fig. 12. Formation of a [Cu3Fe-4S] cluster in *Desulfovibrio africanus* Fd III. Successive cyclic voltammetric cycles of an immobilized film of Fd III prepared at an edge-oriented pyrolytic graphite electrode using neomycin as co-adsorbate, following introduction of the electrode to a solution (0.1 M NaCl, 20 mM HEPES, pH 7.0) containing 10 μM $CuSO_4$. Scan rate 191 mV s^{-1}, temperature 0°C. The voltammogram of the final product is displaced and shown as a dashed line.

solution containing 0 μM Cu (II) [54]. Upon cycling at 0.2 Vs^{-1}, the signal A′ due to $[3Fe\text{-}4S]^{1+/0}$ (reaction 1↔2 in Fig. 9) at $E^{0\prime} = -150$ mV was replaced by a signal (D′) at $E^{0\prime} = +150$ mV. The oxidized and reduced forms of the species giving rise to signal D′ were generated by appropriately designed bulk solution experiments and characterized by EPR and MCD spectroscopy. As a result, it was established that the new signal arises from formation of a $[Cu3Fe\text{-}4S]^{2+/1+}$ cluster exhibiting the "HiPIP" couple, reaction 5↔6. The transformation occurs mainly via reaction 2↔6 with M being either Cu(II) (which is not expected to enter a tetrahedral-type geometry easily) or Cu(I) produced at the electrode surface. The EPR spectrum of $[Cu3Fe\text{-}4S]^{2+}$ (S = 1/2) is sufficiently similar to that of the parent cluster $[3Fe\text{-}4S]^{1+}$ that existence of [Cu3Fe-4S] clusters in natural systems could go undetected by this method. On the other hand, the MCD spectra reveal large changes in the electronic structure; however, MCD is rarely used in primary investigations of biological samples. It was noted that the binding of Cu(I) to $[3Fe\text{-}4S]^{0}$ is much tighter than the binding of Zn(II) or Tl(I) (see below), and whereas EGTA or EDTA are capable of removing M(II) from $[M3Fe\text{-}4S]^{2+}$, this was not observed for the $[Cu3Fe\text{-}4S]^{2+/1+}$ cluster in either oxidation level. Instead, the transformation could be reversed quantitatively by transferring the film to a dilute solution of CN^- and holding at $E_{applied} > +150$ mV, whereupon signals A′ reappeared.

Uptake and release of Fe(II), Zn(II), and Cd(II) are slow-to-intermediate on the normal cyclic voltammetry timescale (case B as defined in section 2.3.1) and sweeps at rates of between 20 and 200 mV s^{-1} revealed the progress of interconversion of signals due to the product (bound) and reactant (unbound) species which have different reduction potentials [53]. New species were assigned by spectroscopic analysis of bulk-transformed samples. By holding the applied potential so that the oxidation levels of reactants and products are well defined, e.g. as $[3Fe\text{-}4S]^{0}$ (i.e. one electron reduced) and $[M3Fe\text{-}4S]^{2+}$ (oxidized) respectively, it was possible to use periodic rapid sweeps to measure the amounts of metallated and non-metallated cluster present at equilibrium and thus determine the appropriate equilibrium constants for reaction 2↔6, free of the complications of redox effects. In this way it was ascertained that Cd(II) and Zn(II) bind to the $[3Fe\text{-}4S]^{0}$ cluster much more tightly than Fe(II), i.e. that $[Zn3Fe\text{-}4S]^{2+}$ is more stable than $[4Fe\text{-}4S]^{2+}$. However, Zn-Fe-S clusters have not so far been detected in natural systems, even though their relative thermodynamic stability with respect to [4Fe-4S] is reasonable on the basis of the relative positions of Zn(II) and Fe(II) in the Irving-Williams order [53]. Against this argument, no consideration is given to the probability that the Fe in typical [4Fe-4S] clusters (i.e. with four cysteines) is stabilized through electron delocalisation (which effectively imparts Fe(III) character). In the case of Da. Fd III, it is possible that the substitution of cysteine by a weaker donor (perhaps carboxylate or water) accounts for the relative instability of the [Fe3 Fe-4S] cluster.

The [3 Fe-4 S] cluster of Da. Fd III also reacts with Tl(I). Binding and release of Tl(I) is much faster than for the 2+ ions, and even at 50 mV s a single redox couple is observed at an electrode potential dependent on the concentration of Tl(I) [54]. This is a fast chemical transformation, and represents case C of section 2.3.1. The system is always equilibrated within the voltammetric timescale (up to 1 V s^{-1}) and the result obtained is effectively the same as the potentiometric reduction potential. Tl(I) binds to $[3Fe\text{-}4S]^0$ with about the same affinity as Zn(II), and to $[3Fe\text{-}4S]^{1+}$ with much lower affinity. In support of this assignment, convincing spectroscopic evidence has recently been obtained for the binding of Tl(I) to both 1+ and 0 oxidations levels of the [3 Fe-4 S] cluster of *Pyrococcus furiosus* ferredoxin [91].

3.1.5. Ligand interchange at clusters: Ligand exchange at Fe-S clusters is of interest if these centres serve to trigger protein conformational change perhaps as a function of redox state, or act as catalysts, as in aconitase [83]. For the purposes of visualizing and analyzing redox-coupled ligand exchange, the binding of 1-hydroxyethane-2-thiolate (the mercaptoethanol anion) to the transformed [4 Fe-4 S] cluster of Da. Fd III has provided a very well-behaved "model" system [55]. The non-cysteine ligated cluster (distinguished hereafter by the formula [Fe 3 Fe-4 S] is first formed by cycling a film of the 7 Fe protein in 0.1 mM Fe(II), and then the electrode is transferred to a cell containing the ligand. The concentration of thiolate (RS^-) is dependent on the thiol concentration and pH (pK = 9.2), and the amount of RS^- available is many orders of magnitude greater than the minuscule quantity of protein on the electrode. 1-hydroxyethane-2-thiolate is a particularly suitable ligand to study because it forms only a weak complex with Fe(II). Thus even with very high levels of thiol in the contacting solution, the [Fe 3 Fe-4 S] cluster remains intact and the supporting level of Fe(II) (present in the thiol solution in order to prevent loss of Fe from the cluster) is not quenched.

The signals (D′) originating from $[Fe3Fe\text{-}4S]^{2+/1+}$ show marked sensitivity to thiol concentration, pH and scan rate. Figure 13 shows how the signals originating from D′ vary with scan rate. The main complication for analysis is that signals D′ overlap signal B′, which is due to the inert [4 Fe-4 S] cluster (see Fig. 10). The results are analyzed by computer simulation using the square scheme model shown in Fig. 14, in which electron transfer is fast for all species, while ligand binding occurs to both oxidized and reduced states of the cluster with differing affinities K_d^{ox}, K_d^{red}, and rate constants k_{on}^{ox}, k_{off}^{ox}, k_{on}^{red}, and k_{off}^{red}. The values resulting from the study are included in Fig. 14. At low scan rate, the species in the cycle remain in Nernstian equilibrium (case C of Fig. 6) and the position of the signal at $E_{obs}^{0'}$ is dependent on the concentration of thiolate. At high scan rates, the species $[Fe3Fe\text{-}4S]^{2+}$-L is "trapped out" (case B) and the resulting signal (termed F′) is at the reduction potential $E_F^{0'}$. From the variation of $E_{obs}^{0'}$ with thiolate

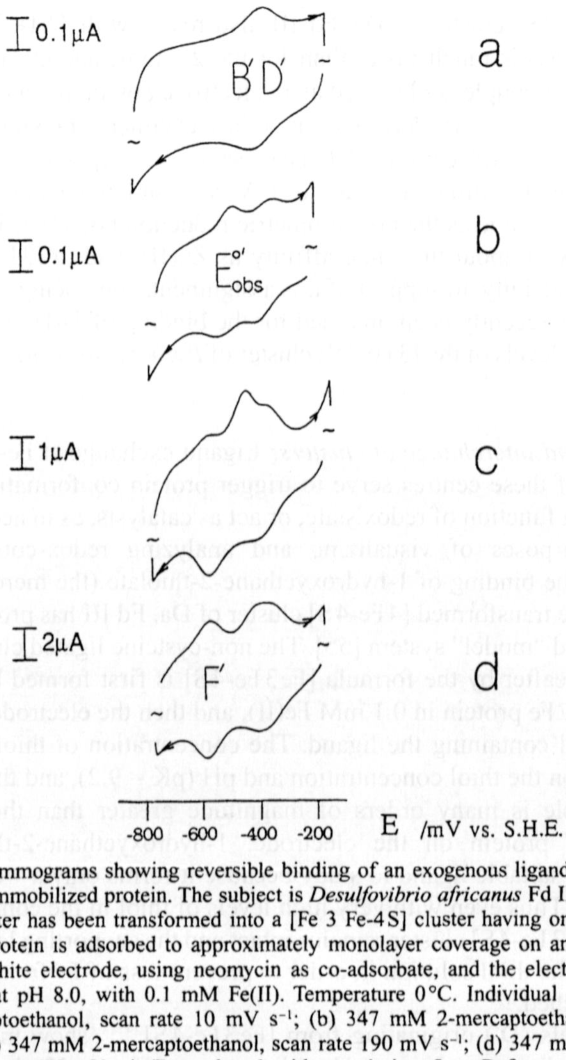

Fig. 13. Voltammograms showing reversible binding of an exogenous ligand to a [4Fe-4S] cluster in an immobilized protein. The subject is *Desulfovibrio africanus* Fd III, in which the [3Fe-4S] cluster has been transformed into a [Fe 3 Fe-4S] cluster having one non-cysteine ligand. The protein is adsorbed to approximately monolayer coverage on an edge-oriented pyrolytic graphite electrode, using neomycin as co-adsorbate, and the electrolyte contains 0.2 M NaCl at pH 8.0, with 0.1 mM Fe(II). Temperature 0°C. Individual conditions are: (a) no mercaptoethanol, scan rate 10 mV s^{-1}; (b) 347 mM 2-mercaptoethanol, scan rate 10 mV s^{-1}; (c) 347 mM 2-mercaptoethanol, scan rate 190 mV s^{-1}; (d) 347 mM 2-mercaptoethanol, scan rate 470 mV s^{-1}. (Reproduced, with permission, from Reference 63. Copyright 1993, Academic Press).

concentration, and knowing that $E^{0'}_{obs} = E^{0'}_{F}$ at the hypothetical high-concentration limit (this would be in the region of 1 M, under which conditions the clusters are destroyed) it was possible to determine K_d^{ox} and K_d^{red}. The kinetics of ligand exchange were analyzed by iterative simulation of the coupled E (Butler-Volmer) and C (first- and second-order) reactions. In this way it was determined that the increased affinity of thiolate for the oxidized cluster $[Fe3Fe\text{-}4S]^{2+}$ is almost solely the result of a much faster rate of ligation. Redox transformations thus appear "gated" by ligand exchange reactions. As indicated also on Fig. 14, the voltammetric experiment readily enables us

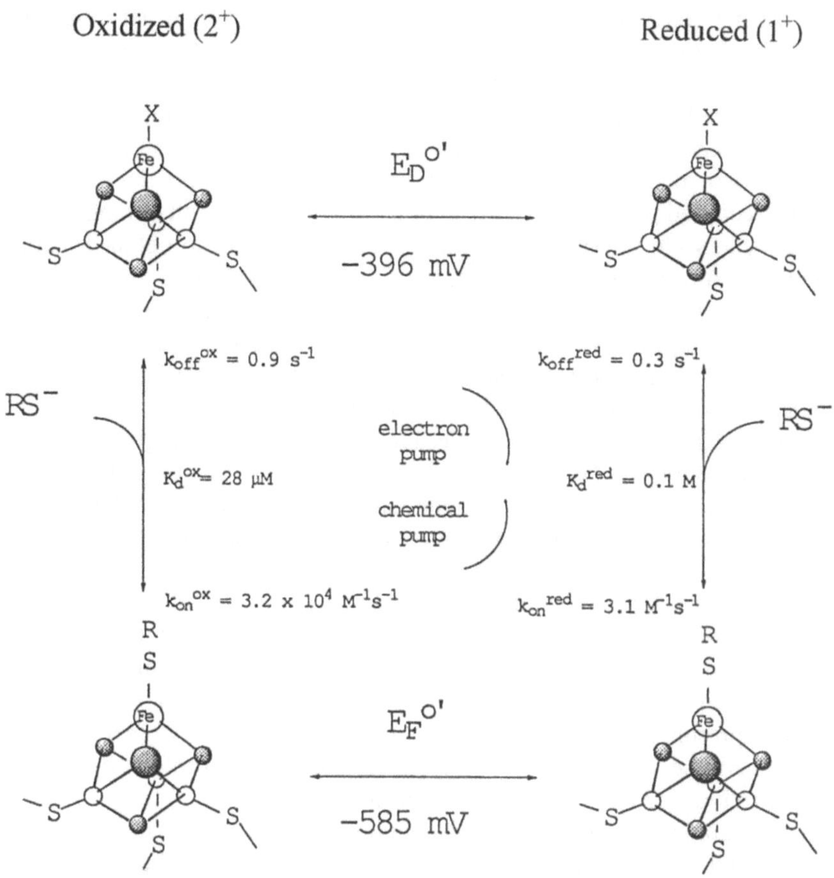

Fig. 14. Square scheme depicting the reversible binding of an exogenous ligand to a [4Fe-4S] cluster. Data are as determined by simulation for binding of 1-hydroxyethane-2-thiolate at the transformed cluster of *Desulfovibrio africanus* Fd III, at 0 °C.

to visualize this gating, since the cycle predominately turns clockwise, with electrochemical energy being used to drive ligand transfer. Hypothetically, if one was able to energize the system instead by enforcing unfavourable ligand binding (for example by mechanically inducing conformational change in a protein, or hydrolyzing ATP etc.) then this energy could be used to increase redox driving forces. In this example, [Fe 3 Fe-4 S]$^{1+}$-L is a much stronger reductant than [Fe 3 Fe-4 S]$^{1-}$, but its formation by chemical means requires the thermodynamically unfavourable binding of L to the reduced cluster.

3.1.6. Kinetics of proton-linked electron transfer: By comparison with electron transfer, proton-transfer reactions in metallo-proteins have received relatively little attention, despite their obvious importance in enzymology and bioenergetics. The transfer centres involved are bound water molecules and side-chain acids and bases, and a good example of the progress being made is the work of Feher and co-workers who have studied a structurally characterized photosynthetic reaction centre and site-directed mutants bearing changes at certain residues (e.g. aspartate, glutamate) which are believed to mediate the transfer of protons to the bound quinone [98]. Redox-coupled proton transfer is another area in which voltammetric techniques can yield important information that is difficult to obtain by other methods. As the example in the previous section showed, provided the E reaction itself can be shown to be sufficiently reversible it becomes relatively straightforward to make simultaneous observations and analyses of the kinetics and thermodynamics of rate-limiting coupled reactions, i.e. to detect and analyze "gated" electron-transfer reactions.

Detailed multi-disciplinary studies have been carried out to determine the properties of the two Fe-S clusters of Av. Fd I [57, 58]. The [3 Fe-4 S] cluster in Fd I (see Fig. 10) is so far unique since the one electron-reduced state exists in two forms that are interconvertable depending on the pH (pK 7.8) [57, 99]. These forms are distinguishable by CD and MCD spectroscopy, whereby it is observed that the alkaline form is similar to $[3\,Fe\text{-}4\,S]^0$ clusters in other proteins, whereas the acid form exhibits very different spectra [99, 100]. Based on a variety of evidence from spectroscopy, crystallography and site-directed mutagenesis, it was concluded that the $[3\,Fe\text{-}4\,S]^0$ cluster binds a single proton, probably at one of the μ_2 sulphido groups. We have already mentioned the evidence for more general, extensive protonation at the highly reduced $[3\,Fe\text{-}4\,S]^{2-}$ clusters in different proteins.

In voltammetric experiments, protonation of the Av. Fd I $[3\,Fe\text{-}4\,S]^0$ cluster is observed as a pH dependence of the reduction potential; the slope above pH 7.8 approaches zero, while the slope below pH 7 increases to a value between −40 and −55 mV (0 °C) [57, 97, 101]. The wider relevance to proton-transfer mechanisms in proteins is that the [3 Fe-4 S] cluster is not directly accessible to solvent molecules but is instead located some 8 Å below the protein surface [95]. However, the intervening medium contains a carboxylate group from aspartate (D) −15. In view of its possible participation in proton mediation, it was therefore of interest to examine the corresponding voltammetry of a mutant (D15N) in which this residue is replaced by asparagine. By studying the protein in the immobilized configuration on a pyrolytic graphite edge electrode, and also (for comparison) as a bulk solution, it was determined that the pK of the $[3\,Fe\text{-}4\,S]^{1+/0}$ couple in the mutant is lowered to 6.9 and the limiting $E^{0\prime}$ value is raised from −430 mV to −409 mV. At pH values sufficiently lower than the respective pK values, both native and D15N were found to have identical reduction potentials. These results are consistent with the removal of a negatively charged group (pK <6) from a site adjacent to the cluster, and provide a numerical gauge of the influence of electrostatics across a crystallographically defined short distance.

As illustrated already in the previous section, a great strength of voltammetric methods lies in their ability to link together thermodynamic and kinetic data. Above pH 8, the waveforms corresponding to the couple [3 Fe-4 S]$^{1+/0}$ were found to be identical for the native and D15N mutant over a range of scan rates, the only difference being the shift in reduction potential. Under conditions in which protonation of the [3 Fe-4 S]^{0} cluster occurs, two cases could be distinguished corresponding to cases C and D of Fig. 6. At slow scan rates, where equilibration of species occurs, native and D15N forms gave indistinguishable voltammetry for the [3 Fe-4 S]$^{1+/0}$ couple. However, at a sufficiently high scan rate, the voltammogram obtained for the D15N mutant became very different to that observed for the native protein. Two result are overlaid in Fig. 15. Increasing the scan rate caused the oxida-

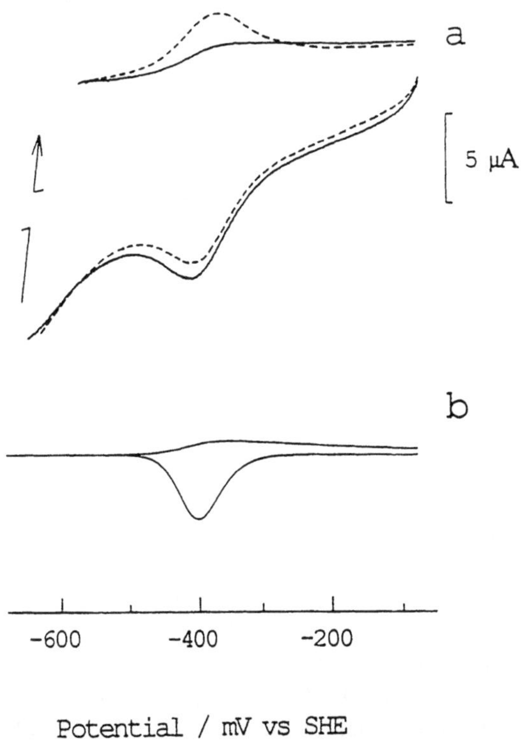

Fig. 15. Voltammograms showing proton-gated electron transfer at the [3 Fe-4 S] cluster in a mutant form of *Azotobacter vinelandii* Fd I (D15N), and comparison with the native form which differs only very marginally in the region of the switched residue, but which shows virtually normal reversible electrochemistry. a) Initial sweep voltammograms of D15N (solid line) and native (dashed line) forms of Fd I, adsorbed at an edge-oriented pyrolytic graphite electrode, using polymyxin as co-adsorbate, at 0 °C, and at scan rate 930 mV s^{-1}. To obtain similar thermodynamic driving forces for cluster protonation, native Fd I was measured at pH 6.5, while the D15N mutant was measured at pH 5.9. b) Simulation of the voltammetry of D15N, pH 5.9, at scan rate 930 mV s^{-1}. Parameters used are shown in Fig. 16. (Figure 15a reproduced, with permission, from Reference 58. Copyright 1993, American Chemical Society.)

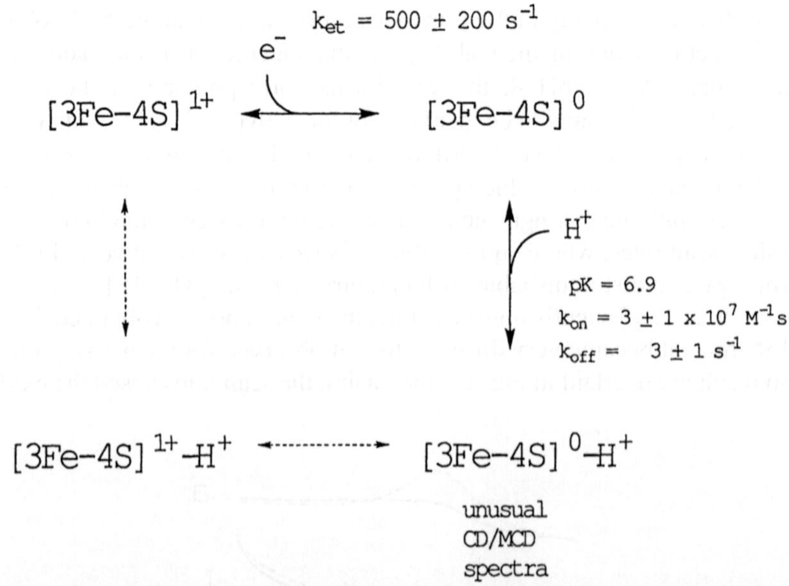

Fig. 16. Square scheme for proton binding to the [3 Fe-4 S] cluster of Av. Fd I, showing parameters giving best fit to data obtained for studies of the D15N mutant, in which proton transfer is retarded. There is no evidence for participation of the species $[3\,Fe\text{-}4\,S]^{1+}$-H^+, thus only two sides of the square are used.

tion wave to diminish in size and take on a drawn-out shape. For the native protein, the effect was much less marked. This behaviour is easily explained on the basis of the square scheme shown in Fig. 16, in which the chemical transformation is a proton-dependent reaction occurring on the intermediate timescale (case D). On sweeping rapidly to high potentials, no evidence was obtained for new signals corresponding to the protonated redox couple $[3\,Fe\text{-}4\,S]^{1+/0}$-$H^+$. It was thus evident that the reactions follow only two sides of the square, i. e. that the pathway for oxidation of $[3\,Fe\text{-}4\,S]^0$-H^- to $[3\,Fe\text{-}4\,S]^{1+}$ involves deprotonation followed by electron loss, and not the other way round. Oxidation is thus gated by deprotonation. Rate constants for proton binding and deprotonation were analyzed by iterative simulation of the coupling of E (Butler-Volmer) and C (first- and second-order) reactions to achieve the best fit to the observed data. The results are included in Fig. 16. It was apparent that as the scan rate is increased, the reduction potential (as measured from the average of E_{pa} and E_{pc}) decreased. This is as expected since shortening the timescale of the experiment results in increased decoupling of the rate-limiting proton transfer reactions. In other words, under conditions of very fast scan rate, the $[3\,Fe\text{-}4\,S]^{+1}$ cluster can gain and then lose an electron before proton transfer can occur; in which case the reduction potential observed is that of the uncoupled $[3\,Fe\text{-}4\,S]^{1+/0}$ couple even under acidic conditions.

A likely rationale for this behaviour is that the rate-limiting step is proton transfer, and that aspartate but not asparagine is active as a transfer site.

While this is not unexpected, in this case the crystal structure is well enough refined to reveal fine details of the aspartate carboxylate environment and its relationship to neighbouring groups, i.e. one carboxylate O is ca. 4.8 Å from the cluster while the other is H-bonded to nearby Lysine-84 [95]. A possibility to be investigated by further mutation studies is whether it is the presence of the salt bridge that enables the proton-transfer reaction to be rapid. The enhanced ability of film voltammetry to measure and compare properties of proteins differing only by single replacements of internal amino acids holds promise for future investigations of this type of reaction.

3.2. Studies of catalytic electron transfer in enzymes

3.2.1. Properties of fumarate reductase and succinate dehydrogenase: Recent voltammetric studies of two particular enzymes, fumarate reductase and succinate dehydrogenase, have revealed properties that have not been detected by conventional methods, and these final parts of Section 3 will focus upon discussion of the results. At appropriate points, we shall mention other studies of enzyme voltammetry that complement the emerging perspectives.

Fumarate reductase (FRD) and succinate dehydrogenase (SDH) are common names for two closely related enzymes that link fumarate reduction or succinate oxidation to redox interconversion of the membrane-confined quinone/quinol (Q/QH_2) pools of respiratory membranes [102].

$$\text{fumarate} + QH_2 \rightleftharpoons \text{succinate} + Q$$

Both enzymes consist of a catalytic domain that "plugs in" to the respiratory chain by binding to "anchor" proteins located in the membrane. This configuration is depicted in Fig. 17. The catalytic domain, which comprises two tightly interlocked subunits α and β, can be fully resolved from the anchor proteins by anaerobic treatment with butanol or perchlorate [102]. Although the anchor proteins are essential for reaction with Q/QH_2, the soluble $\alpha\beta$ domain is still a very active catalyst for redox interconversion of fumarate and succinate by artificial redox partners. Mammalian mitochondrial SDH is particularly important, since as Complex II (i.e. in association with the hydrophobic 'anchor' subunits) it provides a direct link between the citric acid cycle and the respiratory electron-transport chain. In an analogous manner, FRD is located in the cytoplasmic membrane of *E. coli*, when grown under anaerobic conditions with fumarate as the terminal electron acceptor.

In either enzyme, the α subunit (70 kDa) contains one covalently bound FAD, while the smaller β subunit (30 kDa) contains three Fe-S clusters – Centre 1 ($[2Fe\text{-}2S]^{2+/1+}$), Centre 2 ($[4Fe\text{-}4S]^{2+/1+}$) and Centre 3 ($[3Fe\text{-}4S]^{1+/0}$) [102]. Reduction potentials (at pH 7) of centres in *E. coli* FRD and beef heart SDH have been measured. Values for the Fe-S centres

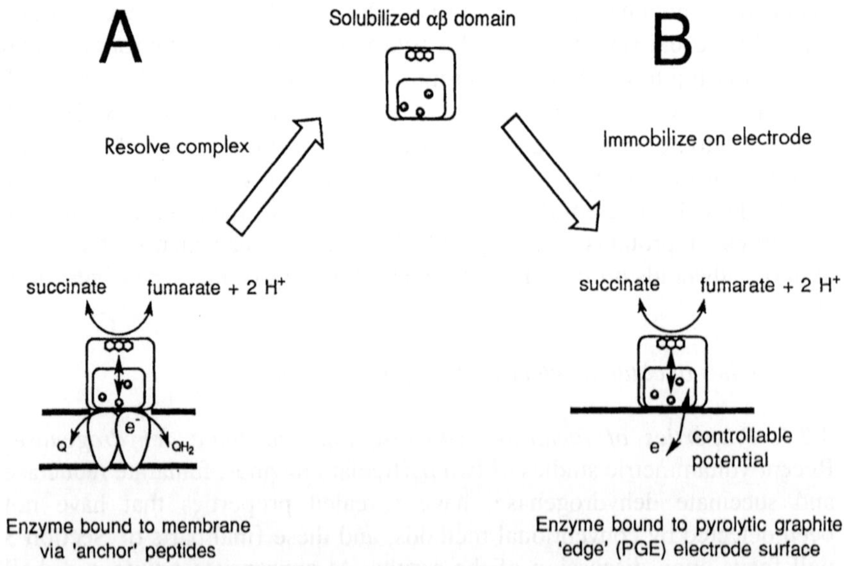

Fig. 17. Cartoons showing fumarate reductase and succinate dehydrogenase, their resolution to give a soluble $\alpha\beta$ domain, and the adsorption of this domain at an electrode.

are: Centre 1: FRD, –79 mV, SDH, 0 mV; Centre 2: FRD, –320 mV, SDH, –260 mV; and Centre 3: FRD, –70 mV, SDH, +60 mV. The FAD is tuned for two-electron reaction (FRD, –55 mV; SDH, –79 mV) the radical being thermodynamically unstable (maximum < 10% in SDH). The FAD and Centres 1 and 3 are appropriate for electron transport between succinate/fumarate (30 mV) and the quinone pool (–80 for menaquinone in *E. coli* and 60 mV for ubiquinone of mitochondria), although there is an obvious problem as to the role of Centre 2. As is the case with other multi-centred redox enzymes, little is established about activity regulation, nor indeed about the pathway and energetics of intramolecular electron transfer (although it is assumed that the FAD is the site of fumarate/succinate interconversion). It is also suggested that Centre 3 is the immediate donor/acceptor site for electron exchange with the quinone pool [102].

3.2.2. Voltammetry of fumarate reductase: A detailed voltammetric study of catalytic electron transport by fumarate reductase has been carried out [61]. The $\alpha\beta$ domain of FRD adsorbs from a submicromolar solution onto a freshly polished edge-oriented pyrolytic graphite (PGE) electrode, yielding a modified electrode that is catalytically active in fumarate/succinate interconversion. The immobilization process can be viewed (Fig. 17 right) in terms of the anchor domain being replaced by an electrode surface. High coverage is obtained in the presence of polymyxin, and this enables direct observation of at least three centres through their voltammetric signals (I, II and III) in the absence of substrate. Of these signals, I and II have been

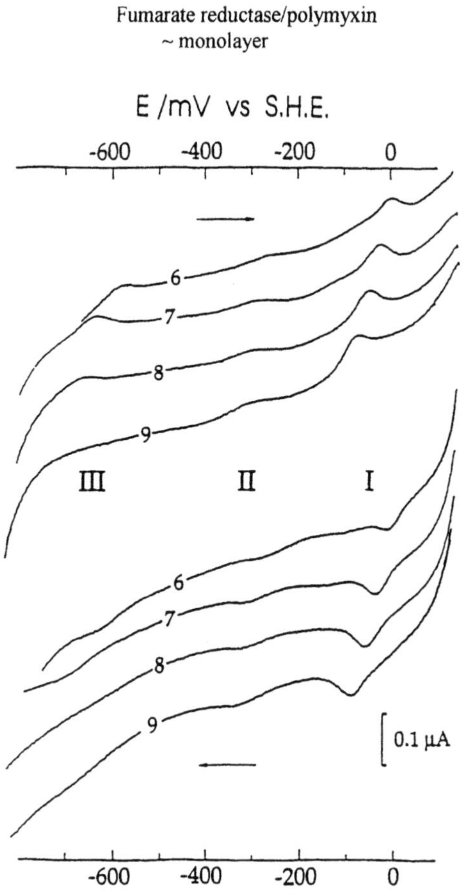

Fig. 18. Voltammograms of an adsorbed film of fumarate reductase at various pH values. The enzyme adsorbs at an edge-oriented pyrolytic graphite electrode from a 0.82 μM solution containing 0.1 M NaCl and 0.1 g/l polymyxin, at 0 °C. Scan rate 50 mV s^{-1} in all cases. Signal I is assigned to FAD, signal II is assigned to the [4 Fe-4 S] cluster, and signal III is tentatively suggested to be the novel couple $[3 Fe\text{-}4 S]^{0/2-}$. (Reprinted, with permission, from Reference 61. Copyright 1993, American Chemical Society.)

tentatively assigned, on the basis of their shape and correspondence of $E^{0'}$ with identified centres. The voltammograms in Fig. 18 were measured at 3 °C and were established to be the same whether measured at a stationary or rotating electrode. The electroactive coverage was estimated using several different determinations of wave area [61]. While no single one of these methods was totally satisfactory, they each yielded a value close to expectations for one electroactive monolayer. The problems in carrying out such a determination are readily apparent. First, even with a high coverage, the faradaic signals are small compared to the double-layer capacitance. Second,

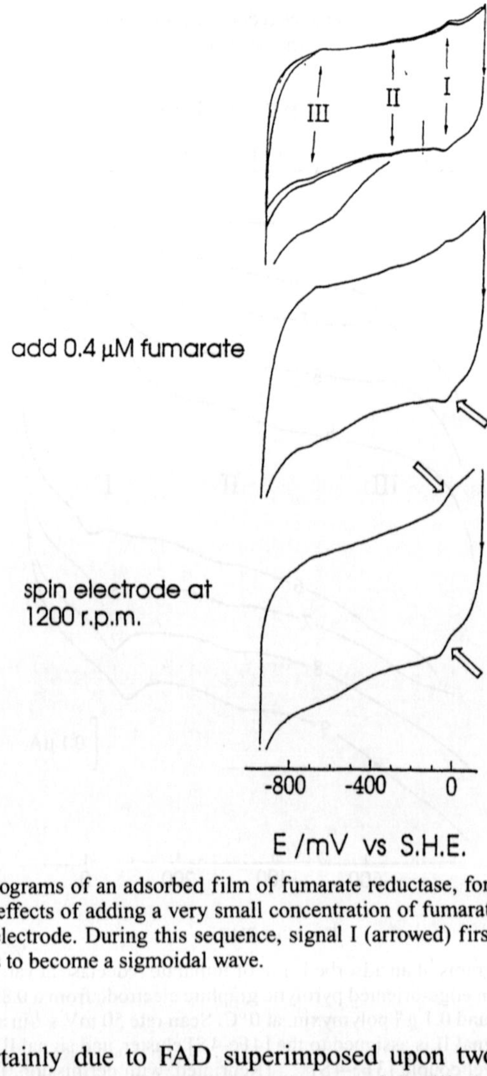

Fig. 19. Voltammograms of an adsorbed film of fumarate reductase, formed as described in Ref. 61, showing effects of adding a very small concentration of fumarate (0.4 μM) and then also rotating the electrode. During this sequence, signal I (arrowed) first becomes distorted, and then collapses to become a sigmoidal wave.

signal I is certainly due to FAD superimposed upon two signals due to Centres 1 and 3. The best estimation made use of the likely probability that the main body of signal I arises from the FAD component. The sharpness of the FAD signals, in particular, indicates that this layer comprises a reasonably homogeneous assembly of non-interacting molecules.

Figure 19 illustrates the subtle effects that can be observed when an extremely active enzyme on an electrode is provided with substrate [61]. At 25 °C, and without substrate, the predominantly FAD signal is observed at $E^{0\prime} = -48$ mV, while the signal assigned to Centre 2 appears at −311 mV. Addition of just a very small amount of fumarate (i.e. a concentration as low as 0.4 μM) to the electrolyte causes the FAD signals to distort, then upon rotating the electrode they collapse to a sigmoidal reduction wave with half-wave potential $E_{1/2} = -32$ mV. This observation identifies the catalytic activity with

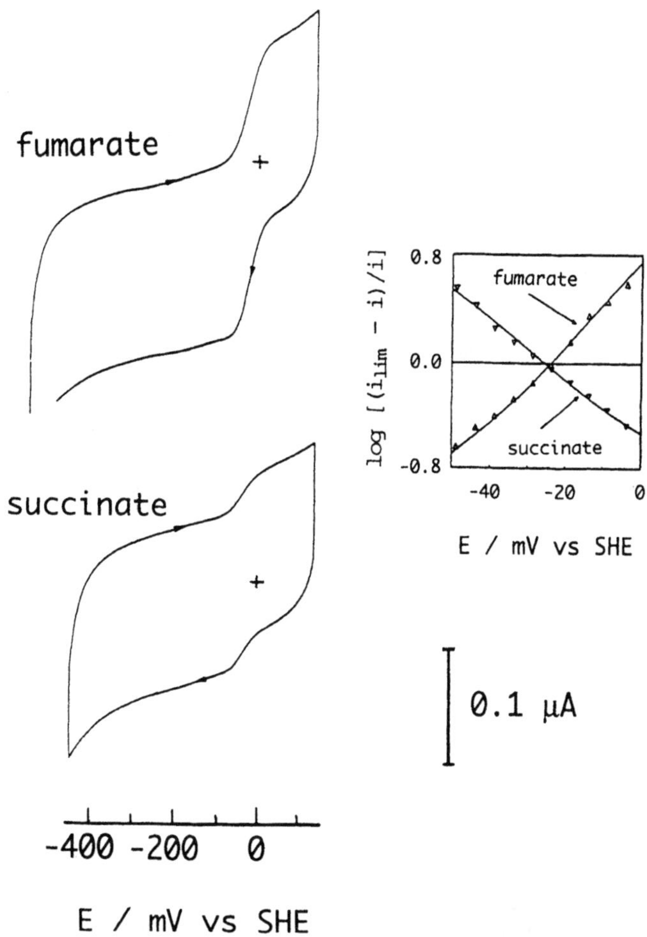

Fig. 20. Rotating disc cyclic voltammograms for adsorbed film of fumarate reductase catalyzing fumarate (2.6 μM) reduction or succinate (13 μM) oxidation. Enzyme adsorbed as described in legend to Fig. 18. Rotation rate 900 r.p.m., temperature 25 °C, pH 7.0. Inset shows corresponding Heyrovsky-Ilkovic plots, and demonstrates that the same value of $E_{1/2}$ is obtained for electrocatalysis in either direction.

a specific signal (redox centre) of the enzyme. As the fumarate concentration is increased this wave becomes larger but appears increasingly distorted, the possible reasons for which are discussed below. As Fig. 20 shows, a sigmoidal response was also obtained using succinate as substrate (albeit at a rather higher concentration) where in this case the rotating disc electrode revealed an oxidation current with $E_{1/2} = -26$ mV. The close correspondence between $E_{1/2}$ values for catalysis in either direction shows that the adsorbed enzyme functions as a 'electrochemically reversible' electrocatalyst, at least under conditions of low electrochemical demand, i.e. at low substrate concentrations.

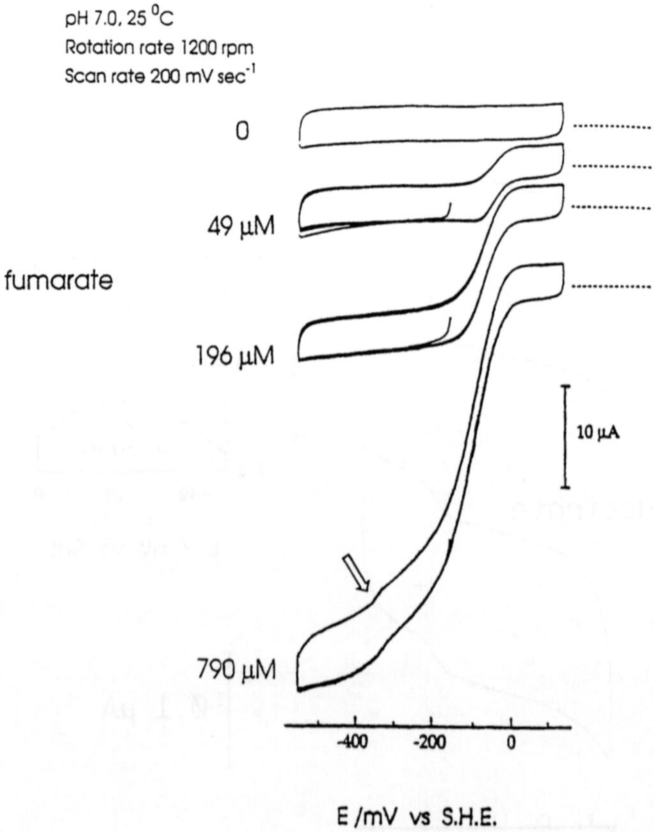

Fig. 21. Catalytic reduction of fumarate by FRD/polymyxin film at rotating PGE disk electrode. Rotating disc cyclic voltammograms for electrocatalysis by an adsorbed film of fumarate reductase at an edge-oriented pyrolytic graphite electrode, showing effect of increasing the fumarate concentration on currents and waveforms. Temperature 25 °C, pH 7.0, rotation rate 1200 r.p.m. Arrow indicates ridge that coincides with position of signal II in the no-turnover voltammetry. (Adapted, with permission, from Reference 61. Copyright 1993, American Chemical Society.)

The kinetics of electrocatalysis of fumarate reduction by FRD were investigated by varying both substrate concentration and rotation rate, and employing the approach outlined in Section 2.3.2 [61]. Figure 21 shows cyclic voltammograms measured at a rotation rate of 1200 r.p.m. for increasing concentrations of fumarate. The data processed for a large number of such experiments are shown in Fig. 22. The catalytic current was measured in the region of potential at which a limiting value (i_{lim}) is reached, thus minimizing or removing the term $1/i_E$. Then the term $1/i_{Lev}$ was removed, by measuring i_{lim} as a function of electrode rotation rate and analyzing the data as a double-reciprocal (Koutecky-Levich) plot. For each fumarate concentration this yielded values of i_{cat}, the equivalent of reaction velocity in conventional steady-state enzyme kinetics. The relationship be-

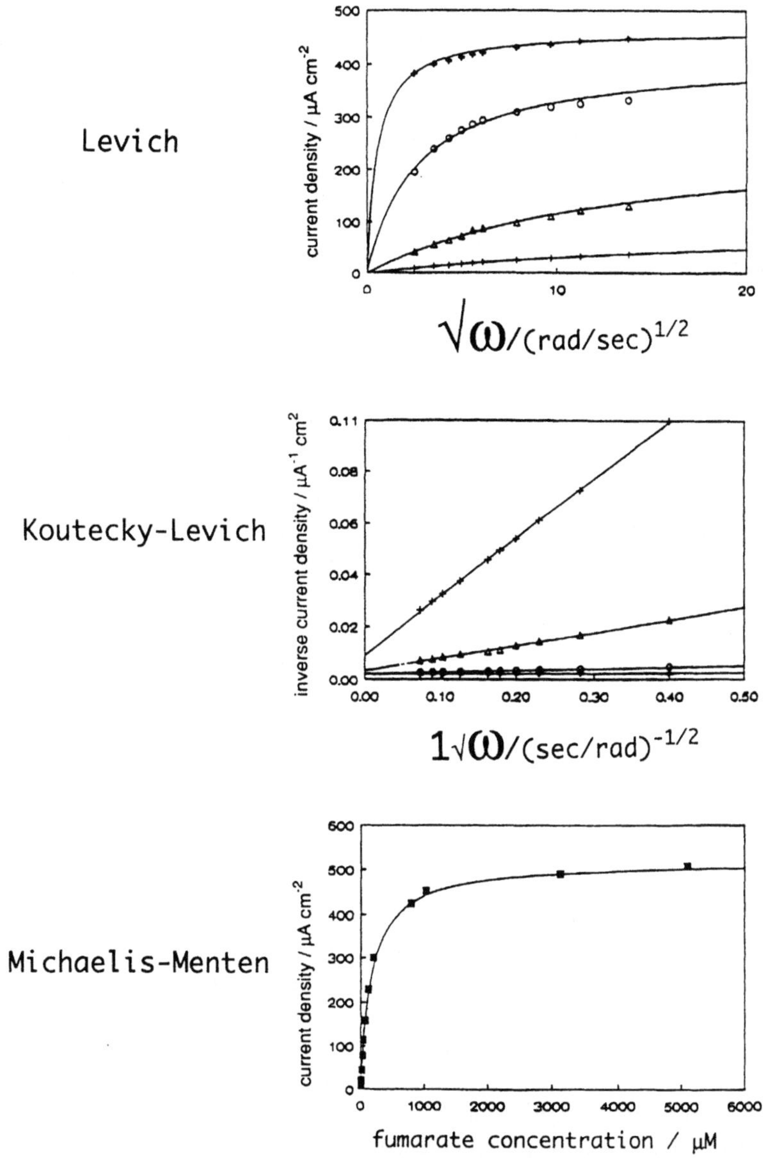

Fig. 22. Levich, Koutecky-Levich and Michaelis-Menten plots for electrocatalysis by an adsorbed film of fumarate reductase at an edge-oriented pyrolytic graphite electrode. Temperature 25 °C, pH 7.0. Other conditions are as given for Fig. 18.

tween i_{cat} and fumarate concentration thus led to the quantities i_{max} and K_M. Finally, k_{cat} was determined from i_{max} by using the calculated surface coverage of electroactive enzyme. Working through these steps, the values k_{cat} = 840 s^{-1} and K_M = 0.16 mM were obtained.

The interesting point about these results is not so much the values themselves but the fact that they are in very good agreement with results obtained by conventional methods [102]. In so testifying to the validity of the voltammetric approach, the data provide confidence to probe more deeply. Referring once again to Fig. 21, an observation of interest is the marked distortion of voltammograms at very high electron flux (see the example of 0.79 mM fumarate). One explanation for this is that a limitation now arises in the first-order rate constant for interfacial electron transfer (the exchange rate constant in the Butler-Volmer equation). In other words i_{cat} has become so large that interfacial electron transfer (i_E) becomes rate limiting unless the electrochemical driving force is increased. Along the same lines, comparing 49 and 196 μM scans, it is also easily seen that the half-wave potential shifts to a more negative value as higher current is drawn (the increasing deviation from reversibility as work is performed is consistent with classical thermodynamic principles). Another observation is the minor wave that appears close to the limiting current region under high electron flux, and which provides a small boost to the catalytic current. This wave coincides with the position of the non-turnover voltammetric signal assigned to Centre 2 and suggests that this site provides a second route into the enzyme to supplement the primary pathway (which probably involves Centres 1 or 3).

3.2.3. Voltammetry of succinate dehydrogenase: For SDH, the analogous voltammetric experiments revealed an interesting effect that may have physiological relevance [59, 60, 62]. As with FRD, the soluble $\alpha\beta$ domain of SDH again adsorbs at PGE from a dilute solution to yield a system capable of electrocatalytically oxidizing succinate, although (unlike FRD) a visible response in the absence of substrate has not so far been obtained. The direction of catalytic action was established to be pH dependent, with catalysis in the oxidation direction being increasingly favoured as the pH is raised. The results shown in Fig. 23 were obtained for experiments performed at 36 °C over a range of pH conditions. At pH 8.3, succinate oxidation is fast, and estimates of kinetic parameters (based upon a maximum of 5% electroactive coverage) indicated k_{cat} = ca. 330 s^{-1} and K_M = 0.22 mM, each broadly in agreement with published values. At pH 7.5, the succinate catalytic current is diminished, but addition of fumarate to the solution produces a strong, peak-type reduction wave which persists irrespective of whether the scan is in the direction of increasing or decreasing potential, and whether the electrode is stationary or rotating. The effect observed at pH 6.5 with increased concentration of fumarate is even more striking. Use of a Au electrode revealed a similar response (though attenuated and less stable) with peak positions matching those observed at graphite [60]. This showed that

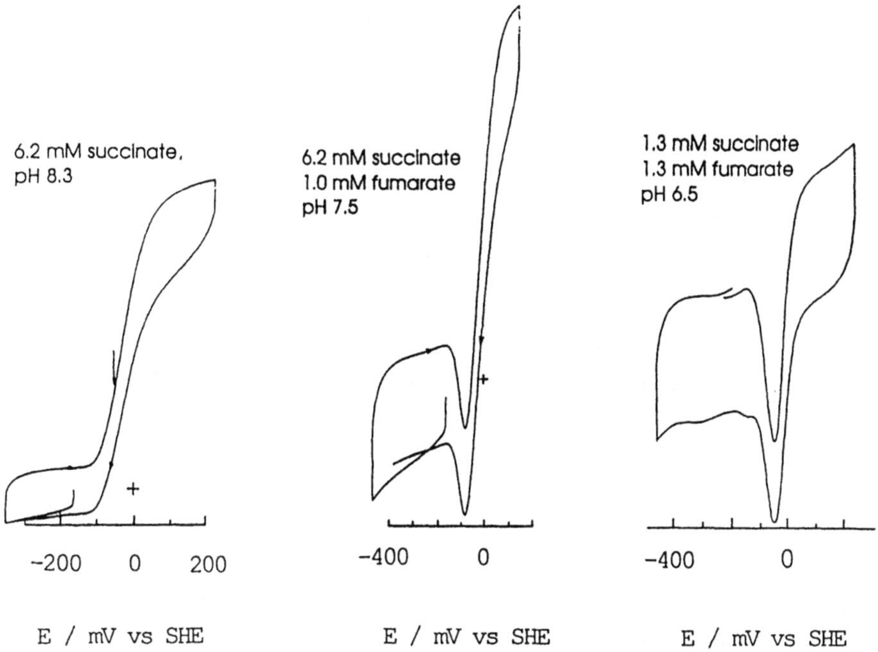

Fig. 23. Voltammograms revealing the characteristic electrocatalytic properties of succinate dehydrogenase. In each case, enzyme is adsorbed at an edge-oriented pyrolytic graphite electrode from a 0.7 μM solution at 36 °C, and scan rate is 10 mV s^{-1}. Other conditions as shown.

the effect does not arise from potential-dependent desorption or re-orientation of the enzyme, which would not be expected to have identical characteristics for two very different electrode materials.

The conclusion reached from these experiments is that reduction of fumarate is catalyzed by SDH only within a narrow potential region. The catalytic electron transport pathway may thus be described as "gated", since the rate is determined not by elementary components of electron transfer theory but by the status of the enzyme (for example, a specific conformation or distribution of electrons among centres) at some critical stage of the turnover process. In the terminology of electronics, the i-E curve shown in Fig. 23 has a region of *negative resistance* and SDH has the properties of a tunnel diode [103].

Independent evidence for the "tunnel diode effect" was subsequently obtained from conventional, steady-state kinetic studies of fumarate reduction by SDH and FRD [60]. These assays are carried out spectrophotometrically by measuring the rate of oxidation of reduced (radical) benzylviologen by fumarate in the presence of the enzyme. When the catalyst is fumarate reductase, a typical decay curve is obtained with the reaction rate decreasing with time, i.e. as reduced benzyl viologen is consumed. This is typical behaviour and consistent with the voltammetry that is observed. However, when SDH is used as the catalyst, the reaction rate *increases* as the reaction proceeds, i. e. as reduced benzylviologen is oxidized. The reaction therefore displays a *negative order* with respect to reactant. Thus, in as far as the driving force is governed by the ratio of oxidized to reduced benzylviologen, the "tunnel-diode" effect is also manifested in conventional methods. Further steady-state studies of a variety of succinate dehydrogenases from different organisms, and in different stages of isolation, including detergent-solubilized Complex II and submitochondrial particles, have confirmed and generalized the observation of negative order with respect to viologen radical and highlighted the contrasting behavior of the enzymes classified as fumarate reductases [60].

The unusual inverse relationship between driving force and current can be easily explained and understood for electrocatalytic reactions, since all that is required is for one of the redox states of the catalyst to be inactive with regard to an essential step such as substrate binding. Indeed what we have termed the tunnel diode effect is apparent in several studies of non-biological electrocatalysis. For example, monomeric molybdenum catecholate complexes catalyse the reduction of oxoanions nitrite, chlorate and bromate [104]. At the electrode, the Mo(VI) state is reduced to Mo(V) or Mo(IV), each of which is substitutionally labile and reoxidized by substrate in an inner-sphere mechanism. However, increasing the electrochemical driving force causes generation of Mo(III) which is substitutionally inert, and the electrocatalytic reaction slows down. Another example is the four-electron reduction of O_2 by iridium porphyrins adsorbed on graphite [105]. The active form of the catalyst is a dimer, but reduction causes formation of an inactive hydrido monomer. Thus, as before, *increasing* the electrochemical driving force causes a *decrease* in the rate of catalysis. These types of behaviour can remain hidden unless one examines the system by electrochemical methods.

For SDH, one mechanistic possibility is illustrated by the simplified catalytic cycle shown in Fig. 24. Here the tunnel-diode effect arises by introduction of rigid requirements for the sequential ordering of electron transfer and of substrate binding or product release. This situation is analogous to substrate inhibition, the substrate in question being the electron. If SDH can bind reactants or release products *only* when a certain centre (for example the FAD) is oxidized, then the enzyme will cease to function when the rate of reduction of this centre exceeds the rate of substrate interchange. In the context of Fig. 8, the FAD would be both catalytic and regulatory site. In terms of function, this property would have the ability to regulate the direction of electron transport. Reverse electron flow (which would act to reverse the direction of the TCA cycle) would be allowed under normal conditions but would be suppressed if the quinone pool potential becomes too negative. It is interesting to speculate that SDH may have evolved from FRD, in effect involving the imposition of a "ratchet-like" restriction on the reverse reaction.

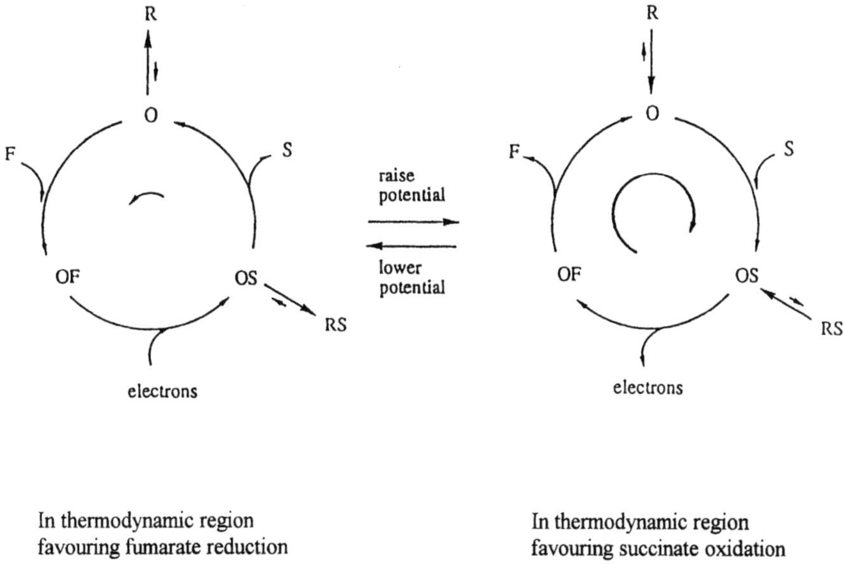

Fig. 24. Scheme showing how catalytic activity can be gated if substrate binding or release is dependent upon the oxidation state of a redox centre in the enzyme. In this example, only the oxidized form O is active, R is inactive. The terms F and S refer to fumarate and succinate respectively, while OF, OS and RS are enzyme-substrate complexes. An excessive (reductive) driving force has the effect of inhibiting reduction. (Adapted, with permission, from Reference 62. Copyright 1992, Nature Publishing Company).

Another example of a tunnel-diode enzyme effect is apparent in a recent report by Ikeda and co-workers, of the catalytic electrochemistry of D-gluconate dehydrogenase (GADH) [70]. This enzyme is a flavometallo-enzyme of 138 kDa, comprising three subunits and containing one FAD, one heme and Fe-S clusters. The enzyme is associated with the cytoplasmic membrane and resembles FRD and SDH in that it also serves to couple an aqueous phase organic redox transformation (D-gluconate to 2-keto D-gluconate) with the redox interconversion of a membrane-bound quinone/quinol pool. The detergent-solubilized enzyme adsorbs at various electrode materials and exhibits electrocatalysis of D-gluconate. As with SDH, the enzyme activity is maximized at a certain electrode potential, resulting in the cyclic voltammogram shown in Fig. 25. One may speculate that effects such as we have observed may be widespread among electron-transport enzymes and could serve to regulate and harmonize activity, yet are subtle enough to remain undetected by conventional methods.

4. Concluding remarks

At the expense of detail, this article has sought to provide a broad spectrum of brief outlines, illustrating how voltammetric methods can provide an

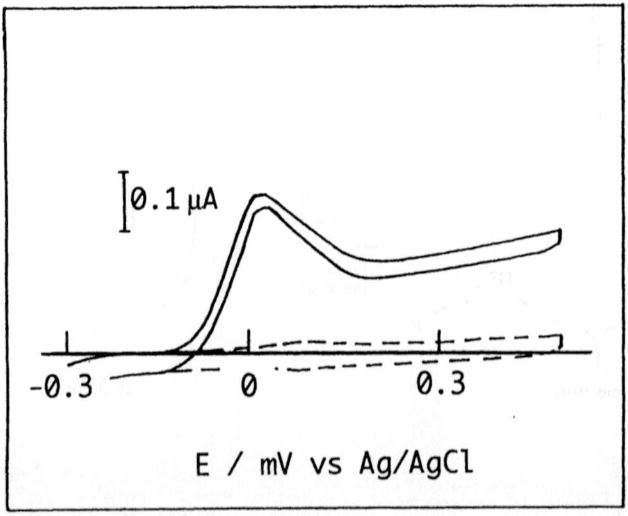

Fig. 25. Cyclic voltammograms for electrocatalytic oxidation of D-gluconate by gluconate dehydrogenase adsorbed at a glassy carbon electrode. Scan rate 5 mV s. Concentration of D-gluconate is 50 µM, temperature 5 °C. (Adapted, with permission, from Reference 70. Copyright 1993, Elsevier Sequoia S. A.)

"electrochemical eye" for visualizing and analyzing complicated, convoluted and even chaotic behaviour of biological redox systems. It may be appreciated that even for the most complex of electron-transport enzymes, the methods can provide a "direct read-out" of characteristic reaction properties. The field is in its infancy, and I expect and indeed hope that this chapter will soon be out of date through improvements in analysis, interpretation and technical sophistication.

Acknowledgments

I would like to thank numerous colleagues for their enthusiastic input throughout the past years, in particular Julea Butt, Artur Sucheta, Barbara Burgess, Brian Ackrell, Andrew Thomson, Simon George, Jacques Breton, Lisa Martin, Alan Bond and most recently, Jill Duff. I am grateful also to Allen Hill who first interested me in this field. Research in the author's laboratory has been supported over the past five years by the UK Science and Engineering Research Council, the Royal Society, the Exxon Education Foundation, the Petroleum Research Fund administered by the American Chemical Society, the United States National Science Foundation, NATO, the UK Engineering and Physical Sciences Research Council, and The Wellcome Trust.

References

1. G Dryhurst, KM Kadish, F Scheller and R Renneberg, *Biological Electrochemistry*, Academic Press, New York and London, 1982.
2. GS Wilson, Meth. Enzymol. 54 (1978) 396–410.
3. K Niki, T Yagi, H Inokuchi and K Kimura, J. Electrochem. Soc. 124 (1977) 1889–1891.
4. MJ Eddowes and HAO Hill, J. Chem. Soc. Chem. Commun. (1977) 771–772.
5. P Yeh and T Kuwana, Chem. Lett. (1977) 1145–1146.
6. MJ Eddowes and HAO Hill, J. Amer. Chem. Soc. 101 (1979) 4462–4464.
7. AJ Bard and LR Faulkner, *Electrochemical Methods: Fundamentals and Applications*, John Wiley, New York (1980).
8. AM Bond, *Modern Polarographic Methods in Analytical Chemistry*, Dekker, New York (1980).
9. Southampton Electrochemistry Group, *Instrumental Methods in Electrochemistry*, Ellis Horwood, Chichester (1985).
10. CMA Brett and AM Oliveira Brett, *Electrochemistry: Principles, Methods and Applications*, Oxford Science Publications, Oxford (1993).
11. RS Nicholson and I Shain, Anal. Chem. 36 (1964) 706–723.
12. AM Bond, KB Oldham and CG Zoski, Anal. Chim. Acta 216 (1989) 177–230.
13. CMA Brett and AMCF Oliveira Brett in *Comprehensive Chemical Kinetics*, CH Bamford and RG Compton (eds), Elsevier. Amsterdam, (1986) Vol 26, Chapter 5.
14. AP Brown and FC Anson, Anal. Chem. 49 (1977) 1589–1595.
15. E Laviron in *Electroanalytical Chemistry*, AJ Bard (ed), Marcel Dekker, New York (1982) Vol 12, pp 53–157.
16. H Matsuda, K Aoki and K Tokuda, J. Electroanal. Chem. 217 (1987) 15–33.
17. FC Anson, Anal. Chem. 38 (1966) 54–57.
18. CED Chidsey, Science 251 (1991) 919–922.
19. JG Osteryoung and RA Osteryoung, Anal. Chem. 57 (1985) 101A–110A.
20. JH Reeves, S Song and EF Bowden, Anal. Chem. 65 (1993) 683–688.
21. WR Hagen, Eur. J. Biochem. 182 (1989) 523–530.
22. ET Smith, DW Bennett and BA Feinberg, Anal. Chim. Acta 251 (1991) 27–33.
23. WJ Albery, MJ Eddowes, HAO Hill and AR Hillman, J. Amer. Chem. Soc. (1981) 103, 3904–3910.
24. J Haladjian, P Bianco and R Pilard, Electrochim. Acta 28 (1983) 1823–1828.
25. PM Allen, HAO Hill and NJ Walton, J. Electroanal. Chem. 178 (1984) 69–86.
26. I Taniguchi, N Higo, K Umekita and K Yasukouchi, J. Electroanal. Chem. 206 (1986) 341–348.
27. T Sagara, K Niwa, A Sone, C Hinnen and K Niki, Langmuir 6 (1990) 254–262.
28. FA Armstrong, Structure and Bonding 72 (1990) 137–230.
29. AM Bond and HAO Hill in *Metal Ions in Biological Systems*, H Sigel and A Sigel (eds), Marcel Dekker, New York (1991) Vol 27, Chapter 12, pp. 431–494.
30. DA Reed and FM Hawkridge, Anal. Chem. 59 (1987) 2334–2339.
31. T Daido and T Akaike, J. Electroanal. Chem. 344 (1993) 91–106.
32. EF Bowden, FM Hawkridge and HN Blount, J. Electroanal. Chem. 161 (1984) 355–376.
33. FA Armstrong, PA Cox, HAO Hill, VJ Lowe and BN Oliver, J. Electroanal. Chem. 217 (1987) 331–366.
34. FA Armstrong, HAO Hill, BN Oliver and D Whitford, J. Amer. Chem. Soc. 107 (1985) 1473–1476.
35. D Datta, HAO Hill and H Nakayama, J. Electroanal. Chem. 324 (1992) 307–323.
36. FA Armstrong, AM Bond, HAO Hill, ISM Psalti and CG Zoski, J. Phys. Chem. 93 (1989) 6485–6493.
37. AM Bond, HAO Hill, DJ Page, ISM Psalti and NJ Walton, Eur. J. Biochem. 191 (1990) 737–742.
38. FN Büchi and AM Bond, J. Electroanal. Chem. 314 (1991) 191–206.
39. AM Bond, Analytical Proceedings 29 (1992) 132–148.
40. W Norde, Adv. Colloid Interfac. Sci. 25 (1986) 267–340.
41. Z Salamon and G Tollin, Bioelectrochem. Bioenerg. 26 (1991) 321–334.
42. Z Salamon and G Tollin, Bioelectrochem. Bioenerg. 27 (1992) 381–391.
43. JL Willit and EF Bowden, J. Electroanal. Chem. 221 (1987) 265–274.
44. JL Willit and EF Bowden, J. Phys. Chem. 94 (1990) 8241–8246.

45. M Collinson and EF Bowden, Langmuir 8 (1992) 2552–2559.
46. M Collinson and EF Bowden, Anal. Chem. 64 (1992) 1470–1476.
47. MJ Tarlov and EF Bowden, J. Amer. Chem. Soc. 113 (1991) 1847–1849.
48. M Collinson, EF Bowden and MJ Tarlov, Langmuir 8 (1992) 1247–1250.
49. S Song, RA Clark, EF Bowden and MJ Tarlov, J. Phys. Chem. 97 (1993) 6564–6572.
50. EF Bowden, RA Clark, JL Willit and S Song, in *Redox Mechanisms and Interfacial Properties of Molecules of Biological Importance*, FA Schultz and I Taniguchi (eds), The Electrochemical Society Inc., (1993) pp 34–45.
51. FA Armstrong in *Advances in Inorganic Chemistry*, R Cammack and AG Sykes (eds), Academic Press, San Diego (1993) Vol. 38, pp. 117–163.
52. FA Armstrong, JN Butt, SJ George, EC Hatchikian and AJ Thomson, FEBS Lett. 259 (1989) 15–18.
53. JN Butt, FA Armstrong, J Breton, SJ George, AJ Thomson and EC Hatchikian, J. Amer. Chem. Soc. 113 (1991) 6663–6670.
54. JN Butt, A Sucheta, FA Armstrong, J Breton, AJ Thomson and EC Hatchikian, J. Amer. Chem. Soc. 113 (1991) 8948–8950.
55. JN Butt, A Sucheta, FA Armstrong, J Breton, AJ Thomson and EC Hatchikian, J. Amer. Chem. Soc. 115 (1993) 1413–1421.
56. JN Butt, J Niles, FA Armstrong, J Breton and AJ Thomson, Nature Structural Biology 1 (1994) 427–433.
57. B Shen, LL Martin, JN Butt, FA Armstrong, CD Stout, GM Jensen, et al., J. Biol. Chem. 268 (1993) 25928–25939.
58. JN Butt, A Sucheta, LL Martin, B Shen, BK Burgess and FA Armstrong, J. Amer. Chem. Soc. 115 (1993) 12587–12588.
59. A Sucheta, BAC Ackrell, B Cochran and FA Armstrong, Nature 356 (1992) 361–362.
60. BAC Ackrell, FA Armstrong, B Cochran, A Sucheta and T Yu, FEBS Lett. 326 (1993) 92–94.
61. A Sucheta, R Cammack, J Weiner and FA Armstrong, Biochemistry 32 (1993) 5455–5465.
62. FA Armstrong, A Sucheta, BAC Ackrell and JH Weiner in *Redox Mechanisms and Interfacial Properties of Molecules of Biological Importance*, FA Schultz and I Taniguchi (eds), The Electrochemical Society Inc. (1993) pp. 184–196.
63. FA Armstrong, JN Butt and A Sucheta, Methods in Enzymology 223 (1993) 479–500.
64. TM Cotton, SG Schultz and RP Van Duyne, J. Amer. Chem. Soc. 102 (1980) 7962–7965.
65. G Smulevich and TG Spiro, J. Phys. Chem. 89 (1985) 5168–5173.
66. K Niki, Y Kawasaki, K Kimura, Y Higura and N Yasuoka, Langmuir 3 (1987) 982–986.
67. P Hildebrandt, J. Molecular Structure 242 (1991) 379–395.
68. C Hinnen, R Parsons and K Niki, J. Electroanal. Chem. 147 (1983) 329–337.
69. A Manjaqui, J Haladjean and P Bianco, Electrochim Acta 35 (1990) 177–185.
70. T Ikeda, S Miyaoka and K Miki, J. Electroanal Chem. 352 (1993) 267–278.
71. DL Scott, RM Paddock and EF Bowden, J. Electroanal. Chem. 341 (1992) 307–321.
72. FA Armstrong, AM Bond, FN Büchi, A Hamnett, HAO Hill, AM Lannon et al., Analyst 118 (1993) 973–978.
73. T Yokuta, K Itoh and A Fujishima, J. Electroanal. Chem. 216 (1987) 289–292.
74. JM Cooper, KR Greenough and CJ McNeil, J. Electroanal. Chem. 347 (1993) 267–275.
75. JF Rusling and A-EF Nassar, J. Amer. Chem. Soc. 115 (1993) 11891–11897.
76. I Taniguchi, K Watanabe, M Tominaga and FM Hawkridge, J. Electroanal. Chem. 333 (1992) 331–338.
77. BC King, FM Hawkridge and BM Hoffman, J. Amer. Chem. Soc. 114 (1992) 10603–10608.
78. L Campanella, MP Sammartino, P Stefanoni, M Brunori and R Santucci, Bioelectrochem. Bioenerg, 21 (1989) 55–62.
79. BN Oliver, JO Egekeze and RW Murray, J. Amer. Chem. Soc. 110 (1988) 2321–2322.
80. AM Bond and KB Oldham, J. Phys. Chem. 87 (1983) 2492–2502.
81. DH Evans and KM O'Connell in *Electroanalytical Chemistry*, AJ Bard (ed), Marcel Dekker, New York (1986) Vol 14, pp 113–207.
82. M Wikström and GT Babcock, Nature 356 (1992) 301–209.
83. H Lauble, MC Kennedy, H Beinert and CD Stout, Biochemistry 31 (1992) 2735–2748.
84. R Cammack in *Advances in Inorganic Chemistry*, R Cammack and AG Sykes (eds), Academic Press, San Diego (1993) Vol 38, pp 281–322.

85. J Kim and DG Rees, Nature 360 (1992) 553–560.
86. MW Hentze and P Argos, Nucleic Acids Research 19 (1991) 1739–1740.
87. JL Smith, EJ Zaluzec, J-P Wery, L Niu, R Switzer, H Zalkin et al., Science 264 (1993) 1427–1433.
88. RH Holm in *Advances in Inorganic Chemistry*, R Cammack and AG Sykes (eds), Academic Press, San Diego (1993) Vol 38, pp 1–71.
89. KK Surerus, E Münck, I Moura, JJG Moura and J LeGall, J. Amer. Chem. Soc. 109 (1987) 3805–3807.
90. KKP Srivastava, KK Surerus, RC Conover, MK Johnson, J-B Park, MWW Adams et al., Inorg. Chem. 32 (1993) 927–936.
91. W Fu, J Telser, BM Hoffman, ET Smith, MWW Adams, MG Finnegan et al., J. Amer. Chem. Soc. 116 (1993) 5722–5729.
92. SJ George, FA Armstrong, EC Hatchikian and AJ Thomson, Biochem. J. 264 (1989) 275–284.
93. WR Hagen in *Advances in Inorganic Chemistry. Fe-S Proteins*, R Cammack and AG Sykes (eds), Academic Press, San Diego (1993) pp 165–222.
94. FA Armstrong, SJ George, R Cammack, EC Hatchikian and AJ Thomson, Biochem. J. 264 (1989) 265–273.
95. CD Stout, J. Biol. Chem. 268 (1993) 15920–25927.
96. AJ Thomson, FEBS Lett. 285 (1991) 230–236.
97. FA Armstrong, SJ George, AJ Thomson and MG Yates, FEBS Lett. 234 (1988) 107–110.
98. G Feher, Annu. Rev. Biochem. 61 (1992) 861–896.
99. SJ George, AJM Richards, AJ Thomson and MG Yates, Biochem. J. 224 (1984) 247–251.
100. PJ Stephens, GM Jensen, FJ Devlin, TV Morgan, CD Stout, AE Martin et al., Biochemistry 30 (1991) 3200–3209.
101. SE Iismaa, AE Vázquez, GM Jensen, PJ Stephens, JN Butt, FA Armstrong et al., J. Biol. Chem. 266 (1991) 21563–21571.
102. BAC Ackrell, MK Johnson, RP Gunsalus and G Cecchini in *Chemistry and Biochemistry of Flavoenzymes*, F Müller (ed), CRC Press, Boca Raton, Florida (1992) Vol 3, pp 229–297.
103. DR Mackenroth and LG Sands in *Illustrated Encyclopedia of Solid-state Circuits and Applications*, Prentice Hall, Englewood Cliffs, New Jersey (1984) Chapter 1.
104. SK Lahr, HO Finklea and FA Schultz, J. Electroanal. Chem. 163 (1984) 237–255.
105. JP Collman and K Kim, J. Amer. Chem Soc. 108 (1986) 7847–7849.

Appendix A. Glossary

A	Electrode surface area
C	Concentration (of species O or R in bulk solution, as appropriate)
D	Diffusion coefficient (of species O or R in bulk solution, as appropriate)
F	Faraday constant
R	Gas constant
T	Absolute temperature
K_M	Michaelis constant
k_{cat}	Limiting catalytic rate (turnover number)
k_s	standard electrochemical rate constant (units of s^{-1} for immobilized system)
n	number of electrons transferred (crossing the transition state)
α	transfer coefficient
Γ	surface concentration (of species O or R)
μ	Scan rate
γ	Kinematic viscosity
ω	Electrode rotation rate (angular frequency in rad $\cdot$ s^{-1})

Bioelectrochemistry of Biomacromolecules
ed. by G. Lenaz and G. Milazzo†

CHAPTER 5
Thiol/disulfide exchange and redox potentials of proteins

Hiram F. Gilbert

Verna and Marrs McLean Department of Biochemistry, Baylor College of Medicine, Houston, Texas, USA

1. Introduction

The sulfhydryl group of the amino acid cysteine plays a pivotal role in the structure and function of numerous proteins. The high nucleophilicity of the thiolate anion coupled with its reasonably low pK_a produces high chemical reactivity in nucleophilic additions and substitutions, accounting for its widespread use as a catalytic group in a variety of enzymes catalyzing hydrolysis, substitution, and exchange reactions. In addition, the reversible two electron oxidation converting the thiol of cysteine to its disulfide form is used not only in electron transfer reactions but also in crosslinking and stabilizing the tertiary and quarternary structure of proteins, the regulation of enzyme activity, and the protection of the intracellular environment. The direct application of electrochemical methods to the study of the redox properties of thiols, either in small molecules or in proteins, is complicated by interactions of the thiol with most electrode surfaces. Consequently, there have been few reliable electrochemical measurements of the redox potentials of cysteine and other biological thiols and disulfides. Nevertheless, alternate methodology involving reversible thiol/disulfide exchange equilibria has allowed the estimation of standard redox potentials of the sulfhydryl groups of biologically important thiols, including those found in proteins. In this chapter, the redox chemistry and biochemistry of the sulfhydryl groups of proteins will be outlined, methodology for applying the thiol/disulfide exchange reaction to the measurement of redox potentials will be described, and examples of the biological function of these redox state changes will be given.

1.1. Redox chemistry of the sulfhydryl group

The sulfhydryl group of cysteine is not the only amino acid side chain which can undergo redox state changes; however, the reversible (and sometimes irreversible) interconversion of protein sulfhydryl groups and disulfides is the most widely recognized redox state change which occurs normally in proteins. Sulfhydryl oxidation can lead to oxidation states which are higher than that of the disulfide [1, 2].

$$\begin{array}{ccccccc} RSH & \rightarrow & RSOH & \rightarrow & RSO_2H & \rightarrow & RSO_3H \\ \downarrow & & & & & & \\ RSSR & \rightarrow & RSOSR & \rightarrow & RSO_2SR & \rightarrow & RSO_2SO_2R \end{array} \qquad (1)$$

The higher oxidation states of sulfur may be significant in the long term description of protein structure since their formation tends to be irreversible; however, the interconversion between thiol and disulfide forms of protein cysteine residues will be the primary focus of this discussion.

The net formation of disulfides from two thiols can be coupled to the reduction of molecular oxygen in a reaction catalyzed by a variety of heavy metals, particularly copper and iron, and which can produce superoxide anion as well as hydrogen peroxide [3].

$$2\ R\text{-}SH + O_2 \rightleftharpoons R\text{-}SS\text{-}R + H_2O_2 \qquad (2)$$

Glutathione (γ-glutamylcysteinylglycine), the major low molecular weight thiol in most cells [4], is converted to its disulfide by the reduction of hydrogen peroxide to water, catalyzed by the seleno-enzyme, glutathione peroxidase [5].

$$2\ GSH + H_2O_2 \rightleftharpoons GSSG + 2\ H_2O \qquad (3)$$

In cells, net disulfide reduction to two sulfhydryl groups is generally coupled to oxidation of pyridine nucleotides (most frequently NADPH) and catalyzed by reductases and dehydrogenases of varying substrate specificity [6]. Glutathione reductase catalyzes the NADPH-dependent reduction of GSSG to GSH:

$$GSSG + NADPH + H^+ \rightleftharpoons NADP^+ + 2\ GSH \qquad (4)$$

while other specific dehydrogenases/reductases catalyze the NADPH-dependent reduction of disulfides of such diverse species as lipoamide and thioredoxin.

In addition of the forementioned coupling of thiol/disulfide redox state changes to oxidants and reductants which do not contain sulfur, the process of thiol/disulfide exchange can reversibly link the oxidization of one thiol to the reduction of another disulfide [7].

$$\begin{array}{lrcl} & R_1SH + R_2SSR_2 & \rightleftharpoons & R_1SSR_2 + R_2SH \quad (5) \\ & R_1SH + R_1SSR_2 & \rightleftharpoons & R_1SSR_1 + R_2SH \quad (6) \\ \hline \text{Net:} & 2\ R_1SH + R_2SSR_2 & \rightleftharpoons & R_1SSR_1 + 2\ R_2SH \quad (7) \end{array}$$

The thiol/disulfide exchange process has the capacity to distribute redox equivalents among the many different protein and non-protein thiol/disulfide species present in cellular environments.

1.2. Biological function of the sulfhydryl group

Thiol groups in proteins have been widely studied. Cysteine sulfhydryl groups are intimately involved in enzyme catalysis, ligand binding, metal binding, maintenance of protein conformation, as well as protection against oxidative stress. The participation of cysteine sulfhydryl groups in the catalytic mechanisms of numerous enzymes has been repeatedly emphasized by numerous, extensive reviews [1, 2, 8–14]. As polarizable nucleophiles with reasonably low pK_as, enzyme thiols participate in a host of acylation and transacylation reactions, as well as other carbonyl additions. Biology is replete with thiol-containing cofactors such as coenzyme A, pantatheine, cysteamine, and glutathione which also participate in a variety of enzyme-catalyzed processes including acyl transfer, carbonyl additions, and the formation of carbon-carbon bonds. The increased acidity of the α-proton of a thiol ester is utilized as a strategy in the formation of carbon-carbon bonds [15]. The contribution of protein thiols to the maintenance of protein structure has been elegantly underlined by the description of the "zinc finger" structural motif of DNA binding proteins in which cysteine residues provide metal binding ligands [16].

The intracellular environment of most organisms contains high concentrations of protein and non-protein thiols including glutathione [17]. A major function of this large pool of intracellular thiols (approaching 1–10 mM in most cells) is to provide protection from xenobiotic chemicals, radiation damage, and biological oxidants such as hydrogen peroxide through mechanisms which utilize the unique nucleophilicity and redox properties of the thiol group [18].

1.3. Biological functions of disulfides

The functional role generally assigned to protein disulfides is one of providing increased stability against denaturation in a hostile extracellular environment [19]. While this view is generally valid, recent studies have suggested some limitations. Protein disulfides undergo β-elimination to dehydroalanine at elevated temperatures and pH, so that the presence of disulfide bonds may, under extreme conditions, be a liability [20]. Like other interactions which stabilize protein structure, the quantitation of the net contribution of individual disulfide bonds to the stability of protein tertiary and quaternary structure is complex [21]. While some disulfides can be demonstrated to contribute significantly to protein stability and structure [22, 23], the free energy of stabilization provided by disulfide bond formation is variable and may be small in some cases [24].

In contrast to the thiol group, there are no examples (to the author's knowledge) in which a disulfide participates chemically in enzymatic catalysis without cycling between the disulfide and thiol redox states. Consequently, the involvement of disulfides in biological catalysis invariably involves catalytic interconversion of thiol and disulfide redox states [6].

1.4. Biological function of thiol/disulfide exchange

The catalytic mechanisms of some enzymes involve dithiol to disulfide redox interconversions as an integral component [6]. With many of these enzymes, even when the substrates themselves are thiols and disulfides, a redox state change involving the interconversion of protein dithiols and disulfides accompanies turnover. Protein thiol/disulfide redox state changes provide a catalytic mechanism for the exchange of reducing equivalents between the substrates and cofactors.

Hopkins and Dixon [25] and later, Guzman-Barron [8], suggested that the regulated interconversion of the thiol and disulfide oxidation states of an enzyme might be used as a biological control mechanism. Changes in the activity of specific enzymes could be linked through thiol/disulfide exchange to the thiol/disulfide redox state of the cellular environment [10–14]. Regulation of enzyme activity by thiol/disulfide exchange operates in plants as a device to reductively activate (and inactivate) key regulatory enzymes in response to light [13, 26]. Although not so well established in the animal kingdom, enzyme regulation by thiol/disulfide exchange has been supported by a variety of somewhat circumstantial experimental evidence [10–14]. There is little doubt that inducing changes in cellular thiol/disulfide status has significant metabolic effects [27].

2. Equilibrium behavior of thiols/disulfides in a redox buffer

Thiol/disulfide exchange equilibria (Eq. 5–7) provide a means of evaluating and discussing the relative redox potential of two thiol/disulfide pairs [1]. In measuring thiol/disulfide exchange equilibria it is often convenient to maintain one of the thiol/disulfide pairs in large molar excess of the other. The thiol disulfide pair in excess provides some protection against the effects of air oxidation and maintains a resonably stable redox environment. For purposes of discussion, the buffer component in excess will be termed a redox buffer. In biological systems, the low molecular weight thiol/disulfide pair of highest abundance is generally the glutathione redox buffer composed of the thiol, GSH and its symmetrical disulfide, GSSG. Consequently, much of the discussion presented here will employ a glutathione redox buffer for illustrative purposes; however, the concepts developed also apply to other thiol/disulfide redox buffers.

2.1. *General disulfide bond formation*

The process of thiol/disulfide exchange can result in effective oxidation of a protein thiol at the expense of another protein or non-protein disulfide. Two types of protein disulfides will be distinguished with respect to redox behavior: disulfides between a protein and a low molecular weight thiol such as glutathione (glutathione mixed disulfides, PSSG) (Eq. 8), and intra- or intermolecular disulfides formed between two protein thiols (PSSP) (Eq. 12). In the protein-SS-protein species, disulfide bonds may be inter- or intramolecular. For our purposes, inter- and intramolecular protein-SS-proteins will be treated the same since many intermolecular disulfide bonds are formed between two protein subunits which are tightly associated and to which the disulfide bond may add little stabilization energy. Additionally, the distinction between inter- and intramolecular disulfides introduces a dependence of the observed dithiol oxidation potential on protein concentration.

2.2. *Formation of protein-SSG mixed disulfides*

The equilibrium formation of protein-glutathione mixed disulfides is described by the expression:

$$\text{P-SH} + \text{GSSG} \rightleftharpoons \text{P-SSG} + \text{GSH} \tag{8}$$

The thermodynamic equilibrium constant (K_{ox}) for this process is simply [28]

$$K_{ox} = \frac{[\text{PSSG}]\,[\text{GSH}]}{[\text{PSH}]\,[\text{GSSG}]} \tag{9}$$

In this case, the ratio of oxidized to reduced protein ([PSSG]/[PSH]) will depend only on the thiol/disulfide ratio of the redox buffer, [GSH]/[GSSG]. Since the total concentration of the protein (P_t) is given by [PSH] + [PSSG], the fraction of the total protein existing as the reduced species will be given by:

$$\frac{[\text{PSH}]}{P_t} = \frac{[\text{GSH}]/[\text{GSSG}]}{K_{ox} + [\text{GSH}]/[\text{GSSG}]} \tag{10}$$

For convenience, the quantity [GSH]/[GSSG] will also be denoted as **R** (for ratio) so that Eq. 10 becomes:

$$\frac{[\text{PSH}]}{P_t} = \frac{\mathbf{R}}{K_{ox} + \mathbf{R}} \tag{11}$$

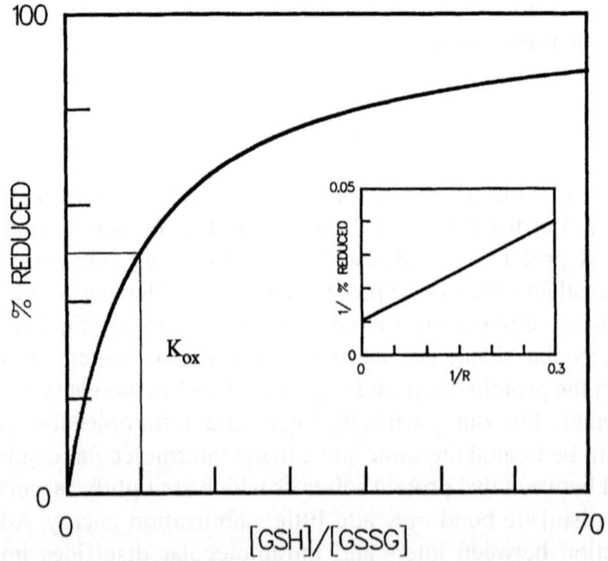

Fig. 1. Dependence of the extent of protein reduction on the [GSH]/[GSSG] ratio at equilibrium for the formation of a protein glutathione mixed disulfide (PSSG). The K_{ox} represents the [GSH]/[GSSG] ratio at which the protein is half reduced. The curve is drawn according to Eq. 10 with a K_{ox} of 7. The inset shows a double reciprocal transformation of Eqn. 11.

The form of Eq. 11 is that of a rectangular hyperbola, similar to that of the Michaelis-Menten equation used to describe saturation behavior in enzyme kinetics. A plot of the fraction of protein present in the reduced state $[PSH]/P_t$ against **R** is hyperbolic. The value of **R** at which the protein is half reduced corresponds to K_{ox} (Fig. 1). A double reciprocal transformation of Eq. 11 leads to a linear expression when $P_t/[PSH]$ is plotted as a function of $1/\mathbf{R}$, with a negative x-intercept equal to $-1/K_{ox}$. Note that for the formation of a mixed disulfide species described by Eq. 8, the value of K_{ox} is unitless.

2.3. *Formation of protein-SS-protein disulfides*

The behaviour of a protein which forms a protein-SS-protein disulfide is characteristically different from that of a protein which forms only PSSG species [29].

$$P\frac{SH}{SH} + GSSG \rightleftharpoons P\frac{S}{S} + 2\,GSH \tag{12}$$

The corresponding equilibrium expression and equations which describe the extent of enzyme reduction at equilibrium as a function of the redox status of the glutathione redox buffer become

$$K_{ox} = \frac{[PSSP]\,[GSH]^2}{[PSH]\,[GSSG]} \tag{13}$$

$$\frac{[PSH]}{P_t} = \frac{[GSH]^2/[GSSG]}{K_{ox} + [GSH]^2/[GSSG]} \tag{14}$$

$$\frac{[PSH]}{P_t} = \frac{\mathbf{R}[GSH]}{K_{ox} + \mathbf{R}[GSH]} \tag{15}$$

When protein thiol oxidation leads to PSSP formation, a plot of the fraction of protein in the reduced form at equilibrium against the composite quantity **R**[GSH] (rather than **R**) will give a simple hyperbola (Fig. 2).

When the protein is half-reduced, the **R**[GSH] will be equal to the K_{ox}. While the K_{ox} for equilibria involving PSSG species will be unitless, the K_{ox} for equilibria involving PSSP species will have units of concentration. Plotting the fraction of protein reduced as a function of **R** will not yield a simple hyperbola. If a series of redox equilibrations are performed by varying **R** at several different, fixed concentrations of GSH, a family of hyperbolas will result, one for each constant [GSH] [29]. The apparent K_{ox} ($\mathbf{R}_{0.5}$ at which the protein is half-reduced) will be a function of [GSH] so that $K_{ox,obs} = K_{ox}/[GSH] = \mathbf{R}_{0.5}$ (Fig. 3). These graphical methods will yield appropriate equilibrium constants for the thiol/disulfide exchange between the protein and redox buffer. The way in which the protein redox state changes with **R** or **R**[GSH] can be used to infer the mechanism of enzyme oxidation, even when the fraction of enzyme in the reduced state is monitored indirectly by activity.

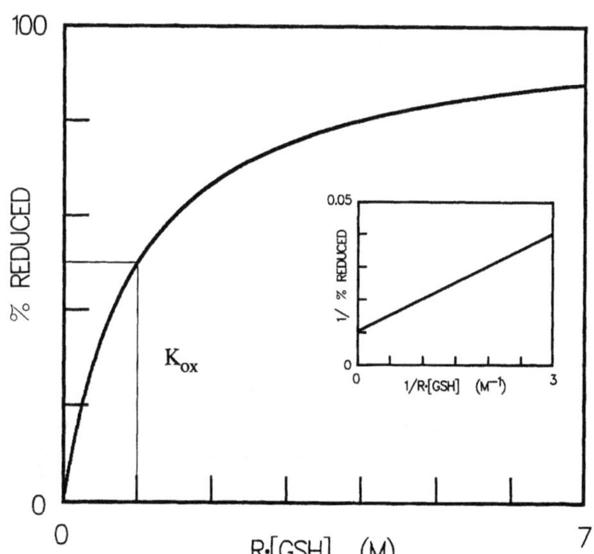

Fig. 2. Dependence of the extent of protein reduction on the quantity $[GSH]^2/[GSSG]$ at equilibrium for the formation of a protein-protein disulfide (PSSP). The K_{ox} represents the $[GSH]^2/[GSSG]$ at which the protein is half reduced at equilibrium. The curve is drawn according to Eq. 14 with a K_{ox} of 1 M. The inset shows a double reciprocal transformation of Eqn. 15.

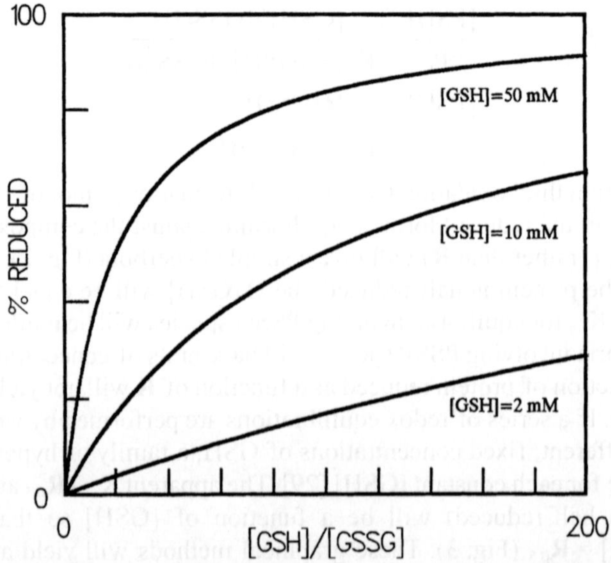

Fig. 3. Dependence of the extent of protein reduction on the quantity [GSH]/[GSSG] at equilibrium for the formation of a protein-protein disulfide (PSSP). The apparent K_{ox} for each curve of constant [GSH] represents the [GSH]/[GSSG] ratio at which the protein is half reduced at equilibrium. As the constant concentration of GSH decreases, the apparent K_{ox} increases illustrating that at a lower GSH concentration, a higher ratio of [GSH]/[GSSG] is required to maintain the protein half-reduced. The curves are drawn according to Eq. 14 with a K_{ox} of 1 M.

2.4. Simultaneous formation of protein-SS-protein disulfides and protein-SSG mixed disulfides

The formation of a protein-SS-protein disulfide from vicinal dithiols proceeds through the initial formation of a mixed disulfide intermediate (Eq. 16).

$$\mathrm{P}\begin{matrix}\diagup \mathrm{SH}\\ \diagdown \mathrm{SH}\end{matrix} + \mathrm{GSSG} \underset{}{\overset{K_1}{\rightleftharpoons}} \mathrm{P}\begin{matrix}\diagup \mathrm{S\text{-}SG}\\ \diagdown \mathrm{SH}\end{matrix} + \mathrm{GSH} \overset{K_2}{\rightleftharpoons} \mathrm{P}\begin{matrix}\diagup \mathrm{S}\\ | \\ \diagdown \mathrm{S}\end{matrix} + 2\,\mathrm{GSH} \tag{16}$$

If K_1 and K_2 have appropriate values, the mixed disulfide intermediate (P(SSG)SH) may accumulate significantly at redox equilibrium [30]. If P_t is the total concentration of all protein species ($P_t = [P(SH)_2] + [P(SSG)SH] + [P(SS)]$), the fraction of the protein existing in the fully reduced form is given by Eq. 17.

$$\frac{[\mathrm{P(SSG)SH}]}{\mathrm{P_t}} = \frac{\mathbf{R}[\mathrm{GSH}]}{\mathbf{R}[\mathrm{GSH}] + K_1[\mathrm{GSH}] + K_2} \tag{17}$$

where K_1 and K_2 are the equilibrium constants for formation of the mixed disulfide intermediate and the equilibrium constant for intramolecular displacement of the glutathione mixed disulfide, respectively. The equilibrium concentration of the mixed disulfide intermediate is given by Eq. 18.

$$\frac{[\mathrm{P(SSG)SH}]}{\mathrm{P_t}} = \frac{\mathbf{R}[\mathrm{GSH}]}{\mathbf{R}[\mathrm{GSH}] + \mathrm{K_1}[\mathrm{GSH}] + \mathrm{K_2}} \tag{18}$$

The mixed disulfide intermediate will comprise a significant fraction of the protein when the quantity K_1 [GSH] is significantly greater than **R** [GSH] + K_2. If either the equilibrium constant for intramolecular displacement of glutathione from the protein (K_2) or the ratio of [GSH]/[GSSG] is significantly greater than K_1, the intermediate mixed disulfide will not be detectable at equilibrium.

2.5. A thermodynamic scale of comparison and standard reduction potentials

Because the formation of new disulfide bonds in proteins, an oxidative process with respect to the protein, can lead to the formation of both mixed disulfides (PSSG) with the redox buffer and to the formation of protein-SS-protein disulfides, there is some difficulty in discussing the thermodynamics of thiol/disulfide exchange using standard reduction potentials (E_o'). While the oxidation of a single thiol species to the symmetrical disulfide (e.g. 2 GSH $\rightleftharpoons$ GSSG or $P(SH)_2 \rightleftharpoons P(SS)$) are two electron oxidations for which an E_o' can be defined (see below), the formation of an asymmetrical mixed disulfide species (PSSG) is a simple exchange reaction for which no simple relation exists between the unitless equilibrium constant (K_{ox}) and the standard reduction potential. In addition, the expression of the reduction potential in volts obscures the arbitrary choice of a standard state of 1 M for reactions in which the equilibrium constants have units of M or M^{-1}. Standard reduction potentials are logarithmic functions of the equilibrium constants involved; for a two electron process such as thiol/disulfide interconversions, a change in reduction potential of 0.03 volts corresponds to a factor of 10 change in equilibrium constant [31].

The tripeptide glutathione (γ-glutamylcystenylglycine) comprises the most abundant thiol/disulfide redox buffer under most physiological situations. The oxidation potential of a particular sulfhydryl-containing molecule may be expressed relative to that of glutathione by determining the equilibrium constant for the thiol/disulfide exchange [1, 10]

$$\mathrm{E_{red}} + \mathrm{aGSSG} \rightleftharpoons \mathrm{E_{ox}} + \mathrm{bGSH} \tag{19}$$

where E_{red} and E_{ox} represent the different thiol and disulfide oxidation states of the protein and a and b represent the stoichiometry coefficients for GSSG and GSH, respectively.

The thermodynamic equilibrium constant for this reaction in the direction shown is

$$K_{ox} = \frac{[E_{ox}][GSH]^b}{[E_{red}][GSSG]^a}. \tag{20}$$

These K_{ox} values represent thermodynamic, thiol/disulfide oxidation potentials relative to glutathione. Larger K_{ox} values reflect an increasingly favorable oxidation of the protein thiol by glutathione disulfide. The equilibrium behavior of a particular protein in a glutathione redox buffer can be predicted if the K_{ox} and the coefficients a and b are known. This scale has been employed by several laboratories [1, 10, 32, 33] and can be conveniently used to compare directly protein thiol/disulfide oxidation potentials of the various type of disulfides formed.

Like other thermodynamic scales, it is possible to determine oxidation potentials using redox buffers other than glutathione (e.g. dithiothreitol) if the oxidation potential of the redox buffer is known relative to glutathione.

$$P\begin{smallmatrix}\diagup SH\\ \diagdown SH\end{smallmatrix} + DTT\begin{smallmatrix}\diagup S\\ |\\ \diagdown S\end{smallmatrix} \xrightleftharpoons{K_{obs}} P\begin{smallmatrix}\diagup S\\ |\\ \diagdown S\end{smallmatrix} + DTT\begin{smallmatrix}\diagup SH\\ \diagdown SH\end{smallmatrix} \tag{21}$$

$$DTT\begin{smallmatrix}\diagup SH\\ \diagdown SH\end{smallmatrix} + GSSG \xrightleftharpoons{K_{ox,R}} DTT\begin{smallmatrix}\diagup S\\ |\\ \diagdown S\end{smallmatrix} + 2\,GSH \tag{22}$$

$$\text{Net:}\quad P\begin{smallmatrix}\diagup SH\\ \diagdown SH\end{smallmatrix} + GSSG \xrightleftharpoons{K_{ox}} P\begin{smallmatrix}\diagup S\\ |\\ \diagdown S\end{smallmatrix} + 2\,GSH \tag{23}$$

The K_{ox} on the glutathione scale will then be given by the product of the observed equilibrium constant in the dithiothreitol redox buffer (K_{obs}, Eq. 21) and the K_{ox} for dithiothreitol ($K_{ox,R}$) so that $K_{ox} = K_{obs}\,K_{ox,R}$.

This same thermodynamic scale can also be used to describe the relative thermodynamic oxidizing power of different redox buffers on a common scale. A dithiothreitol redox buffer ($K_{ox} = 260$ M) [34, 35, 36] with a $[DTT_{red}]/[DTT_{ox}]$ ratio of 0.1 would be thermodynamically equivalent to a glutathione redox buffer with an **R** [GSH] of 26 M. A protein which forms a PSSP in the oxidized state would arrive at the same equilibrium position in either redox buffer.

In order to anchor the glutathione redox scale to the more commonly accepted standard reduction potentials, the equilibrium constant for the glutathione reductase equilibrium can be employed.

$$GSSG + NADPH + H^+ \rightleftharpoons NADP^+ + 2\,GSH \tag{24}$$

At pH 7.0, 25 °C, the equilibrium constant for this reaction has been reported to be 98 M [37] and 200 M [38]. The two electron reduction of GSSG would then be 0.065 v (0.03 × log (150 M)) more positive than the reduction potential of NADPH. Using a reduction potential of $NADP^+$ of −0.315 v [39], the two electron reduction of GSSG corresponding to the glutathione reductase reaction would have a standard reduction potential of −0.25 v.

For proteins that form a protein-SS-protein disulfide, the standard reduction potential for reduction of this disulfide can be calculated from the K_{ox} (with units of M) from the following relationship (Eq. 25) using 0.03 v for the value of ln (10) RT/nF for a two electron reduction.

$$E_o' \text{ (in volts)} = -0.25 \text{ v} - 0.03 \log (K_{ox}) \tag{25}$$

The reduction potentials for the formation of symmetrical disulfides of monothiols such as cysteine can be evaluated by analysis of the equilibrium constant for the reaction [1]

$$2\,RSH + GSSG \xrightleftharpoons{K_{eq}} RSSR + 2\,GSH \tag{26}$$

Consequently, for systems in which a symmetrical disulfide is formed in equilibrium with a glutathione redox buffer, the standard reduction potential can be evaluated using Eq. 25; however, in systems where an asymmetrical disulfide is formed, such as PSSG, it is impossible to calculate a standard reduction potential. By using the glutathione scale it is possible to include mixed disulfide formation on a thermodynamic scale using the unitless K_{ox} for the formation of the mixed disulfide. Since K_{ox} as defined above is actually an oxidation potential, the larger the K_{ox} the more easily the disulfide is formed by oxidation of the dithiols and the more difficult it is to reduce.

2.6. *Practical consequences of equilibrium behavior*

The simple equilibrium expressions given above can be used to describe the behavior of proteins in which thiol/disulfide exchange processes equilibrate the redox state of the protein with the external thiol/disulfide redox buffer. For the formation of a PSSG mixed disulfide species in equilibrium, the fraction of the protein existing in the reduced form will depend only on the quantity **R**. As long as the concentration of the redox buffer is much greater than that of the protein, and only a PSSG species is formed, the extent of protein oxidation at equilibrium will be independent of the concentration of GSH and GSSG as long as **R** remains constant.

The situation is different for proteins which form protein-SS-protein disulfides. Since the equilibrium position is dictated by the quantity $[GSH]^2/[GSSG] = \mathbf{R}\,[GSH]$, the extent of protein oxidation at equilibrium will depend not only on **R**, but on the absolute concentration of GSH as well (Fig. 3). Practically, this means that simply diluting a redox buffer of GSH and GSSG without changing **R** will result in increased protein-SS-protein disulfide formation at equilibrium. This may seem to be a trivial point;

however, the practical consequences are rather important. Virtually all solutions of GSH are contaminated by GSSG formed from air oxidation. Even carefully prepared solutions of GSH generally contain 0.5–2% GSSG. This is present in the solid material so that removing oxygen or adding EDTA to the solution will not prevent the presence of GSSG. Thus, even freshly prepared, carefully neutralized solutions of GSH are redox buffers with a [GSH]/[GSSG] ratio of less than 100. If the K_{ox} is sufficiently large, even this small amount of GSSG will be significant. For example, at a [GSH]/[GSSG] of 100, a protein with a K_{ox} of 10 M will require a GSH concentration of 100 mM to be half-reduced at equilibrium. In both the PSSG and PSSP cases, increased protein oxidation at equilibrium can be observed without a change in the GSSG concentration. A two-fold decrease in GSH imposes a two-fold increase in the oxidizing power of the redox buffer if a PSSG disulfide is formed, but a four-fold increase if a PSSP is formed.

2.7. *Standard state problem*

The differences in behavior of PSSG disulfides and PSSP disulfides in glutathione redox buffers create a problem in comparing the oxidation potentials for the two types of disulfides. In the case of PSSG disulfides, K_{ox} is unitless, but for PSSP disulfides, K_{ox} has units of concentration. Direct comparisons of the two are impossible without defining the GSH concentration at which the comparisons are made. Reporting K_{ox} in units of M assumes an arbitrary standard state in which the concentration of GSH in the redox buffer is 1 M. For example a dithiol such as dithiothreitol ($K_{ox} = 1 \times 10^4$ M) will be half-reduced in a glutathione redox buffer in which the concentration of GSH is 1 M and the ratio of [GSH]/[GSSG] is 1×10^4. An identical redox state could be maintained by 1×10^4 M GSH at a [GSH]/[GSSG] ratio of 1. In comparison, a monothiol with a K_{ox} of 1×10^4 would be half reduced at a [GSH]/[GSSG] of 1×10^4 regardless of the GSH concentration, and the monothiol would be 99.99% oxidized at a GSH/GSSG ratio of 1 even if the GSH concentration was 1×10^4 M. Although the standard state of M (or occasionally mM) will be used in this review, it should be borne in mind that the redox equilibrium distribution of a protein dithiol cannot be determined from a K_{ox} value alone. The equilibrium distribution will be a function of both K_{ox} and the external redox buffer composition and will depend on the type of protein disulfide formed.

3. Measuring thiol/disulfide oxidation potentials

3.1. *Potentiometric*

Electrochemical measurements have generally proven to be unsatisfactory, giving widely variable results with a variety of electrodes. Oxidation of thiols on mercury or mercury-containing electrodes is complicated by the formation of stable mercury-thiol complexes which interfere with the measurement of the oxidation potential [1, 40]. Even "inert" platinum electrodes appear to behave anomalously with regard to thiol redox reactions [41]. As a result of the anomalous behavior of thiols in most electrochemical experiments and the particular problems associated with applying electrochemical methods to the sulfhydryl groups of proteins, the redox potentials of both small molecules and proteins that contain thiols have generally been determined by alternative methods [1].

3.2. *Coupling to NADPH/NADP$^+$*

The enzymes lipoamide dehydrogenase and glutathione reductase can be used to couple the oxidation state of NADPH/NADP$^+$ to that of lipoamide or glutathione [34, 42]. If a test thiol/disulfide is present and will react with either lipoamide or glutathione via thiol/disulfide exchange (even non-enzymatically), the redox state of the test thiol/disulfide can be indirectly coupled to that of NADPH/NADP. By combining the equilibrium constants for the equilibrium between NADPH and the test disulfide and that for the equilibrium between NADPH and GSSG, the value for the K_{ox} for the test thiol on the glutathione scale can be derived so that $K_{ox} = K_{eq} K_{NADPH}$.

$$R\begin{smallmatrix}\diagup SH\\ \diagdown SH\end{smallmatrix} + NADP^+ \underset{}{\overset{K_{eq}}{\rightleftharpoons}} R\begin{smallmatrix}\diagup S\\ |\\ \diagdown S\end{smallmatrix} + NADPH + H^+ \quad (27)$$

$$H^+ + NADPH + GSSG \overset{K_{NADPH}}{\rightleftharpoons} NADP^+ + 2\,GSH \quad (28)$$

$$\text{Sum: } R\begin{smallmatrix}\diagup SH\\ \diagdown SH\end{smallmatrix} + GSSG \overset{K_{ox}}{\rightleftharpoons} R\begin{smallmatrix}\diagup S\\ |\\ \diagdown S\end{smallmatrix} + 2\,GSH \quad (29)$$

The NADPH/NADP couple has been used to measure dithiol/disulfide oxidation potentials relative to glutathione in the range of 1 to 10^4 M [42]; however, this technique is less useful for monothiols or dithiols of lower oxidation potential. If the equilibrium constant for reduction of the disulfide by NADPH is very favourable, the concentration of the disulfide will become very small, presenting analytical problems even at very low NADPH/NADP ratios.

3.3. *Equilibration in redox buffers*

The most widely applicable technique for the measurement of thiol/disulfide oxidation potentials involves establishing equilibrium between the test species and some standard thiol/disulfide redox buffer, and measuring the concentrations of all species involved in the equilibrium at a variety of redox buffer concentrations and ratios [1, 10, 40]. The method is generally applicable, and any analytical technique which can differentiate the species involved can be used. For low molecular weight species, including peptides, HPLC and NMR have been successfully employed.

Oxidation potential measurements of proteins present a variety of problems. Most proteins possess multiple cysteine residues, and multiple redox isomers in a complex equilibrium are possible. Scheraga *et al.* point out that ribonuclease A (8 cysteines, 4 disulfide bonds) has 7193 possible thiol/disulfide redox isomers if one considers the formation of protein-SSG mixed disulfides [43]. Under different redox conditions, the populations of the various redox states will change, and it is possible that the equilibrium mixture may be populated by a much smaller number of species. For some small, well-characterized proteins such as bovine trypsin inhibitor and T4 ribonuclease, Creighton and coworkers [44] have been able to estimate individual oxidation potentials for specific disulfide bonds by rapidly quenching equilibrium mixtures with thiol alkylating agents and resolving and quantitating at least some of the redox isomers by gel electrophoresis or chromatography. However, Weissman and Kim [45] have shown that the rate of thiol alkylation by iodoacetamide is comparable to the rate of thiol/disulfide exchange, so that rearrangement of disulfide bonds may occur during the trapping process. In the case of a single disulfide bond genetically engineered into subtilisin [46], the oxidation potential could be measured by equilibrating the protein in dithiothreitol redox buffers, quenching with alkylating agent, and resolving the disulfide and alkylated thiol forms of the protein by gel electrophoresis.

Incorporation of radiolabelled glutathione into the protein at equilibrium in glutathione redox buffers can be used to quantitate mixed disulfide formation [47, 48]. With radiolabel incorporation it may be possible to detect individual oxidations of multiple protein thiol groups, if the oxidation potentials are sufficiently different.

If the biological activity of a protein changes upon oxidation/reduction, activity measurement can be used to determine the oxidation potential and type of disulfide formed, even for proteins which are not available in large quantities and which may not be pure. Differences in the kinetic properties (K_m, V_{max}, K_i, pK_a, stability, etc.) of an enzyme (or protein) between the oxidized and reduced form can be exploited to monitor redox changes in the thiols whose oxidation affects the kinetic property [10]. While this approach allows an estimation of the oxidation potential and the type of protein disulfide formed, the amount of detailed structural information is limited. It is also possible that non-redox effects (i.e. specific binding of GSH or GSSG to the enzyme) could produce behaviour similar to redox effects. To ensure that redox effects are responsible for the change in the observed property, it is essential to show that GSH has no effect on the measured parameter when a large excess of reduced dithiothreitol (or excess NADPH and glutathione reductase) is also present, and that the effects of the GSH and GSSG are not reversed by dialysis or gel filtration in the absence of reducing or oxidizing agents.

4. Studies with small molecules

4.1. Thermodynamics

4.1.1. Mixed disulfides: The equilibrium constant for the identity thiol/disulfide exchange

$$GSH + GSSG \rightleftharpoons GSSG + GSH \tag{30}$$

must, by the symmetry of the exchange process, be one. For the general thiol/disulfide exchange equilibrium between any monothiol and GSSG

$$RSH + GSSG \rightleftharpoons RSSG + GSH \tag{31}$$

the equilibrium constants at neutral pH will be close to one for most alkyl monothiols which have pK_as near that of glutathione [7, 42]. The K_{ox} will be displaced from one by any interaction which specifically stabilizes RSSG or RSH (or RS^-). For example, if RSH represents an aryl thiol such as the popular thionitrobenzoate (Ellman's anion), resonance stabilization of RS^- (which also results in a significantly decreased pK_a) lowers the apparent K_{ox} for mixed disulfide formation with glutathione to approximately 0.001 at pH 7.0 where RSH is largely ionized [40, 49].

Potentially favorable charge-charge interactions which stabilize the RSSG state seem to have only a small effect on K_{ox}. When R is $^+NH_3{-}CH_2{-}CH_2{-}$ and the interaction of the positively charged amino group might be expected to interact favorably with the predominantly negatively charged glutathione, the K_{ox} is increased from 1 to 4.2 [50]. For alkyl monothiols, K_{ox} can reasonably be expected to be between 0.1 and 10 (Table 1).

The determination of the equilibrium constant for the formation of mixed disulfides between low molecular weight monothiols and glutathione can be complicated by the possibility of multiple thiol/disulfide exchange equilibria [1].

$$RSH + GSSG \xrightleftharpoons{K_{ox}} RSSG + GSH \tag{32}$$

$$RSSG + RSH \xrightleftharpoons{K_2} RSSR + GSH \tag{33}$$

$$\text{Net: } 2\,RSH + GSSG \rightleftharpoons RSSR + 2\,GSH \tag{34}$$

Under conditions where the glutathione concentration is in large excess over the concentration of the other thiol/disulfide pair, glutathione mixed disulfides will predominate unless there are very favorable interactions between the two R groups in the RSSR species or very unfavorable inter-

Table 1. Thiol/disulfide oxidation potentials. For each thiol or dithiol the value of K_{ox} refers to the equilibrium constant for the general reaction $R(SH)n + a\,GSSG \rightleftharpoons RSSR(G) + b\,GSH$. For monothiols, this equilibrium constant will be unitless but for dithiols the equilibrium constant will have units of M.

Thiol	Conditions[a]	K_{ox}	E_0'[b]	Ref
Symmetrical intermolecular disulfides $2\,RSH + GSSG \rightleftharpoons RSSR + 2\,GSH$				
GSH	pH 7.0		−0.25	[37, 38]
Cysteine	pH 7.0		−0.24	[51, 52]
Cysteamine	pH 7.3		−0.24	[53]
N-Ac-cysteamine	pH 7.3		−0.25	[7]
$NADPH + GSSG + H^+ \rightarrow NADP^+ + 2\,GSH$	pH 7.0	150 M[d]		[37, 38]
Intramolecular disulfides $R(SH)_2 + GSSG \rightleftharpoons R(SS) + 2\,GSH$				
Dithiothreitol	pH 7.0	260 M	−0.32	[34–36, 42]
Lipoic acid	pH 7.0	320 M[c]	−0.33	[32]
Glutathione mixed disulfides $RSH + GSSG \rightleftharpoons RSSG + GSH$				
CoASH	pH 8.0	3.1		H.F. Gilbert, unpublished observations
Cysteamine	pH 8.0	1.8		[7]
N-Ac-cysteamine	pH 7.3	2.9		[7]
DTNB (Ellman's Reagent)	pH 8.0	0.001		[54]
Peptides				
YSRCVC	3 M Gdn · HCl			
mixed disulfide	high pH limit[e]	1.3		[55]
loop disulfide	high pH limit	0.003 M	−0.17	[55]
Soybean trypsin inhibitor peptide	3 M Gdn · HCl pH 6.9	0.045 M	−0.21	[33]
Cys_{26}–Cys_{87} in IgG light chain	pH 8.1 8 M urea	0.0035	−0.18	[56]

[a] Unless specified otherwise, the temperature is 25 °C.

[b] E_0' values are based on a standard reduction potential of −0.315 volts at pH 7.0, 25 °C for NADPH [39] and values for the equilibrium constant for the glutathione reductase reaction of 150 M [37, 38]. Equilibrium constants measured by equilibration with DTT redox buffers are converted to K_{ox} on the glutathione scale using a K_{ox} of 260 M for DTT. The K_{ox} of DTT shown in the table and used for the calculations of the K_{ox} for other thiol disulfide systems is based on direct measurement and differs significantly from previous values of approximately 10^4 M.

[c] Calculated from the equilibrium constant of 0.032 for the lipoamide dehydrogenase equilibrium in aqueous solution.

[d] The equilibrium constant for the glutathione reductase reaction links thiol/disulfide redox potentials to standard redox potentials.

[e] The effects of pH on the K_{ox} values will probably be rather small since pK_as are similar to that of glutathione.

[f] Estimated as described in text.

[g] Estimated from the effects of GSSG in the presence of GSH and glutaredoxin on the rate of the ribonucleotide reductase reaction.

[h] Estimated from the kinetics of the reduction of fructose-1,6-bisphosphatase.

[i] Calculated from ratios of rate constants for oxidation of the reduced protein by oxidized dithiothreitol.

[j] Calculated for formation of a disulfide bond between subunits which are already associated non-covalently.

[k] Estimated equilibrium constant for the initial, reversible oxidation event.

Table 1 (continued)

Thiol	Conditions[a]	K_{ox}	$E_o'^{b}$	Ref
Catalytically active dithiols				
Thioredoxin reductase				
with flavin reduced	pH 7.0	2.3 M	−0.271	[57]
with flavin oxidized		0.63 M	−0.254	[57]
Lipoamide dehydrogenase				
$E(SH)_2FAD$	pH 7.0	4.6 M	−0.280	[58]
Mercuric reductase	pH 7.3	2 M	−0.269	[59]
Ribonucleotide reductase	pH 7.6	0.3 M[g]	−0.23	[60]
Redox carriers				
E. coli thioredoxin	pH 7.0	13 M	−0.26	[61]
T4 thioredoxin	ph 7.0	0.14 M	−0.22	[62]
Chloroplast thioredoxin m	pH 7.4	29 M	−0.33	[63]
Chloroplast thioredoxin f	pH 7.4	42 M[h]	−0.31	[64]
Trypanothione	pH 7.0	0.25 M	−0.24	[65]
Disulfides which stabilize proteins				
Bovine trypsin inhibitor				
Cys_{14}–Cys_{38} with other disulfides intact	pH 8.0	1500 M	−0.35	[66]
Cys_5–Cys_{55} with other disulfides intact	pH 8.0	1.1×10^7 M	−0.46	[66]
Ribonuclease T1				
Cys_6–Cys_{103} with other not formed	pH 8.0	34 M[i]	−0.30	[67]
Cys_2–Cys_{10} with other disulfide intact		1000 M	−0.34	[67]
Genetically engineered disulfides				
Subtilisin				
Cys_{22}–Cys_{87}	pH 8.0	120 M	−0.33	[46]
Cys_{24}–Cys_{87}		500 M	−0.35	[46]
λ repressor-intersubunt[j]				
Cys_{85}–Cys'_{85}	pH 8.7	8 M	−0.29	[68]
Cys_{88}–Cys'_{88}		100 M	−0.32	[68]
Thiols which affect enzyme activity				
Monothiols PSH + GSSG ⇌ PSSG + GSH				
Phosphofructokinase				
easiest	pH 8.0	6.1		[47]
hardest		0.7		[47]
Glycogen phosphorylase[k]	pH 8.0	10^{-5}		[50]
Glycogen debrancher	pH 8.0	0.06		[69]
3-hydroxy-3-methylglutaryl-CoA reductase from mevinolin treated rats	pH 8.0	27		[70]

Table 1 (continued)

Thiol	Conditions[a]	K_{ox}	E_0'[b]	Ref
Dithiols $P(SH)_2 + GSSG \rightleftharpoons P(SS) + 2\,GSH$				
Chloroplast				
Fructose bisphosphatase				
easiest	pH 8.0	104 M	−0.31	[64]
hardest		31 M	−0.29	[64]
Chloroplast NADP-Malate Dehydrogenase	pH 7.4	114 M	−0.31	[63]
HMG-CoA Reductase	pH 7.1	0.55 M	−0.25	[71]
with substrates		0.2 M	−0.24	[71]
Fatty acid synthase	pH 8.0	0.015 M	−0.20	[29]
3-ketoacyl thiolase	pH 8.0	0.001 M	−0.17	[52]
Protein disulfide isomerase	pH 8.0	7×10^{-5} M	−0.12	[72]
dsbA	pH 8.0	8×10^{-5} M	−0.12	[73]

actions in the RSSG species. By measuring the individual values for the equilibrium constants K_{ox} (Eq. 32) and K_2 (Eq. 33), the value of the equilibrium constant for reversible formation of the symmetrical disulfide, RSSR (Eq. 34), can be obtained from the product $K_{ox}K_2$, and this value can be used to calculate the standard reduction potential for the symmetrical disulfide according to Eq. 25.

4.1.2. Cyclic disulfides: For small organic dithiols, the extensive studies of Houk and Whitesides [32] have provided insight into the magnitude of entropic and strain effects observed on formation of intramolecular disulfides. These studies were conducted at neutral pH, 25 °C in 50% methanol where the thiols were predominantly in the protonated form.

The formation of cyclic disulfides by thiol/disulfide exchange must proceed through the formation of an inital mixed disulfide intermediate.

$$\mathbf{R}\begin{matrix}/SH\\ \backslash SH\end{matrix} + GSSG \underset{k_2}{\overset{k_1}{\rightleftharpoons}} \mathbf{R}\begin{matrix}/SSG\\ \backslash SH\end{matrix} + GSH \qquad (35)$$

$$\mathbf{R}\begin{matrix}/SSG\\ \backslash SH\end{matrix} \underset{k_4}{\overset{k_3}{\rightleftharpoons}} \mathbf{R}\begin{matrix}/S\\ |\\ \backslash S\end{matrix} + GSH \qquad (36)$$

The intramolecular displacement of GSH from the mixed disulfide intermediate (k_3, Eq. 36) will usually be very favorable compared to the intermolecular displacement of GSH by another molecule of GSH (k_2, Eq. 35) because of the more unfavorable translational entropy barrier to intermolecular reactions [74]. The entropic advantage of an intramolecular reaction over its intermolecular counterpart may be described in terms of the concentration of GSH which would be required to compete effectively with the intramolecular process. Creighton provides a detailed discussion of this concept as applied to disulfide bond formation [75]. For dithiothreitol ($K_{ox} = 1 \times 10^4$ M) the sulfhydryl group in the mixed disulfide intermediate behaves as if its "effective concentration" were 1×10^4 M.

Page and Jencks [74] have estimated that the entropic effects alone can result in "effective concentrations" of up to 10^6–10^8 M in moderately constrained intramolecular reactions and up to 10^{10} M in more highly constrained systems in which most of the translational and rotational entropy barriers have been removed. By extrapolation of the data of Houk and Whitesides [32], a dithiol which forms an unstrained disulfide bond without any loss of conformational freedom (i.e. in a very conformationally constrained situation) would be expected to exhibit a K_{ox} of approximately 10^6–10^7 M.

For smaller rings, such as found in lipoic acid (five-membered ring), the K_{ox} is observed to drop approximately 10-fold compared to a ring size of six, consistent with a moderate ring strain for disulfides in smaller rings. This strain effect increases markedly with a four-membered ring so that the primary oxidation product observed with 1,2-dithiols is generally the more stable intermolecular dimer with a ring size of 8 rather than the intramolecular disulfide of ring size four [32]. For ring sizes above six, the K_{ox} falls rapidly with increasing size by a factor of approximately 5-14 for each additional atom in the ring. As suggested by Houk and Whitesides [32], this factor is in reasonable agreement with the predicted entropic effects of one additional degree of freedom for each additional methylene group. For rings containing more than eight atoms, polymeric products are observed, suggesting that intermolecular disulfide bond formation becomes comparably favorable to intramolecular disulfide bond formation. The complication of polymer formation will of course depend on the molar concentration of the test dithiol. Houk and Whitesides also observed a significant correlation between the K_{ox} and the disulfide dihedral angle in various five-membered rings [32]. Variation in the angle defined by the C_1-C_2-C_3 atoms of various disulfides with 5-membered rings show that between angles of 105°–120°, the K_{ox} decreased by a factor of approximately 1.2 for each degree.

For cyclic organic disulfides formed in reasonably constrained systems in the absence of strain, the K_{ox} may be expected to be determined predominantly by entropic effects. While K_{ox} values as high as 10^9 M might be expected in highly entropically constrained systems, values of 10^6–10^7 M represent the largest K_{ox} values observed.

4.1.3. Peptides: Only a few studies of the oxidation potentials of cysteine-containing peptides are available. Zhang and Snyder, in a series of thorough experiments, have examined the kinetics and thermodynamics of thiol/

disulfide exchange reactions between glutathione and the peptides YSRCVC and YSRCVCA [55]. The equilibrium constants for formation of mixed disulfides between cysteine residues in peptides and glutathione are, as expected, close to one. In the peptide YSRCVC, the K_{ox} values for formation of the two possible PSSG species are 1.4 and 1.2, and the two K_{ox}s observed for formation of the $P(SSG)_2$ species from the two PSSG intermediates are 1.3 and 1.5. The K_{ox} for formation of the cyclic disulfide in the peptide YSRCVC is about 0.003 M, so that glutathione mixed disulfide intermediates are observed to accumulate.

The effect of nearby charged residues appears to be relatively small. Compared to YSRCVCA, the presence of a backbone carboxyl group adjacent to the terminal cysteine in YSRCVC decreased the rates of disulfide bond formation from the glutathione mixed disulfide intermediate about 3-fold; however, the K_{ox} values at pH 6.9, 23° in 3 M guanidinium hydrochloride were identical (0.0035 M). Milburn *et al.* [76] have reported the relative oxidation potentials for a series of hexapeptides of the general structure CXPGYC. A peptide where X and Y were both glycine was 8.2-fold more easily oxidized thermodynamically than a peptide with X and Y both valine. Favorable medium range charge-charge interactions may exhibit a stabilizing effect toward disulfide formation. The peptide CEPCKC was 1.4 times more easily oxidized than CGPCKC, where the more bulky glutamate residue was replaced by glycine. The ability of proline residues to stabilize the formation of cyclic disulfide peptides cannot be directly assessed from the data of Milburn *et al.* [76] since no direct comparisons were made between peptides in which the sole difference was proline; however, CGVVGC was only 1.02 times less easily oxidized than CVPGGC and 1.25 more easily oxidized than CVPGVC. While local steric and charge factors could have significant effects on the oxidation potential of cysteine-containing peptides, these effects appear to be rather small (less than 10-fold) in the limited number of cases examined.

For peptides containing two cysteine residues separated by a variable number of spacer amino acids, a significant factor in determining the thiol/disulfide oxidation potential (K_{ox}) appears to be the differences in conformational entropy which are lost on formation of cyclic, intramolecular disulfides [77]. Considerations of the conformational entropy barriers to the formation of intramolecular disulfide bonds between two cysteine residues separated by n residues suggests that the entropic contribution should decrease [21, 77] as a function of $n^{3/2}$, where n is the number of amino acid residues separating the two cysteines. As the two cysteine residues become more distant, the K_{ox} decreases as it becomes increasingly improbable that the two cysteines will find each other within the correct distance and proper orientation for intramolecular reaction.

From the very limited data available, strain and conformational entropy emerge as the dominant factors governing the oxidation potential of dicysteine peptides. A fragment of soybean trypsin inhibitor (residues

115–181) which contains two cysteine residues (Cys_{136}, Cys_{145}) separated by eight intervening non-cysteine residues, exhibits a K_{ox} (against a N-acetyl-cysteine ethyl ester redox buffer) of 0.0045 M at pH 6.9 [33]. Goto and Hamaguchi [56] report detailed studies of disulfide bond formation with glutathione and a fragment of an immunoglobulin light chain in which the two cysteine residues were separated by 60 residues. The K_{ox} for this peptide is 0.0035 M at pH 8.1 in 8 M urea. The peptide YSRCVC, in which the two cysteines are separated by only one residue, exhibits a K_{ox} of 0.0029 M at pH 6.9 in 3 M guanidine, undoubtedly reflecting strain. YSRCVC also forms significant concentrations of a dimeric species in glutathione redox buffers. Zhang and Snyder [55] have reported the equilibrium constant ($K_{eq} = 190$) for the formation of this dimer by the process

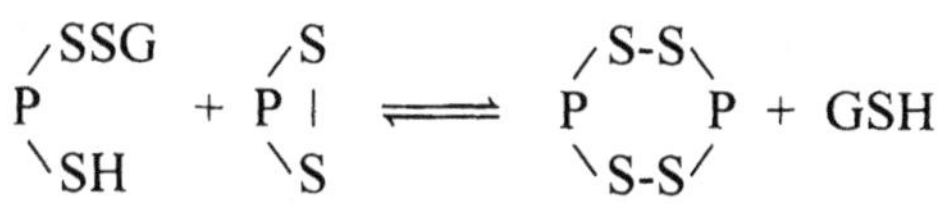
(37)

Making the assumption that the equilibrium constant for the intermolecular reaction of PSSG(SH) with PSSP would be near one, and considering the equilibrium constant for monomeric disulfide formation, makes it possible to estimate the equilibrium constant for the process

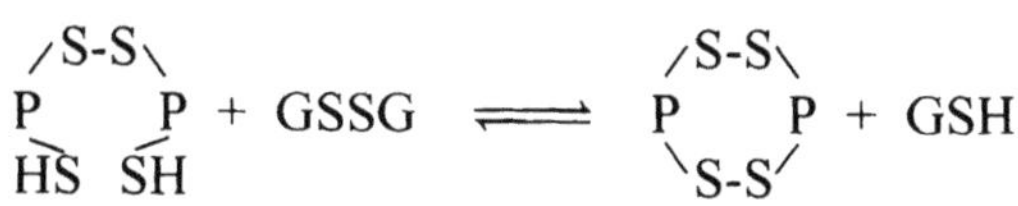
(38)

as approximately 0.400 M. This species would be analogous to a peptide in which the two cysteine residues were separated by 3–4 amino acid residues (depending on how the disulfide bond is counted).

Figure 4 shows a plot of $\log(K_{ox})$ vs. $\log(n)$ where n is the number of residues separating the two cysteines (the number of intervening non-cysteine residues is $n-1$). The line through the data shown in Figure 4 is drawn with a slope of $-3/2$, predicted from considerations of the conformational entropy of an unstrained random coil [21, 77]. For peptides where n is greater than or equal to 5, it is apparent that entropic effects are dominant and can account for much of the variation in K_{ox}. However, for peptides where only one residue separating the cysteine residues, the K_{ox} is at least 100-fold lower than predicted, suggesting significant ring strain in forming this disulfide, perhaps due to the unfavorable requirement for *cis* peptide bond conformations in these small rings [55]. Disulfide bonds between adjacent cysteine residues would be expected to be even less easily formed. Disulfides between adjacent cysteines are exceedingly rare in proteins [19], and only one example has been reported, the acetyl choline receptor [78].

Zhang and Snyder [55] have presented similar entropic arguments with regard to the kinetics of ring closure in small peptides and conclude that in the absence of strain effects, predictions based on conformational entropy considerations are in reasonable agreement with observations. For small

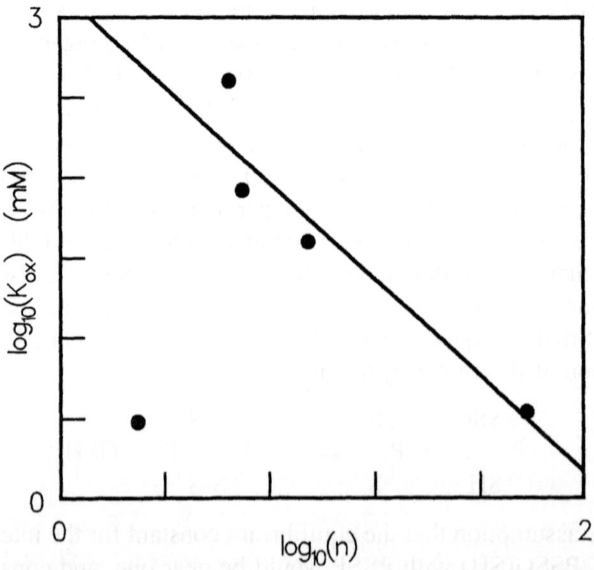

Fig. 4. Dependence of the K_{ox} for formation of cyclic disulfides in peptides on the number of residues separating the two cysteines. The K_{ox} values are in units of mM. The line is drawn with a slope of −3/2 and an intercept of 2 M. Notice that the point for forming a disulfide between adjacent cysteines is well below the line, showing that it is destabilized relative to the others.

rings (one intervening residue) in which the disulfide state is conformationally strained, both ring opening to the mixed disulfide intermediate and ring closure from the mixed disulfide to the cyclic disulfide are significantly faster than expected from entropic considerations alone. The reasons for this deviation, particularly in the direction of ring closure, are not clear.

4.2. Kinetics

Several excellent discussions are available which describe the effects of thiol structure and pK_a on the kinetics of thiol/disulfide exchange reactions of alkyl and aryl thiols [32, 42, 49, 79]. For thiol/disulfide exchange between a variety of alkyl thiols of pK_as near that of glutathione (8.6) and a variety of disulfides, the second-order rate constants for the reaction

$$\text{R-SH} + \text{GSSG} \underset{k_2}{\overset{k_1}{\rightleftharpoons}} \text{E-SSG} + 2\text{GSH} \qquad (39)$$

of a thiol anion (of RSH or GSH) are generally very near $10\ M^{-1}s^{-1}$ (expressed in terms of the thiolate) [55]. Since equilibrium constants for these homologous exchange reactions are frequently near one (if pK_as are comparable), k_1 and k_2 should be approximately equal. For dithiol/disulfide interconversions involving cyclic disulfides, the kinetics of the thiol/disulfide exchange process can be strongly influenced by thermodynamic considerations. For reactions involving intramolecular disulfide formation

$$P\begin{matrix}\diagup SH\\ \diagdown SH\end{matrix} \underset{k_2[GSH]}{\overset{k_1[GSSG]}{\rightleftharpoons}} P\begin{matrix}\diagup SSG\\ \diagdown SH\end{matrix} \underset{k_4[GSH]}{\overset{k_3}{\rightleftharpoons}} P\begin{matrix}\diagup S\\ |\\ \diagdown S\end{matrix} \qquad (40)$$

in which the cyclic disulfide is not appreciably strained, all of the bimolecular rate constants (k_1, k_3, and k_4) should have comparable values (k_1 and k_4 will have values that are 2-fold higher due to statistical effects). Since the equilibrium constant for the overall thiol/disulfide exchange is given by

$$K_{ox} = \frac{k_1 k_3}{k_2 k_4} \qquad (41)$$

and k_1 and k_4 should be comparable, the K_{ox} will be determined mainly by and be approximately equal to the partition ratio k_3/k_4. This partition ratio is a direct measure of the "effective molarity" of the intramolecular thiol in displacing GSH from the P-(SH)-SSG intermediate (k_3) compared to the bimolecular displacement of PSH from this intermediate by GSH (k_2) [75]. The relationship between the K_{ox} and the partition ratio again emphasizes that the thermodynamic K_{ox} will, to a major extent, be dependent on the same conformational entropy effects as the kinetics of ring closure.

Another important concept which can be derived from a consideration of the relationship between K_{ox} and the partition ratio is that the observed rate constant for reduction of an intramolecular disulfide by GSH (reverse reaction of Eq. 40) will depend on K_{ox}. The more thermodynamically unfavorable the reduction (higher K_{ox}), the slower it should be kinetically. For the reduction of high K_{ox} disulfides (> 100 mM) by a monothiol such as GSH, the intermediate mixed disulfide will revert to reactants (P(SS)) much faster than it can be trapped by low concentrations (less than 10 mM) of GSH, i.e. $k_3 \gg k_2[GSH]$. The initial step of the reduction will be essentially at equilibrium, and the observed rate constant will be given by $k_{obs} = k_4 k_2[GSH]^2/k_3$. Since K_{ox} can be approximated by k_3/k_4, the k_{obs} will be approximately $k_2[GSH]^2/K_{ox}$. In this case the observed rate constant for the reduction will be second-order in [GSH] and third-order overall. As K_{ox} increases, k_3 will increase proportionately so that the rate of reduction becomes slower. In intermediate cases where K_{ox} is near the concentration of GSH used in the reduction, k_3 and $k_2[GSH]$ will be comparable so that it should be possible to observe a change in the rate-determining step with increasing GSH concentration from a process which is second-order in GSH to one that is first-order in GSH. The reduction of the disulfides of protein disulfide isomerase displays such kinetics in the single turnover reduction of its disulfide bonds by GSH [80].

In the direction of disulfide reduction, a special situation arises when the reductant is also a dithiol. In this case, the mixed disulfide intermediate should partition more favorably in the direction favored by the thiol group of highest "effective molarity" (highest K_{ox}). For example, reduction of a peptide (or protein) disulfide by DTT will lead to an initial intermediate disulfide between the peptide and DTT. As long as the K_{ox} of the peptide dithiol is significantly less than the K_{ox} of DTT (1×10^4 M), the disulfide of this intermediate will be attacked in an intramolecular reaction more rapidly by the thiol of DTT than the thiol of the peptide. In general, DTT and other cyclic disulfides of high K_{ox} will be more effective reductants both kinetically *and* thermodynamically, a consideration of practical importance when protein disulfides must be reduced for analytical or other purposes. Similar considerations of the partitioning of intermediates in the direction of peptide dithiol oxidation by oxidized DTT show that oxidized DTT will be kinetically a poorer oxidant than GSSG.

5. Factors affecting protein thiol/disulfide oxidation potentials

Studies of the oxidation potentials of dicysteinyl peptides and of proteins under denaturing conditions have revealed that in the absence of significant protein tertiary structure, the oxidation potentials of most dithiols separated by two or more residues should be significantly less than about 0.5 M (Fig. 4). Since the thiol/disulfide oxidation potentials actually observed for some proteins are much larger than this, the net favorable free energy of protein folding must contribute to disulfide bond stability. The K_{ox} for formation of the Cys_5–Cys_{55} disulfide bond of bovine pancreatic trypsin inhibitor (BPTI) is 1.1×10^7 M ($E_o' = -0.46$ v) [66] and the oxidation potential of a dithiol/disulfide which modulates the activity of chloroplast fructose-1,6-bisphosphatase is 6×10^3 M ($E_o' = -0.36$ v) [64]. The K_{ox} values observed for proteins forming intramolecular disulfides vary by more than twelve orders of magnitude. Observed K_{ox} values for protein dithiol oxidation can be significantly lower or higher than the K_{ox} expected for disulfide bond formation in a random polypeptide. While it is impossible to predict protein thiol/disulfide oxidation potentials *ab initio*, several factors have been recognized which undoubtedly contribute to this large variation in oxidation potential.

5.1. *Effects of protein conformation*

The thermodynamically favorable folding of a protein into a stable tertiary structure may bring the sulfhydryl groups of two cysteine residues widely separated in primary sequence into close proximity. This increase in "effective molarity" of one sulfhydryl group relative to the other will significantly increase the observed thiol/disulfide oxidation potential over that expected from a similar peptide in the absence of tertiary structure. If there were no unfavorable enthalpic or entropic barriers to disulfide bond formation in a folded protein, studies with small molecules would suggest that the K_{ox} would be in the range 10^6–10^7 M [42]. Creighton [66] has estimated an effective concentration of Cys_5–Cys_{55} of 1.1×10^7 M in the folded state of bovine pancreatic trypsin inhibitor, illustrating that such thermodynamically favorable dithiol oxidations are actually observed in proteins. If disulfide bond formation is linked to a thermodynamically unfavorable conformational change of the protein, the observed K_{ox} will be significantly lower than 10^6–10^7 M. The oxidation of a dithiol in the species P^*_{red} to P_{ox} (Eq. 42) in which no conformational distortion or further entropy losses are required for disulfide bond formation should exhibit a $K_{ox,N}$ near 10^6–10^7 M.

$$P_{red}\begin{matrix}/SH\\ \backslash SH\end{matrix} \underset{}{\overset{K_{con}}{\rightleftharpoons}} P^*_{red}\begin{matrix}/SH\\ \backslash SH\end{matrix} \underset{}{\overset{K_{ox,N}}{\rightleftharpoons}} P_{red}\begin{matrix}/S\\ |\\ \backslash S\end{matrix} \qquad (42)$$

If P^*_{red} is not the thermodynamically most stable conformation of the reduced form of the protein ($K_{con} < 1$), upon reduction P^*_{red} will relax to the most stable conformation P_{red}, and the observed K_{ox} will measure the transition between P_{red} and P_{ox} ($K_{ox,obs} = K_{con} K_{ox,N}$), and the observed K_{ox} will be smaller than 10^6–10^7 M by the factor K_{con}. A thermodynamically unfavorable change in protein conformation which must accompany disulfide bond formation will result in a corresponding decrease in K_{ox}, making disulfide bond formation less favorable.

5.2. *Specific interactions with the oxidized or reduced state*

If a ligand or effector interacts more strongly with one of the thiol/disulfide redox states of a protein than with the other, the observed K_{ox} will be shifted by the presence of the ligand or effector. An effector which interacts more strongly with the reduced protein will make the oxidation more difficult and lower the K_{ox}. An effector which exhibits a stronger interaction with the oxidized state will stabilize the oxidized state, making the oxidation easier and increasing K_{ox}.

$$\begin{array}{ccccc} E_{red} + GSSG & \overset{K_{ox}}{\rightleftharpoons} & E_{ox} + nGSH & & \\ K_{d,red}\ \Big\Updownarrow \pm S & & K_{d,ox}\ \Big\Updownarrow \pm S & & (43) \\ E_{red}S + GSSG & \overset{K_{ox,S}}{\rightleftharpoons} & E_{ox}S + nGSH & & \end{array}$$

In Eq. 43, $K_{d,ox}$ and $K_{d,red}$ are the dissociation constants of S from E_{ox} and E_{red} respectively, and K_{ox} and $K_{ox,S}$ are the thiol/disulfide oxidation potentials in the absence and presence of an effector. Since the net free energy change around this cycle must be zero, $K_{ox}/K_{ox,S} = K_{d,ox}/K_{d,red}$. If the effector, S, interacts more strongly with E_{red} than with E_{ox} ($K_{d,red} < K_{d,ox}$), then K_{ox} must

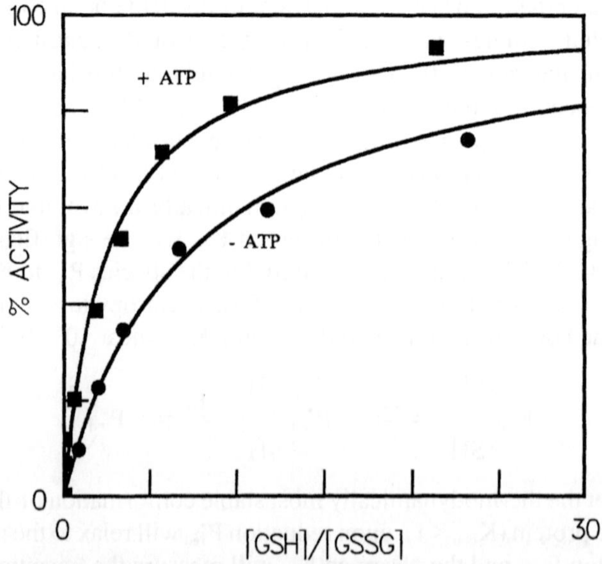

Fig. 5. The effect of the presence of a substrate (ATP) on the observed K_{ox} for equilibration of rabbit muscle phosphofructokinase in glutathione redox buffers. In the absence of ATP (●) the K_{ox} is 7.1, while in the presence of ATP (■) the K_{ox} is 2.5. Taken with permission from ref 28.

be proportionately larger than $K_{ox,S}$, i. e. the protein will be harder to oxidize thermodynamically in the presence of substrate. Such a thermodynamic effect is observed in the oxidation of phosphofructokinase by GSSG in the presence of saturating levels of ATP [28]. The enzyme is approximately 3-fold more difficult to oxidize thermodynamically in the presence of ATD than in the absence of ATP (Fig. 5). The oxidation of the enzyme hydroxymethylglutaryl-CoA reductase is approximately 2-fold more difficult to oxidize in the presence of saturating levels of both substrates.

As discussed above, proteins in which a PSSG mixed disulfide is formed should exhibit K_{ox} values near one. However, in theory, much larger K_{ox} values might be observable if the glutathione moiety of the mixed disulfide were to interact favorably and more strongly with the oxidized form of the protein than the reduced form (Fig. 6). Due to the "chelate effect", the interaction of a covalently attached-SSG moiety with even a weak glutathione binding site ($K_{d,intra}$) on the protein could be significantly more favorable than the interaction of the reduced protein with the free GSH or GSSG in solution (K_d). Covalent immobilization of the glutathione near a binding site would overcome the entropic disadvantage of intermolecular interactions. The subsequent intramolecular binding step ($K_{d,intr}$) would be considerably more favorable. The favorable binding interactions between the glutathione moiety and the glutathione binding site would stabilize the oxidized state relative to the reduced state, raising the K_{ox} and making the

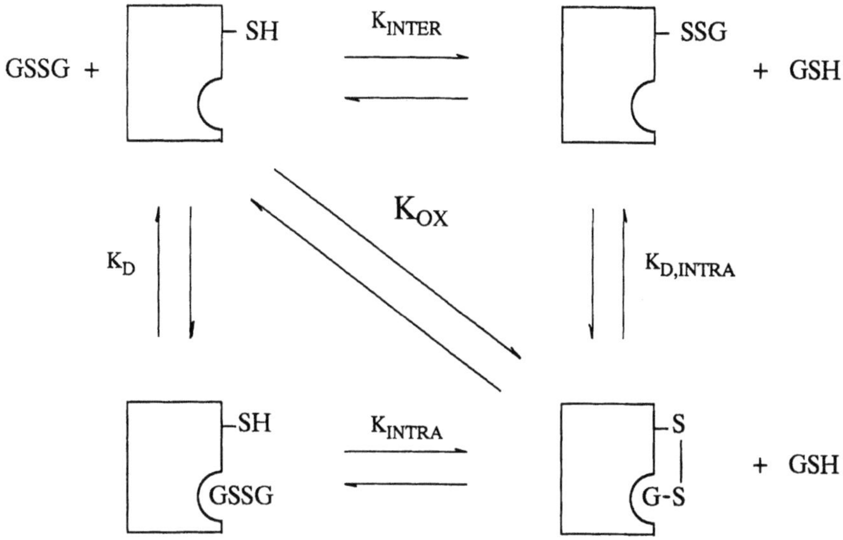

Fig. 6. Favourable interactions between the glutathione moiety and the protein in a protein-glutathione mixed disulfide can increase the observed K_{ox}. If the K_{intra} was the theoretical maximum of 10^7 M and the dissociation constant of GSSG (or GSH) from the glutathione binding site in the reduced protein (K_D) was 0.1 M, the observed K_{ox} (diagonal) would be $K_{intra}/K_D = 10^8$ rather than the expected value of 1.

observed oxidation more thermodynamically favorable. This effect could be quite large [81], up to a factor of 10^8 M depending on how strongly the glutathione is immobilized after interaction.

5.3. Dihedral angle

Disulfide bonds exhibit a preferential dihedral angle of +90° or −90°. Katz and Kossiakoff [82] have analyzed the geometric destabilization due to non-optimal dihedral angle geometries in 84 disulfide bonds in globular proteins of known structure. Their analysis suggests that dihedral angle strain energies in protein disulfide bonds range from 0.5 to 4.7 kcal/mole as a consequence of nonoptimal disulfide bond geometries. This strain energy will make disulfide bond formation less favorable thermodynamically and result in a decreased K_{ox}. Since 4.7 kcal/mole destabilization would amount to a 2900-fold change in any equilibrium constant, it can be estimated that K_{ox} values could be lowered from the maximum of 10^6–10^7 M to as low as 350–3500 M by the normal dihedral angle strain observed in globular proteins.

5.4. Thiol ionization

Acid dissociation of a thiol may provide a mechanism for specific stabilization of the reduced state, the magnitude of which will depend on the solution pH. Szajewski and Whitesides [42], O'Donnell and Williams, [57] and Milburn *et al.* [83] have all provided a thorough analysis of the effects of pH and thiol pK_a on the equilibrium constants for thiol/disulfide exchange. The effects of thiol pK_a on the observed thiol/disulfide oxidation potential will depend on two factors: the effects of electron withdrawal on the stability of the RSSG species, and the stabilization of RS^- by ionization.

The effects of thiol ionization on the observed K_{ox} can be discussed rather simply by defining the observed K_{ox} in terms of total concentrations of thiol and thiolate species

$$[RSH + RS^-] + GSSG \rightleftharpoons RSSG + [GSH + GS^-] \tag{44}$$

$$K_{ox,\,obs} = \frac{[RSSG]\,[GSH + GS^-]}{[RSH + RS^-][GSSG]} \tag{45}$$

By appropriate substitution it can be shown that

$$K_{ox,\,obs} = K_{ox}\,\frac{(K_{a,G} + H^+)}{(K_{a,R} + H^+)} \tag{46}$$

where $K_{ox,\,obs}$ refers to the observed thiol/disulfide oxidation potential at any pH, K_{ox} is the oxidation potential at the low pH limit at which there is no significant thiol ionization, and $K_{a,G}$ and $K_{a,R}$ are the acid dissociation

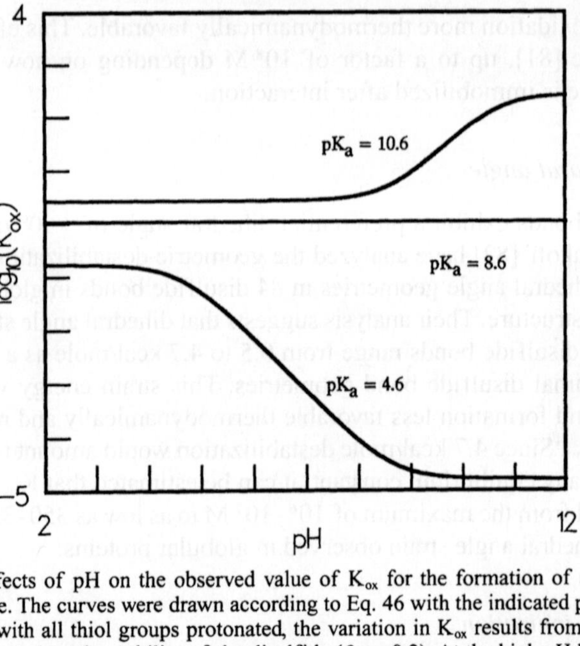

Fig. 7. The effects of pH on the observed value of K_{ox} for the formation of a glutathione mixed disulfide. The curves were drawn according to Eq. 46 with the indicated $pK_{a,R}$s. At the low pH limit, with all thiol groups protonated, the variation in K_{ox} results from the effect of electron withdrawal on the stability of the disulfide ($\beta_{eq} = 0.2$). At the high pH limit, with all thiols ionized, the variation in K_{ox} reflects the relative effects of electron withdrawal on the stabilization of the thiolate anion of glutathione and the test thiol.

constants of glutathione ($pK_a = 8.6$) and the species RSH, respectively. In the low pH limit, electron-donating R groups stabilize the RSSG species relative to RSH. A plot of $\log(K_{ox})$ against $pK_{a,R}$ has a slope (β_{eq}) of approximately +0.2, showing that electron-donating groups in RSH (higher pK_a) stabilize the +0.2 "effective charge" on the central sulfur in the RSSG species [42, 49]. A drop of one pK_a unit of RSH would result in a $10^{0.2}$ (1.6-fold) decrease in K_{ox}.

Examination of Eq. 46 points out that if the pK_a of the thiol is similar to that of GSH (8.6), as most alkyl thiol pK_as are, there will be only a small effect of ionization on the observed thiol/disulfide oxidation potential at any pH (Fig. 7). The stabilization offered by ionization of the RSH state will be exactly compensated for by similar stabilization of GSH. If RSH has a higher pK_a than GSH, the low pH limit $K_{ox,}$ will be somewhat greater than 1 due to the stabilization of RSSG by electron donation (a factor of $10^{0.2}$ for each pK_a unit higher than 8.6). At the high pH limit (pH > 12.6), $K_{ox,obs}$ will approach $K_{ox}(K_{a,G}/K_{a,R})$ which will be greater than K_{ox} ($K_{a,R} < K_{a,G}$). If the pK_a of RSH is lower than that of GSH, the K_{ox} at low pH will be lower because of destabilization of RSSG by the electron-withdrawing group in R. As the pH increases, the high pH limit of $K_{ox}(K_{a,G}/K_{a,R})$ will be again approached but $K_{ox,obs}$ will be significantly less than $K_{ox}(K_{a,G} \ll K_{a,R})$.

Similar effects may be observed on the behavior of dithiol to disulfide oxidation; however, the equations become more complex due to the formation of two equivalents of GSH and the ionization behaviour of both dithiols. The general description is given by

$$K_{ox,obs} = K_{ox} \frac{[(H^+)^2 + 2K_{a,G}H^+ + (K_{a,G})^2]}{[(H^+)^2 + K_{a,R1}H^+ + K_{a,R1}K_{a,R2}]} \tag{47}$$

where $K_{ox,obs}$, K_{ox}, and $K_{a,G}$ are defined as before, and $K_{a,R1}$ and $K_{a,R2}$ refer to the acid dissociation constants of the two thiols of the dithiol species.

Matthews and Williams [58] observed that the pK_a of one of the redox active dithiols of lipoamide dehydrogenase was 4.8. This drop in pK_a can be achieved by locating a positively charged residue nearby the sulfhydryl group which would effectively lower the pK_a by stabilizing an ion pair between the thiolate and the nearby positive charge. Assuming a pH of 7 and values of 4.8 and >8.6 for the two pK_as of the dithiol and 8.6 for the pK_a of glutathione, the effect of lowering the pK_a from 8.6 to 4.8 would be predicted to produce a 200-fold decrease in the observed K_{ox}. An additional 6-fold decrease in K_{ox} would result from the effect of electron withdrawal ($\beta_{eq} = 0.2$) in destabilizing the disulfide. Thus, the ion pair effect in lipoamide dehydrogenase might be expected to decrease the K_{ox} by a factor of approximately 1200. The low pK_a would also enhance kinetic reactivity at pH 7 due to increased concentration of the thiolate anion compared to a thiol of normal pK_a.

6. Thiol/disulfide interconversions during enzyme catalysis

Catalytically essential sulfhydryl groups which participate directly in enzyme catalysis are legion. The utility of sulfhydryl group participation in such diverse reactions as acyl transfer, carbonyl oxidation, acyl group reduction, and pyrimidine methylation has been widely appreciated. However, another group of enzymes requires reversible redox state changes in one or more catalytically essential dithiol/disulfide moities to effect catalysis. All of these enzymes, with the exception of ribonucleotide reductase and ferredoxin-thioredoxin reductase, are known to be flavoenzymes. Generally, this class of enzymes [6] couples the reduction/oxidation of some substrate (S_{ox}/S_{red}) to the reduction/oxidation of NAD(P)/NAD(P)H. The function of the thiol/disulfide redox pair is to transfer reducing equivalents between the flavin cofactor and the substrate while the flavin transfers reducing equivalents between the thiol/disulfide pair and the nicotinamide substrates [6].

$$\mathrm{FlH_2\,E}\begin{matrix}\diagup \mathrm{SH}\\ \diagdown \mathrm{SH}\end{matrix} + \mathrm{S_{ox}} \rightleftharpoons \mathrm{FlH_2E}\begin{matrix}\diagup \mathrm{S}\\ |\\ \diagdown \mathrm{S}\end{matrix} + \mathrm{S_{red}} \tag{48}$$

$$\mathrm{FlH_2\,E}\begin{matrix}\diagup \mathrm{S}\\ |\\ \diagdown \mathrm{S}\end{matrix} \rightleftharpoons \mathrm{FlE}\begin{matrix}\diagup \mathrm{SH}\\ \diagdown \mathrm{SH}\end{matrix} \tag{49}$$

$$\mathrm{FlE}\begin{matrix}\diagup \mathrm{SH}\\ \diagdown \mathrm{SH}\end{matrix} + \mathrm{NADPH} \rightleftharpoons \mathrm{FlH_2E}\begin{matrix}\diagup \mathrm{SH}\\ \diagdown \mathrm{SH}\end{matrix} + \mathrm{NADP^+} \tag{50}$$

6.1. *Thioredoxin reductase*

Thioredoxins are a class of small (9–12 kDa) proteins with a redox active dithiol/disulfide which serve as a source of reducing equivalents for ribonucleotide reductase in the biosynthesis of deoxyribonucleotides and as a potential mediator of other cellular processes (see below) [84]. Thioredoxin reductase catalyzes the reversible, NADPH-dependent reduction of oxidized thioredoxin (disulfide) to the dithiol form.

$$\mathrm{NADPH} + \mathrm{Trx}\begin{matrix}\diagup \mathrm{S}\\ |\\ \diagdown \mathrm{S}\end{matrix} \rightleftharpoons \mathrm{NADP} + \mathrm{Trx}\begin{matrix}\diagup \mathrm{SH}\\ \diagdown \mathrm{SH}\end{matrix} \tag{51}$$

O'Donnell and Williams [58], in a series of careful and comprehensive experiments with *E. coli* thioredoxin reductase, have determined the microscopic redox potentials of the various redox forms of the thiol/disulfide and flavin at a variety of pH values. The oxidation potential of the dithiol is influenced by the redox state of the enzyme-bound flavin. When the flavin is reduced, the K_{ox} for the dithiol is 2.3 M (−0.271 v) while the K_{ox} for the dithiol is 0.63 M (−0.254 v) when the flavin is oxidized. The oxidation state of the dithiol must also affect the oxidation potential of the flavin. In the disulfide form, the K_{ox} for the flavin is 0.27 M (−0.243 v) and in the dithiol form the K_{ox} for the flavin is 1 M (−0.26 v). The redox state independent equilibrium constant between the $FADH_2E(SS) \rightleftharpoons FAD(E(SH)_2)$ (Eq. 49) is 0.43, sufficiently close to 1 that both states are observable in redox titrations. Thermodynamically, the K_{ox} values for the various redox forms of the enzyme are quite similar, differing only by a factor of 8.5 (0.27 M−2.3 M). The redox potentials of the various forms of the enzyme are also similar to the K_{ox} for the thioredoxin substrate.

Holmgren and Fagerstedt [85] have determined that approximately 30−40% of the thioredoxin in *E. coli* exists in the oxidized form. This would set the oxidation state of thioredoxin substrate at a potential of approximately 3.7 M ($RG_{eff} = K_{ox} F_{red}/(1-F_{red})$, where F_{red} is the fraction of the thioredoxin in the reduced form and RG_{eff} is the **R**[GSH] of an equivalent glutathione redox buffer which would maintain the same ratio of reduced to oxidized thioredoxin). At this external redox poise of the thioredoxin, if the active site was at equilibrium with the thioredoxin substrate pool, catalytically active enzyme forms $E(SH)_2FADH_2$ (K_{ox} = 2.3 M) and $E(SH)_2FAD$ (K_{ox} = 0.63 M) would represent the major enzyme species (about 52% and 14%, respectively), ensuring catalytic efficiency for the net reduction of oxidized thioredoxin. The thermodynamic properties of the redox active groups of thioredoxin reductase appear to be appropriately poised to ensure high concentrations of catalytically competent enzyme under physiological conditions.

6.2. Lipoamide dehydrogenase

Lipoamide dehydrogenase, which functions to supply oxidized lipoic acid from reduced lipoic acid and NAD^+ in the pyruvate dehydrogenase complex, exhibits a very similar mechanism to that of thioredoxin reductase. The microscopic redox potentials have not been fully determined; however, two different macroscopic oxidation potentials have been reported [58]. The K_{ox} for oxidation of the two-electron reduced enzyme to the fully oxidized enzyme (most likely involving the thiol/disulfide conversion $E(SH)_2FAD \rightleftharpoons E(SS)FAD$) is 4.6 M (−0.28 volts). However, oxidation of the four-electron reduced enzyme, $E(SH)_2FADH_2$ to the two-electron reduced enzyme, $E(SH)_2FAD$, is considerably more favorable (K_{ox} = 735 M, −0.346 volts) than the comparable oxidation in thioredoxin reductase (K_{ox} = 1 M). The external redox state of NAD/NADH in mitochondria is normally set at a ratio of approximately 8 [86] so that the effective **R**[GSH] of the external NAD/NADH couple is approximately 12.5 M (−0.29 v). This external **R**[GSH] would be insufficient to reduce the enzyme significantly to the inactive, four-electron reduced form. However, if this oxidation potential was more like that of thioredoxin reductase (1 M), the external **R**[GSH] of 12.5 M would tend to reduce the enzyme fully to the inactive state.

The oxidation potentials of the redox active functionalities appear to the balanced toward an optimization of specific redox states of the enzyme which are required for efficient turnover in the physiological direction. Surprisingly, the oxidation potentials for the thiol/disulfide interconversions are remarkably similar in the examples studied. Optimization of specific enzyme redox behaviors tailored to physiological conditions results largely from changes in the oxidation potential of the flavin.

6.3. Glutathione reductase

Glutathione reductase catalyzes the NADPH dependent reduction of glutathione disulfide.

$$GSSG + NADPH + H^{+} \rightleftharpoons NADP^{+} + 2\,GSH \tag{52}$$

The equilibrium constant at pH 7.0 is 150 M [37, 38]. This defines the standard reduction potential of GSSG to be −0.25 V at a standard state of pH 7.0, and 1 M. The oxidation potential of NADPH on the glutathione scale would be 150 M.

Glutathione reductase has be isolated and characterized from a variety of sources [6, 87] and a high resolution X-ray crystal structure is available [88]. Like other members of this class of enzymes glutathione reductase is a flavoenzyme containing a redox active dithiol/disulfide which serves to mediate electron transfer between glutathione and the flavin cofactor. The physiological function of this enzyme is to regenerate GSH from the GSSG formed by the action of glutathione peroxidase on hydrogen peroxide. The glutathione peroxidase/glutathione reductase system constitutes the major pathway for non-peroxisomal destruction of hydrogen peroxide and as such provides a mechanism for protection of cellular constitutents against oxidative damage by hydrogen peroxide and hydroxy radicals [4] (Fig. 8).

In the intracellular environment of most cells, the vast majority of glutathione is in the reduced state [89]. More than 99.5% of the total cellular glutathione pool is in the reduced (GSH) form. This would correspond to a redox poise (**R**[GSH]) of between 3–4 M at a [GSH]/[GSSG] ratio of 300–400. Changes in metabolic state such as fasting/feeding [86, 90, 91] or oxidative stress induced by the direct administration of hydrogen peroxide, hydrogen peroxide-generating drugs [92], or other thiol oxidants such as diamide [12], can cause a significant conversion of cellular GSH to GSSG, resulting in [GSH]/[GSSG] ratios nearing one and **R**[GSH] values < 0.1 M. In the light of these large changes in glutathione status, glutathione reductase must be able to catalyze efficient reduction of GSSG under a wide

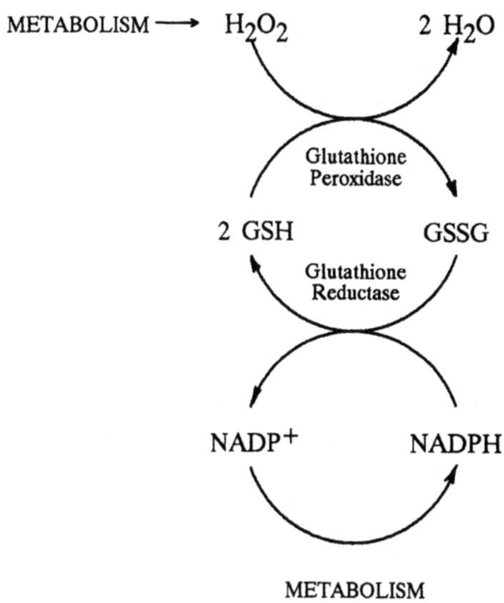

Fig. 8. The glutathione peroxidase, glutathione reductase pathway.

variety of conditions including a 100–400-fold molar excess of the product GSH over the substrate GSSG.

The redox potentials for the thiol/disulfide and FAD moities in glutathione reductase are unknown. The two-electron reduced protein forms tight complexes with NADPH without apparent formation of a significant fraction of the four-electron reduced species $E(SH)_2FADH_2NADP$. Formation of the four-electron reduced species requires removal of the product NADP by NADase [6]. In the absence of NADP(H), a 565 molar excess of GSH over flavin appears to reduce the flavin partially (about 30% reduction) as evidenced by changes in the visible absorption spectrum [93]. Assuming that the ratio of [GSH]/[GSSG] was near 50 in these experiments (not reported), the K_{ox} of the entire system would be about 2 M, near the physiological redox poise of the glutathione system. If the K_{ox} of the enzyme was this close to the physiological **R** [GSH] of cellular glutathione, GSH at high concentrations might be capable of partially reversing the enzyme oxidation by GSSG and could account for the inhibition of the enzyme by the product, GSH [94–95].

6.4. Mercuric reductase

Bacterial mercuric reductase catalyzes the net thermodynamically favorable reduction of Hg^{2+} to elemental Hg° ($E_o' = 0.798$ volts) by NADPH [59]. Like other members of this class of enzymes, mercuric reductase contains a redox active dithiol/disulfide ($K_{ox} = 2$ M, $E_o' = -0.269$ v) and a flavin ($K_{ox} = 316$ M, $E_o' = -0.335$ v). The flavin redox potential is sufficiently negative compared to that of NADPH that no significant four-electron reduced species is observed during catalysis, similar to the observations with lipoamide dehydrogenase. The two-electron reduced enzyme contains a dithiol which forms a tight complex with Hg^{2+}; however, in the absence of NADPH there is no reduction of the mercury bound to the enzyme. Hg^{2+} reduction is only observed in the presence of NADPH.

6.5. Trypanothione reductase

Trypanosomes utilize the low molecular weight dithiol, trypanathione, in much the same way that most other organisms utilize glutathione [96]. Trypanothione, a recently discovered dithiol in these species, is a bis-glutathione adduct of spermidine [65]. A flavoprotein, trypanothione reductase, has been isolated which specifically reduces oxidized trypanothione and the monoglutathionyl spermidine disulfide in an NADPH-linked process. The enzyme has been shown to belong to the class of flavin/dithiol enzymes; however, the oxidation potentials of the functional groups of this enzyme are not known [97].

6.6. Ferredoxin-thioredoxin reductase

During the light-induced, reductive activation (or inactivation) of specific enzymes in plant chloroplasts (see below), reducing equivalents must be transferred from the photoreduced ironsulfur center of ferredoxin to protein dithiol redox carriers (the plant thioredoxins, see below), and ultimately to the target enzymes. The enzyme which catalyzes the initial step of this reductive process, ferredoxin-thioredoxin reductase [98], has a catalytically essential dithiol located on the M_r 13,000 subunit. Presumably the function of this dithiol is to mediate electron transfer between the ferredoxin and thioredoxin. The redox potential of this protein dithiol is not known.

6.7. Ribonucleotide reductase

Ribonucleotide reductase, a mechanistically complex enzyme, catalyzes the reduction of ribonucleotides to deoxyribonucleotides [99–100]. Unlike the enzymes discussed above, ribonucleotide reductase is not a flavoprotein. The enzymes from various species possess either a Vitamin B_{12} cofactor or a unique tyrosyl radical in the active enzyme. A mechanism involving initial abstraction of the 2′ hydrogen atom of the substrate by an enzyme free radical has been proposed [100].

The overall reduction of ribonucleotides requires a dithiol reductant to regenerate the oxidized enzyme at the end of a catalytic cycle. While dithiothreitol will serve in this capacity, the physiological reductants are thioredoxin and glutaredoxin [84]. Oxidized thioredoxin is reduced by

NADPH through thioredoxin reductase while glutaredoxin may be reduced by GSH. The GSSG produced is then reduced by NADPH and glutathione reductase. Through both pathways, the ultimate source of reducing equivalents is NADPH.

Thelander [102] and Lin *et al.* [102] have found that the B1 subunit of ribonucleotide reductase from various species contains multiple dithiols/disulfides. The reduced enzyme will reduce three equivalents of CDP to deoxyCDP in the absence of any additional reductant. Lin *et al.* [102], on the basis of positional incorporation of radiolabelled iodoacetamide into the reduced enzyme, have suggested that two redox active dithiol/disulfides per active site could comprise a redox shuttle between thioredoxin and the active site and that sequential (or parallel) transfer of reducing equivalents could be accommodated.

7. Thiol/disulfides as redox carriers

The reversible nature of the thiol/disulfide exchange process can be utilized biologically in the transfer of reducing equivalents between different molecular species. For example, the net biological reaction catalyzed by ribonucleotide reductase

$$\text{NDP} + \text{NADPH} \rightleftharpoons \text{deoxyNDP} + \text{NADP}^+ \tag{53}$$

is actually mediated by a sequence of thiol/disulfide exchange processes involving the mobile, redox active, thiol/disulfide cofactors, thioredoxin and glutaredoxin [58].

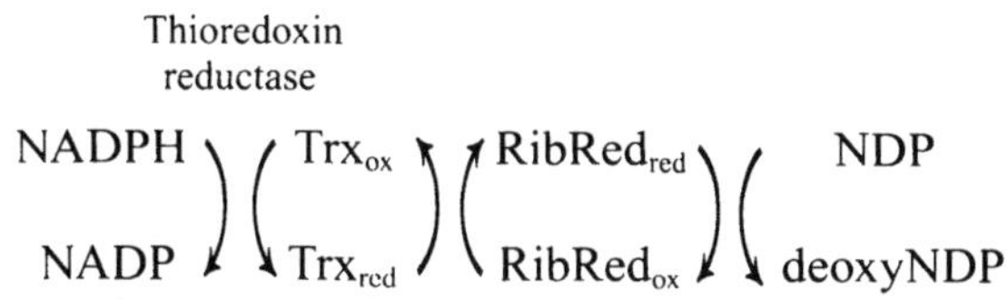

(54)

Cellular thiol/disulfide redox buffers such as thioredoxin, glutaredoxin, trypanothione, and glutathione can serve to transport reducing equivalents between electron donors and acceptors including proteins.

7.1. *Thioredoxins and glutaredoxins*

Thioredoxins and glutaredoxins are small (M_r 10,000–12,000), soluble proteins which have a redox active dithiol/disulfide. In bacterial and mammalian cells, these molecules appear to be primarily involved in providing reducing equivalents to ribonucleotide reductase; however, other potential functions have been suggested [84]. The thioredoxin from *E. coli* has been best characterized. A high resolution X-ray structure of the disulfide form is available [103] and an NMR structure of the reduced form has been deduced [104]. The redox active dithiol/disulfide is located on a protuberance, and upon reduction of the Cys_{32}–Cys_{35} disulfide bond, the distance between the two cysteines increases from 2.05 A to 6.8 A, and a new interaction is noted between the cysteine and a nearby lysine residue. There are no large conformation changes which accompany disulfide reduction; however, small local effects are observed [105]. The equilibrium constant for the reaction catalyzed by thioredoxin reductase is approximately 50 at pH 7.0, so that the K_{ox} for *E. coli* thioredoxin is 2 M on the glutathione scale [61]. Replacing the Pro_{34} residue of *E. coli* thioredoxin with a His to mimic the active site sequence of the related protein disulfide isomerase results in a 35 mV decrease in the E_o' of the dithiol (a 15-fold decrease in the K_{ox}) and a 7-fold reduction in the catalytic activity of the protein in the reduction of insulin [106].

Bacteriophage T4 encodes a phage thioredoxin as well as its own ribonucleotide reductase. In contrast to other thioredoxins, the T4 thioredoxin does not have a proline residue within the disulfide loop. Berglund and Holmgren [107] showed that the T4 thioredoxin is 10-fold more easily reduced than *E. coli* thioredoxin. Bacteriophage T7 utilizes the *E. coli* thioredoxin as an obligate cofactor of the phage DNA polymerase. While the reduced form of thioredoxin is specifically required for activation and increased processivity of the enzyme [108], site-directed mutagenesis of the cysteines to serines has shown that the activation does directly require a redox active role for the thioredoxin [62].

The plant thioredoxins (see below for a discussion of the biological role of these thioredoxins) are particularly fascinating in at least two aspects – the multiplicity of different structures, and the very high oxidation potential in comparison to the *E. coli* protein. In plants, the primary function of the various thioredoxin species is to modulate the activity of key regulatory enzymes by reduction (see below). Two primary classes of thioredoxin species have been recognized by their specificity toward reduction of their target enzymes [13]. The "m" class thioredoxins are more effective in the reduction of NADP-malate dehydrogenase while the "f" class thioredoxins are more effective in the reduction of chloroplast fructose-1,6-bisphosphatase.

Three iso-proteins of thioredoxin m have been isolated (b, c, and d) which have recently been shown to differ only by the loss of one or two amino acids from the amino terminus [109]. Two apparent iso-proteins of the f class have been observed [110]; however, the molecular differences between the two are not known. The oxidation potentials of spinach thioredoxin f may be estimated from the data of Clancey and Gilbert [64] as about 63 M. Rebeille and Hatch [63] have determined the equilibrium constant for the oxidation of thioredoxin m by oxidized dithiothreitol at pH 7.4 ($K_{eq} = 0.11$) which corresponds to a K_{ox} of 29 M on the glutathione scale. The structural basis for the 10^3 difference in oxidation potential between the *E. coli* thioredoxin and the chloroplast thioredoxins is not apparent from sequence differences around the redox active dithiol/disulfide.

In a mutant strain of *E. coli* lacking thioredoxin, Holmgren [111] observed that ribonucleotide reduction still occurred at a rate capable of maintaining cell growth. These observations led to the isolation of another redox active dithiol/disulfide protein, glutaredoxin, which was similar in size to thioredoxin and capable of supporting reduction of ribonucleotide reductase in the presence of glutathione, glutathione reductase, and NADPH. On a molar basis, glutaredoxin was kinetically a superior reductant for ribonucleotide reductase. An *E. coli* strain deficient in glutaredoxin has recently been isolated [113]. This strain is deficient in sulfate and ribonucleotide reduction, suggesting that glutaredoxin, rather than thioredoxin, is the primary reductant of ribonucleotide reductase *in vivo* even though the levels of glutaredoxin are significantly lower than those of thioredoxin [113]. In addition to their function as redox carriers, alternative thioredoxins and glutaredoxins have been suggested to play roles in catalysis of thiol/disulfide exchange reactions [84].

7.2. *Trypanothione*

A novel dithiol species has been found in trypanosomes. Termed trypanothione [65] the molecule is a bis-glutathione adduct of spermidine in which two molecules of glutathione are covalently linked by an amide bond to the terminal amino groups. The conjugation of glutathione to spermidine has also been shown to occur in *E. coli* [114]. During logarithmic growth, glutathione represents the major thiol species; however, during stationary phase the vast majority of the glutathione is conjugated to N-1 of spermidine forming a mono-glutathionyl adduct. The most abundant glutathione conjugate in the trypanosome *Crithida fasciculata* is N_1, N_8-bisglutathionylspermidine; however, significant levels of the mono conjugate are observed [115]. The function of these glutathione conjugates appears to be similar to that of glutathione in that most of the destruction of hydrogen peroxide by these organisms can be attributed to thiol-dependent pathways [96]. The oxidized trypanothione generated from peroxide reduction can be reduced by the NADPH-linked trypanothione reductase.

The redox potential of trypanothione has been reported as −0.242 volts, corresponding to a K_{ox} of 0.25 M on the glutathione scale [116]. The biological function of this conversion of glutathione to an intramolecular dithiol is not known.

8. Role of oxidation potential of a dithiol in protein folding

Since the classic experiments of Anfinsen and his colleagues [117] on the regeneration of the native structure of ribonuclease A from the unfolded, reduced protein, it has generally been held that the native conformation of a folded protein (including the disulfide bonds) is the thermodynamically most stable one under the prevailing physiological conditions. All of the information required for the attainment of the proper tertiary structure (including disulfide bonds) appears to be resident in the primary sequence. This implies that given the proper conditions, an unfolded protein will spontaneously fold into the correct structure – given enough time. However, post-translational covalent modification (such as proteolysis, cofactor incorporation, glycosylation, etc.) may alter this conclusion somewhat.

Relatively few proteins with disulfide bonds have been studied in detail, but in appears that the disulfide bonding patterns observed are dictated by the overall thermodynamic stability of the various possible disulfide isomers. Under the appropriate oxidizing conditions, one of the numerous possible disulfide isomers is generally found to predominate at equilibrium [118]. While the final thermodynamic stability dictates which disulfide bonds will eventually form, the mechanistic pathway between the fully reduced protein and the correct redox isomer of the fully oxidized protein is littered with a variety of intermediates [23, 41].

8.1. Mechanistic pathway

The attainment of the correct disulfide bonds by oxidation of protein sulfhydryls to disulfides is far from straightforward. Disulfide bonds do not appear to be formed or broken in a simple, sequential manner when a reduced, unfolded protein is exposed to an excess of oxidant (such as GSSG). The protein thiols are oxidized much more rapidly than the native structure and biological activity is regained. This phenomenon can be accounted for by the initial formation of either a large [43] or relatively small [23] number of "incorrect" disulfide bonds (including glutathione mixed disulfides) which slowly rearrange to the native structure (Fig. 9). The initial distribution of disulfide bonds which form rapidly upon oxidation does not appear to be random (based on closeness in sequence), suggesting that even in the "mis-oxidized" state there are significant effects of protein structure which affect the kinetics and thermodynamics of disulfide bond formation [23]. While the formation of non-native disulfide bonds during the folding process has been accepted for some time, Weissman and Kim's [45] reexamination of the intermediates involved in the folding of bovine trypsin inhibitor using alternative procedures for stopping thiol/disulfide rearrangements has suggested that the population of non-native disulfides is significantly lower than previously believed.

The initial set of "productive" and "nonproductive" intermediates may be converted to the most stable, native structure by one or more processes which may involve 1) trapping a productive intermediate by oxidation, 2) removing a non-productive intermediate by oxidation or reduction, 3) conversion of a non-productive intermediate to a productive intermediate by rearrangement of disulfide bonds, or 4) conformation changes not involving thiol/disulfide

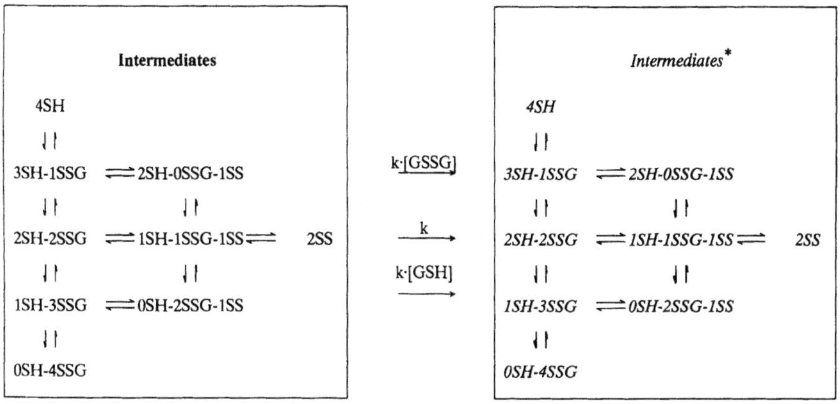

Fig. 9. A generalized scheme for protein folding pathways. The various oxidation states of a species with four cysteine residues, all of which are in disulfide bonds in the native structure, are shown. SH, SSG, and SS refer to thiol, mixed-disulfide and disulfide species respectively. N is the native protein, in which there is only one arrangement of disulfide bonds. Note that each of the species shown in the figure is a collection of redox isomers involving different cysteine residues. While some proteins such as BPTI follow a relatively straightforward route to the native states, other proteins such as RNAse involve many more intermediates with multiple, most likely nonnative, disulfide bonds (see text).

rearrangements. Mechanisms 1 and 2 will involve an oxidative or reductive transition between two different protein oxidation states while mechanisms 3 and 4 will not involve any net redox state change [43].

Creighton and his colleagues have suggested that oxidation of bovine pancreatic trypsin inhibitor (BPTI), which contains three disulfides, proceeds through a mechanism involving rate-determining, intramolecular rearrangements of a limited number of non-native, two-disulfide species through reasonably well-defined intermediates [23]. Interestingly, an intermediate two-disulfide species with native disulfide bonds (Cys_{30}–Cys_{51}, Cys_{14}–Cys_{38}) must rearrange (to Cys_{30}–Cys_{51}, Cys_{5}–Cys_{55}) before the native structure can be attained by formation of the final Cys_{14}–Cys_{38} disulfide. Recent studies of BPTI folding in which the Cys_{14} and Cys_{38} were replaced by alanines or threonines suggest that multiple pathways may be a available for disulfide bond rearrangement [24]. When treated with oxidants, this mutant protein did attain an active conformation with two correct disulfide bonds, although the correct disulfide bonds formed approximately 3-fold more slowly than the wild-type protein. While not making a large contribution to protein stability, Cys_{14} and Cys_{38} can kinetically facilitate (by a relatively small 3-fold) the attainment of the correct disulfide bonded state. However, the recent observations of Weissmann and Kim [45] suggest that interpretation of the folding pathway, at least for trypsin inhibitor, may be complicated by rearrangements that occur after quenching the reaction. Further experiments will be required to evaluate the role of non-native disulfide bonds in the mechanism of protein folding.

Disulfide bond formation that accompanies the folding of a protein from its reduced, denatured state to the native structure is generally more facile when the oxidation is performed in a thiol/disulfide redox buffer [119]. Dithiothreitol, glutathione, and mercaptoethanol redox buffers have been utilized. For must of the proteins examined, the correct formation of disulfide bonds occurs under reasonably well-defined optimum concentrations of thiol and disulfide. For ribonuclease A, optimum activity (at a fixed time of 30 min) is regained at a GSH = 2.3 mM and GSSG = 0.8 mM [102]. Surprisingly, the formation of disulfide bonds from a reduced protein, frequently requires what might be considered reducing conditions. However, the oxidation potentials of the disulfides in the native protein are generally very high relative to glutathione (e.g. the K_{ox} for Cys_{14}–Cys_{38} in BPTI is about 1500 M) [66] so that most readily attainable glutathione redox buffers would be thermodynamically capable of oxidizing BPTI (or ribonuclease).

Sheraga *et al.* [43] have proposed a "folding" mechanism for disulfide bond formation in native ribonuclease A from the reduced form in the presence of glutathione redox buffers. Ribonuclease A (4 disulfides) has a total of 7163 different redox isomers if all possible combinations of free thiols, glutathione mixed disulfides, and protein disulfides are considered. According to this mechanism, addition of the reduced protein to a glutathione redox buffer results in the rapid establishment of an equilibrium distribution of "intermediates" which are then slowly converted, via six parallel kinetic processes, to another set of "intermediates". The "intermediates" then rapidly attain the native disulfide bonded structure.

The distribution of "intermediates" after establishment of the initial, rapid equilibrium depends on the redox status of the folding buffer and the oxidation potentials of the various protein disulfide species. The distribution of redox isomers of the same net oxidation state will be fixed by the relative stabilities of the various redox isomers and will be independent of the external state of the glutathione redox buffer [43, 66]. Changes in the concentrations of GSH and/or GSSG will result in changes in the equilibrium population of the 25 different "intermediates". Kinetically, the "intermediates" are converted to the native state (after rapidly passing through the "intermediate" population) by only a few (actually 6) kinetically significant processes. The possible rate-limiting steps appear to be of three types – rearrangement of disulfides without change of redox state, formation of a disulfide, and reduction of a disulfide. The probability of a given pathway contributing to formation of the native state will depend on the relative rates of the various pathways under different redox buffer conditions. In turn, this rate will depend on the concentration of the "intermediate" involved and the concentration of any external reductant (GSH) or oxidant (GSSG) required for the step [43]. Due to the ability to generate a complex kinetic dependence on the concentrations of GSH and GSSG via both thermodynamic and kinetic considerations, this mechanism (or one like it) should be capable of describing the behavior of most protein oxidations in different redox buffers. However, for different proteins the magnitude of rate and equilibrium constants would be expected to be significantly different, so that the populations of "intermediates" and the chemical nature of the rate-limiting reactions will undoubtedly be different.

In a recent study on the oxidative folding of ribonuclease A, Wetlaufer *et al.* [121] emphasize that rate and yield of native protein do not correlate simply with the redox status of a mercaptoethanol redox buffer. These authors also analyze the difficulty of maintaining a constant redox buffer status during the folding process, and provide a source of excellent practical considerations which should be followed in experiments of this type. Although the mechanism and nature of the intermediates in protein in disulfide formation undoubtedly are influenced by thermodynamic effects, the process of protein oxidation is a complex kinetic process, the rate of which will depend on both thermodynamic and kinetic factors.

8.2. "Cooperativity" in disulfide bond formation

It might be expected that disulfide bond formation would become more favorable as successive disulfide bonds are formed, the tertiary structure of the protein becomes more native-like and cysteine residues are brought into close proximity by protein folding. While there are only a few proteins with multiple disulfide bonds for which the oxidation potentials of individual disulfide bonds are known, there appears to be no consensus regarding the cooperativity of disulfide bond formation.

Increasing oxidation potential with increasing completeness of folding would provide evidence of cooperativity in disulfide bond formation. For BPTI, the gross oxidation equilibrium constants (statistical averages of all intermediates involved) for formation of the I, II, and III disulfide-bonded species are approximately 8.7 M (0 disulfide – 1 disulfide), 2.3 M (1-disulfide to incorrect 3-disulfide intermediates), 4000 M (1-disulfide to the native-like 2-disulfide intermediate), and 1500 M for the formation of the native 3-disulfide structure from the native-like 2-disulfide intermediate [66]. The increase in observed oxidation potential with increasing similarity to the native structure implies that the net favorable interactions which drive the formation of native structure are increasingly utilized to stabilize disulfide bond formation. A similar behavior has been observed for RNAse T1 in which both native and non-native single disulfides exhibit K_{ox} values between 2.6 M and 22 M while the final disulfide exhibits a K_{ox} of 1100 M [67].

This tendency toward a higher K_{ox} (more stable disulfide) for disulfide bonds formed late in the folding process is not apparent in the oxidation of ribonuclease A. Creighton [122] observed an increasing resistance to disulfide bond formation as successive disulfide bonds are formed during the regeneration of RNAse A from the reduced form in the presence of oxidized DTT. In fact, oxidized DTT was observed to be thermodynamically incapable of oxidizing the 3-disulfide intermediate to the native structure, yet oxidized intermediates could be formed. For RNAse A, this implies that the formation of the final disulfide bonds must overcome increasingly unfavorable interactions in the native-like structure. A recent review of disulfide formation during protein folding is available [123].

9. Proteins in which a disulfide bond stabilizes structure

Proteins with disulfide bonds are more likely to be found in the extracellular environment in the form of secreted and extracellular-membrane bound proteins. The stability of proteins to denaturation is increased substantially by the introduction of disulfide bonds [19]. For example, disulfide bond formation has been estimated to contribute approximately 19 kcal/mole toward the stability of ribonuclease A [120].

9.1. Wild-type structures

The most straightforward demonstration that disulfide bond formation contributes to net protein stability and structure is the loss of structure and biological activity upon reduction of the disulfides. Ribonuclease A, [122] ribonuclease T1 [67], lysozyme [119], and bovine pancreatic trypsin inhibitor [66] all lose a large fraction of their biological activity on reduction. However, the stability of the native-like structure in the absence of disulfide bonds may be significant [65].

The contribution of disulfide bond formation to protein stability is generally discussed in terms of the effects of covalent crosslinking on the stability of the unfolded state [77]. In the transition between the unfolded state and the folded state, the largest single unfavorable energetic barrier can be attributed to the loss of chain conformational entropy which must accompany the formation of a discrete, compact tertiary structure. The formation of covalent crosslinks restricts the conformational freedom of the peptide chain in the unfolded state [21, 77]. As a consequence of this lower entropy of the unfolded state, the transition to the folded state is accompanied by a smaller entropy loss and the net transition to the folded state is correspondingly more favorable.

Poland and Scheraga [77] have applied polymer theory to estimate the entropy loss resulting from crosslink formation between two cysteines separated by n intervening residues. They estimated the entropy loss to be $\Delta S = -6.9 - (3/2)\,R\ln(n)$. Using a somewhat larger volume area defined by a sphere of closest approach of two sulfur atoms, Pace *et al.* [21] arrived at estimated entropy losses of $\Delta S = -2.1 - (3/2)\,R\ln(n)$, which more closely matched a variety of experimental measurements on the stability of disulfide and chemically crosslinked proteins. From these predictions, it may be concluded that crosslinks between more remote cysteine residues (larger n) would contribute more to the apparent stability of the folded state than crosslinks between nearby residues. Pace *et al.* [21] have suggested that the entropic contribution to the conformational stability of a protein will be raised by 3.0, 4.0, and 5.0 kcal/mole by the introduction of crosslinks of 15, 45, and 135 residues, respectively. Since most protein disulfides are formed between residues separated by less than 24 residues, Pace *et al.* argue that the individual contribution of a disulfide bond to protein stability is likely to be less than 5 kcal/mole (a factor of 5000). The stability of ribonuclease T1 (2 disulfides, Cys_2–Cys_{10} and Cys_6–Cys_{103}), was decreased by 3.4 kcal/mole by reduction and alkylation of the Cys_2–Cys_{10} disulfide and by 7.2–9.3 kcal/mole when both disulfides were reduced. Most of the large decrease in stability upon reduction could be accounted for by the expected increase in the conformational entropy of the unfolded protein. The single Cys_{32}–Cys_{35} disulfide bond in *E. coli* thioredoxin stabilizes the molecule toward denaturation by about 2.4 kcal/mole which is within a factor of two of the 1.6 kcal/mole stabilization expected from chain entropy effects [102, 124].

In the simple case involving one disulfide bond, protein stability and oxidation potential are linked through the thermodynamic cycle shown in Fig. 10 [124]. In this diagram, the species designated U represents the unfolded state of the protein while N represents the native, folded conformation. The equilibrium constants designated K_1 and K_2 represent folding equilibria ($K_1 = [N]/[U]$) while the equilibrium constants K_u and K_n

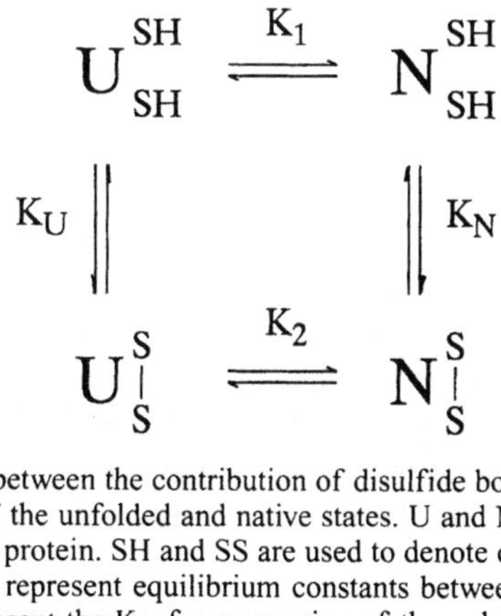

Fig. 10. The relationship between the contribution of disulfide bonds to protein stability and the oxidation potentials of the unfolded and native states. U and N refer to the unfolded and folded conformations of a protein. SH and SS are used to denote oxidation states of the sulfhydryl groups. K_1 and K_2 represent equilibrium constants between the unfolded and folded states and K_n and K_u represent the K_{ox} for conversion of the reduced to oxidized states. The effects of a disulfide bond on the stability of the protein is represented by the ratio K_2/K_1 which must, by nature of this thermodynamic cycle, be equal to K_n/K_u. Disulfide bonds which show the largest increase in K_{ox} between the unfolded and folded states should increase the stability of the protein the most.

represent thiol/disulfide oxidation potentials of the unfolded and native structures. The designation of SH or S-S represents the reduced and oxidized species for both U and N forms. The native state with the disulfide bond reduced may be entirely hypothetical and unobservable experimentally if K_1 is significantly less than one, since reduction of the native disulfide will be accompanied by unfolding. The relationships of this thermodynamic cycle require that $K_2/K_1 = K_n/K_u$.

The ratio K_2/K_1 reflects the relative stabilities of the protein with and without disulfide bonds so that disulfide stabilization against denaturation occurs when K_2/K_1 is large. This stabilization toward denaturation must also be reflected in the ratio K_u/K_n which defines the oxidation potential for dithiol oxidation in the denatured and native states. If K_2/K_1 is large and the protein is stabilized against denaturation, the ratio K_n/K_u must also be large, and the dithiol will be easier to oxidize in the folded state than in the denatured state. This relationship predicts that dithiols which exhibit large changes in oxidation potential between the folded and unfolded state should provide more stability against denaturation. Anything which decreases the oxidation potential, such as dihedral, conformational, or steric strain, will detract from the stabilizing effects of the disulfide bond. An interesting example of the link between oxidation potential and thermodynamic stability of a folded protein is provided by dsbA, a periplasmic protein oxidant in *E. coli*. In this case, the single disulfide bond that forms actually destabilizes the protein toward urea denaturation. The reduced protein is more stable than the oxidized protein. Consistent with this, oxidation of the dithiols is much more favorable for the denatured protein than for the folded protein [73]. Thus, when disulfide bond formation is coupled to the induction of strain or the loss of favorable interactions in the reduced state, both the disulfide bond and the protein are less stable.

9.2. *Genetically engineered disulfides*

Site-directed mutagenesis has been employed in an effort to engineer increased protein stability by introducing disulfide crosslinks at defined locations [125]. The number of potential sites for engineered disulfide bonds in globular proteins appears to be rather limited if major distortions of the structure are to be avoided [126]; however, disulfide bonds have been engineered into dihydrofolate reductase [127], T4 lysozyme [128], subtilisin [46] and the λ repressor [68]. In all cases, the disulfide crosslinked enzymes exhibited essentially the same enzymatic activity as the wild type enzymes.

Cys_{22}–Cys_{87} and Cys_{24}–Cys_{87} disulfide bonds have been introduced into subtilisin in an effort to enhance stability. The thiol/disulfide oxidation potentials (measured with dithiothreitol redox buffers) were 120 M and 500 M respectively on the glutathione scale [46]. Based on the arguments presented above concerning the relationship between K_{ox} and the effect of the disulfide bond on protein stability, it would be expected that the Cys_{24}–Cys_{87} mutant would be approximately four times more stable (assuming that K_u, Figure 10, was comparable for the two mutants). Irreversible thermal denaturation of subtilisin occurs mainly through autolysis. Differences in stability will reflect both the effects of conformation on enzyme activity and the suitability of the protein as a substrate for proteolysis. Since both mutant enzymes exhibited very similar catalytic activities, the major contribution to differences in autolytic stability is probably local or global unfolding of the protein making it a better substrate for itself. With the disulfide intact, the Cys_{24}–Cys_{87} mutant was 2.1–2.7 times more stable than the Cys_{22}–Cys_{87} mutant, in reasonable agreement with the prediction made from the relative K_{ox} values. Somewhat surprisingly, almost all of the change in K_n/K_u (or equivalently K_2/K_1) can often be attributed to differences in the stabilities of the reduced state or to unfavorable strain accompanying the formation of the disulfide bond [129].

In a study of five disulfide-containing mutants of T4 lysozyme, Matsumura *et al.* [130] observed that the introduction of cysteine residues decreased the thermal stability of the mutant T4 lysozyme (measured under reducing conditions) by 2–6 °C. Formation of the disulfide bond then increased thermal stability (compared to the reduced form) by 3–14 °C, depending primarily on the size of the loop formed with the engineered disulfide. The most stable of the disulfide-crosslinked mutants (Cys_{21}–Cys_{142}) was 2.8 kcal/mole more stable than the reduced form of the protein. The K_{ox} for formation of this disulfide was approximately 200 M on the glutathione scale [129]. The stabilization produced by the introduction of multiple disulfide bonds into the T4 lysozyme molecules is additive [130]. Three disulfide bonds increased the melting temperature of T4 lysozyme by 32 °C relative to the reduced molecule and by 23 °C relative to the wild-type lysozyme. In general, the stability imparted by a disulfide bond is greatest when the size of the loop formed by the crosslink is as large as possible, the destabilizing effect of the introduction of cysteine into the native structure is as small as possible, and the strain introduced by disulfide bond formation is as low as possible [129]. The last of these criteria would suggest that the K_{ox} for disulfide formation should be as large as possible since the effect of strain would be to decrease the oxidation potential for disulfide bond formation.

10. Redox active thiol/disulfides which affect activity

A very large number of proteins possess one or more cysteine residues, the modification of which by oxidation, alkylation, or other chemical means results in some change in the properties of the protein [1, 2, 8, 11–13]. Increases or decreases in V_{max}, K_m, the K_is for inhibitors, stability, interactions with other proteins, and sensitivity to proteolysis have been observed after modification of cysteine residues. While these "essential" cysteine residues may play a direct role in catalysis, often they do not.

The conversion of protein thiols to disulfides by thiol/disulfide exchange with another thiol/disulfide pair provides a mechanism for reversible interconversion of protein thiol/disulfide oxidation states. If the biological activity (or another property) of the reduced and oxidized forms of a protein are different, the potential for regulation of protein function in response to cellular changes in thiol/disulfide redox status exists. Hopkins and Dixon [25] and later Guzman Baron [8] alluded to a relationship between the redox state of glutathione, the predominant low molecular weight thiol in most cells, and the activity of biological processes.

Regulation of enzyme activity by thiol/disulfide exchange in response to changes in the thiol/disulfide redox status of some intracellular redox buffer has become an established control mechanism in the light-induced activation of several chloroplast enzymes [13, 26]. In other biological systems, including mammalian systems, an increasing body of circumstantial evidence suggests that this mechanism of regulation deserves serious consideration [12–14].

10.1. Light regulation in plants

The most compelling evidence for the modulation of enzyme activity by reversible thiol/disulfide redox state changes can be found in the effects of light on the activities of key enzymes in chloroplasts [13, 131]. Several enzymes of the reductive pentose phosphate cycle (fructose-1,6-bisphosphatase, sedoheptulose bisphosphatase, NADP-glyceraldehyde-3-phosphate dehydrogenase, and phosphoribulokinase) as well as other chloroplast enzymes (NADP-malate dehydrogenase, phenylalanine ammonia lyase) exist in the dark in inactive, disulfide states. Upon illumination, reducing equivalents from non-cyclic electron transport at photosystem I reduce ferredoxin. In turn, these reducing equivalents are passed from ferredoxin to a series of mobile, chloroplast thioredoxins through the action of ferredoxin-thioredoxin reductase. Like their bacterial counterparts, these thioredoxins are small proteins with a redox active dithiol/disulfide center. The reduced forms of the various chloroplast thioredoxins then specifically reduce the disulfide bonds of the target enzymes (Fig. 11). *In vitro*, the reductive activation of these enzymes can be achieved with DTT in the presence of the appropriate thioredoxin.

10.1.1. Malate dehydrogenase: While it had been argued convincingly that thioredoxin-mediated thiol/disulfide exchange regulates enzyme activity in response to light, a key observation was provided by experiments of Scheibe with pea chloroplasts [132]. Because only reduced chloroplast thioredoxin m

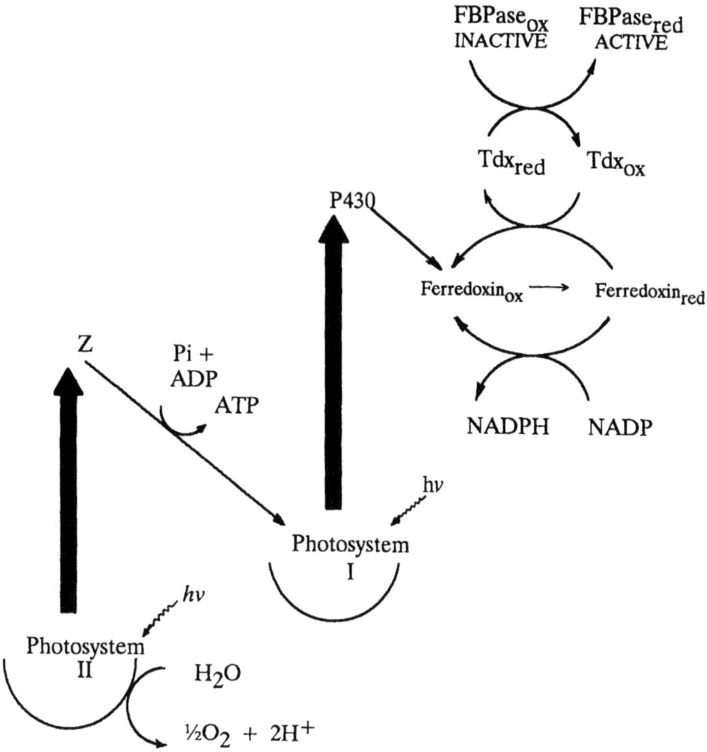

Fig. 11. Light-induced reductive modulation of the activity of chloroplast enzymes. Reducing equivalents generated in the light are passed to ferredoxin which in turn generates reduced thioredoxins. The thioredoxins, in turn, reduce specific target enzymes, activating or inactivating them.

could be inactivated by alkylation, the sensitivity of thioredoxin to inactivation after different periods of illumination was used to determine changes in the ratio of reduced to oxidized thioredoxin m in response to light. In the dark, the majority of the thioredoxin "m" was in the oxidized state and the activity of malate dehydrogenase was low. Upon illumination, thioredoxin m became reduced as malate dehydrogenase became more active. The thioredoxin m reduction occurred in a kinetically competent manner at a rate equivalent to or slightly faster than reductive activation of the malate dehydrogenase.

In a series of comprehensive experiments, Rebeille and Hatch [63, 133] measured the activity of NADP-malate dehydrogenase at equilibrium with thioredoxin m at varying fractions of reduced thioredoxin. Replots of the data from Rebeille and Hatch [63] suggest that the enzyme is half-reduced (half active) when the ratio of thioredoxin$_{red}$/thioredoxin$_{ox}$ is approximately 4; however, the reduction of the enzyme appears to be strongly cooperative, exhibiting a Hill coefficient of 3.6–4 (Fig. 12). This

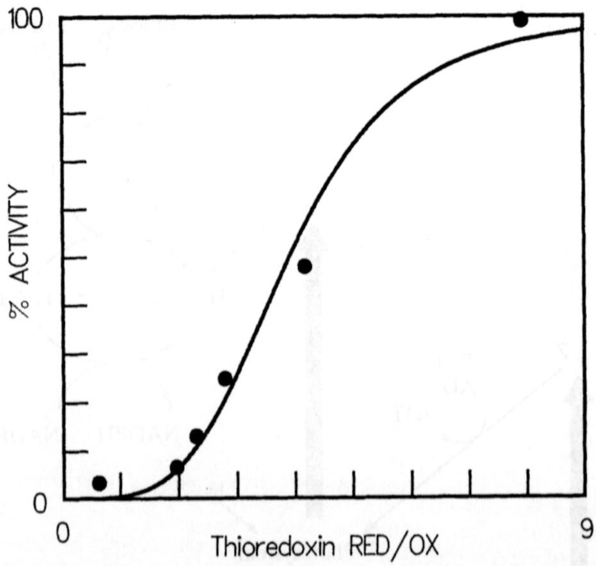

Fig. 12. Activity of chloroplast NADP-malate dehydrogenase in equilibrium with thioredoxin m redox buffers. Data for this plot are taken from ref 64. The curve is drawn as a positively cooperative reduction according to the relationship % activity = $100 \cdot \mathbf{R}^n/(K_{ox}^n + \mathbf{R}^n)$, where n has the value 4 and K_{ox} has the value 3.9.

behavior results in a sharp transition between oxidized and reduced enzyme with a small change in the redox state of the thioredoxin m [63]. From their data, the effective K_{ox} of the malate dehydrogenase can be calculated to be about 114 M on the glutathione scale. The effective K_{ox} was also dependent on the presence of NADP and NADPH, being raised as the NADP/NADPH ratio decreased.

In intact chloroplasts the redox states of NADP-malate dehydrogenase, NADP, and thioredoxin m changed during the light to dark transition and in the presence of alternative electron acceptors such as oxaloacetate [133]. In agreement with the measurements of Scheibe [132], about 20% of the thioredoxin was in the reduced state in the dark, but greater than 90% of the thioredoxin was in the reduced state after illumination. Given that the K_{ox} of thioredoxin m is about 29 M (see above), the effective redox potential which could be maintained by thioredoxin m would change from 7.2 M in the dark to >260 M in the light. At equilibrium with the thioredoxin, the malate dehydrogenase would be expected to increase from 6% to >70% in the dark-light transition, consistent with observations.

10.1.2. Fructose-1,6-bisphosphatase: Spinach chloroplast fructose-1,6-bisphosphatase is activated in the dark by thioredoxin f in the presence of dithiothreitol [134]. The reaction is reversible, and an excess of oxidized DTT will oxidatively inactivate the enzyme [62]:

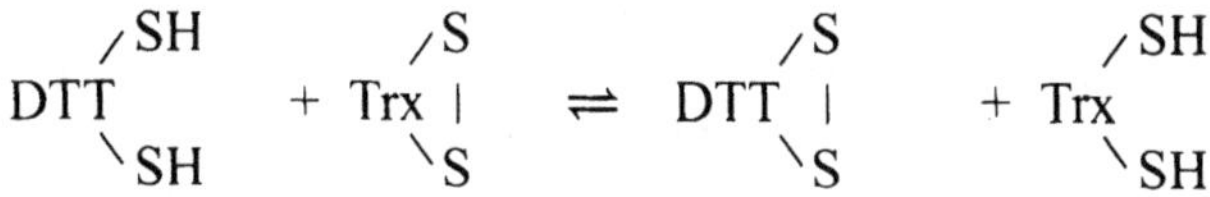
(55)

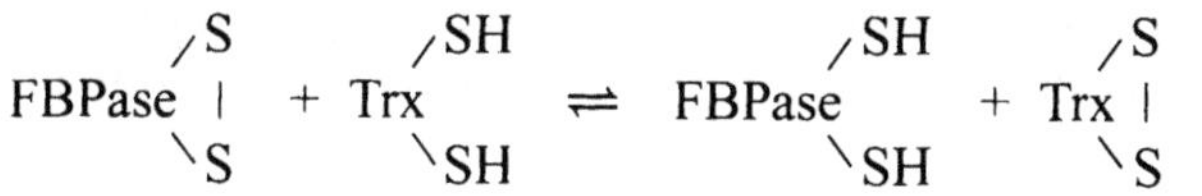
(56)

Kinetically, the overall reduction of FBPase is zero order in DTT concentration and first-order in thioredoxin f and FBPase so that the measured second-order rate constants for the forward and reverse processes correspond to the rate constants for the reversible reduction of the enzyme by thioredoxin (Eq. 56). Since the overall redox state of the enzyme is dictated by the ratio of DTT_{red}/DTT_{ox}, the equilibrium constant determined from the

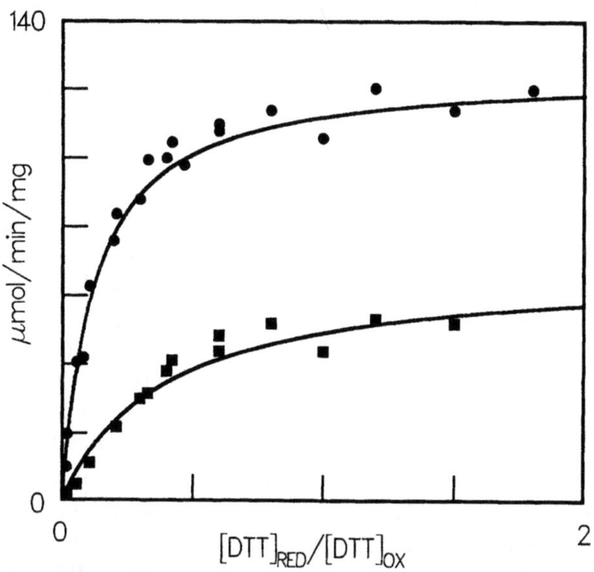

Fig. 13. Redox equilibrium between dithiothreitol redox buffers and the activity of spinach chloroplast fructose-1,6-bisphosphatase. Different equilibrium constants are observed depending on the concentrations of magnesium and fructose-1,6-bisphosphate used in the assay for activity (no substrates were present in the equilibration mixture). A reduction event which drops the apparent K_m for fructose-1,6-bisphosphate is observed by assaying at low fructose-1,6-bisphosphate and high magnesium concentration: 10 mM Mg^{2+}, 0.4 mM fructose-1,6-bisphosphate – (●). Another reduction event which drops the apparent K_m for magnesium is observed by assaying at low magnesium and high fructose-1,6-bisphosphate concentrations: 1 mM Mg^{2+}, 3 mM fructose-1,6-bisphosphate, (■). The two $K_{Eq.}$ values are 0.12 and 0.39 respectively. The two different assays reach two different limiting velocities when fully reduced, 120 units/mg and 68 units/mg respectively. Reproduced, with permission, from ref 64.

variation in enzyme activity with DTT_{red}/DTT_{ox} measures the K_{ox} of the thiol/disulfide pairs of the FBPase.

In reality, the situation is somewhat more complex because not one, but two different redox processes which affect activity are observed [64]. The two equilibrium and two kinetic processes can be observed relatively independently because they have different effects on the apparent K_ms of Mg^{2+} and fructose-1,6-bisphosphate. The K_{ox}s (on the glutathione scale) for the two dithiols of the enzyme are approximately 3-fold different, 31 M and 104 M (Fig. 13).

The ratios of the rate constants for the forward and reverse reaction of Eq. 56 can be used to estimate the two different equilibrium constants for the oxidation of the enzyme by thioredoxin since the kinetics in both directions are measurements of the reaction of thioredoxin with the enzyme. By knowing the equilibrium constants for the reaction of the enzyme with thioredoxin and the equilibrium constants for the reaction of the enzyme with DTT, the equilibrium constant for the reaction of thioredoxin with DTT can be estimated. If the reduction mechanism is kinetically simple, and the ratios of the rate constants do correspond to the equilibrium constant for the process, the two essentially independent estimates of the K_{ox} for thioredoxin should agree. In fact, both estimates agree reasonably well. The K_{ox} for thioredoxin f estimated in this manner is 42 ± 10 M on the glutathione scale.

10.1.3. Lack of coupling between NADPH/NADP and GSH/GSSG in chloroplasts: The available data on the thiol/disulfide oxidation potentials of chloroplast thioredoxin m and f and chloroplast NADP-malate dehydrogenase and fructose-1,6-bisphosphatase show that these proteins are thermodynamically extremely sensitive to dithiol oxidation. The plant thioredoxins m and f have similar oxidation potentials (29 M and 42 M, respectively) which are comparable to or somewhat below the oxidation potential for NADP-malate dehydrogenase (114 M) and the two oxidation potentials observed for spinach chloroplast fructose-1,6-bisphosphatase (31 M and 104 M). These oxidation potentials are comparable to or greater than those observed for extracellular mammalian proteins, and enormously greater than any conceivable redox state which could be maintained by cellular glutathione redox buffers which, at least in mammalian cells and most likely in chloroplasts, are poised at an effective **R**[GSH] of about 3–4 M. In fact, to keep these proteins reduced would thermodynamically require an NADPH/NADP ratio of at least 6 (equivalent to an **R**[GSH] of 900 M at pH 8.0), which is comparable to the observed NADPH/NADP ratio in the light [135].

Reduction of these two chloroplast enzymes by glutathione would require a [GSH]/[GSSG] ratio of 2.7×10^6 at a concentration of GSH of 5 mM. Halliwell and Foyer [136] have measured GSH/GSSG ratios of about 10–40 in chloroplasts. While this ratio appears to be suspiciously low, it is unlikely that GSSG levels in chloroplasts could be as low as 2×10^{-9} M and still support the necessary flux through glutathione reductase. The conclusion is that if these enzyme systems were allowed to equilibrate with chloroplast $[GSH]^2/[GSSG]$, the enzymes would be oxidized both in the light and dark.

How are these enzymes maintained in a reduced state in opposition to the thermodynamically favorable oxidation by either the NADPH/NADP system or the glutathione system in chloroplasts? The redox potential of ferredoxin [137] (–0.42 volts, 2.2×10^5 M on the glutathione scale) is sufficient to drive reduction of these chloroplast systems thermodynamically even at ratios of $ferredoxin_{red}/ferredoxin_{ox}$ of 0.06. What must be avoided is any kinetically significant communication between the thioredoxins (and enzymes) and NADPH/NADP or glutathione redox buffers. Chloroplasts do not have an NADPH-dependent thioredoxin reductase activity [139]. The absence of this enzyme provides kinetic insulation from the NADPH/NADP system. At least for thioredoxin f and fructose-1,6-bisphosphatase, the oxidation of the reduced species by GSSG is surprisingly slow. The second-order rate constant for GSSG oxidation of reduced fructose-1,6-bisphosphatase in the presence of thioredoxin f is approximately 0.1 $M^{-1}s^{-1}$ at pH 8.0 [64], which is roughly 20-fold slower than the rate constant for a typical thiol/disulfide exchange involving thiols of pK_a 8.6. Even if the chloroplast GSSG concentration was as high as suggested by Halliwell and Foyer [136] (0.5 mM), the half-life for oxidation of reduced fructose-1,6-bisphosphatase by GSSG would be 3.8 hours. Since light activation and dark inactivation occur with half-times of the order of a few minutes, the slow rate constant for reaction of the enzymes with GSSG effectively insulates them against equilibration with the glutathione redox buffer.

10.2. Mammalian systems

A large body of relatively circumstantial *in vivo* and *in vitro* evidence suggests that alteration of the cellular thiol/disulfide redox state of mammalian cells may have an effect on a variety of metabolic processes [10, 11, 14]. Such diverse cellular processes as Ca^{2+} homeostasis [139], protein synthesis [140], protein degradation [141], activity of the hexose monophosphate shunt [142–143], glycogenolysis [144], and the regulation of transcription factor binding to DNA [145–147] have been suggested to be dependent on the cellular thiol/disulfide status.

10.2.1. Variation in the thiol/disulfide redox status of cells: In most cells, the major thiol/disulfide redox buffer, glutathione, is found predominantly in the reduced state [4]. There are numerous experimental difficulties encountered in obtaining reliable measurements of the GSSG levels in biological materials, the most serious of which is the tendency to convert small amounts of GSH to GSSG by artefactual, post-mortem oxidation [148].

With regard to thiol homeostasis, the most extensively studied tissue is rat liver. Careful measurements of total levels of GSH and GSSG in rat liver have shown that in a normal, fed animal, the ratio of [GSH]/[GSSG] is typically 300–400 [89, 149]; greater than 99.7% of the cellular glutathione is in the reduced form. Because of intracellular compartmentation, the measured total cellular levels of GSSG provide only an upper limit on the cytoplasmic concentration so that the [GSH]/[GSSG] of 300–400 actually represents a lower limit. In addition to an approximately 2-fold diurnal variation in GSH, fasting for 24–72 h may decrease GSH levels by 2–3-fold [90, 91], the result of a faster turnover of liver glutathione [151]. Normal total levels of GSSG are approximately 0.025–0.04 mM in rat liver [87, 152] in both the fed and fasted state.

The [GSH]/[GSSG] ratio which governs the thermodynamic response of proteins which can form glutathione mixed disulfides may range, in liver, from 350 to as low as 120 under normal metabolic conditions, approximately a 3-fold change [87, 89]. The quantity **R** [GSH], which governs the thermodynamic response of proteins capable of forming PSSP disulfides, changes from about 3 M in the livers of fed rats to about 0.3 M in livers from fasted rats, a 9–10-fold change. The larger change in the quantity **R** [GSH] results predominantly from the squared dependence of this quantity on the GSH concentration.

More severe changes in the glutathione status of cells can be induced by chemical means. A variety of therapeutically useful drugs can drastically affect the glutathione redox status. Several of these drugs, menadione, diaquat, paraquat, nitrofurantoin, adriamycin, and others, induce the net oxidation of GSH to GSSG by redox cycling (see ref [3, 152] for review). Others such as diethylmaleate [153] simply decrease GSH by chemical alkylation of the sulfhydryl group, and some, such as diamide [12], stoichiometrically oxidize GSH to GSSG. Inhibitors of glutathione synthesis, such as buthionine sulfoximine [154], can be used to deplete glutathione levels to 20% of their normal values. Hydrogen peroxide, various alkyl hydroperoxides and various peroxide and superoxide generating systems can be utilized to induce large increases in cellular GSSG through the increased flux through the glutathione peroxidase system [5]. Under such conditions, the [GSH]/[GSSG] can fall to values less than one (most of the glutathione as GSSG) and R [GSH] can plummet to the mM range.

The physiological consequences of these perturbations of the glutathione status are remarkably varied – ranging from no obvious short-term effect to rapid cell death. Animals will survive conditions in which the liver glutathione levels are depleted to 20% of control values for at least 15 days [155], and mutant strains of *E. coli* which lack glutathione but grow normally have been described [156]. However, oxidation of cellular glutathione to GSSG can induce rapid loss of cellular integrity [157]. The processes by which changes in glutathione redox status may be causally related to such multifaceted terminal responses such as cell death are far from clear. Nevertheless, large changes in glutathione (and other cellular thiol/disulfide species) redox state are observed, and it is reasonable and important to pose questions regarding the effects of these thiol/disulfide redox state changes on cellular processes.

10.2.2. Relationsips between K_{ox} and redox buffer status: Under normal physiological conditions and even under conditions of moderate oxidative stress, the large excess of GSH over GSSG imposes significant kinetic and thermodynamic constraints on the ability of proteins to undergo changes in thiol/disulfide oxidation state. The oxidation of protein sulfhydryl groups to disulfides by thiol/disulfide exchange processes is often (though not always) reversible.

$$E\begin{matrix}\diagup SH \\ \diagdown SH\end{matrix} \underset{k_2[GSH]}{\overset{k_1[GSSG]}{\rightleftharpoons}} E\begin{matrix}\diagup SSG \\ \diagdown SH\end{matrix} \underset{k_4[GSH]}{\overset{k_3}{\rightleftharpoons}} E\begin{matrix}\diagup S \\ | \\ \diagdown S\end{matrix} \quad (57)$$

When redox state changes are reversible, and the rate of the exchange process is relatively fast in comparison to the change in the external redox state, the fraction of the protein in the reduced from will depend on the external redox status (set by the redox buffer) and the thermodynamic oxidation potential of the specific protein. Significant changes in the thiol/disulfide redox state (and activity) of the protein will only be observed when the thermodynamic oxidation potential of the critical sulfhydryl group(s) is close to the redox poise of the environment.

Consider the equilibrium formation of a biologically inactive protein-SSG mixed disulfide from a protein sulfhydryl group in a glutathione redox buffer (Fig. 14). In the absence of oxidative stress, the normal [GSH]/ [GSSG] ratio is thought to vary from approximately 100 to 300. If the K_{ox} for this protein thiol/disulfide exchange were 1.0, the biological activity of the protein would change by less than 0.7% over this range – hardly significant. If the K_{ox} were 200, the amount of the enzyme in the reduced (active) form at equilibrium would decrease from 60% to 33% as the [GSH]/[GSSG] ratio changed from 300 to 100 – a two-fold change in activity. In the other extreme, with a K_{ox} of 1000, the activity would change from 23% to 9%, a larger fractional change but at the expense of reduced catalytic capacity under normal conditions. Similar arguments apply to enzyme systems in which the oxidized form is more active. Protein-SSG species are found *in vivo* [151, 158–161]. Under normal conditions, the levels are comparable to those of GSSG. Under redox stress, PSSG levels increase linearly with increased GSSG [158]. Collison *et al.* [48] have shown that oxidative stress induced in cardiac cells by t-butylhydroperoxide or diamide results in the accumulation of several specific protein-SSG disulfides. The changes in [GSH]/[GSSG] (a drop to 3–5) precede the accumulation of protein-SSG, but only by a few minutes. Upon removal of the oxidative stress, specific PSSG species were observed to disappear at very different rates and to different extents. Two of the major

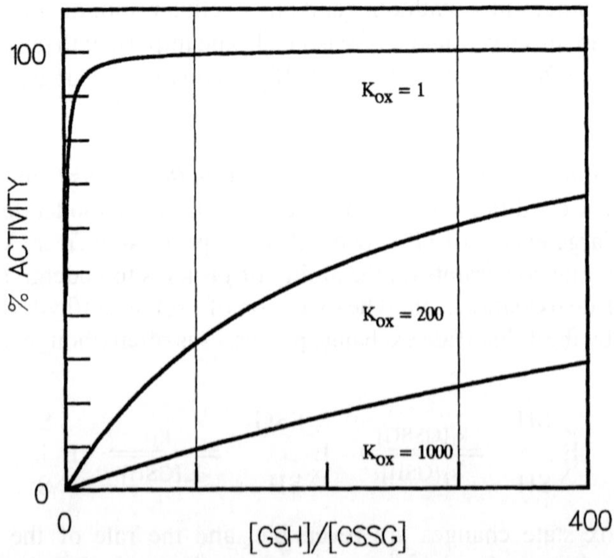

Fig. 14. The specificity of the response of changes in the glutathione redox buffer status can be provided by differences in K_{ox}. The changes in the extent of protein reduction in response to changes in the glutathione redox buffer status are illustrated for three hypothetical proteins of the indicated K_{ox} which form protein-SSG mixed disulfides. The vertical lines represent normal physiological limits on the [GSH]/[GSSG] ratio. Small changes in the fraction of the protein which is reduced are obtained if K_{ox} is either too high or too low. Similar arguments apply to proteins which form protein-SS-protein disulfides.

proteins observed to be specifically S-thiolated by formation of a protein-glutathione mixed disulfide upon oxidative stress of myocytes were creatine kinase and phosphorylase b [162]. Interestingly, S-thiolation, in the presence of superoxide and peroxide generating systems, proceeded faster than could be accounted for by the rate of simple thiol/disulfide exchange with the GSSG present in the reaction mixture. In hepatocytes, a 30 kDa protein [163], subsequently identified as carbonic anhydrase [164], was observed to be specifically modified by glutathione mixed disulfide formation in response to t-butyl hydroperoxide, diamide, or menadione administration under conditions which did not lead to significant cell death.

Redox interconversions in which the oxidized state of the protein is a protein-SS-protein disulfide are restricted in the same way. However, in this case the extent of protein oxidation will be governed by the quantity $[GSH]^2/[GSSG]$ (i.e., **R**[GSH]) rather than [GSH]/[GSSG]). Normally, **R**[GSH] might vary from 3 M to 0.3 M in response to fasting, a 10-fold change in the applicable redox buffer parameter rather than a 3-fold change in the [GSH]/[GSSG] ratio.

Three major conclusions emerge from considerations of the thermodynamics of protein thiol/disulfide exchange with glutathione redox buffers. 1) In order to observe significant changes in the redox state of the protein under normal physiological conditions, the K_{ox} for the protein must lie within limits set by the redox state of cellular redox buffers. 2) Proteins which form protein-SS-protein disulfides should be more responsive to changes in the cellular thiol/disulfide redox status since the response is governed by **R**[GSH] rather than just [GSH]/[GSSG]. 3) Since not all proteins have the same K_{ox}, redox state changes in proteins will be specific, i.e. limited to those proteins with K_{ox} near the cellular thiol/disulfide status.

The kinetics of thiol/disulfide exchange with some proteins can be very slow on the biological time scale due to steric and/or charge effects so that they would never undergo even thermodynamically favorable redox state changes. Other proteins do undergo rapid thiol/disulfide exchange with glutathione even in the absence of any catalyst for the process. Non-equilibrium, steady-state kinetic effects dominate biology and may produce significant deviations from the behaviour expected from thermodynamic considerations. However, special energy-requiring mechanisms other than simple reversal of the thiol/disulfide exchange process would be required to prevent the attainment of thiol/disulfide equilibrium under likely *in vivo* conditions. Mechanisms of this kind have been demonstrated in the light regulation of some enzyme activities (see above); however, in non-photosynthetic cells no such mechanism has yet been confirmed, although its existence would be of considerable significance.

10.2.3. Reversible behavior: The thermodynamic and kinetic behavior of rabbit muscle phosphofructokinase (PFK) [28, 47] toward oxidative inactivation in glutathione redox buffers illustrates some of the principles which should be considered in evaluating the possibility of regulation of enzyme activity by a thiol/disulfide exchange mechanism. DTNB (Ellman's reagent) had been shown to cause inactivation of PFK by oxidation of multiple sulfhydryl groups [165]. This inactivation is reversible by thiols such as DTT. As an oxidizing agent, DTNB is approximately 1000-fold more potent thermodynamically than GSSG while the reducing power of DTT is equivalent to that of a glutathione redox buffer of [GSH]/[GSSG] = 100 at a GSH concentration of 150 M (Table 1). Thus, the inactivation of the enzyme by DTNB and reactivation by DTT only shows that the enzyme is inactive under extremely oxidizing conditions and active under extremely reducing conditions. Activity changes caused by such large fluctuations in the thermodynamic oxidation potential of the medium cannot be used to assess the possibility of activity changes under the more restrictive conditions likely to be found in the cell.

The observed equilibrium behavior of PFK activity in glutathione redox buffers can be adequately described by the formation of a protein-SSG mixed disulfide with a K_{ox} of 6.1 (Fig. 5) [28]. On more detailed examination [47], the redox behavior of PFK is more complex. Kinetically, the inactivation of PFK by GSSG is biphasic. Approximately 65% of the

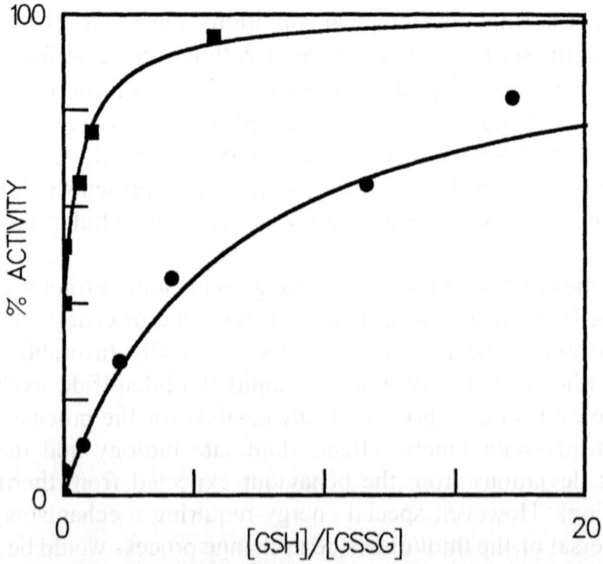

Fig. 15. Thiol/disulfide exchange equilibria between glutathione redox buffers and rabbit muscle phosphofructokinase. The thiol/disulfide exchange equilibria involving the formation of two protein-SSG mixed disulfides are shown. The oxidation of the more easily oxidized group, $K_{ox} = 6.5$ (●), results in complete inactivation while oxidation of the less easily oxidized group, $K_{ox} = 0.7$ (■) results in the loss of 60% activity. Taken, with permission, from ref 47.

enzyme activity is lost rapidly while the remaining activity is lost more slowly. The kinetics of the reverse reaction, reduction of oxidized PFK by GSH, could also be shown to depend on the reduction of two different protein-SSG species.

The two protein-SSG mixed disulfide species formed at equilibrium exhibit individual K_{ox} values of 0.7 and 6.1, and the oxidation of the two groups occurs randomly rather than sequentially (Fig. 15). The sulfhydryl group which is more easily oxidized kinetically is the one which is more difficult to oxidize thermodynamically. The relative kinetic sensitivity of protein sulfhydryl groups to oxidation provides little useful information on which to base conclusions concerning the potential regulation of enzyme activities by thiol/disulfide exchange. Kinetically, a rapid oxidation by even low concentrations of GSSG will be opposed *in vivo* by reduction of the oxidized protein by GSH. For example, the rapidly reacting group of PFK ($K_{ox} = 0.7$) is oxidized by GSSG with a second-order rate constant of 110 M^{-1} min^{-1} at pH 8.0, and the protein-SSG that is formed is reduced by GSH with a second-order rate constant of 150 M^{-1} min^{-1}. Under even reasonably oxidizing *in vivo* conditions ([GSH]/[GSSG] = 100, [GSH] = 5 mM, [GSSG] = 0.05 mM), the observed rate constant for the oxidation of PFK by GSSG (0.0055 min^{-1}) would be opposed by an observed rate constant of 0.75 min^{-1} for reduction by GSH. The system would come to thermodynamic equilibrium with an observed rate constant of ($0.75\ min^{-1}$ + $0.0055\ min^{-1}$) giving a half-time for reaching equilibrium of 0.9 min. However, at equilibrium only 0.7% of the enzyme would be in the oxidized form. In this case the conclusion reached by considering either kinetic and thermodynamic factors is the same – this protein sulfhydryl group would not become significantly oxidized unless the [GSH]/[GSSG] ratio drops to a value approaching the K_{ox}.

The example of PFK illustrates three major points with respect to regulation of enzyme activity by thiol/disulfide exchange. 1) The sulfhydryl groups which are more easily oxidized kinetically need not be the ones which are more easily oxidized thermodynamically. 2) Redox titrations of multiple groups which affect enzyme activity will be dominated by the most easily oxidized group (highest K_{ox}). 3) Rabbit muscle PFK would become significantly oxidized only at [GSH]/[GSSG] ratios of 20 or lower – ratios rarely achieved *in vivo* except under rather severe oxidative stress.

HMG-CoA reductase from both rat liver and yeast is extremely sensitive to oxidative inactivation by a variety of disulfides, including GSSG and the mixed disulfide of CoASH and GSSG (CoASSG) [167, 168]. The yeast enzyme is so sensitive to oxidation by CoASSG that μM concentrations of CoASSG will inactivate the enzyme even in the presence of 10 mM dithiothreitol [162]. Dotan and Shechter [168] suggested that the cellular ratio of GSH to GSSG might effect the activity of this enzyme *in vivo*.

With the rat-liver microsomal enzyme, Roitelman and Shechter [169] found that the activity and apparent $S_{0.5}$ for NADPH was dependent on the concentration of GSH present in the assay. Higher concentrations of GSH were associated with increased enzyme activity. These results could be explained either by an "allosteric" effect of GSH on the enzyme activity or by redox equilibrium effects. Although Roitelman and Shechter did not report measurements of GSSG levels in their experiments, GSSG is always present in GSH solutions unless specific precautions are taken to eliminate it. The effect of increasing glutathione concentration on the activity of HMG-CoA reductase could be explained as a redox titration performed at a constant [GSH]/[GSSG] ratio and increasing GSH. If the oxidized form of HMG-CoA reductase was protein-SS-protein disulfide, increasing GSH concentrations at a constant [GSH]/[GSSG] ratio would lead to increased enzyme reduction at equilibrium.

Further examination of the behavior of HMG-CoA reductase in glutathione redox buffers [72] revealed that the effects of glutathione on the activity of the enzyme could be accounted for by redox equilibrium behavior. In the presence of excess (20 mM) DTT, even high concentrations of GSH (100 mM) had no significant effect on enzyme activity, ruling out an "allosteric" effect of GSH. Incubation of the enzyme in glutathione redox buffers until equilibrium was attained shows that the activity behavior is consistent with the formation of an inactive PSSP species in equilibrium

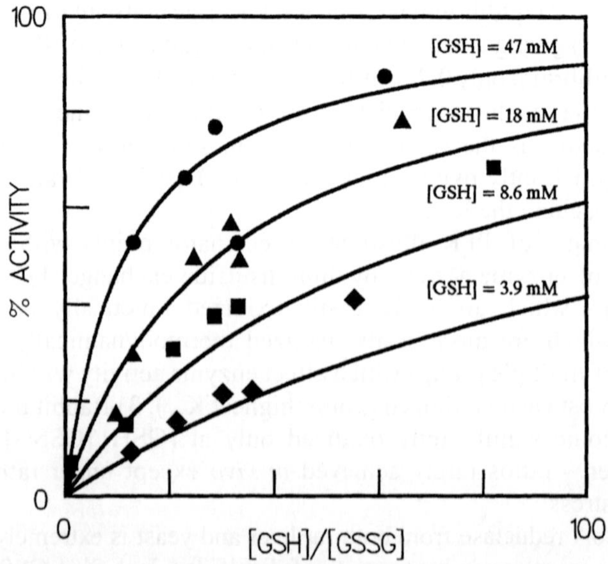

Fig. 16. Thiol/disulfide exchange equilibrium between glutathione redox buffers and rat liver microsomal 3-hydroxy-3-methylglutaryl-coenzyme A reductase. Equilibrium formation of a protein-SS-protein disulfide with $K_{ox} = 0.55$ M accounts for the observed behavior of the enzyme activity. Curves are drawn according to Eq. 15 at different constant GSH concentrations, as indicated. Taken, with permission, from ref 71.

with an active dithiol species and is inconsistent with the formation of protein-SSG disulfides which affect activity (Fig. 16). The K_{ox} is sufficiently large (0.55 M at pH 7) that with only 2% of the glutathione present as GSSG ([GSH]/[GSSG] = 50), 10 mM GSH is required to half-reduce the enzyme, and 30 mM GSH would be necessary to achieve 75% reduction. The approach of HMG-CoA reductase to redox equilibrium with glutathione redox buffers is sufficiently fast, even in the presence of the saturating levels of both substrates, that significant activity changes can be observed during the assay for enzyme activity. By using relatively high concentrations of GSH and GSSG in the assay and long assay times (30–60 min), it was possible to measure the K_{ox} for HMG-CoA reductase in the presence of substrates – 0.22 M [72].

In contrast to the kinetically slow reaction of chloroplast fructose-1,6-bisphosphatase with GSSG, the reaction of HMG-CoA reductase with GSSG is much faster than expected for a typical thiol/disulfide exchange. For a typical thiol of pK_a 8.6, the second-order rate constant for reaction with GSSG at pH 7.1 would be approximately 20 M^{-1} min^{-1}. The observed rate constant for reaction of GSSG with HMG-CoA reductase at pH 7.1 is 1700 ± 200 M^{-1} min^{-1}, a factor of 85 faster than the reaction of a typical thiol with GSSG. This large difference cannot be totally explained by a lower pK_a for the thiol of HMG-CoA reductase. While the fraction of the thiol present as the thiolate anion at pH 7.1 will increase as the pK_a decreases, the nucleophilicity of the thiolate will also decrease by a factor of 3 for each unit decrease in pK_a ($\beta_{nuc} = 0.5$) [42, 49].

With a K_{ox} of 0.55 M, the potential for changes in the redox state of HMG-CoA reductase with changes in the cellular glutathione redox status is significant. Normal, unstressed changes in the R[GSH] of the cellular environment between 3 M and 0.3 M would result in changes in the oxidation state of the enzyme from 84% reduced to 39% reduced. Under conditions of oxidative stress, even more extensive enzyme oxidation should occur.

Ness *et al.* [170] discovered that oxidation of microsomal HMG-CoA reductase by GSSG resulted in the formation of a covalent disulfide crosslink between subunits. SDS gel electrophoresis under non-reducing conditions showed the presence of a M_r 200,000 form of HMG-CoA reductase which could be converted to a M_r 100,000 monomer on reduction with DTT. In isolated microsomes from the livers of fasted rats, the fraction of enzyme in the crosslinked, M_r 200,000 species was increased compared to control animals, suggesting an increased oxidation of the enzyme in response to fasting – precisely the effect expected from the observed changes in the glutathione status with fasting. Although this observation is somewhat qualitative, it provides evidence that protein disulfide formation may be possible even in an environment of predominantly reduced glutathione. The potential for thiol/disulfide regulation of this enzyme becomes even more intriguing with the isolation of a protein (reductase activating protein) from rat liver which catalyzes the reductive activation of HMG-CoA reductase by NADPH in the absence of added thiols [171].

10.2.4. Irreversible behavior: The oxidation of protein thiols and reduction of protein disulfides are not always reversible. Thiol/disulfide redox state changes may be coupled to irreversible conformation changes, enhanced proteolytic modification, aggregation, or precipitation. For several proteins such as pyruvate kinase [172], aldolase [173], phosphoenolpyruvate carboxy kinase [137], and liver fructose-1,6-bisphosphatase [174], the sensitivity of the protein to proteolytic inactivation increases dramatically with protein sulfhydryl oxidation. The increased formation of disulfide bonds in the lens crystallins is associated with formation of an opaque protein precipitate in cataract formation [175].

Consequently, the suggestion has been made that changes in the thiol/disulfide redox status of the cellular environment could have an effect on the turnover (and hence steady-state levels) of specific proteins. Oxidation of glycogen phosphorylase a and b [50] proceeds through an initially reversible oxidation of sulfhydryl groups which is followed by a slower, irreversible process which ultimately leads to precipitation of the protein from solution. Lysozyme, when fully reduced, also becomes less soluble and will aggregrate and precipitate [117].

With phosphorylase, the initial oxidation of 4 ± 1 sulfhydryl groups per subunit with the disulfide cystamine is thermodynamically unfavorable. The K_{ox} for this initial, reversible oxidation can be estimated to be ~ 1×10^{-4}, so unfavorable that the small amount of cysteamine formed from the reaction of the enzyme with cystamine can prevent complete oxidation even in the presence of 100 mM cystamine [50]. The initial, reversible activity loss is followed by a slower, irrevesible inactivation. Thiol oxidation must also be involved in the irreversible step because the rate of the irreversible process depends on the square of the cystamine concentration. This type of behavior is consistent with the rapid establishment of an unfavorable equilibrium in which the extent of enzyme oxidation depends on the concentration of cystamine. If this small amount of oxidized enzyme is then kinetically trapped by an irreversible oxidation by cystamine in the rate-limiting step, second-order kinetics result.

If an initial thiol/disulfide exchange equilibrium, even an unfavorable one, is followed by some irreversible process, the overall rate of irreversible loss of activity and structure will exhibit a dependence on the initial redox state of the protein – even if only very small amounts of the oxidized species are present at equilibrium.

$$E_{red} + aGSSG \underset{}{\overset{K_{ox}}{\rightleftharpoons}} bGSH + E_{ox} \xrightarrow{k_{irr}} E^{*}_{ox} \tag{58}$$

The observed rate of the irreversible loss in activity will depend on the concentration of the oxidized species at equilibrium and the rate of the irreversible process.

$$v = k_{ir}[E_{ox}] = K_{ox}\mathbf{R}\,k_{ir} \tag{59}$$

Even when the fraction of the protein present in the oxidized form is small, changes in **R** will lead to an increased rate of the irreversible process. If, as is the case with phosphorylase, the irreversible step also depends kinetically on the concentration of the disulfide in the redox buffer, a "cooperative" squared dependence of the rate of activity loss on the level of disulfide in the medium will be observed. However, if the irreversible step is fast relative to the initial oxidation event, the observed rate will depend only on the concentration of oxidant present.

11. Catalysis of thiol/disulfide exchange

Thiol/disulfide exchange reactions are catalyzed by several enzymes in different cellular compartments. These catalysts generally exhibit broad substrate specificity and will accept various protein and non-protein thiols and disulfides as substrates. As with other enzyme catalysts, under conditions when the concentration of substrate is in excess of the concentration of enzyme, thiol transferases will accelerate the approach to thermodynamic equilibrium but will not change the equilibrium position. Thus, the existence of these catalysts increases the importance of thermodynamic effects in governing the oxidation state of proteins.

11.1. Thiol transferases

Axelsson *et al.* purified a cytoplasmic thioltransferase (M_r 12,000) from rat liver which exhibited broad substrate specificity, catalyzing the net glutathione-dependent reduction of a variety of disulfides including insulin, ribonuclease, and the glutathione mixed disulfides of lysozyme [176]. The amino acid composition [177], kinetic properties, immunological cross-reactivity [178] and sequence homology [179] suggest that thioltransferase is identical to glutaredoxin, which is known to catalyze similar thiol/disulfide exchange reactions and may play a role in ribonucleotide reductase reduction [180]. The thioltransferase, in its reduced form, is inactivated by iodoacetamide suggesting the involvement of an enzyme sulfhydryl group of unusually low pKa (3.8) in catalysis [181]. The mechanism by which the enzyme efficiently catalyzes thiol/disulfide exchange with such a variety of substrates appears to involve a sequential interaction with glutathione and the disulfide substrate [182]. The thioltransferase isolated from red blood cells has been shown to catalyze the reduction of the mixed disulfide between hemoglobin and glutathione and to reactivate oxidatively inactivated phosphofructokinase [183].

11.2. Protein disulfide isomerase

Protein disulfide isomerase (PDI) is a thiol/disulfide exchange catalyst, M_r 56,000, which catalyzes thiol/disulfide exchange reactions including the unscrambling of mis-oxidized ribonuclease A [184–185]. The same protein will also catalyze the GSH-dependent reduction of a variety of disulfide-containing proteins and peptides, including insulin [186]. This broad specificity [187] initially led to some controversy concerning the biological role and identity of these activities. However, conclusive evidence suggests that the two activities belong to the same protein [188].

PDI activity is associated with the microsomal fraction of liver; however, the protein does not appear to be an integral membrane protein. Much of the activity is latent in microsomes and disruption of the microsomal membrane increases the availability of the active enzyme [189]. The rat liver [190] and human [191] genes for PDI have been cloned and sequenced, and the presence of an amino terminal signal sequence directing the enzyme to the lumen of the endoplasmic reticulum has been suggested. That the enzyme is involved in catalysis of disulfide bond formation in proteins destined to be secreted has been suggested by the finding that PDI can be covalently crosslinked to nascent immunoglobulin light chains *in vivo* [192]. Interestingly, PDI has been found to be identical to the β subunit of prolyl hydroxylase [193], and copurifies as a component of the microsomal triglyceride transfer complex [194]. Photoaffinity labels have identified PDI as a thyroid hormone binding protein [195] and as a component of the glycosylation site binding protein of the glucosyl transferase of the endoplasmic reticulum [197]. The myriad functions of this protein have raised some intriguing questions regarding its biological role.

Creighton *et al.* [197] have examined the mechanism of PDI catalysis of the thiol/disulfide exchange reaction involved in folding of bovine trypsin inhibitor and shown that the protein catalyzes intramolecular rearrangements of disulfide bonds more effectively than net oxidations or reductions. In catalysis of the formation and rearrangements of disulfide bonds in reduced, denatured RNAse A, PDI appears to catalyze disulfide bond formation effectively and to accelerate the rate of thiol/disulfide exchange involved in regeneration of the native structure from misfolded or partially folded species without substantially affecting the nature of the rate-limiting step in the overall folding process [72]. From variation in the activity of PDI with the composition of a glutathione redox buffer [72], and by measuring the sensitivity of PDI to inactivation by alkylation after incubation with a glutathione redox buffer [198], the K_{ox} for the active site dithiols of PDI have been estimated at $4–7 \times 10^{-5}$ M. PDI would appear to be the most strongly oxidizing disulfide yet observed as a protein-SS-protein disulfide. DsbA, a protein of the periplasmic space required for efficient disulfide bond formation in *E. coli* [199], also has significant homology to thioredoxin and PDI. The dithiol/disulfide active site of dsbA has a K_{ox} of 8×10^{-5} M, comparable to that of PDI [73].

PDI is not a particularly effective catalyst [187], frequently requiring near stoichiometric amounts of PDI to achieve significant rate enhancements for protein oxidation. However, simple calculations based on the reported purification of the enzyme from bovine liver [184] suggest that very high concentrations of the protein would be expected in the lumen of the endoplasmic reticulum. Because of its high local concentration, PDI may, in fact, be a major redox buffer in the endoplasmic reticulum. The thiol/disulfide redox status in the lumen of the endoplasmic reticulum and the identity of the disulfide oxidant in that compartment are not known. Ziegler and Poulsen [200] have suggested that a novel mixed-function oxidase situated on the endoplasmic reticulum could provide oxidizing equivalents for protein disulfide bond formation in the form of cystamine. Wells *et al.* [201] have recently found that PDI also has significant dehydroascorbate reductase activity, catalyzing the GSH-dependent reduction of dehydroascorbate.

11.3. Thioredoxin

Thioredoxin is an effective thiol/disulfide exchange catalyst. In the presence of thioredoxin reductase and NADPH, thioredoxin catalyzes the net reduction of disulfide bonds in a variety of proteins [61]. Reduced thioredoxin reduces insulin about 10,000 times more rapidly than dithiothreitol reduces insulin. Not all disulfides are reduced as effectively by thioredoxin, implying some specificity in the thioredoxin-insulin interaction. Thioredoxin will also catalyze the attainment of the correct structure and disulfide bonding in ribonuclease [203], albeit at a much slower rate than with PDI on a molar basis [203].

12. Conclusions

Thiol/disulfide exchange processes are intimately involved in fundamental cellular processes – enzyme catalysis, maintenance of protein structure, transport of redox equivalents, and metabolic regulation. In each of these roles, both the thermodynamic and kinetic behavior of protein thiols and disulfides play an important part. Thiol/disulfide oxidation potentials cover an extremely large range which spans the physiological thiol/disulfide redox state of most cells. The interior of most cells is not sufficiently reducing to preclude the existence of disulfides in the intracellular environment, nor are the oxidation potentials of all proteins sufficiently high or low to preclude changes in the thiol/disulfide redox states of proteins in response to cellular changes in the glutathione redox buffer.

The equilibrium behavior of a thiol/disulfide exchange redox system in response to an external redox buffer will depend on the type of protein disulfide species involved in the equilibrium. The quantities [GSH]/[GSSG] and **R**[GSH] and their relationship to the thiol/disulfide oxidation potential (K_{ox}) of a specific protein will dictate the equilibrium position of a protein thiol oxidation. While equilibrium may not always be attained for kinetic reasons, thermodynamic factors should be considered when dealing with thiol/disulfide exchange events.

Acknowledgments

Work from the author's laboratory was supported by grants from the National Institutes of Health (HL-28521 and GM-40379) and the National Science Foundation (DMB-8411783).

References

1. PC Jocelyn, *Biochemistry of the Sulfhydryl Group,* Academic Press, New York, 1972.
2. Yu M Torchinskii, *Sulfhydryl and Disulfide Groups of Proteins*, Plenum, Nex York, 1974.
3. H Sies, Angew. Chem. Int. Ed. Engl. 25 (1986) 1058–1064.

4. A Meister and ME Anderson, Annu. Rev. Biochem. 52 (1983) 711–760.
5. B Chance, H Sies and A Boveris, Pysiol. Rev. 59 (1979) 527–605.
6. CH Williams in *The Enzymes,* PD Boyer (ed), Academic Press, New York, 1976, Vol 13, pp. 89–174.
7. L Eldjan and A Phil, J. Amer. Chem. Soc. 79 (1957) 4589–4593.
8. ES Guzman Barron, Adv. Enzymol. 11 (1951) 201–266.
9. M Freidman, *The Chemistry and Biochemistry of the SH Group,* Pergamon Press, Elmsford, New York, 1973.
10. HF Gilbert, Adv. Enzymol. 63 (1990) 69–172.
11. NS Kosower and EM Kosower, Int. Rev. Cytol. 54 (1978) 109–160.
12. DM Ziegler, Annu. Rev. Biochem. 54 (1985) 305–329.
13. BB Buchanan, Ann. Rev. Plant Physiol. 31 (1980) 341–374.
14. HF Gilbert, Methods Enzymol. 107 (1984) 330–351.
15. H Eggerer and A Klette, Eur. J. Biochem. 1 (1967) 447–453.
16. A Klug and D Rhodes, Trend. Biol. Sci. 12 (1987) 464–469.
17. RC Fahey in *Protein Crosslinking: Biochemical and Molecular Aspects,* part A, Plenum Publishing Corp., New York, 1977, pp. 1–30.
18. DJ Reed and PW Beatty in *Reviews in Biochemical Toxicology,* E Hodgson, JR Bend, and RM Philpot (eds), Elsevier, New York, 1980, pp. 231–242.
19. JM Thornton, J. Mol. Biol. 151 (1981) 261–287.
20. DB Volkin and AM Klibanov, J. Biol. Chem. 262 (1987) 2945–2950.
21. CN Pace, Trends Biochem. Sci. 15 (1990) 14–17.
22. CB Anfinsen and HA Scheraga, Adv. Protein Chem. 29 (1975) 205–225.
23. TE Creighton, Prog. Biophys. Mol. Biol. 33 (1978) 231–297.
24. CA Marks, H Naderi, PA Kosen, ID Kuntz and S Anderson. Science 235 (1987) 1370–1373.
25. FG Hopkins and M Dixon, J. Biol. Chem. 54 (1922) 527–563.
26. LE Anderson in *Thioredoxin and Glutaredoxin Systems: Structure and Function,* A Holmgren (ed), Raven Press, New York, 1986, pp 26–37.
27. H Sies, R Brigelius and P Graf, Adv. Enzyme Reg. 26 (1987) 175–189.
28. HF Gilbert, J. Biol. Chem. 257 (1982) 12086–12091.
29. DW Walters and HF Gilbert, J. Biol. Chem. 261 (1986) 12135–13143.
30. RE Cappel and HF Gilbert, J. Biol. Chem. 265 (1990) 15464–15470.
31. F Daniels and RA Alberty, *Physical Chemistry,* 3rd Edition, John Wiley and Sons, New York, 1966, p. 241.
32. J Houk and GM Whitesides, J. Amer. Chem. Soc. 109 (1987) 6825–6836.
33. GH Snyder, Biochemistry 26 (1987) 688–694.
34. WW Cleland, Biochemistry 3 (1964) 480–483.
35. DM Rothwarf and HA Scheraga, Proc. Natl. Acad. Sci. USA 89 (1992) 7944–7948.
36. MH Chau and JW Nelson, FEBS Lett. 291 (1991) 296–298.
37. EM Scott, IW Duncan and V Ekstrand, J. Biol. Chem. 238 (1963) 3928–3933.
38. G Gorin, A Esfandi and GB Guthrie, Arch. Biochem. Biophys. 168 (1975) 450–454.
39. WM Clark, *Oxidation-Reduction Potentials of Organic Systems,* The Williams and Wilkins Co., Baltimore, 1960, p. 486.
40. IM Kolthoff and C Barnum, J. Amer. Chem. Soc. 62 (1940) 3061–3066.
41. GT Rogers and DN Mallett, Bioelectrochem. Bioenerg. 10 (1983) 269–277.
42. RP Szajewski and GM Whitesides. J. Amer. Chem. Soc. 102 (1980) 2011–2026.
43. HA Scheraga, Y Konishi and T Ooi, Adv. Biophys. 18 (1984) 21–41.
44. TE Creighton, Meth. Enzymol. 107 (1984) 305–329.
45. JS Weissmann and PS Kim, Science 253 (1991) 1386–1393.
46. JA Wells and DB Powers, J. Biol. Chem. 261 (1986) 6564–6570.
47. DW Walters and HF Gilbert, J. Biol. Chem. 261 (1986) 15372–15377.
48. MW Collison, D Beidler, LM Grimm and JA Thomas, Biochim. Biophys. Acta 885 (1986) 58–67.
49. DJ Hupe and ER Pohl, Isr. J. Chem. 26 (1986) 395–399.
50. RE Cappel and HF Gilbert, J. Biol. Chem. 261 (1986) 15378–15384.
51. PC Jocelyn, Eur. J. Biochem. 2 (1967) 327–333.
52. IM Kolthoff, W Stricks and RC Kapoor, J. Amer. Chem. Soc. 77 (1955) 4733–4741.

53. L Eldjarn and A Phil, J. Amer. Chem. Soc. 79 (1957) 4589–4595.
54. P Eyer and D Podhradsky, Anal. Biochem. 153 (1986) 57–61.
55. R Zhang and GH Snyder, Biochemistry 27 (1988) 3785–3794.
56. Y Goto and K Hamaguchi, J. Mol. Biol. 146 (1981) 321–327.
57. ME O'Donnell and CH Williams, J. Biol. Chem. 258 (1983) 13795–13805.
58. RG Matthews and CH Williams, J. Biol. Chem. 251 (1976) 3956–3964.
59. SM Miller, DP Ballou, V Massey, CH Williams and CT Walsh, J. Biol. Chem. 261 (1986) 8081–8084.
60. A Holmgren, J. Biol. Chem. 254 (1979) 3672–3678.
61. A Holmgren, Meth. Enzymol. 107 (1984) 295–300.
62. HE Huber, M Russel, P Model and CC Richardson, J. Biol. Chem. 261 (1986) 15006–15012.
63. F Rebeille and MD Hatch, Arch. Biochem. Biophys. 249 (1986) 164–170.
64. CJ Clancey and HF Gilbert, J. Biol. Chem. 262 (1987) 13545–13551.
65. AH Fairlamb, P Blackburn, P Ulrich, BT Chait and A Cerami, Science 227 (1985) 1485–1487.
66. TE Creighton, *Functions of Glutathione: Biochemical, Physiological, Toxicological, and Chemical Aspects*, A Larsson, S Orrenius, A Holmgren and B Mannervik (eds), Raven Press, New York, 1983, pp. 205–213.
67. CN Pace and TE Creighton, J. Mol. Biol. 188 (1986) 477–486.
68. RT Sauer, K Hehir, RS Stearman, MA Weiss, A Jeitler-Nilsson, EG Suchandek et al., Biochemistry 25 (1986) 5992–5998.
69. RE Cappel, JW Bremer. TM Timmons, TE Nelson and HF Gilbert, J. Biol. Chem. 261 (1986) 15385–15389.
70. RE Cappel and HF Gilbert, J. Biol. Chem. 264 (1989) 9180–9187.
71. RE Cappel and HF Gilbert, J. Biol. Chem. 263 (1988) 12204–12212.
72. MM Lyles and HF Gilbert, Biochemistry 30 (1991) 613–619.
73. A Zapun, JCA Bardwell and TE Creighton, Biochemistry 32 (1993) 5083–5092.
74. MI Page and WP Jencks, Biochem. Biophys. Res. Commun. 57 (1974) 887–889.
75. TE Creighton, Meth. Enzymol. 131 (1986) 83–104.
76. PJ Milburn, YC Meinwald, S Takahashi, T Ooi and HA Scheraga, Int. J. Peptide Protein Res. 31 (1988) 311–321.
77. DC Poland and HA Scheraga, Biopolymers 3 (1965) 379–399.
78. PN Kao and A Karlin, J. Biol. Chem. 261 (1986) 8085–8088.
79. JM Wilson, RJ Bayer and DJ Hupe, J. Amer. Chem. Soc. 99 (1977) 7922–7926.
80. HF Gilbert, Biochemistry 28 (1989) 7298–7305.
81. WP Jencks, Proc. Natl. Acad. Sci. USA 78 (1981) 4046–4050.
82. BA Katz and A Kossiakoff, J. Biol. Chem. 261 (1986) 15480–15485.
83. PJ Milburn, Y Konishi, YC Meinwald and HA Scheraga, J. Amer. Chem. Soc. 109 (1987) 4486–4496.
84. A Holmgren in *Thioredoxin and Glutaredoxi Systems: Structure and Function*, A Holmgren (ed), Raven Press, New York, 1986, pp. 1–9.
85. A Holmgren and M Fagerstedt, J. Biol. Chem. 257 (1982) 6926–6933.
86. EA Newsholm and C Start, *Regulation in Metabolism*, John Wiley, London, 1973, p. 325.
87. I Carlberg and B Mannervik, Meth. Enzymol. 113 (1985) 484–489.
88. RA Karplus and GE Schulz, J. Mol. Biol. 195 (1987) 701–729.
89. AJ Alpert and HF Gilbert, Anal. Biochem. 144 (1985) 553–562.
90. N Nitateishi, T Higashi, S Shinya, A Naruse and Y Sakamoto, J. Biochem. (Tokyo) 75 (1974) 93–103.
91. RJ Jaeger, RB Connolly and SD Murphy, Res. Commun. Chem. Pathol. Pharmacol. 6 (1973) 465–467.
92. H Sies, *Oxidative Stress*, Academic Press, London, 1985.
93. V Massey and CH Williams, J. Biol. Chem. 240 (1965) 4470–4480.
94. B Mannervik, Acta Chem. Scand. 23 (1969) 2912–2914.
95. PM Chung, RE Cappel and HF Gilbert, Arch. Biochem. Biophys. 288 (1991) 48–53.
96. PG Penketh, WPK Kennedy, CL Patton and AC Sartorelli, FEBS Lett. 221 (1987) 427–432.
97. SL Shames, AH Fairlamb, A Cerami and CT Wash, Biochemistry 25 (1986) 3519–3526.

98. M Droux, M Miginiac-Maslow, JP Jacquot, P Gadal, NA Crawford, NS Kosower et al., Arch. Biochem. Biophys. 256 (1987) 372–380.
99. L Thelander and P Reichard, Ann. Rev. Biochem. 48 (1979) 133–202.
100. J Stubbe, Adv. Enzymol. 63 (1990) 349–419.
101. L Thelander, J. Biol. Chem. 249 (1974) 4858–4862.
102. AN Lin, GW Ashley and J Stubbe, Biochemistry 26 (1987) 6905–6909.
103. A Holmgren, BO Soderberg, H Klund and CI Branden, Proc. Natl. Acad. Sci. USA 72 (1990) 2305–2309.
104. HJ Dyson, GP Gippert, DA Case, A Holmgren and PE Wright, Biochemistry 29 (1990) 4129–4135.
105. RF Kelley, W Shalongo, MV Jagannadham and E Stellwagen, Biochemistry 26 (1887) 1406–1411.
106. G Krause, J Lundstron, JL Barea, CP de la Cuesta and A Holmgren, J. Biol. Chem. 266 (1991) 9494–9500.
107. O Berglund and A Holmgren, J. Biol. Chem. 250 (1975) 2778–2782.
108. S Tabor, HE Huber and CC Richardson, J. Biol. Chem. 262 (1987) 16212–16223.
109. K Maeda, A Tsugita, A Dalzoppo, F Vilbois and P Schurmann, Eur. J. Biochem. 154 (1986) 197–203.
110. J Buc, M Riviere, B Gontero, P Sauve, J-C Meunier and J Ricard, Eur. J. Biochem. 140 (1984) 199–202.
111 A Holmgren, J. Biol. Chem. 254 (1979) 3664–3670.
112. B Kren, D Parsell and JA Fuchs, J. Bacteriol. 170 (1988) 308–315.
113. JO Hishoishog, H von Bahr-Lindstrishom, H Jhornvall and A Holmgren, Gene 43 (1986) 13–21.
114. H Tabor, J. Biol. Chem. 250 (1975) 2648–2658.
115. AH Fairlamb, GB Henderson and A Cerami, Mol. Biochem. Parasitol. 21 (1986) 247–257.
116. AH Fairlamb and GB Henderson in *Host-Parasite Cellular and Molecular Interactions in Protozoal Infections,* KP Chang and D Snary (eds), Springer-Verlag, New York, 1987, pp. 29–40.
117. CB Anfinsen, E Haber, M Sela and FH White, Proc. Natl. Acad. Sci. USA 47 (1961) 1309–1314.
118. TE Greighton, *Proteins, Structures and Molecular Properties,* WH Freeman, New York, 1984, Chap. 7.
119. VP Saxena and DB Wetlaufer, Biochemistry 9 (1970) 5015–5023.
120. Y Konishi, T Ooi and HA Scheraga, Biochemistry 21 (1982) 4734–4780.
121. DB Wetlaufer, PA Branca and G-X Chen, Protein Engineering 1 (1987) 141–146.
122. TE Creighton, *J. Mol. Biol.* 113 (1977) 329–341.
123. HF Gilbert in *Mechanisms of Protein Folding,* RH Pain (ed) Oxford: RL Press (1994) pp. 104–136.
124. TY Lin and PS Kim, Biochemistry 28 (1989) 5282–5287.
125. R Wetzel, Trend. Biol. Sci. 12 (1987) 478–482.
126. CO Pabo and EG Suchanek, Biochemistry 25 (1987) 5987–5991.
127. JE Villafranca, EE Howell, DHY Voet, MS Strobel, RC Ogden, JN Abelson et al., Science 222 (1983) 782–788.
128. R Wetzel, LJ Perry, WA Baase and WJ Becktel, Proc. Natl. Acad. Sci. USA 85 (1988) 401–405.
129. M Matsumura, WJ Becktel, M Levitt and BW Matthews, Proc. Natl. Acad. Sci. USA 86 (1989) 6562–6566.
130. M Matsumura, G Signor and BW Matthews, Nature 342 (1989) 291–293.
131. GE Edwards, H Nakamoto, JN Burnell and MD Hatch, Annu. Rev. Plant Physiol. 36 (1985) 255–286.
132. R Scheibe, FEBS Lett. 133 (1981) 301–304.
133. R Rebeille and MD Hatch, Arch. Biochem. Biophys. 249 (1986) 171–179.
134. BB Buchanan, P Schurmann and PP Kalberer, J. Biol. Chem. 246 (1971) 5952–5959.
135. C Foyer and B Halliwell, Planta 133 (1976) 21–25.
136. B Halliwell and CH Foyer, Planta 139 (1978) 9–17.
137. CJ Batie and H Kamin, J. Biol. Chem. 261 (1986) 11214–11223.

138. RA Wolosiuk, NA Crawford, BC Yee and BB Buchanan, J. Biol. Chem. 254 (1979) 1627–1632.
139. D Di Monte, G Bellomo, H Thor, P Nicotera and S Orrenius, Arach. Biochem. Biophys. 235 (1984) 343–350.
140. NS Kosower, GA Vanderhoff and EM Kosower, Biochim. Biophys. Acta 272 (1972) 623–630.
141. GL Francis and FJ Ballard, Biochem. J. 186 (1980) 581–586.
142. H Sies and KH Summer, Eur. J. Biochem. 57 (1975) 503–512.
143. R Brigelius, Hoppe Seyler's. Z. Physiol. Chem. 364 (1983) 989–996.
144. M Usami, H Matsushita and T Shimazu, J. Biol. Chem. 255 (1980) 1928–1934.
145. C Abate, L Patel, FJ Rauscher and T Curran, Science 249 (1990) 1157–1161.
146. KA Hutchison, G Matic, S Meshinchi, EH Bresnick and WB Pratt, J. Biol. Chem. 266 (1991) 10505–10509.
147. TH Rushmore, MR Morton and CB Pickett, J. Biol. Chem. 266 (1991) 11632–11639.
148. ME Anderson, Meth. Enzymol. 113 (1985) 548–554.
149. TPM Akerboom, M Bilzer and H Sies, J. Biol. Chem. 257 (1982) 4248–4252.
150. BH Lauterburg, JD Adams and JR Mitchell, Hepatology 4 (1984) 586–590.
151. R Brigelius in *Oxidative Stress*, Academic Press, London, 1985, pp. 243–270.
152. DJ Reed and MW Fariss, Pharmacol. Rev. 36 (1984) 25S–32S.
153. JL Plummer, BR Smith, H Sies and JR Bend, Meth. Enzymol. 77 (1981) 50–59.
154. A Meister, Meth. Enzymol. 113 (1985) 571–580.
155. OW Griffith, RJ Bridges and A Meister, Proc. Natl. Acad. Sci. USA 76 (1979) 6319–6322.
156. P Apontoweil and W Berends, Biochim. Biophys. Acta 399 (1975) 10.
157. S Orrenius, AA Jewell, H Thor, G Bellomo, L Eklow and MT Smith in *Isolation, Characterization, and Use of Hepatocytes*, RA Harris and NW Cornell (eds), Elsevier, New York, 1983, pp 333–340.
158. R Brigelius, C Muckel, TPM Akerboom and H Sies, Biochem. Pharm. 32 (1983) 2529–2534.
159. MJ Meredith, Anal. Biochem. 131 (1983) 504–509.
160. T Higashi, M Furukawa, A Hikita, N Naruse, N Tateishi and Y Sakamoto, J. Biochem. (Tokyo) 257 (1983) 1661–1667.
161. MF Lou, LL Poulsen and DM Ziegler, Meth Enzymol. 143 (1987) 124–129.
162. E-M Park and JA Thomas, Biochim. Biophys. Acta 964 (1988) 151–160.
163. K Rokutan, JA Thomas and H Sies, Eur. J. Biochem. 179 (1989) 233–239.
164. YC Chai, CH Jung, CK Lii, SS Ashraf, S Hendrich, B Wolf et al. Arch. Biochem. Biophys. 284 (1991) 270–278.
165. RG Kemp, Meth. Enzymol. 42 (1972) 71–80.
166. HF Gilbert and MD Stewart, J. Biol. Chem. 256 (1981) 1782–1785.
167. CF Tormanen and TJ Scallen, Circulation 64 (1980) VI271.
168. I Dotan and I Shechter, Arch. Biochem. Biophys. 226 (1983) 401–410.
169. J Roitelman and I Shechter, J. Biol. Chem. 259 (1984) 870–877.
170. GC Ness, SJ Eales, LC Pendleton and M Smith, J. Biol. Chem. 260 (1985) 12391–12393.
171. I Dotan and I Schechter, J. Biol. Chem. 262 (1987) 17058–17064.
172. TJC Van Berkel, JF Koster and GEJ Staal, Biochim. Biophys. Acta 321 (1973) 496–501.
173. JS Bond, J. Biol. Chem. 259 (1984) 886–891.
174. S Pontremoli, E Melloni, M Michetti, F Salamino, B Sparatore and BL Horecker, Arch. Biochem. Biophys. 213 (1982) 731–733.
175. Y Ozaki, A Mizuno, K Itoh and K Iriyama, J. Biol. Chem. 262 (1987) 15445–15551.
176. K Axelsson, S Eriksson and B Mannervik, Biochemistry 17, 2978, 1978.
177. ZR Gan and WW Wells, J. Biol. Chem. 261 (1986) 996–1001.
178. ZR Gan and WW Wells, J. Biol. Chem. 263 (1988) 9050–9054.
179. Y Yang and WW Wells, J. Biol. Chem. 265 (1990) 589–593.
180. M Luthman and A Holmgren, J. Biol. Chem. 257 (1982) 6686–6690.
181. ZR Gan and WW Wells, J. Biol. Chem. 262 (1987) 6699–6703.
182. YF Yang and WW Wells, J. Biol. Chem. 266 (1991) 12766–12772.
183. JJ Mieyal, DW Starke, SA Gravina, C Dothey and JS Chung, Biochemistry 30 (1991) 6088–6097.
184. DA Hillson, N Lambert and RB Freedman, Meth. Enzymol. 107 (1984) 281–300.

185. JE Morin and JE Dixon, Meth. Enzymol. 113 (1985) 541–547.
186. PT Varandani in *Mechanisms of Oxidizing Enzymes,* TP Singer and RN Ondarza (eds), Elsevier, Amsterdam, 1978, pp. 29–35.
187. RB Freedman and DA Hillson in *The Enzymology of Post-translational Modification of Proteins,* RB Freedman and HC Hawkins (eds), Academic Press, New York, 1980, Vol I, pp. 157–212.
188. S Bjelland, K Wallevik, J Kroll, JE Dixon, JE Morin, RB Freedman et al., Biochim. Biophys. Acta 747 (1983) 197–199.
189. RB Freedman, Trend Bioch. Sci. 9 (1984) 438–441.
190. JC Edman, L Ellis, RW Blacher, RA Roth and WJ Rutter, Nature 317 (1985) 267–270.
191. JI Morris and PT Varandani, Biochim. Biophys. Acta 949 (1988) 169–180.
192. R Roth and S Pierce, Biochemistry 26 (1987) 4179–4182.
193. J Koivu and R Mullyla, J. Biol. Chem. 262 (1987) 6159–6164.
194. JR Wetterau, KA Combs, SN Spinner and BJ Joiner, J. Biol. Chem. 265 (1990) 9800–9807.
195. T Obata, S Kitagawa, QH Gong and I Pastan, J. Biol. Chem. 263 (1988) 782–785.
196. M Geeth-Habib, R Novia, HA Kaplan and WJ Lennarz, Cell 54 (1988) 1053–1060.
197. TE Creighton, DA Hillson and RB Freedman, J. Mol. Biol. 142 (1990) 43–48.
198. HC Hawkins, M deNardi and RB Freedman, Biochem. J. 275 (1991) 1991–1998.
199. JC Bardwell, K McGovern and J Beckwith, Cell 67 (1991) 581–589.
200. DM Ziegler and LL Poulson, Trend Biol. Sci. 2 (1977) 79–81.
201. WW Wells, DP Xu, Y Yang and PA Rocque, J. Biol. Chem. 265 (1990) 153–157.
202. VP Pigiet and BJ Schuster, Proc. Natl. Acad. Sci. USA 83 (1986) 7643–7647.
203. HC Hawkins, EC Blackburn and RB Freedman, Biochem J. 275 (1991) 349–353.

Bioelectrochemistry of Biomacromolecules
ed. by G. Lenaz and G. Milazzo†

CHAPTER 6
Electrochemistry of iron-sulfur proteins

Benjamin A. Feinberg and Michael D. Ryan

Department of Chemistry, University of Wisconsin-Milwaukee, WI, USA, and Department of Chemistry, Marquette University, Milwaukee, WI, USA

1. Introduction to iron-sulfur proteins

1.1 Structure to iron-sulfur proteins

Iron-sulfur proteins and enzymes are unique because they possess one or more iron-sulfur clusters. They are ubiquitous in nature, playing key roles in photosynthesis, nitrogen fixation, and terminal respiration, not to mention a host of other biologically important reactions including CO oxidation to CO_2, interconversion of H_2 and H^+, and sulfite/nitrite reduction. Historically, most studies focused on the electron-transferring properties of iron-sulfur proteins, but now several enzymes have been discovered that clearly use iron-sulfur clusters for catalysis. The iron-sulfur clusters in these proteins and enzymes are usually coordinated to the polypeptide by the amino acid cysteine in its thiolate form, $-CH_2S^-$. Inorganic sulfide (S^{2-}, "acid labile sulfide") is required for the formation of the cluster, except for the special case of $Fe(SCH_2R)_4$-containing proteins. While these are single iron proteins which do not contain a metal cluster, they are generally included in the iron-sulfur proteins. Iron-sulfur proteins can then be characterized by considering the type and quantity of iron-sulfur clusters they possess. The most commonly encountered cluster types found in iron-sulfur proteins are shown in Fig. 1.

These include:

a) $Fe(SCH_2R)_4$ which is coordinated to the polypeptide by four cysteinyl residues; a member of this class is rubredoxin, with a reduction potential of −60 mV;
b) $[2Fe\text{-}2S](SCH_2R)_4$ where the S^{2-} is inorganic sulfide; spinach ferredoxin is a typical member of this class;
c) $[4Fe\text{-}4S](SCH_2R)_4$ and $2[4Fe\text{-}4S](SCH_2R)_4$ ferredoxins (Fd) where the cluster has a cubane structure; an example of a double clustered iron-sulfur protein is *Clostridium pasteurianum (Cp)* Fd; and

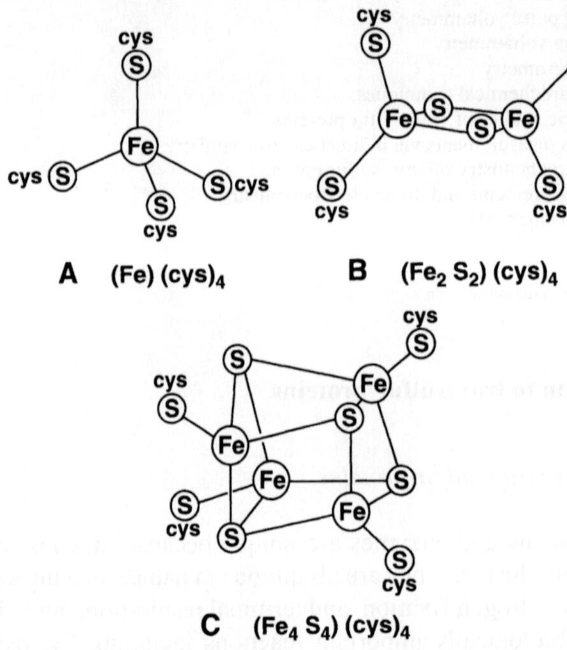

Fig. 1. Different types of iron-sulfur clusters found in iron-sulfur proteins and enzymes. From W.H. Orme-Johnson and N.R. Orme-Johnson in *Iron-Sulfur Proteins,* T.G. Spiro, ed., John Wiley and Sons, New York, 1982, Chapter 2, pp. 67–96. Reprinted with permission of John Wiley and Sons, New York.

d) $[3Fe\text{-}4S](SCH_2R)_4$ which are like the [4 Fe-4 S] proteins but either do not have a fourth iron under any conditions or have lost an iron; an example is *Desulfovibrio gigas (Dg)* Fd II. In another [3Fe-4S] protein to be discussed, one of the cysteinyl ligands is apparently replaced with the weaker ligand aspartic acid, with the amino acid side chain $-CH_2CO_2^{1-}$.

The [4 Fe-4 S] proteins exist in a wide variety of structures. One large class is the 2[4 Fe-4 S] Fd (double clustered ferredoxin), which was discovered in *Clostridium pasteurianum* by Mortenson *et al.* in 1962 [1]. These Fd have a molecular mass of about 6200 Daltons and are generally acidic. Single clustered low potential [4 Fe-4 S] Fds have also been discovered [2]. In addition, there is a class of [4 Fe-4 S] Fds which have a high redox potential ($E^{\circ\prime} \approx +350$ mV), and are called high potential iron-sulfur proteins, HiPIP.

In this chapter we would like to introduce some key iron-sulfur proteins, several important and useful electrochemical methods, and then explore in some detail the electrochemistry of iron-sulfur proteins, which includes both the practical methods used and the kinds of information that can be obtained. This chapter is *not* intended to be a comprehensive review of the literature but rather an introduction to the subject that gives the reader a feeling for the range of experiments that have been done and which may encourage the reader to explore electrochemical approaches in their own research. Several general reviews on iron-sulfur proteins, including their electrochemistry, have been published [3–8]. The electrochemistry of iron-sulfur proteins is ultimately an examination of their clusters in specific and unique protein environments. Interesting and useful electrochemical methods have been used to examine a wide variety of iron-sulfur proteins, but most of the electrochemical experiments done to date are on the [4 Fe-4 S] and [3 Fe-4 S] proteins. In addition, while obtaining equilibrium reduction potentials under various conditions, such as changing pH, has been an important area of research, more recent work focuses on the follow-up chemistry of the cluster after a primary electrochemical reaction has been initiated.

The X-ray crystallographic structures of a number of ferredoxins have been completed, including *P. asaccharolyticus* (formerly *P. aerogenes*) [9], a 2 [4 Fe-4 S] ferredoxin, *A. vinelandii* ferredoxin I (a 7 Fe ferredoxin) [10], *S. platensis* ferredoxin, [2 Fe-2 S] [11], *C. vinosum* HiPIP [12], and *D. gigas* [3 Fe-4 S] ferredoxin II [13]. The structure of *P. asaccharolyticus* is shown in Fig. 2, and we will discuss this class of ferredoxins in some detail, as it is probably the most widely studied class of iron-sulfur proteins.

In *Pa* ferredoxin, the two [4 Fe-4 S] clusters are about 12 Å apart. The pH 7 equilibrium reduction potential of *Cp* Fd, $E^{\circ\prime}_{7.0}$, (also called the midpoint potential, $E_{m,pH}$) of both clusters is about −415 mV [14, 15]. The various biological functions of ferredoxins, including *Cp* Fds clusters as a storage form of Fe, have been reviewed [16, 17].

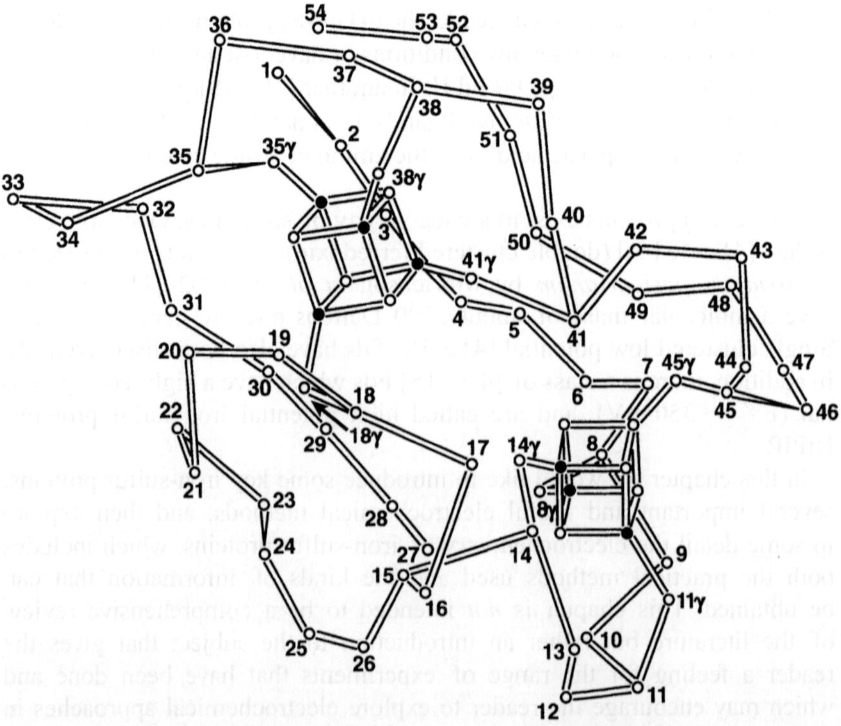

Fig. 2. The α-carbon configuration of *P. asaccharolyticus* (formerly *Peptococcus aerogenes*) ferredoxin. From E. T. Adman, L. C. Sieker and L. H. Jensen, J. Biol. Chem. 248 (1973) 3987–3996. Reprinted with the permission of the *Journal of Biological Chemistry*.

1.2. Equilibrium reduction potentials of iron-sulfur proteins

The comparison of *Cp* Fd and *Cv* HiPIP provides a helpful introduction into differences in redox behavior induced by the protein tertiary structure. When Fd is compared to HiPIP, it is found that they have nearly identical cluster structures but vastly different reduction potentials with $E^{\circ\prime}_{HiPIP} = +350$ mV and $E^{\circ\prime}_{Fd} = -415$ mV. In addition to the differences in reduction potential, Fd and HiPIP are further differentiated from each other by the oxidation levels between which they can interconvert. These differences are explained by the three-state-hypothesis [18] which says that the structural constraints imposed on the cluster by the surrounding polypeptide determine which one of two pairs of oxidation states (a redox couple) is accessible to the protein. The structural and spectroscopic features of the three-state-hypothesis are summarized in Table 1.

Considering the clusters without cysteine (Cys) ligands, the three oxidation states for just the clusters are 3+, 2+, and 1+, respectively. With Cys ligands, there is a total of three formal oxidation states: −1, −2, and −3, which were calculated by taking into account the 8 minus charges of 4 S^{2-}

Table 1. Formal and other representation of Fd and HiPIP redox states

3 Fe(III) 1 Fe(II)		2 Fe(III) 2 Fe(II)		1 Fe(III) 3 Fe(II)
$HiPIP_{ox}$ + e	⇌	$HiPIP_{red}$		
		Fd_{ox} + e + e	⇌	Fd_{red}
$[4Fe\text{-}4S]^{3+}$		$[4Fe\text{-}4S]^{2+}$		$[4Fe\text{-}4S]^{1+}$
$[4Fe\text{-}4S](SCys)_4^{1-}$		$[4Fe\text{-}4S](SCys)_4^{2-}$		$[4Fe\text{-}4S](SCys)_4^{3-}$
Paramagnetic		Diamagnetic		Paramagnetic
EPR active		EPR inactive		EPR active

Table 2. Cluster types, redox couples, and redox states of iron-sulfur proteins

Cluster type	Redox couple	Typical redox potential	Example
$Fe(SCH_2R)_4^{2-/3-}$	Fe^{2+}/Fe^{3+}	−60 mV	*Cp* rubredoxin
$[2Fe\text{-}2S]^{1+/2+}$	$Fe^{2+}Fe^{3+}/Fe^{3+}Fe^{3+}$	−420 mV	Spinach Fd
$[4Fe\text{-}4S]^{2+/3+}$	$3Fe^{3+}1Fe^{2+}/2Fe^{3+}2Fe^{2+}$	+350 mV	*Cv* HiPIP
$[4Fe\text{-}4S]^{1+/2+}$	$2Fe^{3+}2Fe^{2+}/1Fe^{3+}3Fe^{2+}$	−430 mV	*Cp* Fd
$[3Fe\text{-}2S]^{0/1-}$	$1Fe^{2+}2Fe^{3+}/3Fe^{3+}$	−130 mV	*Dg* Fd II

and the 4 minus charges of the four RCH_2S^- ligands. $HiPIP_{red}$ and Fd_{ox} are isoelectronic and diamagnetic, while $HiPIP_{ox}$ and Fd_{red} are strongly paramagnetic. Formal representations of iron-sulfur clusters and their redox states are summarized in Table 2.

Based primarily on X-ray crystallographic data on *Cv* HiPIP and *Pa* Fd [19], a number of suggestions have been made concerning the possible sources of discrimination between the two redox couples. These include differential exposure of the cluster to the solvent [20], geometrical or electronic distortion of the cluster [21, 22], and differential stabilization of cluster charge by N···H hydrogen bonds [19, 23]. Attempts to generate the $HiPIP_{ox}$ state with *Cp* Fd by oxidation with ferricyanide appear to lead to the formation of either [3 Fe-3 S] [4 Fe-4 S] or possibly 2 [3 Fe-4 S] clusters [24]. The electrochemistry of this reaction will be discussed later. For a recent and very interesting comparison of ferredoxin and HiPIP structure and properties, see Backes *et al.* [25].

Having examined some basic iron-sulfur redox chemistry, it is worth noting a few important enzymatic reactions that make use of iron-sulfur clusters since a complete understanding of how these clusters carry out their function in these enzymes may indeed be facilitated by electrochemical studies. Aconitase is an example of an iron-sulfur enzyme which, in the absence of substrate, coordinates OH or H_2O at one of the cubane irons [26]. The substrate, citrate, displaces this ligand to form the active complex. Upon oxidation of the cluster, this iron is lost, a [3 Fe-4 S] cluster is formed, and the enzyme is deactivated. Using EPR, ENDOR and Mössbauer

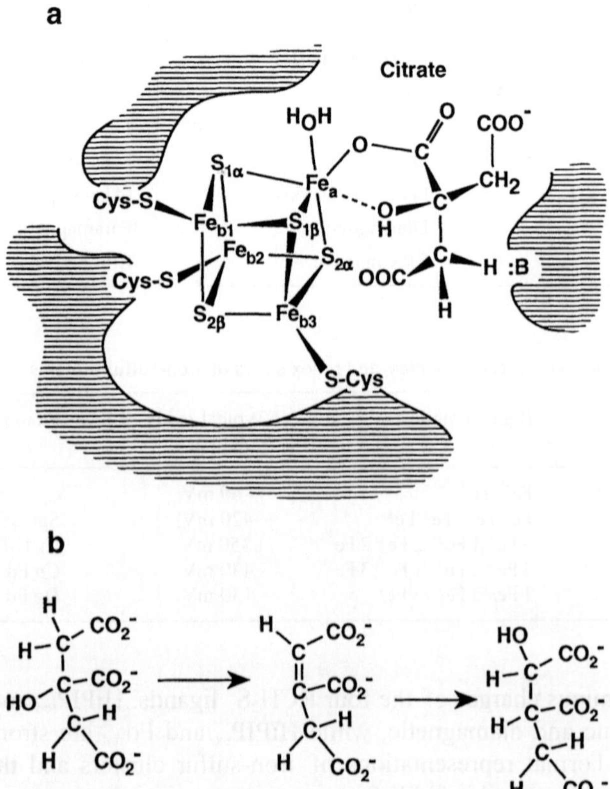

Fig. 3. (a) Representation of ENDOR-derived information abut the $[4Fe\text{-}4S]^{1+}$ cluster of the aconitase enzyme-substrate complex, showing the two pairs of sulfurs ($S_{1\alpha}$ and $S_{1\beta}$; $S_{2\alpha}$ and $S_{2\beta}$) in relationship to the four inequivalent iron sites, Fe_a, Fe_{b1}, Fe_{b2}, Fe_{b3}, along with the bound substrate. From M.M. Wert, M.C. Kennedy, A.L.P. Houseman, H. Beinert and B.M. Hoffman, Biochemistry 29 (1990) 10533. Reprinted with the permission of *Biochemistry*. (b) From left to right: the substrate, citrate; the intermediate *cis*-aconitate; and the product, isocitrate.

spectroscopy, it has been shown that two ligand sites on the citrate molecule chelate the iron atom that is not ligated to the protein [27, 28]. Citrate coordinates to the iron via the geminal hydroxyl and carboxylate groups that are attached to the middle carbon atom shown in Fig. 3. A five-membered ring containing two functional groups from the citrate and an iron atom is formed.

The reaction catalyzed by aconitase is the transformation of citrate to isocitrate (see Fig. 3b) with cis-aconitate as an intermediate [28].

Another iron-sulfur enzyme, *Cp* hydrogenase I, catalyses the reversible interconversion of H^+ and H_2: $2H^+ \rightleftharpoons H_2$. Present evidence indicates that it has four ferredoxin-like $[4Fe\text{-}4S]^{1+/2+}$ clusters and a putative catalytic iron-sulfur H cluster, whose structure has not yet been determined. Apparently, in

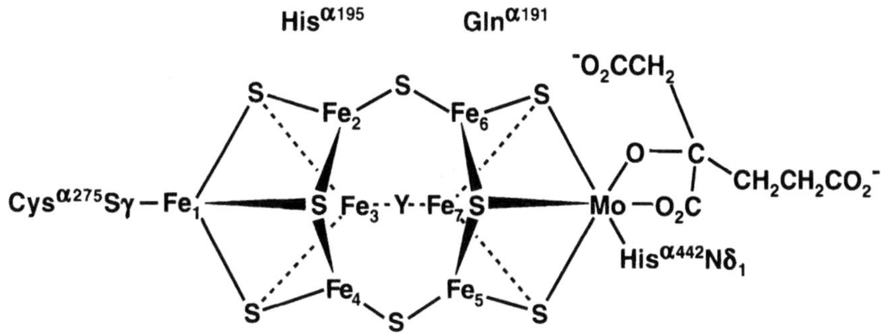

Fig. 4. Schematic representation of the FeMo-cofactor. Y represents the bridging ligand with relatively light electron density. From J. Kim and D.C. Rees, Science 257 (1992) 1677–1682. Reprinted with permission of *Science*.

the reduction of H^+ to H_2, the four [4Fe-4S] clusters donate electrons to the H cluster. All the [4Fe-4S] clusters are isopotential with an $E^{\circ\prime}$ of about −410 mV, which is close to the thermodynamic potential for the reduction of protons at pH 7 [29].

An important enzyme that has one additional metal atom besides iron, is the iron-sulfur-molybdenum enzyme, nitrogenase. The net reaction for the reduction of dinitrogen is:

$$N_2 + 8\,e^- + 18\,H^+ + 16\,MgATP \rightarrow 2\,NH_3 + H_2 + 16\,MgADP + 16\,P_i.$$

Vanadium and all-iron nitrogenase have been discovered [30, 31], but the molybdenum-containing nitrogenase is the most widely studied. It consists of two proteins: the molybdenum-iron protein (MoFe) and the iron (Fe) protein, which donates electrons to the MoFe protein. The MoFe protein has six iron-sulfur clusters, with the two of them being referred to as the iron-molybdenum cofactor (FeMoco) and forming the probable catalytic site for dinotrogen reduction. The other four iron-sulfur clusters are referred to as the P clusters, which are thought to serve as electron storage and transfer sites. The structure for both has recently been determined at 2.7 Å [32] and the FeMoco is shown in Fig. 4.

The clusters in FeMoco are fundamentally different from any of the other clusters discussed in this chapter. The cofactor consists of a [4Fe-3S] and [1Mo, 3Fe-3S] cluster bridged by three non protein ligands. Six of the seven Fe sites in the cofactor have trigonal coordination geometry, including one ligated to the side chain of $Cys^{\alpha 275}$. While details of the reduction of dinitrogen to ammonia are not understood, the FeMoco cluster has been examined electrochemically by Schultz and coworkers [33, 34].

2. Electrochemical techniques

2.1. Introduction

A large variety of electrochemical methods is available for the study of iron-sulfur redox proteins such as ferredoxins. The choice of methodology depends upon the type of information desired, the exact redox characteristics of the protein, and, for spectroelectrochemical methods, the spectral characteristics of the protein. A careful analysis of these factors is critical in order to determine the best method. We will examine these three factors individually.

2.2. Type of information desired

Electrochemical methods can be used to obtain information on the thermodynamic and kinetic properties of redox proteins. The simplest thermodynamic parameter to obtain is the formal potential, $E^{\circ\prime}$. From variations in the formal potential, other equilibrium constants can also be measured. For example, the variation in the $E^{\circ\prime}$ with pH can be used to calculate specific $pK_a s'$ that are linked to electron transfer. The formation constants for the binding of ligands to metal centres can also be determined from $E^{\circ\prime}$ changes.

The second type of information that may be obtained is kinetic data. Kinetic parameters such as the rate of electron exchange with the electrode surface, k° (heterogeneous rate constant), and the rates of homogeneous reactions between proteins and substrates, pseudosubstrates or other redox active materials can be probed with electrochemical methods.

2.2.1. Redox characteristics of the protein: The redox properties of the protein can greatly affect one's ability to obtain the type of information discussed above. For example, except for spectroelectrochemical methods, thermodynamic information can only be obtained if the protein will exchange electrons directly with the electrode surface. If a mediator is required for reduction or oxidation of the protein, one can only provide limits on the value of the redox potential. As will be discussed in more detail later, electron exchange is a necessary, but not sufficient condition for obtaining thermodynamic information. If the electron exchange is too slow, the thermodynamic information may be severely degraded or unusable.

2.2.2. Spectral characteristic of the protein: The combination of electrochemical and spectroscopic techniques (spectroelectrochemistry) provides a powerful method for examining the redox behavior of proteins. While almost every spectroscopic technique has been combined with electrochemistry, the most popular methods with regard to ferredoxins and other redox proteins are UV/visible, EPR and resonance Raman spectroscopy. For UV/visible spectroscopy, there must be a significant change in the molar absorptivity upon electron transfer. EPR requires that the redox protein have a non-zero spin state, and the solution must generally be frozen to obtain the spectra. Resonance Raman spectroscopy requires a strongly absorbing chromophore, and vibrations that are sensitive to the redox states. Because of this, there have been few resonance Raman studies of iron-sulfur proteins, and none that have involved resonance Raman spectroelectrochemistry.

2.3. *Summary of selected electrochemical techniques*

With these characteristics of the protein in mind, it is important to choose an electrochemical method that will enable one to obtain the desired information. There are literally hundreds of voltammetric and amperometric methods, but in this work we will focus on four techniques, three voltammetric and one amperometric, that have shown themselves to be useful in redox protein studies: cyclic voltammetry, differential pulse voltammetry, square wave voltammetry and chronoamperometry. Each of these methods has its own strengths and weaknesses, and it is easy to extend these concepts to the variety of other electroanalytical methods available. In this work, we will first summarize the strengths and weaknesses of each technique for protein studies, and then discuss the details of each technique along with some typical applications. Complete details of these techniques can be obtained from textbooks such as Bard and Faulkner [35].

All of these techniques are based on controlling the potential in some periodic fashion, and measuring the current that flows. The individual techniques are derived from the wave form that is used to control the potential. These techniques have been utilized in the study of iron-sulfur proteins, and an extensive body of literature encompasses the theoretical response for a wide variety of electrochemical mechanisms. The choice of technique to be used on a particular system will depend upon the concentration of the protein and the type of information that is desired. From the analysis of voltammetric data, it is possible to calculate the $E_{1/2}$ for a redox process, which for proteins differs from the $E^{\circ\prime}$ value by less than 1 mV. In general, differential pulse voltammetry and square wave voltammetry have the best analytical sensitivity, and must be used when the protein concentration is low, especially below 10 μM. However, determination of the formal redox potential from the voltammetric wave, may be difficult unless the electron transfer rate with the electrode surface is fast. The analytical sensitivity for cyclic voltammetry is not as good as for the other techniques, but it is much easier to determine thermodynamic parameters. No thermodynamic information can be obtained from chronoamperometry, but it is the best technique to determine solution kinetic values.

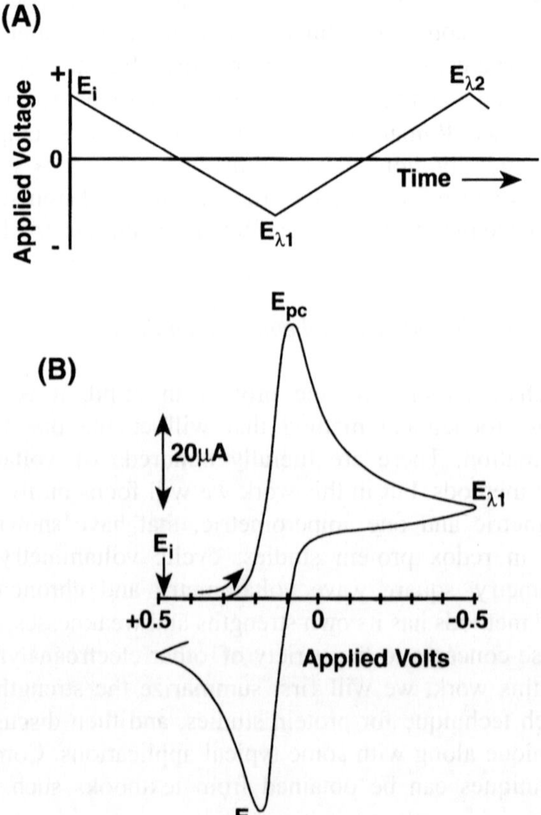

Fig. 5. Waveform and response for cyclic voltammetry. (A) Applied potential scan. $E_{\lambda 1}$ and $E_{\lambda 2}$ switching potentials. (B) Typical cyclic voltammogram of 1 mM $K_3Fe(CN)_6$ at a platinum electrode in aqueous 0.1 M KCl solution. From G.A. Mabbott, J. Chem. Ed. 60 (1983) 697–701. Reprinted with permission of *Journal of Chemical Education.*

2.3.1. Cyclic voltammetry: In cyclic voltammetry (see Fig. 5), the applied potential varies linearly with time. The rate of potential change with time is called the scan rate, v, which has units of volts per second: Vs^{-1}. As the potential passes the $E^{\circ\prime}$ value, current will flow as material is oxidized or reduced. When the potential reaches some arbitrarily defined potential (E_λ switching potential), the direction of the scan is reversed and the scan continues until one reaches a final potential, usually equal to the starting potential. If the electron transfer rate is fast, and the redox species are stable, peaks are observed in the currents for both the forward and reverse scans as shown in Fig. 5.

The current that flows at the peak potentials (i_{pf} for the forward scan and i_{pr} for the reverse scan) depends upon the bulk concentration of the electroactive material, C_o^*, and the scan rate. The peak current varies linearly with C_o^*, and increases with the square root of scan rate, $v^{1/2}$. The combination of these parameters, $i_{pf}/C_o^* v^{1/2}$, should be equal to a constant regardless of scan rate if the current is due to material diffusing to the electrode surface; this is called diffusion control. If the peak current is linearly proportional to scan rate, the observed current is probably due to material adsorbed on the electrode surface, rather than diffusing solution species. One must be careful about drawing conclusions from the redox potentials of adsorbed proteins because significant conformational changes may have occurred, thus altering their properties. While the current increases with scan rate, the greatest analytical sensitivity is obtained at slow scan rates because the background current, which is mainly due to charging the electrode surface (charging current), increases linearly with scan rate. A good summary of the application of cyclic voltammetry to a variety of electrochemical mechanisms is given by Evans *et al.* [36].

The shape of the waves and the relative positions of the forward (E_{pf}) and reverse (E_{pr}) peak potentials are dependent upon the rate of electron exchange between the electrode surface and the redox protein, as well as on chemical reactions that occur prior to or after electron transfer. If the electron transfer rate is fast, the E_{pf} is 28.5 mV negative of the $E_{1/2}$ value for a reduction wave. The strength of cyclic voltammetry is that, upon reversing the potential scan at the switching potential, the voltammetric wave for the reverse electron transfer can be observed. If the reverse reaction is an oxidation, the E_{pr} value is 28.5 mV positive of the $E_{1/2}$ value, and the $E_{1/2}$ is halfway between the E_{pf} and E_{pr} values. If the electron transfer is not rapid, the $E_{1/2}$ values and the rate of electron transfer can be determined from the observed peak potentials [37]. In most cases, the formal redox potential can be estimated from the average of the E_{pf} and E_{pr} values. If a reverse wave is not observed, it is not possible to determine the $E_{1/2}$ value accurately, and no thermodynamic information can be obtained. Cyclic voltammetry has been used by several workers to examine iron-sulfur proteins. For example, Landrum *et al.* [38] showed that spinach ferredoxin can be reduced directly as a surface-modified gold minigrid electrode. Other modified surfaces were used by Hill *et al.* [39] to examine spinach ferredoxin, *Cp* ferredoxin and *Cp* rubredoxin, as shown in Fig. 6.

Because of the digital nature of the computer-generated waveforms from the latest instruments, the voltage scan consists of a series of steps rather than the linear scan that is assumed in cyclic voltammetry (CV). This technique is formally called cyclic staircase voltammetry (CSCV). In practice, these terms (CV and CSCV) are used interchangeably, and the results are generally equivalent under normal experimental conditions [40].

Under conditions where a forward and reverse wave is observed, cyclic voltammetry provides an accurate method for the determination of $E^{\circ\prime}$ values, even in cases where the electron transfer process is slow. From the variation in the $E^{\circ\prime}$ values with pH, it is possible to determine the pK_a of acidic groups that are sensitive to the redox state of the protein. Using the Nernst equation, the variation of the $E^{\circ\prime}$ with pH is equal to:

$$E^{\circ\prime} = E^{\circ\prime}_{O/R} + (RT/nF)\ln((K_{a,O} + [H^+])/(K_{a,R} + [H^+])) \quad (1)$$

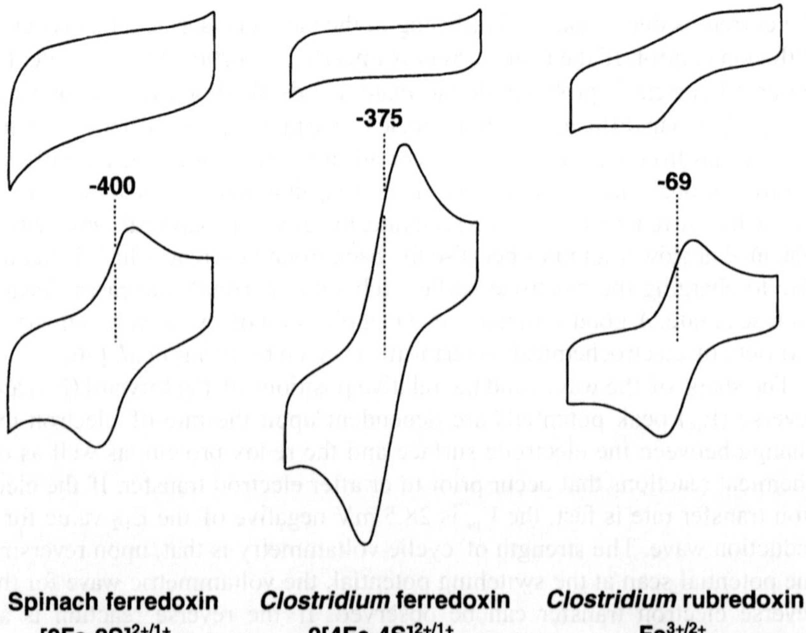

Fig. 6. Cyclic voltammetry of negatively charged iron-sulfur proteins, *Cp* Fd, *Cp* Rb, and spinach Fd, at the pyrolytic graphite "edge" electrode. Promoters used in the lower cyclic voltammograms: $Cr(NH_3)_3$ or decamethonium, $(CH_3)_3N(CH_2)_{10}N(CH_3)_3^{2+}$, and no promoters used in the upper voltammograms. From F. A. Armstrong, H. A. O. Hill and N. J. Walton, Accounts of Chemical Research 21 (1988) 407–413. Reprinted with permission of the *Journal of the American Chemical Society*.

where $E°'$ is the redox potential at a given pH, n the number of electrons transferred, F is Faraday's constant, $E°'_{O/R}$ is the formal potential for the protonated forms of the protein, and $K_{a,O}$ and $K_{a,R}$ are the K_a values of the oxidized and reduced form of the protein, respectively. Similar analysis can be done with the complexing of ligands at the redox site. If a formation constant for ligand binding is dependent upon the redox state of the protein, usually binding at the redox site, the formal potential as a function of ligand concentration is given by:

$$E_K = E°' + RT/nF \ln((1 + K_{f,O}[L])/(1 + K_{f,R}[L])) \quad (2)$$

where E_K is the formal redox potential in the presence of a ligand substrate, [L] is the concentration of the ligand, and $K_{f,O}$ and $K_{f,R}$ are the formation constants for the binding of the ligand to the oxidized and reduced form of the protein, respectively.

In addition to thermodynamic information such as redox potentials, K_a, and K_f values, kinetic information can also be determined using cyclic voltammetry. If the enzyme is reduced in the presence of substrate, turnover may occur, and the enzyme may be re-oxidized by the substrate in a catalytic mechanism.

$$A + e^- \rightleftharpoons A \quad (3)$$

$$A + Z \rightarrow A + Z' \quad (4)$$

where A and A^- are the oxidized and reduced form of the electroactive material, respectively, and Z and Z′ are the oxidized and reduced forms of the material that was reduced by the solution reaction. Either A or Z may be the protein. If the protein, A, reacts directly at the electrode surface, Z would be a substrate or pseudo-substrate. If the protein, Z, is not electroactive, then A is a mediator that can indirectly reduce the protein. Notice that A is regenerated by the second reaction, and may be reduced again at the electrode surface. This will lead to an increase in current relative to the diffusion-controlled case; i_d is the diffusion-controlled current in the absence of Z; and i_k is the kinetic current observed in the presence of Z. If the reaction is fast enough, the concentration of A reaches a steady-state value near the electrode surface and the voltammetric current will become independent of scan rate. Once the current has at least doubled, the ratio of i_k/i_d, under pseudo-first order conditions ($[Z] \gg [A]$, or, if Z is H^+, as in a buffered solution) is given by:

$$i_k/i_d = 0.359\ (k_f[Z]/v)^{1/2} \quad (5)$$

where k_f is the rate of Reaction 4, and [Z] is the concentration of substrate [41]. The analysis becomes considerably more difficult if the conditions are not pseudo-first-order. Under second-order conditions, the current ratio is sensitive to the diffusion coefficients of both A and Z, as well as the [Z]/[A] ratio.

2.3.2. Differential pulse voltammetry: Differential pulse polarography (at a mercury electrode) or voltammetry (at a solid electrode) uses a waveform that was designed to maximize the current relative to the background (charging current) signal. Differential pulse voltammetry/polarography is one of the most sensitive electroanalytical methods, and has been applied widely to the study of redox proteins because of the low concentrations generally utilized. The waveform is shown in Fig. 7A. This waveform consists of a linearly varying potential on which a series of pulses of equal height (ΔE) are superimposed. The current (i_1) is measured just before the pulse is applied (t_1), and then again (i_2), just before the end of the pulse (t_2). The current that is displayed is the current difference, Δi, where $\Delta i = i_2 - i_1$. The difference is largest when the current is strongly dependent upon potential, especially near the $E^{\circ\prime}$, and becomes negligibly small on either side of the $E^{\circ\prime}$. The current-potential waveform has a peak shape with the peak potential, E_p, that is equal to $E_{1/2} - \Delta E/2$, if the electron transfer is fast; ΔE is negative if the scan is from positive to negative potentials.

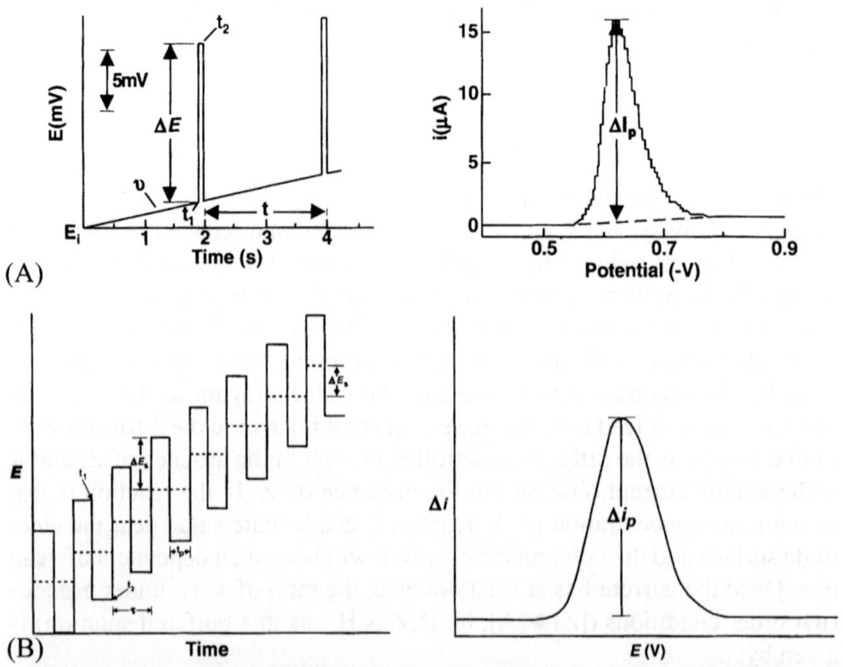

Fig. 7. Waveform and response for differential pulse and square wave voltammetry. (A) A voltage ramp for differential pulse polarography. The current is measured just prior to the application of the pulse at t_1 and typically during the last 17 ms of the pulse at t_2. The difference $i_2 - i_1$ is plotted in the polarogram. ΔE, the potential pulse amplitude (25 mV in this case; E_i, initial potential, and the scan rate, v, 2 mV/s. B) A differential pulse polarogram of 1.2 mM cadmium chloride in an aqueous 0.1 M KCl solution. v = 2 mV/s; ΔE = 5 mV. SCE reference electrode. From R. D. Braun, *Introduction to Instrumental Analysis*, McGraw-Hill Book Co., New York, 1987, pp. 782–784. Reprinted with permission of McGraw-Hill Book Company.

The current at the peak potential, the peak current Δi_p, which depends upon the concentration of the protein, the pulse time (t), and the pulse height (ΔE), is given by:

$$\Delta i_p = \frac{nFAD^{\frac{1}{2}} C_o^* [\exp(nF\Delta E/2RT) - 1]}{\sqrt{\pi t}\, [\exp(nF\Delta E/2RT) + 1} \qquad (6)$$

where D is the diffusion coefficient and C_o^* is the bulk concentration. Pulse heights generally vary from 104–100 mV, with the greatest sensitivity occurring for the largest pulse size. If the electron transfer rate is not fast, the wave will be broader than the reversible case, the peak current will decrease, and the wave will shift to more negative potentials for reductions. Unfortunately, unless one already knows the formal potential, it is difficult to detect the presence of a slow electron transfer because the reverse reaction is not measured directly, as in cyclic voltammetry. The only effect that one can observe is a wave that is broader than expected for a fast electron transfer. For a fast electron transfer, the full-width of the wave at half-height is 90 mV, but this value depends somewhat on the value of ΔE. If the wave is significantly broader than this value, the $E_{1/2}$ value, calculated from the peak potential, is unreliable and cannot be used for thermodynamic calculations. This is significantly different from cyclic voltammetry where deviations from reversible conditions could be directly observed from the difference between the forward and reverse peak potentials. If the differential pulse voltammogram is consistent with fast electron transfer, the $E_{1/2}$ values can be used in the same manner as in cyclic voltammetry. The order of magnitude increase in sensitivity of differential pulse voltammetry enables one to study proteins at μM or lower concentrations, which is a significant improvement over cyclic voltammetry. Examples of the use of differential pulse voltammetry to study iron-sulfur proteins are the work of van Dijk *et al.* [42] on the study of ferredoxins and rubredoxins from spinach and *M. elsdenii.*

The differential pulse voltammetric waveform is much more complex than the cyclic voltammetric waveform. As a result, theories for various electrochemical mechanisms are not as easily obtained and applied as in cyclic voltammetry. The effect of heterogeneous and homogeneous kinetic parameters on the voltammetric response is a function of the pulse height, the length of the pulse and the interval between pulses [43, 44]. Differential pulse voltammetry has been used to study the electrocatalytic reduction of

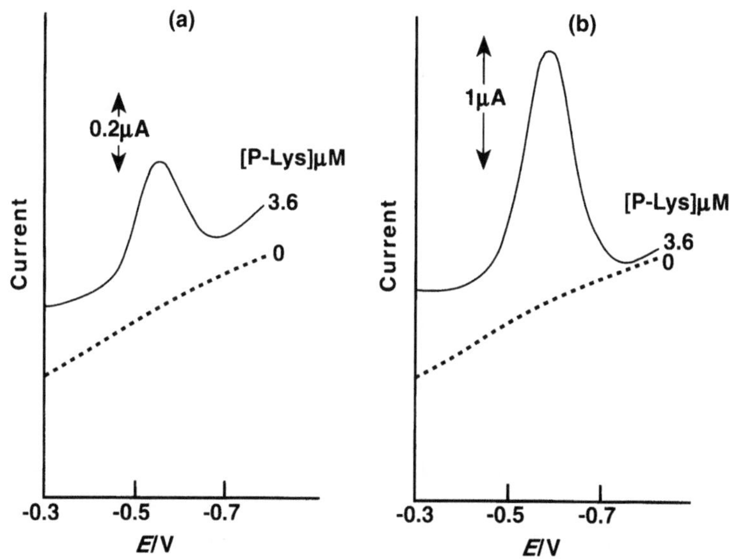

Fig. 8. Differential pulse voltammograms of (a) 54 μM *Desulfovibrio desulfuricans (Dd)* ferredoxin I at pH 5.8; (b) 38 μM *Dd* ferredoxin II at pH 7.4 with 3.6 μM polylysine. 0.05 M Tris buffer with 0.05 M sodium acetate and 0.4 M sodium chloride. From J. Haladjian, P. Bianco and L. Asso, Electrochim. Acta 31 (1986) 1513–1517. Reprinted with permission of *Electrochemical Acta.*

cytochrome c with ferrous-EDTA [44]. In this case, cytochrome c did not reduce directly at the electrode surface, and a mediator, Fe^{III}(EDTA), was added to reduce ferricytochrome c indirectly. In this scheme, species A is the ferric-EDTA complex and ferricytochrome c is Z.

2.3.3. Square wave voltammetry: Square wave voltammetry is closely related to differential pulse voltammetry, but has a number of important differences. Square wave voltammetry consists of a series of square wave pulse superimposed on a step scan (Fig. 7B). The potential is pulsed E_{sw} mV negative of the scan potential for a time t_p. Then, a reverse pulse is applied $-E_{sw}$ mV of the scan potential for the same length of time. The current is measured at two points: the first measurement (i_1) is made at the end of the forward pulse, and a second current value (i_2) is obtained at the end of the reverse pulse. At the end of each pair of pulses, the scan potential is increased by E_s mV, and a new pair of pulses occurs. The difference in current, $\Delta i = i_1 - i_2$, is plotted, but most modern instrumentation also record the individual current i_1 and i_2. A peak shape is obtained, as was observed for differential pulse voltammetry. The peak current, Δi_p, is given by:

$$\Delta i_p = \frac{nFAD^{1/2} C_o^* \Delta \Psi}{\sqrt{\pi t_p}} \quad (7)$$

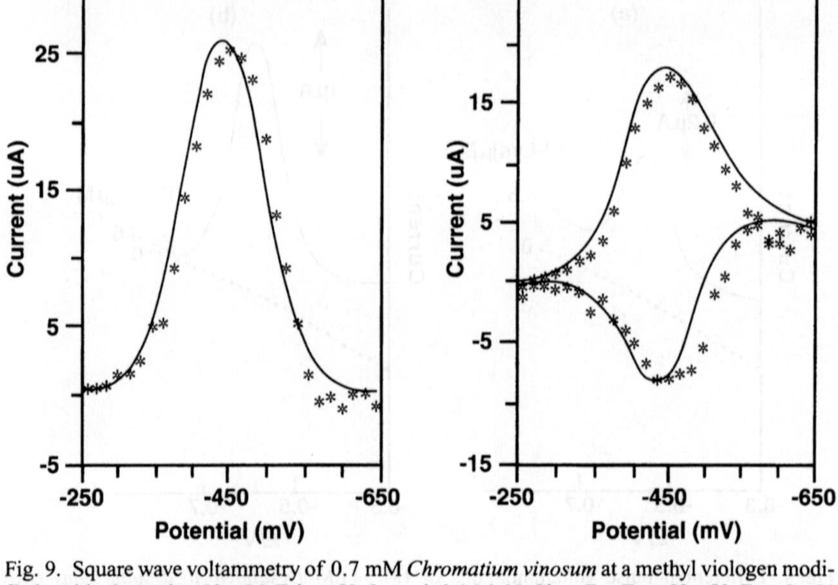

Fig. 9. Square wave voltammetry of 0.7 mM *Chromatium vinosum* at a methyl viologen modified gold electrode, 50 mM Tris, pH 8, and 0.1 M NaCl; $\Delta E = E_p = 50$ mV, $E_s = 5$ mV. Experimental points (*) for the net current (A) and the forward and reverse currents (B). From E. T. Smith and B. A. Feinberg, J. Biol. Chem. 265 (1990) 14371. Reprinted with permission of *Journal of Biological Chemistry*.

where $\Delta\Psi$ is a dimensionless function that depends on the values of E_s and E_{sw} [45]. In particular, the shape of the individual current-potential curve is a sensitive measure of the rate of electron transfer, and the stability of the product. Armstrong *et al.* [46] used square wave voltammetry to examine the stability of *Cp* ferredoxin. Smith and Feinberg used square wave voltammetry to examine the redox properties of several 2[4Fe-4S] ferredoxins [47–49]. An example of the square wave voltammetry of *Chromatium vinosum* Fd [47] is shown in Fig. 9.

Unlike differential pulse voltammetry, one can observe the re-oxidation current, for an initial reduction, during the reverse pulse. Like cyclic voltammetry, it is possible to measure both the reduction and oxidation currents directly. If the electron transfer is slow, the reverse current, i_2, will not reverse sign, and it will be impossible to determine the $E_{1/2}$ value. Often, though, by increasing the pulse height, E_{sw}, the oxidation wave (for an initial reduction) can still be observed. As with differential pulse voltammetry, it must be verified that the electron transfer rate is fast if the $E_{1/2}$ values determined are to have any fundamental thermodynamic significance. The effects of slow electron transfer on the square wave voltammogram have been described by O'Dea *et al.* [50]. Square wave voltammetry can also be used to examine the catalytic mechanism, as described in Reactions 3 and 4 [50]. The voltammetric response is a complex function of the pulse time, pulse height, and scan rate, and the original reference must be examined in order to relate the observed currents to the rate of Reaction 4.

2.3.4. Chronoamperometry: Chronoamperometry has the simplest waveform as shown in Fig. 10. The potential is instantaneously stepped from an initial potential to a fixed, final potential.

After a period of time, the potential may be returned to the initial potential (or some other potential); this technique is called cyclic chronoamperometry. Since the potential is independent of time, potential related information such as $E^{\circ\prime}$ values is lost; however, because the potential is generally stepped well beyond the $E^{\circ\prime}$ of the protein, the rate of electron transfer from the electrode to the protein will be fast, and slower homogenous solution kinetics can now be studied without the interference of the heterogeneous kinetics. For the three other techniques above, the current is measured in the potential region around the $E^{\circ\prime}$, and the rates of solution reactions must be deconvoluted from any slow electron transfer kinetics. If a material is reduced directly at the electrode surface, the current that will be observed in chronoamperometry is given by the Cottrell equation:

$$i_d = nFAD^{1/2}C_o^*/(\pi t)^{1/2} \tag{8}$$

where all terms are as defined before and t is the time from the imposition of the pulse. A typical diffusion-controlled current decay is shown in Fig. 10, curve A. If the observed current depends only on the rate at which the material reaches the electrode surface (diffusion control), $i_d t^{1/2}/C_o^*$ will

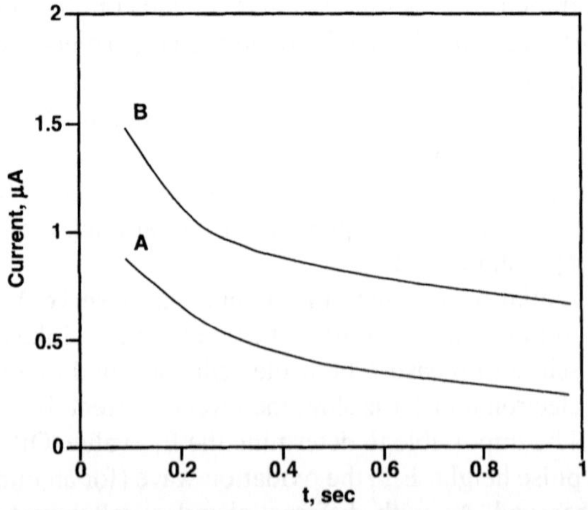

Fig. 10. Chronoamperometric decay of V(II)-EDTA in the presence (B) and absence (A) of spinach ferredoxin. $C_{V(II)\text{-}EDTA} = 0.10$ mM; $C_{Fd} = 0.20$ mM; pH 5, 0.1 M acetate buffer. Average value of $k_f = 3.2 \times 10^4\ M^{-1}s^{-1}$. Adapted from data from S. Dasgupta and M.D. Ryan, Bioelectrochem. Bioenerg. 7 (1980) 587.

be a constant. If additional electroactive material is generated by Reaction 4, the current that will flow, i_k, will be greater than i_d (see Fig. 10, curve B). The kinetic current (i_k) is given by:

$$i_k = nFAD^{1/2}C_o^* \left\{(k_f[Z])^{1/2} erf\,(k_f[Z]t)^{1/2} + \frac{\exp(-k_f[Z]t)}{\sqrt{\pi t}}\right\}. \quad (9a)$$

If i_k is more than twice i_d, Eq. 9a reduces to:

$$i_k = nFAC_o^*(Dk_f[Z])^{1/2} \quad (9)$$

if pseudo-first-order conditions apply. The ratio of i_k to i_d is equal to:

$$\frac{i_k}{i_d} = \sqrt{k_f[Z]t}\left\{\pi^{1/2} erf\,(k_f[Z]t)^{1/2} + \frac{\exp(-k_f[Z]t)}{\sqrt{k_f[Z]t}}\right\}. \quad (10a)$$

When the current ratio is greater than 2, Eq. 10a reduces to:

$$i_k/i_d = (\pi k_f[Z]t)^{1/2}. \quad (10b)$$

The theoretical relationship between i_k and i_d has also been developed for the case where Reaction 4 is second order [51]. Under those conditions, the current ratios depend upon the concentration and diffusion coefficient ratios of the indirectly reduced material, Z, to the electroactive material. Fortunately, the theoretical results are not a strong function of the diffusion coefficient if the ratios differ significantly from unity. Second-order kinetics were used to determine the homogeneous electron transfer rates between a europium(II) complex and spinach ferredoxin [52].

2.3.5. Spectroelectrochemical techniques: Spectroelectrochemical techniques have been used for almost three decades to obtain kinetic and thermodynamic information on redox proteins. Initially, the techniques were based on the mediated reduction/oxidation of the protein, but because unmediated systems have been discovered, the direct reaction at the electrode surface can also be used. The vast majority of the spectroelectrochemical studies have been used to obtain redox potentials or spectral properties of the reduced or oxidized forms of a protein.

The most commonly used methods for protein studies are UV/visible and EPR spectroscopy. For visible spectroscopy, small volume or thin-layer electrochemical cells have been used, and the solution potential is brought into equilibrium with the electrode potential. Once equilibrium has been achieved, a spectrum is then obtained. By changing the electrode potential, a series of spectra are obtained, and the spectral changes are related to the electrode (or solution) potential. A modestly large volume (1.0–2.0 ml) spectroelectrochemical cell has been designed and can be made with some ease [53]. In carrying out spectroelectrochemical (SEC) titrations, a mediator is used which has its $E^{\circ\prime}_{mediator}$ within ± 50 mV of the iron-sulfur protein of interest. For example, in the SEC titration of Fd, the mediator used was methyl viologen, since its $E^{\circ\prime}$ is –410 mV and in the vicinity of what was expected for the Fd. The advantage of these cells is that they are closed systems with low oxygen leak rates and provide an environment where low ambient potentials can be established and maintained. The potential is applied with a potentiostat and, at *each potential*, the following equilibrium prevails for the oxidized (O) and reduced (R) species of interest:

$$E_{app} = E^{\circ\prime}_{med} - 0.059/n \log R_{med}/O_{med} = E^{\circ\prime}_{Fd} - 0.059/n \log R_{Fd}/O_{Fd}. \quad (11)$$

The log R_{Fd}/O_{Fd} are obtained from the spectra and the subsequent Nernst plot provides the $E^{\circ\prime}_{Fd}$ and the number of electrons transferred per cluster, which in case of two identical *one* electron redox centers is n = 1 [54, 55]. Rickard *et al.* [56] determined the redox potential of spinach ferredoxin using visible spectroelectrochemistry, and Taniguchi *et al.* [57] examined a variety of ferredoxins and HiPIP proteins using similar methodology.

More recently a spectroelectrochemical cell was designed for low temperature electron paramagnetic resonance of very low potential, oxygen-sensitive redox proteins [58]. The electrochemical cell was fabricated from quartz and a quartz EPR tube was mounted directly to the cell body. In a typical experiment, the protein solution with mediators is poised at particular potential, the cell tilted to allow the solution into the EPR tube and then an EPR spectrum taken. The frozen solution in the EPR tube was permitted to thaw, tilted back into the small electrolysis chamber of the cell and the experiment repeated at several new E_{app}, and a Nernst plot prepared. One must have spectra of the fully oxidized and fully reduced protein. The working electrode was gold foil. The key advantage of this cell is that the leakage of oxygen into the solution is greatly minimized, e.g. 0.005 nmol/min,

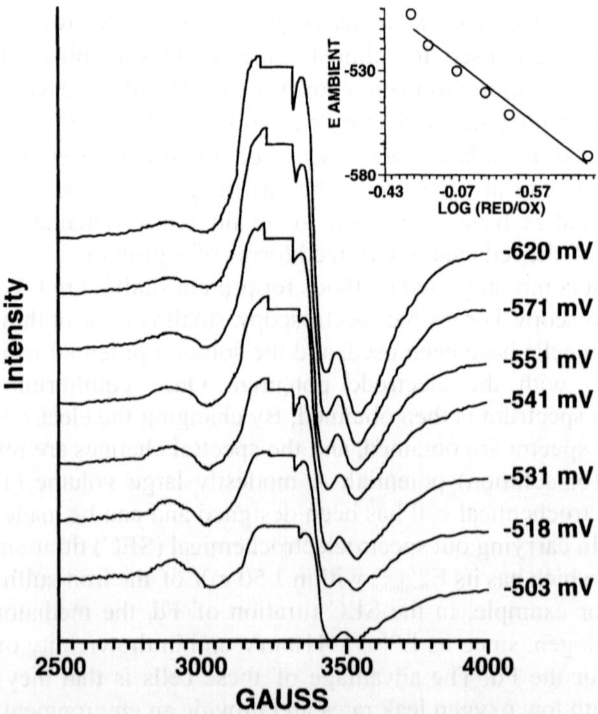

Fig. 11. Determination of the midpoint potential of the $[4Fe\text{-}4S]^{1+/2+}$ cluster of the iron-sulfur/corrinoid protein from *Clostridium thermoaceticum* by EPR SEC. The corrinoid/iron-sulfur protein was 150 μM, 50 mM Tris, pH 7.6. From S.R. Harder, W.P. Lu, B.A. Feinberg and S.W. Ragsdale, Biochemistry 28 (1989) 9080–9089. Reprinted with the permission of *Biochemistry*.

and a drift of ±2 mV after equilibrium was reached. EPR SEC was carried out for the very low potential iron-sulfur cluster of the iron-sulfur/corrinoid protein of *Clostridium thermoaceticum* CO dehydrogenase, shown in Fig. 11.

3. Voltammetric studies of iron-sulfur proteins

3.1. Equilibrium measurements via indirect electrochemistry

Determination of the equilibrium reduction potential of metalloproteins has always been an important goal for bioelectrochemistry and has generally been done by redox titration. For iron-sulfur proteins, the low reduction potentials create a significant challenge in maintaining strictly anaerobic conditions during titration. Additionally, electron transfer pathways and mechanisms are often unintelligible without the sequence of E°'s for the members of a given electron transport chain. Using a dye-mediated spectroelectrochemical titration [47], the low potential ferredoxin from *Chromatium vinosum* was examined. For very low potential titrations of this kind, methyl viologen ($E^{\circ\prime} \approx -445$ mV) needs to be present, usually with other low potential mediators such as: 2,2′-trimethylene-2,2′-bipyridyl ($E^{\circ\prime} = -540$ mV) and 4,4′-dimethylene-2,2′-trimethylene-2,2′-bipyridyl ($E^{\circ\prime} = -650$ mV). These two mediators can readily be prepared following the procedure described by Salmon and Hawkridge [59]. Depending on the pH of the solution, these potentials are in the region where H^+ is reduced to hydrogen and this reaction defines the lower limit that may be reached in this kind of bulk electrolysis reaction. With a new redox protein with an unknown $E^{\circ\prime}$, one may have to search for an appropriate mediator with a formal potential near the value for the iron-sulfur protein being investigated.

3.2. Direct electrochemistry of iron-sulfur proteins

In their study of the low potential [2Fe-2S] ferredoxins form *Halobacterium* of the Dead Sea (MW 14,000), Miller and Werber [60] carried out cyclic voltammetry with a dropping mercury electrode. Sweep rates of 0.1 to 10 Vs^{-1} were used and the observed current responses were nearly identical. The reduction potential of the Fd was determined to be $E^{\circ\prime}_{7.0} = -345$ mV vs SHE, which was identical to the potential measured by Werber and Mevareck using a solution redox titration [61]. Their work also demonstrated that it was the adsorbed Fd molecules that actually underwent electrochemical reduction, and at longer times this adsorption coverage increased to a maximum. The advantage of the adsorption process was that concentrations of Fd of 5 μg mL^{-1} or below 5×10^{-7} M could be detected. The disadvantage of a mercury electrode was that, at higher pH, the cysteines tended to adsorb on the mercury surface. Subsequently, the Hg-cysteinates were reduced at a potential which was close to the reduction of the [2Fe-2S] cluster; fortunately, at pH 7, the protein was not denatured so as to permit Hg-S-Cys formation.

Van Dijk *et al.* [42] investigated the equilibrium reduction potentials of three different iron-sulfur proteins using pulse polarography at a dropping mercury electrode. The iron-sulfur proteins examined were *M. elsdenii* Fd, rubredoxin, and spinach Fd. This work took into account the problem of electrostatic repulsion of the protein at the electrode surface. When reducing Fd such as these at negative potentials, the ferredoxins are apparently repelled from the electrode surface as the potential is scanned to negative potentials, making the electrode more negatively charged. By using polylysine, van Dijk *et al.* were able to increase Fd reduction efficiencies greatly and obtain constant values for the $E^{\circ\prime}$ vs SHE at pH 7.4. Their results are summarized in Table 3. The $E^{\circ\prime}$ for spinach Fd was in good agreement with the value obtained by Tagawa and Arron [62] via solution methods. "Charge compensation" of the proteins was achieved by the binding of polylysine to

Table 3. Pulse polarographic results for three iron-sulfur proteins at the dropping mercury electrode

Fd	Cluster type	MW (kD)	E°′ vs SHE
Me Fd	$2[4Fe\text{-}4S]^{1+/2+}$	6	−370 mV
Me Fd	$Fe^{2+/3+}(S\text{-}Cys)_4$	6	−108 mV
Spinach Fd	$[2Fe\text{-}2S]^{1+/2+}$	11	−405 mV

the electrode surface; the negatively charged Fds were then able to approach the electrode surface freely. This effect was optimum when the net charge of the protein concentration was about equal to the same concentration of opposite charges provided by the added polylysine polymer. See Figure 8 for another dramatic example.

Even earlier experiments by Hill and coworkers were concerned not only with charge compensation at the electrode surface but also with new electrode surfaces. Armstrong, Hill and Walton [39] carried out a series of novel experiments where electrode surfaces were modified. In earlier work, the electrodes such as gold could be effectively modified by having an electrochemically inactive "electrode promoter" such as 4,4′-biypridyl or pyridine-n-carboxaldehyde thiosemicarbazone. There is evidence that these species cover the electrode surface with pyridine groups pointing out perpendicularly to interact with the redox species of interest. In the case of thiosemicarbazone, the sulfur probably interacts with the metal electrode and the pyridine moiety serves as the "facilitator", speeding up heterogeneous electron transfer.

In 1984 Hill and Armstrong [63, 64] introduced the pyrolytic graphite edge (PGE) electrode to the study of iron-sulfur proteins. The discovery of pyrolytic graphite as an electrode material for direct unmediated electron transfer for biological redox systems has been an important development. Pyrolytic graphite is anisotropic in its structure and for this reason either the basal plane or the edge can be used. The PGE is the most widely used and apparently provides the most sensitive response. When polished in air, it possesses a surface layer of various carbon oxide functionalities which can be further modified by selective reagents. Recognizing that most iron-sulfur proteins are generally negative in overall charge, Hill and Armstrong used $MgCl_2$ or $Cr(NH_3)_6^{3+}$ [65], in much the way polylysine was used. In a further development for dealing with electrostatic repulsions and at the same time introducing a level of specificity for various iron-sulfur protein interactions with the electrode surface, they introduced the use of aminoglycoside antibiotics such as neomycin and tobramycin. One side of the antibiotic interacts with the electrode surface, and the other side, with $-NH_3^+$ groups directed to the solution, interacts with the negatively charged Fd and can also be used to produce a co-adsorbed aminoglycoside/Fd film on the electrode surface [66]. For more details, see two reviews by Armstrong [67, 68].

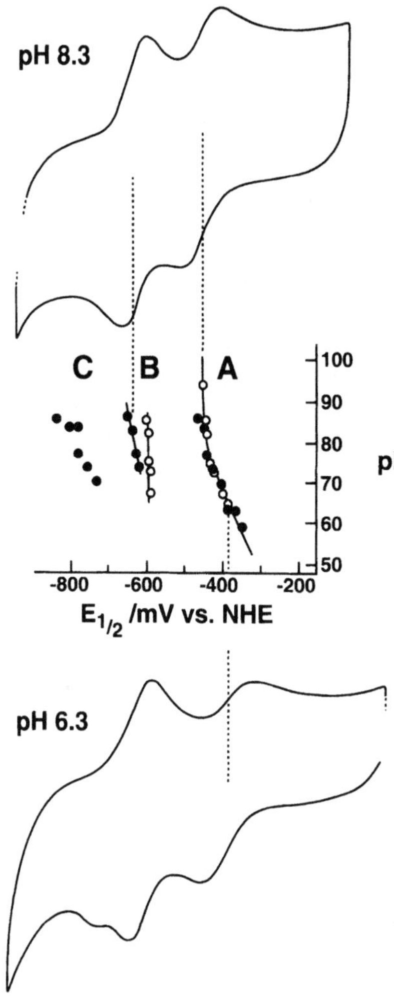

Fig. 12. Cyclic voltammetry of *Azotobacter chroococcum* 7Fe Fd I, 84 μM at pH 8.3 and pH 6.3 with neomycin, tobramycin or $Cr(NH_3)$ as promoters. From F.A. Armstrong, S.J. George, A.J. Thomson and M.G. Yates, FEBS Letters 234 (1988) 107–110. Reprinted with permission of Elsevier Science Publisher B.V., Biomedical Division.

Continuing their work with the PGE, Armstrong *et al.* [69] explored the 7Fe Fd I of *Azotobacter chroococcum*, which was known to have one [4Fe-4S] and one [3Fe-4S] cluster. While this work was unmediated, the aminoglycosides, tobramycin and neomycin, were added to the buffers and were used to promote direct interaction of the protein at the PGE electrodes. It is interesting to note that because the aminoglycosides deprotonate at about pH 8.5, $Cr(NH)_3^{3+}$ was used as a promoter at higher pH. Cyclic voltammograms and their pH dependencies were determined. Armstrong *et al.* found that the reduction of the 3Fe center was directly linked to H^+ uptake.

As seen in Fig. 12, Couple A is pH-dependent and couple B is pH-independent.

For the A cluster/redox couple with E°′ = –460 vs NHE, the following assignment was made:

$$[3\,Fe\text{-}4\,S]^{1+} + e^- \rightleftharpoons [3\,Fe\text{-}4\,S]^0, \quad (12)$$

$$[3\,Fe\text{-}4\,S]^0 + H^+ \rightleftharpoons [3\,Fe\text{-}4\,S]^0\cdots H^+. \quad (13)$$

Comparison of the acidic and basic MCD spectra supported the formation of a protonated [3 Fe-4 S] cluster. Couple B (E°′ = –645 mV) corresponds to the $[4\,Fe\text{-}4\,S]^{1+/2+}$ cluster; it is pH-independent: $[4\,Fe\text{-}4\,S]^{2+} + e^- \rightleftharpoons [4\,Fe\text{-}4\,S]^{1+}$. These assignments were confirmed by X-band EPR spectroscopy of the material generated by anaerobic bulk reduction of the Fd at –835 mV; it was found, for example, that the EPR spectrum of the reduced Fd closely resembled the known $[4\,Fe\text{-}4\,S]^{1+}$ cluster EPR of the 7Fe Fd of *Thermus thermophilus*. Additionally, spin integrations indicated a total of two elec-

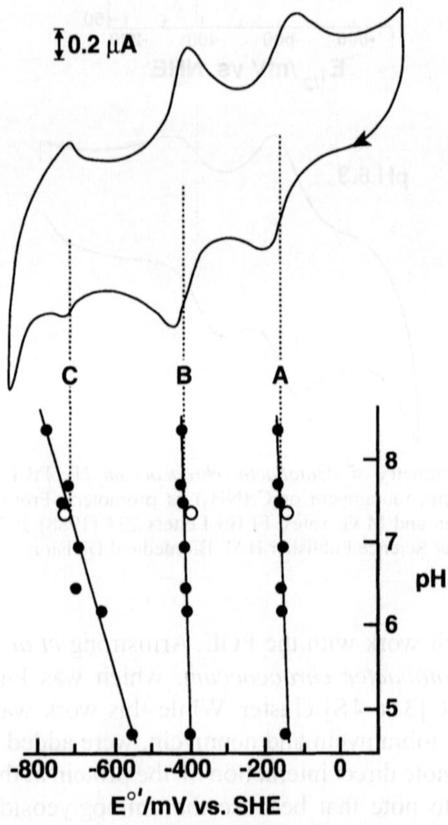

Fig. 13. Cyclic voltammogram (a) of *D. africanus* Fd III and (b) pH dependence of E°′ values. From F. A. Armstrong, S. J. George, R. Cammack, E. C. Hachikian and A. J. Thompson, Biochem. J. 264 (1989) 265–273. Reprinted with permission of *Biochemical Journal*.

trons transferred for the reduction of the fully oxidized Fd, that is one electron per cluster.

Continuing the work in 7Fe Fds, Armstrong *et al.* [70, 71] explored *Desulfovibrio africanus* Fd III which was known to have only 7 cysteines and was thus a good candidate to be a 7Fe Fd. Figure 13 above is a typical cyclic voltammogram of *Da* Fd III.

Through a combination of direct electrochemistry and an EPR and MCD spectroscopic examination of electrolyzed Fd III, they showed that this Fd III contained one [3Fe-4S] and one [4Fe-4S] cluster with E°′ values of −140 and −410 mV, respectively. While redox couple A (E° = −140 mV) was shown to be primarily diffusion-controlled at lower scan rates, adsorbed ferredoxin dominated at high scan rates. Couple B (E° = −410) was well behaved (diffusion-controlled) up to 800 mVs^{-1}. Couple C was pH-dependent with E° values of −670 mV at about pH 6.4 to −726 mV at pH 8.4. EDTA or one of its relatives was used in buffers to eliminate any metal ions in solution, and upon exhaustive electrolysis at −605 mV about $2e^-$ per protein monomer were observed. EPR and MCD spectroscopic studies of electrolyzed solution were carried out to identify A and B. EPR spectroscopy of the oxidized Fd gave a line shape similar to those observed for a known oxidized $[3Fe\text{-}4S]^{1+}$ cluster. The complete, unambiguous characterization of the electrolysis product must be done in order to assign redox states to the electrogenerated protein convincingly.

Electrolyzing at −260 mV, which is between the E°′ of the A and B couples, gave a one electron per monomer reduced product. They observed that the g = 2.01 signal, due to the $[3Fe\text{-}4S]^{1+}$ cluster, nearly disappeared and a new EPR signal with a weak broad feature typical of the reduced $[3Fe\text{-}4S]^0$ cluster took its place. These results confirmed the assignment of this electrochemistry to couple A. The two electron electrolysis at −605 mV provided an EPR signal consistent with a known $[4Fe\text{-}4S]^{1+}$ cluster and thus was assigned to redox couple B. To associate couple A unequivocally with its EPR spectrum, they carried out a dye-mediated potentiometric titration of the Fd over the potential range of −300 mV to +100 mV and followed the g = 2.01 EPR signal of the $[3Fe\text{-}4S]^{+1}$ cluster. They found that one electron was transferred at an E°′ = −130 mV. This titration showed how closely the direct electrochemistry provided the same E°′ as the obtained through the homogenous SEC potentiometric titration.

Low temperature MCD spectroscopy of the native (fully oxidized) and one electron reduced Fd III were similar to Dg Fd II clusters, $[3Fe\text{-}4S]^{1+}$ and $[3Fe\text{-}4S]^0$, respectively, that is, couple A. The two electron reduced Fd III gave a MCD spectrum similar to that for $[4Fe\text{-}4S]^{1+}$ and thus confirmed that B is the $[4Fe\text{-}4S]^{2+/1-}$ redox couple. In this Fd, the 3 Fe cluster appears to be bound by Cys 11, 17 and 51 with a possible fourth ligand in aspartic acid 14 taking the place of Cys 14; there is indirect evidence that aspartic acid may be a ligand to iron-sulfur clusters. Figure 14 rationalizes the very different

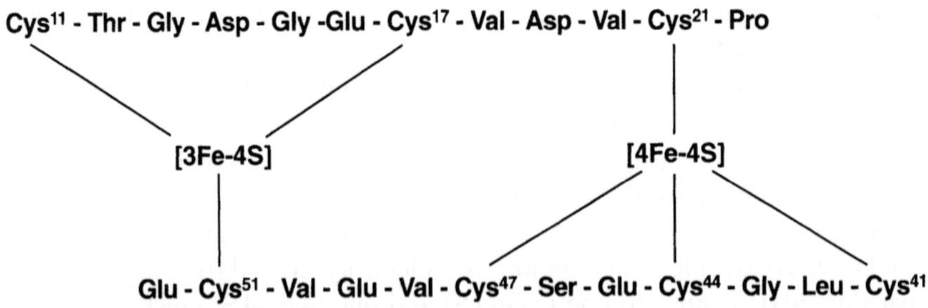

Fig. 14. Proposed assignment of ligands for the [3 Fe-4 S] and [4 Fe-4 S] cluster of *D. africanus* Fd III. From F. A. Armstrong, S. J. George, R. Cammack, E. C. Hatchikian and A. J. Thompson, Biochem. J. 264 (1989) 265–273. Reprinted with permission of *Biochemical Journal*.

structures of the two clusters, which results in the dramatically different redox properties discussed above.

Armstrong *et al.* used similar methods to show that, when reduced, the $[3\,Fe\text{-}4\,S]^0$ cluster reacts rapidly with Fe (II) ion to form a $[4\,Fe\text{-}4\,S]^{2+}$ cluster. This was the beginning of a novel investigation of the uptake of metals by iron-sulfur clusters. This latter cluster, $[4\,Fe\text{-}4\,S]^{2+}$, can then be reduced to $[4\,Fe\text{-}4\,S]^{1+}$. These reactions are summarized in Scheme I. The original Fd III was a 7 Fe Fd, and the final product is a $2\,[4\,Fe\text{-}4\,S]^{2+/1+}$ Fd. These results are seen in the direct electrochemistry shown in Fig. 15.

The cyclic voltammogram was reversible and diffusion-controlled at low scan rates. In the scan to +200 mV, Fe (II) was added in sufficient quantities to provide one equivalent of Fe (II) to the Fd III. Stirring was stopped and the cyclic voltammetric scan continued cathodically to −260 mV and returned to the anodic potential of +100 mV. This is the region of the $[3\,Fe\text{-}4\,S]^{0/1+}$ couple and is shown in Fig. 15 b. Note that the anodic wave of Fig. 15 b is significantly diminished. Upon subsequent cyclic voltammetric scans over this range, both the anodic and cathodic peak currents of couple A diminished in value. Finally, when the cyclic voltammetry was extended to the full range of −100 mV to −850 mV, the current due to the B couple, $[4\,Fe\text{-}4\,S]^{1+/2+}$, increased by 1.5 to 2-fold. The product of the reaction shown in Fig. 15 c gave only the redox couple of $E^{\circ\prime} = -400$ mV and its peak currents were linear with the square root of scan rate, without change in the peak position or shape, and a peak separation of approximately 60 mV at 16 mVs^{-1}, indicating n = 1. When the protein was reduced by bulk electrolysis at −610 mV, additional current was obtained upon adding 1 mole of iron (II) per mole of the [3 Fe-4 S] cluster. Again, the additional current arises from the follow-up chemistry shown in Scheme I. In addition, voltammetry of this final electro-

Scheme 1

$$[3\,Fe\text{-}4\,S]^{1+} + 1\,e^- \rightleftharpoons [3\,Fe\text{-}4\,S]^0$$
$$[3\,Fe\text{-}4\,S]^0 + 1\,Fe^{2+} \rightleftharpoons [4\,Fe\text{-}4\,S]^{2+}$$
$$[4\,Fe\text{-}4\,S]^{2+} + 1\,e^- \rightleftharpoons [4Fe\text{-}4S]^{1+}$$

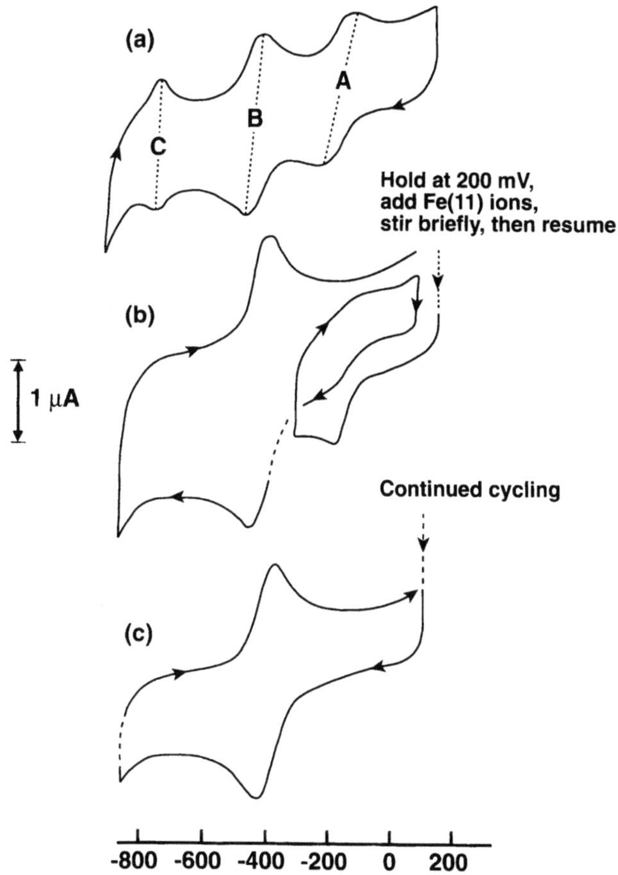

Fig. 15. Cyclic voltammograms of *Da* Fd III, showing the effect of added Fe(II) ions. v = 16 mV/s, T = 2 °C, pH 7.4, Fd = .11 mM, and 1.1 mM neomycin. From S.J. George, F.A. Armstrong, E.C. Hatchikian and A.J. Thomson, Biochem. J. 264 (1989) 275–284. Reprinted with permission of *Biochemical Journal*.

lysis solution (Fig. 15c) gave the cyclic voltammogram of the [4Fe-4S]$^{1+/2+}$ couple. These results are particularly novel because a relatively stable [4Fe-4S] was formed even though four Cys ligands were not available. From Fig. 14, aspartic acid, with its $-CH_2CO_2^{1-}$ side chain, is the most likely candidate for a ligand for the newly introduced Fe(II). In other cases where there is no ligand, as in aconitase, metal uptake can take place anyway, but the [4Fe-4S] cluster is not stable upon oxidation. In this work, the new 8 Fe Fd III can be fully oxidized and is apparently stable as shown by cyclic voltammetry. Other ligands that are possible are H_2O and HO .

As a final addendum to this work [67], the assignment of the C redox couple remains unclear at this time. Nonetheless, Armstrong notes that it appears moderately stable since the oxidation wave is observed even after holding the potential of the electrode below −850 mV for several minutes. A possible explanation proposed by Armstrong, and one which takes into account the observed pH dependence, is:

$$[3\,Fe\text{-}4\,S]^0 + 2\,e^- + nH^+ \rightleftharpoons [3\,Fe\text{-}4\,S]^{2-}\cdots nH^+. \quad (14)$$

Lastly, in their most recent work Armstrong *et al.* [67] have demonstrated the $[3\,Fe\text{-}4\,S]^0$ cluster of *Da* Fd III will take up other metals such as Zn(II) and Co(II) to form mixed-metal clusters, [M-3 Fe-4 Fe] [66, 72, 73]. More specifically, they made use of their earlier discovery that Fds can be made to co-adsorb with aminoglycosides like neomycin at the PGE to form a film on the electrode surface. The advantage of the Fd/neomycin film is that one

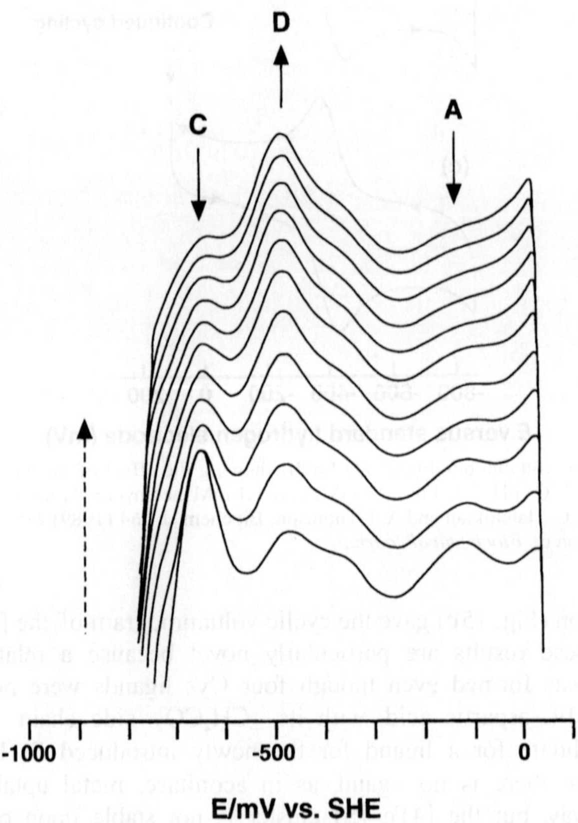

Fig. 16. Oxidative scan (successive cycles) of a film of 7 Fe Fd III upon transfer to a Zn^{2+} solution. Wave D appears and increases with concomitant disappearance of waves A and C. Scanning between 0 and −800 mV vs SHE at 470 mVs^{-1}, temperature 0 °C, pH 7. From F. N. Butt, F. A. Armstrong, J. Breton, S. J. George, A. J. Thomson and E. C. Hatchikian, J. Amer. Chem. Soc. 113 (1991) 6663–6670. Reprinted with the permission of the *Journal of the American Chemical Society*.

can use very small amounts of precious protein and carry out multiple experiments with a single sample. The reactions examined in detail are:

$$[3\,Fe\text{-}4\,S]^0 + M^{2+} \rightleftharpoons [M\text{-}3\,Fe\text{-}4\,S]^{2+}. \quad (15)$$

Experimentally, the preformed Fd/neomycin film on the PGE is dipped into a pH 7 EGTA-containing buffer solution to remove all extraneous metals; three waves were observed in the cyclic voltammograms as shown in Fig. 16. In this figure, A couple belongs to the $[3\,Fe\text{-}4\,S]^{0/1+}$ redox couple. Next, upon dipping the Fd/neomycin film PGE into a solution with one of the metals, Fe^{2+}, Zn^{2+} or Cd^{2+}, at modestly high concentrations (to assure approximately 90% reaction), a new D couple was observed with successive cyclic voltammetry scans between 0 mV to −850 mV, while couples A and C disappeared.

The rates of appearance of D and disappearance of A and C waves were shown to follow a first order kinetics. The affinity of the $[3\,Fe\text{-}4\,S]^0$ cluster for M^{2+} was qualitatively found to be: $Cd^{2+} > Zn^{2+} > Fe^{2+}$. Aiding in the detection of the D couple was the fact the $E^{\circ\prime}_D$ values in the presence of the metals were: −393 mV for Fe^{2+}; −492 mV for Zn^{2+}; and −560 mV for Cd^{2+}. The $E^{\circ\prime}_D$ in the presence of Fe^{2+} overlaps, of course, with the original B couple, but the other two are strikingly different. The potentials of these heterometal clusters, $E^{\circ\prime}_D$, were also determined at the PGE with the Fds in bulk solution and compared to the results in films. Though we shall not outline the details, bulk electrolysis experiments followed by EPR and MCD spectroscopy assigned the new D couple as: $[4\,Fe\text{-}4\,S]^{1+/2+}$, $[Zn,3\,Fe\text{-}4\,S]^{1+/2+}$, and $[Cd,3\,Fe\text{-}4\,S]^{1+/2+}$, respectively.

As noted earlier a qualitative scheme of affinities could be detected visually from the data, of which only the case of Zn^{2+} is shown in Fig. 16. In more detailed experiments, various concentrations of metal were added to the solution into which the Fd/aminoglycoside film PGE electrode was dipped. These solutions were permitted to come to equilibrium at a potential (e.g. for Zn^{2+}, −306 mV) where the $[3\,Fe\text{-}4\,S]^0$ cluster was formed. It was then possible for these proteins to take up metals. A rapid 470 mVs^{-1} oxidative scan was carried out to measure the current due to the equilibrium concentration of both $[3\,Fe\text{-}4\,S]^0$ and $[M\text{-}3\,Fe\text{-}4\,S]^{2+}$, that are the A and D waves, as shown in Fig. 17.

The dissociation equilibrium constant, K_d, for the reverse of Reaction 15 (metal uptake reaction) could then be calculated by:

$$K_d = ([M^{2+}] \cdot [3\,Fe\text{-}4\,S]^0])/[M,3\,Fe\text{-}4\,S]^{2+}. \quad (16)$$

The currents from Fig. 17 were used to calculated the fraction $[3\,Fe\text{-}4\,S]^0$ as f_{3Fe}, which means that $[M,3\,Fe\text{-}4\,S]^{2+}$ is $f_{M3Fe} = 1 - f_{3Fe}$. Since in Eq. 16 both $[3\,Fe\text{-}4\,S]^0$ and $[M\text{-}3\,Fe\text{-}4\,S]^{2+}$ in the film are very low compared to the amount of M^{2+} in solution, the concentration of M^{2+} is effectively constant. The expression was then rearranged to calculated the K_d for these three metals:

$$1/K_d \cdot [M^{2+}] = [3\,Fe\text{-}4\,S]^0/[M,3\,Fe\text{-}4\,S]^{2+}. \quad (17)$$

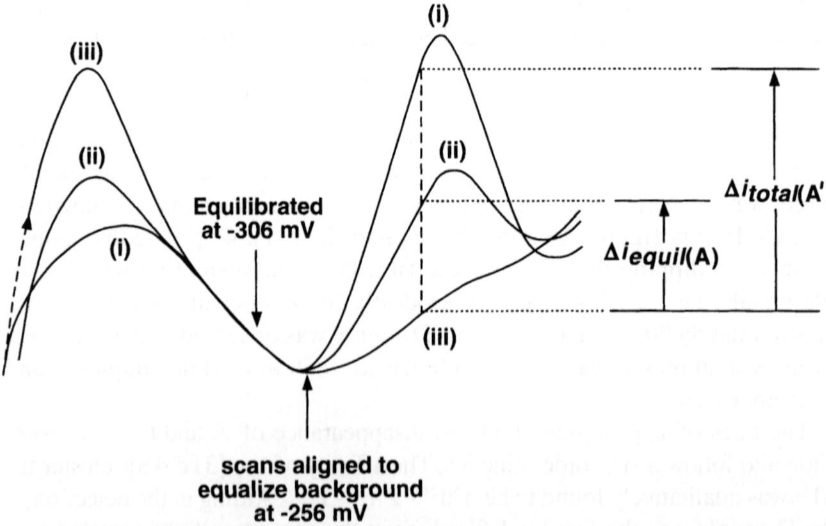

Fig. 17. Example illustrating the measurement of $\Delta i_{equil, A}$ and $\Delta i_{total, A'}$ with oxidative scans (470 mVs^{-1}) showing appearance of waves A and D after equilibration at 0 °C from which the population ratio (1-f_{Fe})/f_{Fe} was determined for Zn^{2+}. For i, no Zn^{2+}; ii, 1 μM Zn^{2+} and iii, 20 μM Zn^{2+}. The potential held at −306 mV for 2 min; currents were measured and compared at −131 mV for [3 Fe-4 S] and −481 mV for [Zn-3 Fe-4 S]. From F. N. Butt, F. A. Armstrong, J. Breton, S. J. George, A. J. Thomson and E. C. Hatchikian, J. Amer. Chem. Soc. 113 (1991) 6663–6670. Reprinted with the permission of the *Journal of the American Chemical Society*.

In summary, they found that these K_d values were: $K_{d, Fe2+} = 30$, $K_{d, Zn2+} = 1.6$, and $K_{d, Cd2+} = 08$, providing quantitative support for the trend of *stabilities* being $Cd^{2+} > Zn^{2+} \gg Fe^{2+}$. The reasons that *Dg* Fd III behaves so elegantly are that both the [3 Fe-4 S] and [4 Fe-4 S] clusters are *stable* after formation, they are both *stable* for considerable times after the reaction comes to equilibrium, and they *are able* to undergo transformation. Their examination of the [3 Fe-4 S] [4 Fe-4 S] form *Azotobacter chroococcum* demonstrated clearly that its [3 Fe-4 S] cluster was not reactive at all!

3.3. *Kinetic measurements and direct electrochemistry*

In early work on *Cp* Fd, Armstrong *et al.* carried out an oxidation study of *Cp* Fd using the PGE [46]. Using differential pulse voltammetry and square wave voltammetry, two oxidation waves were observed at −723 mV and +1040 mV vs SCE, respectively. The 2 [4 Fe-4 Se] derivative was prepared and cyclic voltammetric peaks of both shifted, as expected for Se-substituted clusters, indicating that the clusters were indeed involved in the observed peaks for both S and Se Fds. Square wave voltammetry showed no reverse current after the oxidation of the $[4Fe\text{-}4S]^{2+}$. They proposed the super-oxidation of Fd_{ox}:

$$[4Fe\text{-}4X]^{2+} \rightleftharpoons [4Fe\text{-}4X]^{3+} + e^-. \qquad (18)$$

The most anodic peak was ascribed to localized X = S, Se oxidation. The unstable species, $[4\,Fe\text{-}4\,X]^{3+}$, was seen to decompose rapidly at a homogeneous rate k_f to electrochemically inactive products, Z. Via the Nicholson and Shain treatment, a minimum value for this rate of decomposition was determined to be 425 s^{-1} or $t_{1/2} < 1.6$ ms. For comparison, HiPIP has been rapidly reduced in pulse radiolysis experiments and the resulting spectrum of the very unstable product resembled that of Fd_{red}; this new short-lived HiPIP was referred to as $HiPIP_{super\text{-}reduced}$ [74]. We know from our earlier discussion that the final products here are the [3 Fe-4 S] clusters.

3.4. Kinetic measurements

Reactions 3 and 4 can be used to study reactions between physiological donors/mediators and enzymes. A variation on the use for mediators is typified in work by Haladjian *et al.* [75], in which the rate of the reduction of *Desulfovibrio desulfuricans* Norway cytochrome c_3 (A in Reaction 3) by its physiological partner hydrogenase (Hase, Z in Reaction 4) was determined in the presence of hydrogen. In a solution saturated with hydrogen, the following reactions took place:

$$H_2 + Hase_{ox} \rightleftharpoons 2\,H^+ + Hase_{red} \tag{19}$$

$$2\,Hase_{red} + cyt\ c_{3,ox} \xrightarrow{k_f} 2\,Hase_{ox} + cyt\ c_{3,red} \tag{20}$$

$$cyt\ c_{3,red} \rightleftharpoons cyt\ c_{3,ox} + 4\,e\ . \tag{21}$$

In the absence of hydrogenase, a reverse, diffusion-controlled cyclic voltammetric wave for cyt c_3 was observed. Upon addition of hydrogenase, the cyclic voltammetric current increased due to the regeneration of cyt $c_{3,red}$ by Reaction 20. The pseudo-first order rate constant, k_f, was proportional to the hydrogenase concentration over the range of 50 to about 300 nM. This experiment could only be carried out under pseudo-first order conditions because Reaction 19 was much faster than Reaction 20. Otherwise, one would have depleted the concentration of $Hase_{red}$, and the observed rate constant would not have been proportional to [Hase]. The calculated second-order rate constant was found to be $3 \times 10^7\,M^{-1}s^{-1}$ [76]. This high rate reflects the molecular specificity of cyt c_3-hydrogenase electron transfer reaction, and an extremely high rate for dihydrogen oxidation. A similar study was carried out by Hoogvliet *et al.* [77, 78] using a series of viologens as mediators (A in Reaction 3) in the reduction of *Desulfovibrio vulgaris* hydrogenase (Z in Reaction 4), and chronoamperometry or cyclic voltammetry. Similar results were obtained by both techniques, but the chronoamperometric results were much more precise and reproducible [78]. A typical set of $k_f[Z]t$ values, calculated from the i_k/i_d ratio using Eq. 5, is shown in Fig. 18. A rate constant of $0.5 \times 10^6\,M^{-1}s^{-1}$ was obtained for the reaction between methyl viologen radical cation and hydrogenase. Similar to

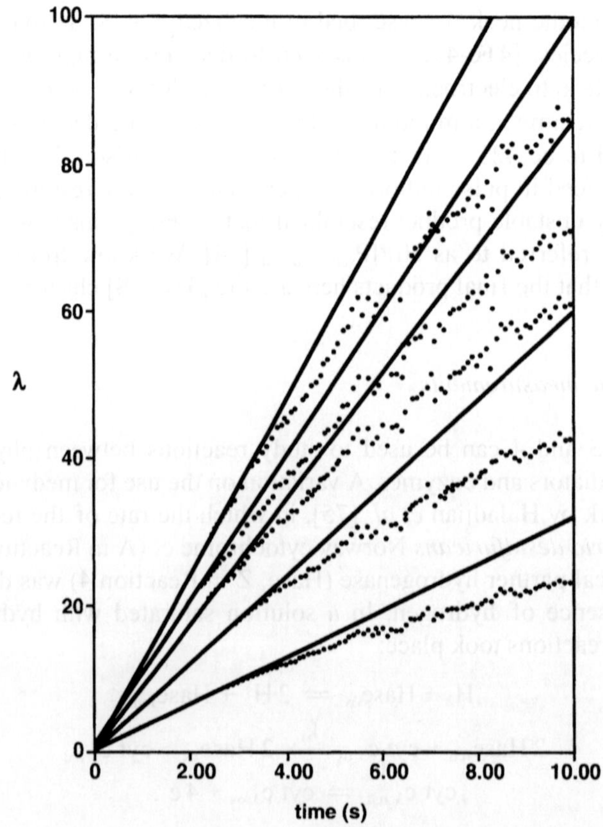

Fig. 18. Plots of the kinetic parameter λ (= k_f[Hase]t) versus time for increasing hydrogenase concentration. [Hydrogenase]/μM = (from lower to upper lines): 0.15, 0.18, 0.31, 0.35, and 0.43. [di(n-aminopropyl)viologen] = 0.25 mM in 0.25 M Tris buffer, pH 7.5, I = 0.2 M. From J.C. Hoogvliet, L.C. Lievense, C. van Dijk and C. Veeger, Eur. J. Biochem. 174 (1988) 281–285. Reprinted with the permission of *European Journal of Biochemistry*.

the above study, pseudo-first-order conditions were obtained by maintaining the solution pH. As a result, one can maintain a steady-state concentration of the enzyme without using a large excess. This approach can be generally applied to any enzyme that can turn over a non-electroactive substrate. Direct reduction of hydrogen ion is minimized by the choice of electrode material.

4. Conclusions

In this chapter we have examined the applications of several electrochemical techniques to explore what turns out to be the complex and exciting electrochemistry of several iron-sulfur proteins. While with few exceptions the work was done with small iron-sulfur proteins, work with larger iron-sulfur proteins and enzymes is in progress in various laboratories. Also, with many important iron-sulfur enzymes, such as nitrogenase, so poorly understood with regard to their electron transfer mechanisms and detailed redox behavior, it is clear that there are many exciting problems to study and new methods to be developed to understand them better.

References

1. LE Mortenson, RC Valentine and JE Carnahan, Biochem. Biophys. Res. Commun. 7 (1962) 448–452.
2. K Fukuyama, H Matsubara, T Tsukihara and Y Katsube, J. Mol. Biol. 210 (1989) 383–398.
3. W Lovenberg (ed) *Iron-Sulfur Proteins*. Volume 1, Academic Press, New York (1973).
4. W Lovenberg (ed) *Iron-Sulfur Proteins*. Volume 2, Academic Press, New York (1973).
5. W Lovenberg (ed) *Iron-Sulfur Proteins*. Volume 3, Academic Press, New York (1977).
6. TG Spiro (ed) *Iron-Sulfur Proteins*. John Wiley and Sons, New York (1982).
7. R Cammack and AG Sykes (eds) *Advances in Inorganic Chemistry, Volume 38: Iron-Sulfur Proteins*, Academic Press, New York (1992).
8. H Matsubara, Y Katsube and K Wada, *Iron-Sulfur Protein Research*, Japan Scientific Societies Press, Tokyo (1987).
9. ET Adman, LC Sieker and LH Jensen, J. Biol. Chem. 248 (1973) 3987–3996.
10. GH Stout, S Turley, LC Sieker and LH Jenson, Proc. Natl. Sci USA 85 (1988) 1020–1022.
11. T Tsukihara, K Fukuyama, M Nakamura, Y Katsube, N Tanaka, M Kakudo, et al., J. Biochem. 90 (1981) 1763–1773.
12. CW Carter, Jr, J Kraut, ST Freer, N-H Xuong, RA Alden and RG Bartsch, J. Biol. Chem. 249 (1974) 4212–4225.
13. CR Kissinger, ET Adman, LC Sieker, LH Jensen and J LeGall, FEBS Lett 244 (1989) 447–450.
14. K Tagawa, HY Tsujimoto and DI Arnon, Proc. Natl. Acad. Sci. USA 49 (1963) 567–572.
15. RC Prince and MWW Adams, J. Biol. Chem. 262 (1987) 5125–5128.
16. LE Mortenson and G Nakos in *Iron-Sulfur Proteins*, W Lovenberg (ed), Academic Press, New York (1973) Vol. 1, p. 37.
17. RK Thauer and P Schonheit in *Iron-Sulfur Proteins*, TG Spiro (ed) John Wiley and Sons, New York (1982) pp. 329–341.
18. CW Carter, Jr, J Kraut, ST Freer and RA Alden, J. Biol. Chem. 249 (1974) 6339–6349.
19. CW Carter, Jr in *Iron-Sulfur Proteins*, W Lovenberg (ed), Academic Press, New York (1977) Volume 3, pp. 157–204.
20. RJ Kassner and W Yang, J. Am. Chem. Soc. 99 (1977) 4351–4355.
21. CW Carter, Jr., J. Biol. Chem. 252 (1977) 7802–7811.
22. EJ Laskowski, RB Frankel, WO Gillum, GC Papefthymiou, J Renaud, JA Ibers et al., J. Am. Chem. Soc. 100 (1978) 5233–5337.
23. RP Sheridan, LC Allen and CW Carter, Jr, J. Biol. Chem. 256 (1981) 5052–5057.
24. MK Johnson, TG Spiro and LE Mortenson, J. Biol. Chem. 257 (1982) 2447–2452.
25. G Backes, Y Mino, TM Loehr, TE Meyer, MA Cusanovich, WV Sweeney, et al., J. Am. Chem. Soc. 113 (1991) 2055–2064.
26. AH Robbins and CD Stout, Proc. Natl. Acad. Sci. USA 86 (1989) 3639–3643.
27. H Beinert and MC Kennedy, Eur. J. Biochem. 186 (1989) 5–15.
28. MH Emptage in *Metal Clusters in Proteins*, L Que, Jr (ed), ACS Symposium Series 372, ACS, Washington, DC (1988) pp. 343–371.
29. MWW Adams, J. Biol. Chem. 262 (1987) 15054.
30. RR Eady, TH Richardson, RW Miller, M Hawkins and DJ Lowe, Biochem. J. 256 (1988) 189–196.
31. RR Eady, Polyhedron 8 (1989) 1695–1700.
32. JS Kim and DC Rees, Science 257 (1992) 1677–1682.
33. WE Newton, SF Gheller, BJ Feldman, WR Dunham and FA Schultz, J. Biol. Chem. 264 (1989) 1924–1927.

34. FA Schultz, SF Gheller and WE Newton in *Redox Chemistry and Interfacial Behavior of Biological Molecules*, G Dryhurst and K Niki (eds), Pro. Int. Symp. Redox Mech. Interfacial Prop. Mol. Biol. Importance, 3rd (1987) Plenum Press, New York (1988), pp. 203–216.
35. AJ Bard and LR Faulkner, *Electrochemical Methods: Fundamentals and Applications*, John Wiley and Son, New York (1980).
36. DH Evans, KM O'Connell, RA Petersen and MJ Kelly, J. Chem. Ed. 60 (1983) 290–293.
37. RS Nicholson, Anal. Chem. 37 (1965) 1351–1355.
38. HL Landrum, RT Salmon and FM Hawkridge, J. Am. Chem. Soc. 99 (1977) 3154–3158.
39. FA Armstrong, HAO Hill and NJ Walton, Acc. Chem. Res. 21 (1988) 407–413.
40. M Seralathan, RA Osteryoung and JG Osteryoung, J. Electroanal. Chem. 22 (1987) 69–100.
41. RS Nicholson and I Shain, Anal. Chem. 26 (1964) 706–723.
42. C van Dijk, JW van Leeuwen, C Veeger, JPGM Schreurs and E Barendrecht, Bioelectrochem. Bioenerg. 9 (1982) 743–759.
43. M-H Kim and RL Birke, Anal. Chem. 55 (1983) 522–527.
44. S Dasgupta and MD Ryan, Bioelectrochem. Bioenerg. 7 (1980) 587–594.
45. JH Christie, JA Turner and RA Osteryoung, Anal. Chem. 49 (1977) 1899–1903.
46. FA Armstrong, HAO Hill and NJ Walton, FEBS Lett. 150 (1982) 214–218.
47. ET Smith and BA Feinberg, J. Biol. Chem. 265 (1990) 14371–14376.
48. ET Smith, DW Bennett and BA Feinberg, Anal. Chim. Acta 251 (1991) 27–33.
49. ET Smith, JM Tomich, T Iwamoto, JH Richards, Y Mao and BA Feinberg, Biochemistry 30 (1991) 11669–11676.
50. JJ O'Dea, J Osteryoung and RA Osteryoung, Anal. Chem. 53 (1981) 695–701.
51. MD Ryan, J-F Wei, BA Feinberg and Y-K Lau, Anal. Biochem. 96 (1979) 326–333.
52. S Dasgupta and MD Ryan, Biochim. Biophys. Acta 680 (1982) 242–249.
53. MT Stankovich, Anal. Biochem. 109 (1980) 295–308.
54. T Kuwana and WR Heineman, Acc. Chem. Res. 9 (1976) 241–248.
55. WR Heineman, BJ Norris and JF Goetz, Anal. Chem. 47 (1975) 79–84.
56. LH Rickard, HL Landrum and FM Hawkridge, Bioelectrochem. Bioenerg. 5 (1978) 686–696.
57. VT Taniguchi, N Sailasuta-Scott, FC Anson and HB Gray, Pure Appl. Chem. 52 (1980) 2275–2281.
58. SR Harder, BA Feinberg and SW Ragsdale, Anal. Biochem. 181 (1989) 283–287.
59. RT Salmon and FM Hawkridge, J. Electroanal. Chem. 112 (1980) 253–264.
60. IR Miller and MM Werber, J. Electroanal. Chem. 100 (1979) 103–110.
61. MM Werber and M Mevarech, Arch. Biochem. Biophys. 187 (1978) 447.
62. K Tagawa and DI Arnon, Nature 195 (1962) 537–543.
63. FA Armstrong, HAO Hill and BN Oliver, J. Chem. Soc., Chem. Commun (1984) 976–977.
64. FA Armstrong, PA Cox, HAO Hill, VJ Lowe and VN Oliver, J. Electroanal. Chem. 217 (1987) 331–366.
65. JE Frew and HAO Hill, Eur. J. Biochem. 172 (1988) 261–269.
66. JN Butt, FA Armstrong, J Breton, SJ George, AJ Thomson and EC Hatchikian, J. Am. Chem. Soc. 113 (1991) 6663–6670.
67. FA Armstrong in *Advances in Inorganic Chemistry,* R Cammack and AG Sykes (eds), Academic Press, New York (1992) pp. 117–163.
68. FA Armstrong, Struct. Bonding 72 (1990) 137–221.
69. FA Armstrong, SJ George, AJ Thomson and MG Yates, FEBS Lett. 234 (1988) 107–110.
70. FA Armstrong, SJ George, R Cammack, EC Hatchikian and AJ Thomson, Biochem. J. 264 (1989) 265–273.
71. SJ George, FA Armstrong, EC Hatchikian and AJ Thomson, Biochem. J. 264 (1989) 275–284.
72. I Moura, JJG Moura, E Münck, V. Papaefthymiou and J LeGall, J. Am. Chem. Soc. 108 (1986) 349–351.
73. KK Surerus, E Münck, I Moura and JJG Moura, J. Am. Chem. Soc. 109 (1987) 3805–3807.
74. J Butler, AG Sykes, GV Buxton, PC Harrington and RG Wilkins, Biochem. J. 189 (1980) 641–644.

75. J Haladjian, P Bianco, F Guerlesquin and M Bruschi, Biochem. Biophys. Res. Commun. 147 (1987) 1289–1294.
76. KK Pandey and M Massoudipour, Inorg. Chim. Acta 138 (1987) 169–170.
77. JC Hoogvliet, LC Lievense, C van Dijk and C Veeger, Eur. J. Biochem. 174 (1988) 273–280.
78. JC Hoogvliet, LC Lievense, C van Dijk and C Veeger, Eur. J. Biochem. 174 (1988) 281–285.

Appendix A. Glossary

Fd	ferredoxin
Cp	*Clostridium pasteurianum*
HiPIP	high potential iron-sulfur protein or high potential iron protein
SEC	spectroelectrochemical
PGE	pyrolytic graphite edge electrode
n	the number of electrons transferred
F	Faraday's constant
D	the diffusion coefficient of the electroactive species of interest
C_o^*	the bulk concentration of the oxidized species
i_d	the diffusion-controlled current
i_k	the kinetically-controlled current
erfc	error function c
EPR	electron paramagnetic resonance
MCD	magnetic circular dichroic
EDTA	ethylene diamine tetraacetic acid
EGTA	1,2-bis(2-aminoethoxy)ethane-N,N,N′,N′-tetraacetic acid
SHE	standard hydrogen electrode

Bioelectrochemistry of Biomacromolecules
ed. by G. Lenaz and G. Milazzo†

CHAPTER 7
Glycosaminoglycans: Sulphated polysaccharides of the cell surface and extracellular matrix

David A. Pye

Cancer Research Campaign Department of Medical Oncology, and Cancer Research Campaign Department of Drug Development, Paterson Institute for Cancer Research, Christie Hospital NHS Trust, Manchester, United Kingdom

1. Introduction

Proteoglycans (PGs) are a large family of glycoconjugates comprising a protein core attached to which are one or more glycosaminoglycan (GAG) chains. Many core proteins have been identified to date, and these vary widely in both size and sequence. This variation is reflected in their biological functions and distribution. At present most information concerning PG biological activity has come via studies of the negatively charged GAG components. PGs as a whole will not be discussed in this chapter; instead only one element, the GAGs, will be described. Detailed information on the biochemistry of PGs can be found in the following reviews [1–4].

GAGs can be placed in four general subsets: three (heparan sulphate (HS), chondroitin sulphate/dermatan sulphate (CS/DS), and keratan sulphate (KS)) are found attached to core proteins with the fourth, hyaluronic acid (HA), existing as free GAG chains. HA is an unsulphated GAG consisting of a backbone of GlcA-GlcNAc disaccharides and will not be discussed further in this article. The remaining subsets of sulphated GAGs can initially be characterized by their repeating disaccharide backbone HexA-GlcNAc/NSO_3 (HS), HexA-GalNAc (CS/DS) and Gal-GlcNAc (KS). These backbones are enzymatically modified to introduce O-sulphate groups into the disaccharides, which produces complex mixtures of molecules with very diverse structures. This leads to a great challenge in the determination of molecular organization and sequence analysis of protein binding sites on the chain.

1.1. Heparan sulphate

HS is initially synthesized from repeating -4GlcAβ1-4GlcNAcα1- units with chain lengths varying form 50 to 150 disaccharides. Following synthesis, this structure is extensively modified by a series of stepwise sulphation and epimerization reactions. Variation in these modification steps results in the synthesis of a heterogeneous population of chains, which is thought to give HS its wide range of functions [6]. The primary modification is the removal of N-acetyl groups with the concurrent addition of sulphate. Other modifications then follow including the epimerization of GlcA to IdoA, selected 2-O-sulphation of IdoA residues, and 6-O-sulphation of GlcNAc/NSO_3. More infrequent modifications include 2-O-sulphation of GlcA and 3-O-sulphation of glucosamine. These modifications can lead to the presence of as many as 24 different disaccharide structures, (as shown in Table 1), making HS the most struc-

Table 1. Heparan sulphate/heparin disaccharides

N-Sulphated disaccharides	N-Acylated disaccharides
1 GlcAβ1-4GlcNSO$_3$	13 GlcAβ1-4GlcNAc
2 GlcA(2S)β1-4GlcNSO$_3$	14 IdoAα1-4GlcNAc
3 IdoAα1-4GlcNSO$_3$	15 IdoA(2S)α1-4GlcNAc
4 IdoA(2S)α1-4GlcNSO$_3$	
5 GlcAβ1-4GlcNSO$_3$(6S)	16 GlcAβ1-4GlcNAc(6S)
6 GlcA(2S)β1-4GlcNSO$_3$(6S)	17 IdoAα1-4GlcNAc(6S)
7 IdoAα1-4GlcNSO$_3$(6S)	18 IdoA(2S)α1-4GlcNAc(6S)
8 IdoA(2S)α1-4GlcNSO$_3$(6S)	
9 GlcAβ1-4GlcNSO$_3$(3S)	Disaccharide sequences that either do not exist or are very infrequent
10 IdoAα1-4GlcNSO$_3$(3S)	
11 GlcAβ1-4GlcNSO$_3$(3S,6S)	19 GlcA(2S)β1-4GlcNAc(6S)
12 IdoAα1-4GlcNSO$_3$(3S,6S)	
	20 GlcA(2S)β1-4GlcNSO$_3$
	21 GlcA(2S)β1-4GlcNSO$_3$(3S)
	22 IdoA(2S)α1-4GlcNSO$_3$(3S)
	23 GlcA(2S)β1-4GlcNSO$_3$(3S,6S)
	24 IdoA(2S)α1-4GlcNSO$_3$(3S,6S)

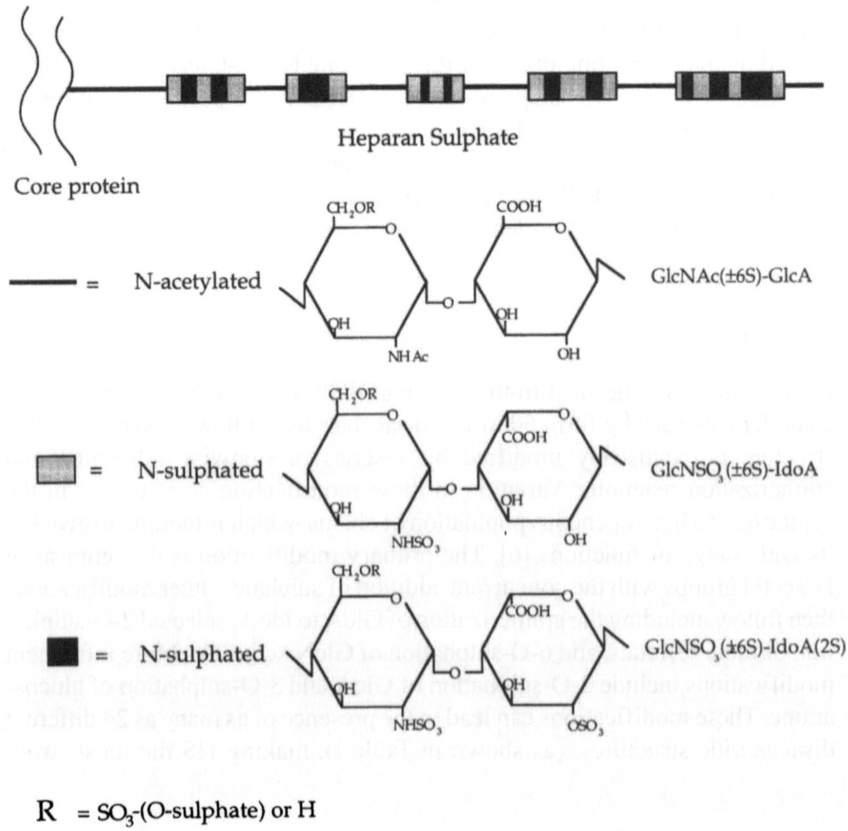

Fig. 1. General structure of fibroblast heparan sulphate chains showing typical domain structure.

turally diverse species of the GAG family. The enzymes act at many sites, and do not modify the entire molecule. Instead, a seemingly well-ordered domain structure is favoured in which sulphated regions are separated by regions containing relatively unsulphated GlcAβ1-4GlcNAc units [7–9], Figure 1. The sulphated domains are generally found to consist of repeating IdoAα1-4GlcNSO$_3$($\pm$6S) units one to eight disaccharides in length. Within these sulphated sequences are 2-O-sulphated IdoA units which are generally located centrally, Figure 2 [8, 9]. Hence HS chains are synthesized with varying chain lengths of mixed disaccharide composition with a complex domain structure. HS chains are covalently attached to the core protein via a linkage tetrasaccharide, which is followed by a sequence of N-acetyl-rich disaccharides of varying lengths [7, 10, 11]. In fibroblast HS for example, the first N-sulphated disaccharide is found an average of 10 disaccharides down from the protein linkage region, the first IdoA(2S)-containing disaccharide being about 16 disaccharides downstream. These two features are highly variable and show microheterogeneity not only in HS from different sources, but also in HS

HexA-GlcNSO$_3$-IdoA-GlcNSO$_3$-IdoA(2S)-GlcNSO$_3$-IdoA(2S)-GlcNSO$_3$-IdoA-GlcNSO$_3$-IdoA-GlcNSO$_3$

HexA-GlcNSO$_3$-IdoA-GlcNSO$_3$-IdoA(2S)-GlcNSO$_3$-IdoA(2S)-GlcNSO$_3$-IdoA-GlcNSO$_3$

Fig. 2. Distribution of IdoA(2S) units within N-sulphated domains of heparan sulphate.

synthesized by a single cell type. More variability is brought about by differences in HS sulphation levels from one cell type to another; some are highly sulphated, e.g. skin fibroblasts [12], whereas others have relatively low levels of sulphation, e.g. endothelial cells [7, 9]. It is this complexity between mixtures of HS chains synthesized by one cell type and the variation between different cells that causes great problems in terms of both structural analysis and structure activity studies.

1.2. Heparan

Heparan is synthesized only in mast cells. Initially attached to a protein core, it is removed from the core and packaged into granules for storage. It is closely related to HS, the difference being that in heparan the series of modification steps have gone almost to completion, yielding a molecule consisting mainly of IdoA(2S)α1-4GlcNSO$_3$(6S) disaccharides with little or no domain structure. The availability of heparin in relatively large quantities compared to that of HS has led to its use in a great many biological studies into the way HS interacts with other molecules. This in turn has led to the designation of the so-called heparan binding proteins and heparan binding domains within them. These are in fact binding sites for HSPGs at the cell surface and within the extracellular matrix. For example in the case of activation of bFGF, heparan is not the natural activator of the growth factor; this role is played by cell surface HSPGs. Heparan is hence destined to be used for the foreseeable future in studies relating the role of HSPGs in modulating biological process.

1.3. Chondroitin sulphate/dermatan sulphate

CS is synthesized as a repeating sequence of -4GlcAβ1-3GalNAcβ1-disaccharides. Again as with HS/heparin, the backbone is subjected to a series of enzymatic modifications [13, 14]. Disaccharide units are modified by sulphatransferases, introducing sulphate groups at positions C-4 and/or C-6 on the GalNAc and C-2 of GlcA/IdoA. DS is formed by epimerization of GlcA at position C-5 to produce IdoA-containing disaccharides, hence chains are produced which have a co-polymeric nature, i.e. containing both CS and

Table 2. Chondroitin/dermatan sulphate disaccharides

Chondroitin sulphate disaccharides	Dermatan sulphate disaccharides
1 GlcAβ1–3GalNAc	13 IdoAα1–3GalNAc
2 GlcA(2S)β1–3GalNAc	14 IdoA(2S)α1–3GalNAc
3 GlcA(3S)β1–3GalNAc	
	15 IdoAα1–3GalNAc(4S)
	16 IdoA(2S)α1–3GalNAc(4S)
4 GlcAβ1–3GalNAc(4S)	17 IdoAα 1–3GalNAc(6S)
5 GlcA(2S)β1–3GalNAc(4S)	18 IdoA(2S)α1–3GalNAc(6S)
6 GlcA(3S)β1–3GalNAc(4S)	
	19 IdoAα1–3GalNAc(4S,6S)
7 GlcAβ1–3GalNAc(6S)	20 IdoA(2S)α1–3GalNAc(4S,6S)
8 GlcA(2S)β1–3GalNAc(6S)	
9 GlcA(3S)β1–3GalNAc(6S)	21 IdoAα1–3GalNSO_3
	22 IdoA(2S)α1–3GalNSO_3
10 GlcAβ1–3GalNAc(4S,6S)	23 IdoA(2S)α1–3GalNSO_3(6S)
11 GlcA(2S)β1–3GalNAc(4S,6S)	
12 GlcA(3S)β1–3GalNAc(4S,6S)	

DS regions. The structural diversity of CS/DS chains is less pronounced than for HS due to the lack of N-sulphation of the hexosamine. Yet 23 disaccharides have been identified [15], the structures of which are shown in Table 2, but the majority are not present in significant amounts. The three most predominant disaccharides are the structural units for CS type A (GlcAβ1–3GalNAc(4S), CS type B (dermatan sulphate) (IdoA(±2S)α1–3GalNAc(4S)), and CS type C (GlcAβ1–3GalNAc(6S)). Problems have arisen in designating a GAG chain as either CS or DS because of their co-polymeric nature. Specification in general deems that CS is devoid of IdoA, therefore any GAG chain with a measurable proportion of the epimer is designated as being DS. Composition is determined by comparing the degradation of the GAG by two enzymes, chondroitinase ABC and AC1. This will be looked at in more detail later in this chapter. Digestion of DS GAG chains with these enzymes showed that GlcAβ1–3GalNAc(±4S/6S) disaccharide repeats were distributed periodically throughout predominantly IdoA(±2S)α1–3GlcNAc(4S)-containing chains. It has also been revealed that the sugar next to the –3Galβ1–3Galβ 1–4Xylβ-serine/threonine linkage region is always GlcA. CS chains are terminated with either GalNAc or GlcA and each terminal residue can be sulphated [16, 17]. At the moment techniques do not exist for elucidation of further CS/DS structural characteristics, such as location of IdoA(2S) within domains or distribution of other sulphates. Hence only general structures based on distribution of IdoA and GlcA can be proposed for CS/DS chains (Fig. 3).

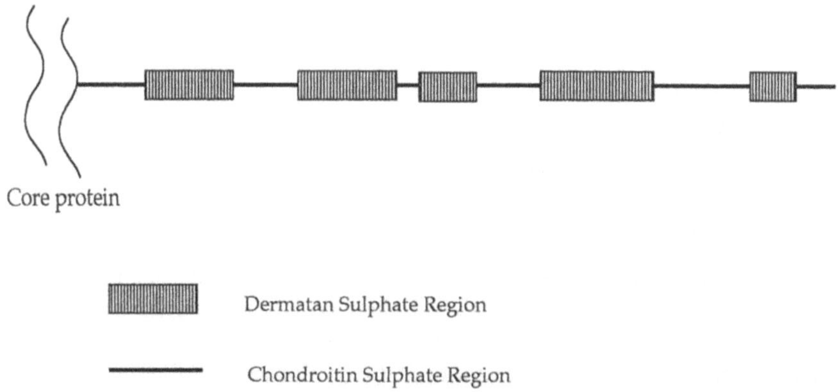

Fig. 3. Representation of typical chondroitin/dermatan sulphate chain domain structure.

1.4. Keratan sulphate

First isolated from cornea [18] Keratan sulphate was originally characterized by its link to the protein core, as N-linked (corneal KS) KS-I and O-linked (skeletal KS) KS-II. KS-II has been further subdivided into groups KS-II-A and KS-II-B on the basis that KS-II-A contains 1–3 linked Fuc and 2–6 linked NeuAc[19]. It is perhaps best to think of KS chains as consisting of three domains: (i) a capping region at the non-reducing terminus of the chain, (ii) a linkage region which connects the chains to the core protein, and (iii) an intervening repeat sequence (Fig. 4).

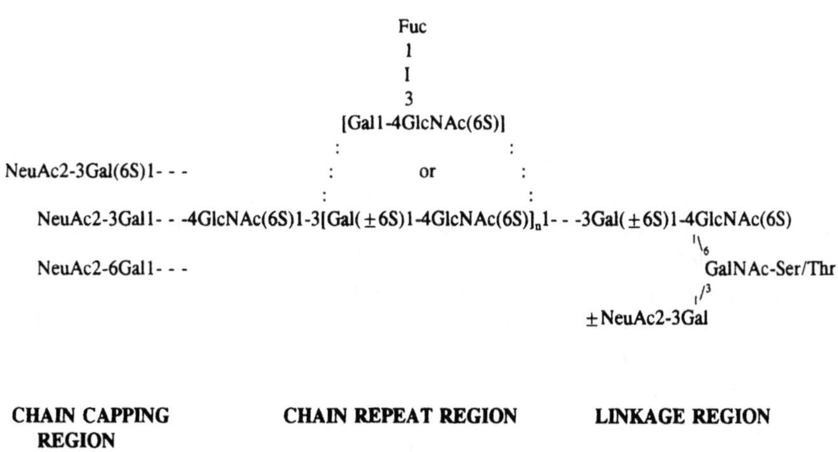

Fig. 4. General structure of skeletal keratan sulphate.

GlcNAc(6S)ß1-6GalNAc-ol | | GlcNAc(6S)ß1-6GalNAc-ol
3 | | 3
I | or | I
1 | | 1
NeuAcα2-3Gal | | Gal

Fig. 5. O-linked keratan sulphate linkage regions, isolated by keratanase digestion of cartilage keratan sulphate.

The non-reducing terminal of N- and O-linked KS has been shown to contain NeuAcα2-3 and NeuAcα2-6 linked to Gal (± 6S). This is described as the capping region [20, 21]. Asialo caps have been observed with no end terminal NeuAc but these are thought to be the result of neuraminidase action. KS attachment to the protein core is through either an N-linked GlcNAc via asparagine or O-linked GalNac to serine/threonine. In O-linked KS the disaccharide NeuAcα2-3Gal may also be attached to position C-3 of the GalNac linkage residue to form a short branch with the main KS chain extending from position C-6 of GalNAc. Alternatively the NeuAc unit may be absent, leaving Gal 1-3 linked to GalNAc [22, 23] (Fig. 5). The α2-3-linked NeuAc units found in both the linkage and capping regions are a common structural feature of both KS-II-A and KS-II-B. The repeat region of KS consists mainly of repeating -3Galβ1-4GlcNAcβ1-disaccharides with O-sulphation occurring at position C-6 of GlcNAc or Gal [24]. In general, unsulphated GlcANAc is not found, hence the predominant disaccharides are Gal-GlcNAc(6S) and Gal(6S)-GlcNAc(6S). Distribution of Gal(6S) within the repeat region is regularly broken by unsulphated Gal [25], with higher levels of Gal(6S) found towards the non-reducing end of the chain. At first sight this makes the KS molecule appear much less structurally diverse; however, the presence of NeuAc [26] in the cap and linkage regions and Fuc [27, 28] in the repeat region greatly enhances structural diversity within synthesized chains.

2. Glycosaminoglycan biosynthesis

GAG synthesis is performed in the rough endoplasmic reticulum and golgi, the mechanisms being similar for all the GAGs. Heparan/HS biosynthesis begins with the formation of a non-sulphated precursor heparan with the GlcAβ1-4GlcNAc disaccharide repeat. The polymerization process is initiated at linkage regions on the core protein via the Serine-Xyl-Gal-Gal sequence which acts as a primer from which chain elongation occurs. The process proceeds by the stepwise addition of the relevant UDP-sugars (UDP-GlcA and UDP-GLNAc) to the non-reducing terminal of nascent chains [29–31]. The first modification of this backbone is a deacylation step followed by N-sulphation to form $GlcNSO_3$ units, which occurs along with or at the end of polymerization. Both N- and O-sulphation are performed by specific sulphatransferases with differing substrate specificities, with each subsequent enzyme recognizing the product of the previous step as its preferred substrate [32, 33]. In effect, these enzymes are carrying out a programmed sequence of events which lead to the formation of a heterogeneous population of GAG chains. The N-sulphated disaccharides (GlcAβ1-4GlcNSO$_3$) initially formed are then modified by a specific epimerase which acts at position C-5 of GlcA to form IdoA, and requires the N-sulphate group for substrate recognition. The next step is the introduction of ester linked O-sulphate groups by a string of specific sulphatransferases, the major variant being the introduction of O-sulphate at position C-2 of IdoA [34]. This is followed by 6-O-sulphation of GlcNAc/SO$_3$ [35, 36] and other minor modifications including 3-O-sulphation GlcNAc/SO$_3$ [37, 38] and 2-O-sulphation of GlcA [39]. In heparan the major modifying steps go nearly to completion yielding large amounts of the tri-sulphated disaccharide IdoA(2S)α1-4GlcNSO$_3$(6S), comprising up to 90% of the total disaccharides [40]. HS however is not extensively modified; instead, domains of predominantly IdoA($\pm$2S)α 1-4GlcNSO$_3$ disaccharides (approximately 50% of the intact structure) occur which are separated by mainly N-acylated regions.

The mechanisms at work in HS/heparan biosynthesis are broadly applicable to that of CS/DS in that again the backbone is built from the -3Gal β1-3Galβ1-4Xylβ-serine linker by addition of UDP-sugars. Yet they differ in the mechanisms involved in sulphation, as little or no initial N-sulphation occurs, a marked difference to HS/heparan biosynthesis where N-sulphation acts as the template for further modifications. Even though GalNSO$_3$ is not formed in significant amounts the conversion of GlcA to IdoA still takes place indicating a different substrate specificity for this epimerase in the conversion of CS to DS. O-sulphation is also initiated at various locations within the chains, noticeably at positions C-4 and C-6 of GalNAc, C-2 or C-3 of IdoA and C-2 or C-3 of GlcA. This lack of N-sulphation reduces the possible combinations of disaccharide structures but there is still scope for many diverse structures due to conversion to IdoA and subsequent O-sulphation patterns.

KS chains are generated by O-sulphation of the basic polylactosamine structure Gal-GlcNAc with O-sulphation occurring at C-6 of Gal and C-6 GlcNAc by specific sulphatransferases. Other additions follow, including capping at the non-reducing terminal of chain and addition of Fuc to GlcNAc units [41]. This leads to the generation of chains with great structural diversity within what seems to be a well-ordered structural pattern of linkage region, repeat region and capping region. However, KS does have less structural diversity than that seen in HS and lacks the extended classical alternating domain structures associated with HS and CS/DS.

Many questions still remain unanswered in PG/GAG biosynthesis, such as why do the modifications in heparin go almost to completion, whereas in the case of HS there are fewer leading to a more diverse domain structure. Also, what controls the type of GAG chain synthesized onto a certain core protein? Is it due to primary amino acid consensus sequences or a consequence of secondary structural features? It may be that determination of chain type is dependent on whether the first unit added to the serine-Xyl-Gal-Gal-GlcA is either GlcNAc or GalNAc. These and many more questions will no doubt be answered in the coming years, but at the moment the whole biosynthesis story remains incomplete.

3. Glycosaminoglycan interactions

Many GAG interactions are currently being studied, covering a wide range of biological functions, and a thorough review of these studies is beyond the scope of this article. Instead some of the more fully explored interactions will be dealt with including the interaction of HS/heparin with basic fibroblast growth factor, heparin with antithrombin III and DS/HS/heparin with heparin cofactor II.

3.1. Fibroblast growth factors

The fibroblast growth factor (FGF) family currently consists of 9 related polypeptides ranging in mass between 16 and 26kDa [42, 43]. All have been shown to bind to heparin, which has facilitated their purification by heparin affinity chromatography. They influence a variety of cellular functions including proliferation, morphogenesis and differentiation [44–46]. The most widely studied members of this family are basic fibroblast growth factor (bFGF) and acidic fibroblast growth factor (aFGF) [47, 48]. Both aFGF and bFGF have been detected in the extracellular matrix where they are associated with HSPGs [43, 49]. It has been demonstrated that binding to HS is a distinct requirement for these growth factors to bind to their high affinity tyrosine kinase receptor [50, 51]. The FGF tyrosine kinase receptors (FGFRs) are a family of four receptors designated FGFR1-4, with an unusually large number of splice variants [52]. X-ray studies of aFGF and bFGF have located a large number of positively charged amino acids near the carboxy terminus, but site-directed mutagenesis studies suggests other regions are involved in HS/heparin binding [53, 54].

Several methods have been applied to the problem of the role played by HS in bFGF activation, including the use of heparin as a competitive inhibitor against endogenous HS and degradation of HS by enzymatic treatment. The most successful approach, however, was taken by Yayon, who used a FGFR1-transfected Chinese hamster ovary (CHO) mutant cell line deficient in HS biosynthesis. This cell line could not bind bFGF unless exogenous HS/heparin was added, whereas wild type CHO cells expressing normal levels of HS and transfected with FGFR1 were able to bind bFGF. This was the first hard evidence of the role of HS in mediating growth factor activity

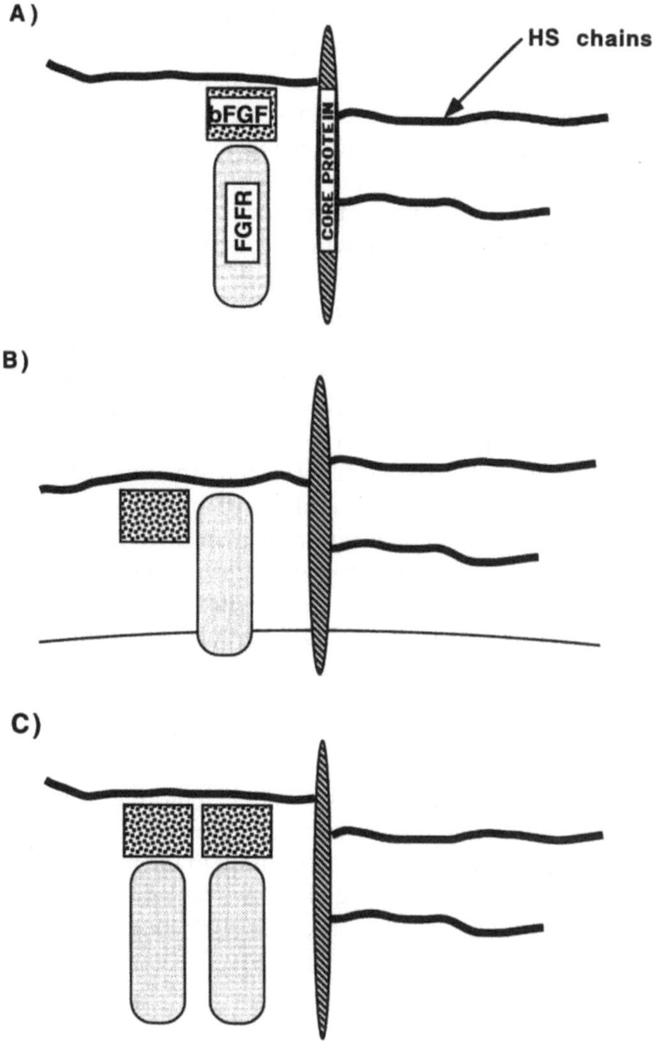

Fig. 6. Proposed models for activation of bFGF by heparan sulphate/heparin.

by assisting binding to tyrosine kinase receptors [50]. Many models have been proposed for the role HS plays in binding to bFGF and the subsequent binding of this complex to the FGFRs. The three main models are shown in Fig. 6. Figure 6 A shows the 1 : 1 : 1 complex of bFGF/HS/FGFR in which HS confers a conformational change in the bFGF molecule subsequently allowing it to interact with the FGFR [50, 55]. An alternative, if not complementary, view, (Fig. 6 B) is that HS binds to both bFGF and the FGFR with HS acts as a template for correct spatial alignment of the other two components. This is based on the observation of a weak interaction between the extracellular domain of FGFR2 and heparin [52]. The final mechanism, Fig. 6 C, depicts the interaction as a 2 : 1 : 2 bFGF/HS/FGFR complex with HS acting as a template bringing two bFGF/FGFR complexes together to form a functional dimer [51]. This final model is based on the belief that FGFR dimerization is required to induce tyrosine kinase activity [56, 57]. This is allowed for in the previous models, by suggesting that initial complexes between bFGF/HS/FGFR could subsequently dimerize to initiate kinase activity. This requirement for HS in growth factor function has also been demonstrated for other HS-binding growth factors, such as vascular endothelial growth factor, aFGF and K-FGF [58, 59].

Both N- and 2-O-sulphate groups have been shown to be required for binding to bFGF by treating HS/heparin chemically in order to remove sulphate groups selectively [60, 61]. Further investigations were aimed at elucidating bFGF binding sequences within the HS/heparin chains. These studies employed specific heparin lyases and chemical treatment to produce a mixture of size-defined HS/heparin oligosaccharides, which were then resolved on the basis of their affinity for bFGF [60, 62]. Analysis of disaccharide composition of a specific high affinity binding sequence yielded the following proposed structure: GlcAβ1-4GlcNSO$_3\alpha$1-4-[IdoA(2S)α1-4GlcNSO$_3$]$_5$-α1-4IdoAα1-4GlcNAc [62]. Smaller oligosaccharides (5-6 disaccharides in length) with lower affinity for bFGF were also analyzed and found to contain small amounts of IdoA(2S) with increasing levels of 6-O-sulphation. A pentasaccharide HexA-GlcNSO$_3$-HexA-GlcNSO$_3$-IdoA(2S) has been implicated as a minimal bFGF binding sequence [60]. These results suggest a specific role for 2-O-sulphation in bFGF/HS interactions, presumably through ionic interactions with basic amino acids on the protein, although there is evidence to suggest that the ionic interactions impart a small contribution to the binding energy and are only involved in the initial docking of the two molecules. The main stabilizing factors are non-ionic interactions, such as H-bonding and van der Waals packing forces.

3.2. Hepatocyte growth factor

Hepatocyte growth factor (HGF) is produced by mesenchymal cells and is a mitogen, motility and morphogenic factor for many cell types including hepatocytes, melanocytes, epithelial and endothelial cells [63]. HGF is a large protein compared to most growth factors, of molecular weight approximately 82kDa. It is a heterodimer consisting of a large α-chain and a smaller β-chain. The α-chain consist of a hairpin loop and four kringle domains. The target for HGF at the cell surface is a tyrosine kinase receptor, the Met receptor [64–66]. HGF has been shown, by use of natural truncated HGF and deletion mutants within the α-chain, to bind to Met via both the first hairpin loop and the second kringle domain [67–70]. Initially no direct interaction between HGF and HS could be established. Indirect evidence came from the study of cell surface HGF interactions, which were found to be sensitive to high ionic strengths and competition from heparin. Binding sites seen in the extracellular matrix laid down by certain cell types were also found to be sensitive to HS degrading enzymes. Recently a direct interaction between HS and HGF has been reported in which liver HSPGs and free HS chains were shown to bind to HGF with an affinity comparable to that of heparin [71]. The main HS binding site was also located and resides on the α-chain, but there may be some weak interaction with the β-chain.

Chemical modification of HS and disaccharide analysis of enzyme-generated oligosaccharides defined by both size and affinity for HGF, were used to define HS structural requirements for HGF binding. This showed a distinct requirement for repeating IdoA-containing disaccharides. The smallest fragments which bound to HGF were found to be of the general structure GlcAβ1-4GlcNSO$_3\alpha$1-4[IdoAα1-4GlcNSO$_3$]$_{1-2}\alpha$1-4IdoAα1-4GlcNAc. Oligosaccharides of 5–7 disaccharides had a high affinity for HS, which increased with fragment size. The disaccharide composition of these larger fragments showed an increase in 6-O-sulphation which correlated with increases in size and HGF affinity. The results suggest that both N-sulphation and 2-O-sulphation of IdoA are not required for HGF binding, yet there is a requirement for specific 6-O-sulphate substitution [71]. This is in stark contrast to the situation for the interaction of HS with bFGF, which implicated N-sulphate groups and IdoA(2S) as playing a major role in binding, with no appreciable role for 6-O-sulphation. These differences in structure/function relationships add weight to the argument that structural diversity within the HS molecule exists in order to allow specific binding to a larger number of distinct factors.

3.3. Glycosaminoglycans and proteinase inhibitors

The blood coagulation process entails the sequential activation of a number of serine proteinases, which results in the generation of thrombin and the subsequent conversion of fibrinogen into fibrin [72]. Modulation of the process is of great physiological importance and is achieved by inactivation of the enzymes by a family of serine proteinase inhibitors known as serpins. Another inactivation pathway involves a GAG-containing component thrombomodulin, a protein C activation cofactor, which leads to the inactivation of coagulation proteins factor V and VIII [73].

GAGs have been found to influence the inhibitory mechanisms of serine proteinases involved in the blood coagulation cascade, the major inhibitors being antithrombin (AT) and heparin cofactor II. Heparin/HS was found to accelerate the inactivation of coagulation enzymes by interacting with the serpin AT. A more complex situation arose with heparin cofactor II, which is activated by both heparin and DS, resulting in selective inactivation of thrombin.

3.4. Antithrombin

Antithrombin (AT) is the major inhibitor of a variety of serine proteinases, such as thrombin in plasma. It is a glycoprotein synthesized in the liver with a molecular weight of approximately 58kDa and inhibits serine proteinases by formation of strong equimolar complexes [74]. AT was found to inhibit thrombin in the absence of other cofactors, but at a very slow rate, the process being greatly accelerated by the presence of heparin. This acceleration of AT-proteinase interaction was found to depend on the presence of a unique pentasaccharide sequence within the heparin chain, "the AT binding region" of the structure GlcNAc(6S)α1-4GlcAβ1-4GlcNSO$_3$(3S)α1-4IdoA(2S)-α1-4-GlcNSO$_3$(6S) [75–78]. Structure/function studies revealed that both N-sulphate groups [79], the 3-O-sulphate group of the internal GlcNSO$_3$ [80], and the non-reducing terminal 6-O-sulphate are essential for biological activity

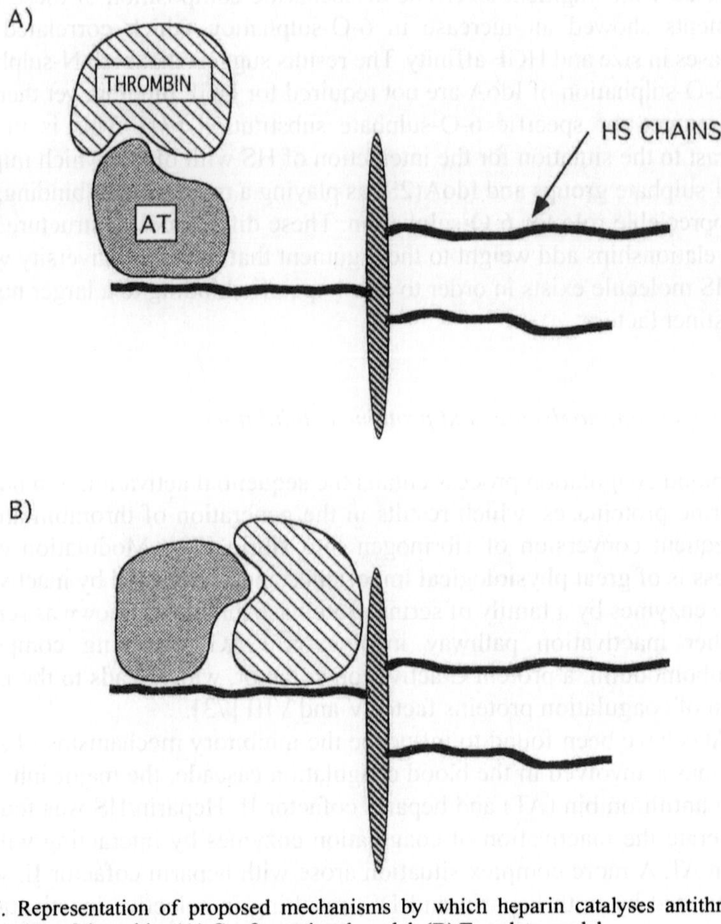

Fig. 7. Representation of proposed mechanisms by which heparin catalyses antithrombin inactivation of thrombin. (A) Conformational model, (B) Template model.

[81]. It was observed that only a few commercial heparin fragments exhibit high affinity AT binding; this correlates with the low occurrence of 3-O-sulphate which is seen in only 30–40% of chains from these preparations [82]. It is now generally accepted that the 3-O-sulphate group acts as an indicator for the presence of anticoagulant regions within heparin/HS chains. The binding of both intact heparin and the pentasaccharide to AT has been found to involve a two-step process. Initially a weak interaction is formed, which is then followed by a conformational change within AT, resulting in high affinity binding [83, 84]. Confusion still exists over the mechanism of heparin-induced AT-proteinase interactions. Initially the conformational change in AT by heparin was assumed to be the important factor, presumably by making an active ligand on AT available to the active site of the proteinases, Fig. 7A [85, 86]. Other evidence which seems to back this theory, is the similarity in stimulation by both intact heparin and the pentasaccharide of AT interactions with Factors Xa and XIIa. However, this mechanism has been challenged following studies on AT interactions with factors IXa and XIa, which showed no appreciable influence by the pentasaccharide. In this case it was found that a chain length of approximately 18 saccharides was required to enhance rates to a level comparable to that of intact heparin [87, 88]. Hence an alternative mechanism was proposed, in which heparin acts as a template or bridge between AT and the proteinases [89, 90, 91], Fig. 7B. This has since been supported by a large amount of data from chemical modification and kinetic studies [90, 92–95]. Divisions, however, still exist, with some investigators insisting that the bridging mechanism is just a secondary event, the main basis for the effect of heparin being the conformational change. A suggested compromise is that both mechanisms contribute to the acceleration effect of heparin, with the relative contributions of each varying with different target proteinases.

3.5. Heparin cofactor II

Heparin cofactor II (HCII) is another plasma serine proteinase inhibitor, consisting of a single peptide chain 480 amino acids in length. HCII inactivates thrombin by formation of a stable complex, yet is ineffective against Factor Xa. Interestingly, both heparin and DS increase inhibition of thrombin by HCII with no effect seen with CS-type A or CS-type C [96], although some activity has been observed for di-sulphated CS-4,6-sulphate [97]. Differences have been seen in the heparin and DS binding sites on HCII, suggesting a positively charged Arg is involved in binding of DS but not of heparin, hence indicating the existence of two distinct binding sites [98]. Site-directed mutagenesis of basic residues of HCII showed the binding sites to be overlapping but not identical [98–101]. Heparin-governed inhibition of thrombin by HCII was found to be independent of specific saccharide sequences but

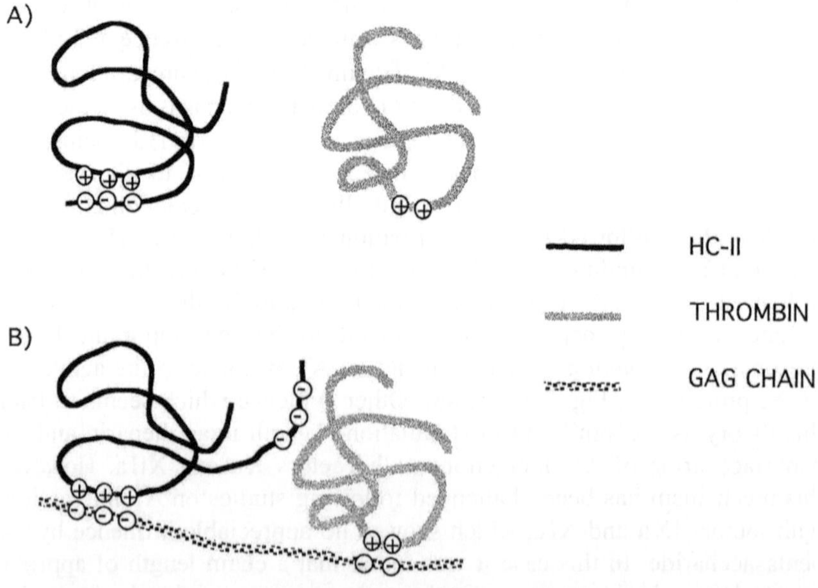

Fig. 8. Proposed model for role of dermatan sulphate and heparin in catalyzing the formation of heparin cofactor II-thrombin complexes. (A) Intramolecular binding of GAG site within heparin cofactor II, (B) GAG-mediated activation of heparin cofactor II-thrombin complexes.

is instead dominated by charge density [102]. This is revealed by an increase in catalytic efficiency of heparin with increasing chain length [103–105]. Dermatan sulphate activity on the other hand appeared to be related to the occurrence of IdoA(2S)α1-3GalNAc(4S) units [106]. The smallest oligosaccharide fragment prepared by chemical degradation of DS capable of displaying anticoagulant activity and high affinity binding for HCII was found to be six disaccharides in length [107]. This suggests a bridging mechanism as seen previously with AT for DS-mediated HCII inhibition of thrombin. The smallest DS high affinity HCII binding oligosaccharide isolated by affinity chromatography consisted of three repeating IdoA(2S)α1-3GalNAc(4S) disaccharides [108]. Detailed information on the role of GAGs in HCII interactions with thrombin has been obtained from recombinant HCII mutants [100, 109] and the following model has been proposed. Figure 8 A shows that in the absence of GAG the N-terminal domain of HCII binds intramolecularly to the GAG binding site, which results in a low rate of HCII-thrombin complex formation. On binding to GAGs the N-terminal region is displaced from the internal binding site, allowing it to interact with thrombin with a subsequent stimulation of HCII-thrombin formation. There is also a simultaneous binding of the GAG chain to thrombin which further stabilizes the complex (Fig. 8 B). The validation of this model, however, will require detailed information from sources such as X-ray crystallography and NMR spectroscopy.

4. Glycosaminoglycan sequencing

Approaches to sequencing GAGs are at the moment far from adequate, the task being complicated by the heterogeneous populations of chains produced and the problems associated with the impossible task of isolating a single species of chain. Approaches to date therefore involve extraction of information from a mixture of GAG chains. So far structural investigations have been limited to general investigations into domain structure, assessment of total disaccharide composition, and spectroscopic characterization of oligosaccharides derived from nascent chains. The procedures involved require methods for depolymerizing GAGs prior to further analysis. Two general methods of scission are employed, chemical and enzymatic.

The common method used for chemical cleavage of HS/heparin is treatment with low pH nitrous acid [110]. This cleaves the chain at $GlcNSO_3$ units with the concurrent removal of sulphate; resistant linkages have the general structure UAα/β1-4[GlcNAc($\pm$6S)α1-4GlcA]$_n\beta$1-4$GlcNSO_3$. Chemical depolymerization of KS is performed by hydrazinolysis followed by nitrous acid deamination [111]. Hydrazinolysis removes the acetyl groups from GlcNAc leaving a free amino group which is then lost on treatment with nitrous acid, with subsequent ring rearrangement and cleavage of the O-glycosidic link. This reaction scheme, if allowed to go to completion, would result in the complete reduction of the chain to disaccharides, therefore in order to generate oligosaccharides, the first hydrogenolysis step is not allowed to go to completion. Although chemical methods of depolymerization are relatively quick and inexpensive, they can lack specificity and introduce artifacts into oligosaccharides which are not present in the parent chains, and so these are not the methods of choice.

Enzymes on the other hand are very specific and act under mild conditions, with the resulting oligosaccharides closely resembling the parent molecule from which they are derived. The reactions can be monitored simply by measuring increases in absorbance, allowing generation of controlled mixtures of products. There are a number of enzymes with varying specificities which act on HS/heparin including the heparin lyases, I, II and III (Hep I, II, III) [112]. Hep I cleaves HS/heparin between Glc-NSO_3($\pm$6S) and IdoA(2S), the absolute substrate requirement being 2-O-sulphation of IdoA. Resistant linkages have the general structure IdoA(2S)α1-4[GlcNR($\pm$6S)α1-4UA]α/β1-4$GlcNSO_3$($\pm$6S), R being either Ac or SO_3. Hep III acts on linkages of the type GlcNR($\pm$6S)α1-4GlcA with resistant linkages being of the type GlcAβ1-4[$GlcNSO_3$($\pm$6S)α1-4IdoA($\pm$2S)]$_n\alpha$1-4GlcNR, where R is either Ac or SO_3. These latter resistant oligosaccharides are the domains which contain the majority of sulphate groups and are the most flexible due to the IdoA content, making them the most likely sequences for interacting with other molecules. Hep III is therefore the enzyme of choice for the study of structure/function relationships with HS/heparin and other components such as growth factors. Hep II has a much broader substrate specificity than I and III with cleavage occurring within IdoA-containing domains regardless of the presence of 2-O-sulphate.

CS/DS structure has been investigated using various chondro/dermatolyase enzymes (chondroitinase ABC, AC and B) which cleave hoxosaminidic bonds at positions depending on the type of hexuronic acid involved [113–115]. Chondroitinase ABC cleaves CS/DS between GalNAc(±4S, 6S)β1-4GlcA(±2S)/IdoA(±2S), generally degrading all material to disaccharides. AC lysase acts on disaccharides of the type GalNAc(±4S, 6S)βl-1GlcA(±2S), resistant linkages having the general structure GlcAβ1-3[GalNAc(4S)-β1-4IdoA(±2S)]$_n$$\alpha$1-3GalNAc(±4S, 6S) which are the DS regions of the chain. Chondroitinase B degrades DS-containing regions of the GAG chain i.e. disaccharides of the structure GlcNAc(4S)β1-4IdoA(±2S). Resistant sequences have the general structure IdoA(±2S)α1-3[GalNAc(±4S,6S)β1-4GlcA (±2S)]$_n$$\beta$1-3GalNAc(4S).

KS is degraded by two enzymes, with keratanase treatment most commonly used cleaving Galβ1-4GlcNAc(6S) units [116], resistant fragments typically having the structure GlcNAc(±6S)β1-3[Gal(6)β1-4GlcNAc(±6S)(±Fuc)]$_n$$\beta$1-3Gal. A more recently identified enzyme, keratanase II, has been reported to cleave the β1-3 glycosidic bond of GlcNAc(6S) [117] and in contrast to keratanase the presence of α1-3 Fuc residue on GlcNAc(6S) does not prevent keratanase II cleavage at this site [118]. It should be noted that the conclusions drawn with respect to structural details rely heavily on the specificity of the method used.

Generally the first approach in structural studies is complete degradation of the GAG chains produced by various cell types. The chains are usually biosynthetically labelled with [^{3}H]-glucosamine and/or [^{35}S]-sulphate to aid detection. The resulting disaccharides are identified by either ion-exchange chromatography [119] or capillary electrophoresis [120, 121]. In this way GAG chains from different sources can be compared on the basis of their total disaccharide composition. This can be useful, but does not provide any information regarding organization of disaccharides within the chain. Hence approaches comparable to peptide mapping are employed, whereby oligosaccharides are produced by a particular depolymerization method and resolved on the basis of their size and sulphate content by gradient PAGE [122] or anion exchange and gel filtration. For example, Hep III degradation of HS followed by sizefractionation by gel filtration results in a typical picture of resistant oligosaccharide size distribution (Fig. 9). The large disaccharide peak is indicative of the presence of large amounts of GlcA-GlcNAc disaccharides with the resistant oligosaccharides size tetra to hexadecasaccharide being IdoA-containing disaccharide repeats, if digestion is complete. Elution profiles of this type yield information on both the size and distribution of these domains within the chains. Two further approaches now follow for further analysis of these oligosaccharides. First they can be cleaved by other methods and the products examined, to yield information for example on distribution of

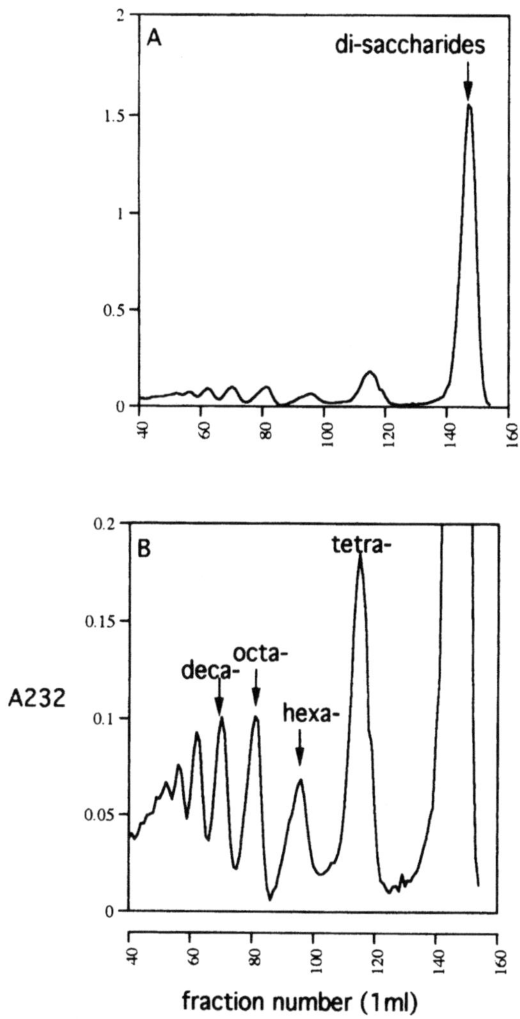

Fig. 9. Biogel P-6 gel filtration profile of Hep III degraded porcine mucosal heparan sulphate. (A) Complete profile, (B) Expansion.

IdoA(2S) residues within different size domains. The second approach, utilizing conventional NMR and MS analysis, is potentially more daunting and is aimed at structural characterization of single oligosaccharide species. This has proved successful in determining structural attributes of heparin [123] and KS [124], yet its use with HS is hampered by the complexity of oligosaccharide mixtures produced on degradation [125]. Problems associated with the isolation of HS oligosaccharides are illustrated in Fig. 10. These HPLC elution profiles show that it is possible to resolve the smaller sequences but overlap in the elution position of the larger fragments makes it extremely difficult to obtain isolated species of this type.

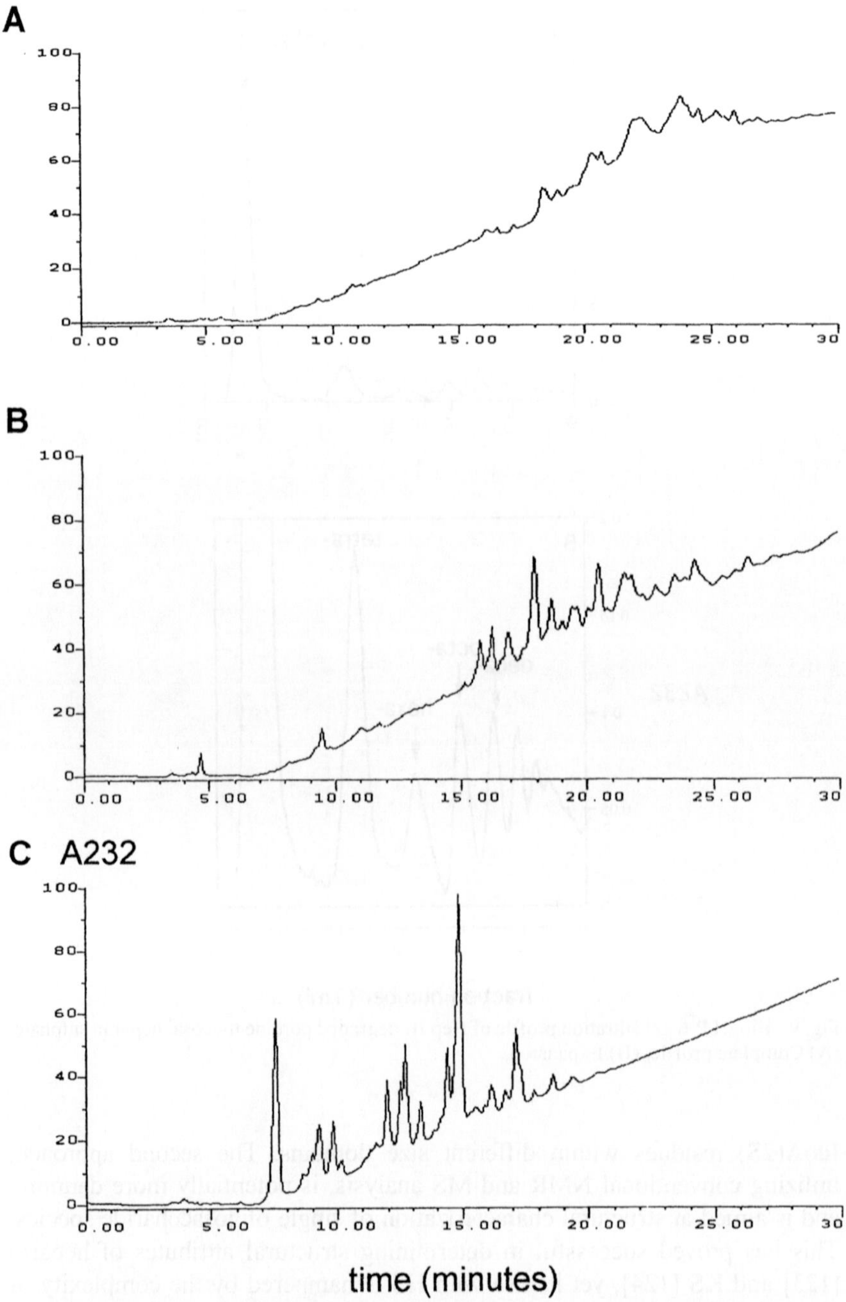

Fig. 10. Weak anion exchange (Hydropore) elution profiles of size-defined Hep III resistant oligosaccharides, 0–0.3 M ammonium bicarbonate in 30 mins. (A) Tetradecasaccharides, (B) Decasaccharides, (C) Hexasaccharides.

The above strategies are currently the methods of choice for structural investigation of GAGs, with combinations of these methods also being employed in identifying structural features of ligand binding regions. Never the less they suffer from many limitations, both in their application, and in interpretation of data. The future of GAG sequence analysis must lie in methods aimed at reading sequence data directly, as is the case for proteins and nucleic acids. Initially applied to isolated oligosaccharides, it may be possible in the future to reach the goal of reading the entire sequence of a single GAG chain.

5. Clinical application of glycosaminoglycans

Since GAGs both bind and regulate a wide variety of proteins, there is extensive scope for their application both as native molecules and synthetic analogues in drug therapy. Heparin has been used clinically for many years as an anticoagulant [126], but problems arose due to the heterogeneous nature of the chains being synthesized, with complications developing in about 10–20% of patients [127]. It was realized that only a small proportion of heparin chains synthesized contained the pentasaccharide sequence required for AT binding and this proportion was found to vary extensively when different preparations were used. Low molecular weight heparin was found to have a significant antithrombotic effect with the best bioavailability. Yet the different methods of depolymerization used during its production lead to further difficulties, since both chemical and enzymatic degradation of heparin can introduce artefacts into the structure leading to loss of activity. Recent approaches have been aimed at producing synthetic antithrombotic agents to eliminate the constraints encountered in using heparin [128]. These agents contain two synthetic molecules, which bind specifically to antithrombin and non-specifically to thrombin. These are linked by spacers to act as a template for alignment of AT and thrombin, so mimicking the role of heparin. This allows for stringent control in production of agents.

Growth factors also make an attractive target for drug development, with both inhibitors and activators of growth factor function finding clinical application in a number of disease states, including malignant cancers, wound healing and various eye disorders. Recently HS/heparin-derived oligosaccharides, five or more disaccharides in length, have been shown to be activators of bFGF in an *in vitro* assay system with hexasaccharides acting as inhibitors [129]. These oligosaccharides may be useful in their own right in certain clinical situations, but perhaps the best chance for drug development lies in a structure/function approach. This would require a more detailed molecular profile of the mechanisms involved in HS/heparin-mediated activation than exists at present, but future developments may yield this information.

It is possible to detect GAGs in both blood and urine, although the high levels of proteins present in serum made this difficult at first. Fortunately bioassays with great sensitivity have been developed, although they are still vulnerable to interference from some components [130]. Recent developments in GAG research have opened up the possibility of exploiting these bioassays in a diagnostic role. One such role may be in radiotherapy treatment of malignant tumours, where the major limiting factor is the damage inflicted on normal tissue during treatment. If methods could be found to determine the extent of this damage during therapy it would be possible to modify the treatment during the radiotherapeutic regime, leading to maximal tumour destruction with minimal normal tissue injury. The initial targets for ionizing radiation in tissue are the blood vessels and in particular the endothelium [131]. It has been observed that X-irradiation of endothelial cells results in a stimulation of both GAG synthesis and secretion [132]. Hence, increased GAG levels in serum may be exploited as an indicator to assess the progress of radiation-induced tissue injury. Skeletal KS has recently attracted attention as a potential early marker for the disease osteoarthritis [133–135], since it is released in both normal turnover of cartilage and the disease state. KS proteoglycans are first secreted into the synovial fluid and subsequently released into the blood supply. A bioassay has been developed utilizing a monoclonal antibody which recognizes a sulphated KS hexasaccharide, to assess levels in patient serum as a possible marker for the disease. These are just a few of the uses of GAGs presently under study for their potential in clinical situations; many more applications are undoubtedly awaiting discovery.

References

1. M Höök, L Kjellén and S Johansson, Ann. Rev. Biochem. 53 (1984) 847.
2. ER Ruoslahti, Ann. Rev. Cell Biol. 4 (1988) 229.
3. JT Gallagher, Curr. Opin. Cell Biol. 1 (1989) 1201.
4. L Kjellén and U Lindahl, Ann. Rev. Biochem. 60 (1991) 443.
5. TE Hardingham and AJ Fosang, FASEB J. 6 (1992) 861.
6. M Yanagishita and VC Hascall, J. Biol. Chem. 267 (1992) 9451.
7. A Lindblom, G Bengtsson-Olivercrona and LA Fransson, Biochem. J. 279 (1991) 821.
8. JE Turnbull and JT Gallagher, Biochem. J. 273 (1991) 553.
9. DA Pye and S Kumar, Biochim. Biophys. Acta 1266 (1995) 235.
10. JE Turnbull and JT Gallagher, Biochem. J. 277 (1991) 297.
11. M Lyon, JA Deakin and JT Gallagher, J. Biol. Chem. 269 (1994) 11208.
12. JE Turnbull and JT Gallagher, Biochem. J. 265 (1990) 715.
13. LA Fransson in *The Polysaccharides*, GD Aspinall (ed) Academic Press New York (1985) Vol. 3, p. 337.
14. L Rodén in *The Biochemistry of Glycoproteins and Proteoglycans*, WJ Lennarz, editor. Plenum, New York (1980) p. 267.
15. NK Karamanos, A Syrokou, P Vanky, M Nurminen and A Hjerpe, Anal. Biochem. 221 (1994) 189.
16. M Tsuji, Y Nakanishi, H Habuchi, K Ishihara and S Suzuki, Biochim. Biophys. Acta 612 (1980) 373.

17. K Otsu, H Inour, Y Tsuzki, H Yonekura, Y Nakanishi and S Suzuki, Biochem. J. 227 (1985) 37.
18. K Meyer, A Linker, EA Davidson and B Weismann, J. Biol. Chem. 205 (1953) 611.
19. IA Nieduszynski, TN Huckerby, JM Dickenson, GM Brown, GH Tai, HG Morris et al., Biochem. J. 271 (1990) 243.
20. JM Dickenson, TN Huckerby and IA Nieduszynski, Biochem. J. 278 (1991) 779.
21. JM Dickenson, TN Huckerby and IA Nieduszynski, Biochem. J. 282 (1992) 267.
22. JJ Hopwood and HC Robinson, Biochem. J. 141 (1974) 57.
23. JM Dickenson, TN Huckerby and IA Nieduszynski, Biochem. J. 269 (1990) 55.
24. VP Bhavanandan and K Meyer, J. Biol. Chem. 243 (1968) 1052.
25. M Oeben, R Keller, HW Stuhlsatz and H Greiling, Biochem. J. 248 (1987) 85.
26. LS Lohmander, S De-Luca, B Nilsson, VC Hascall, CB Caputo, JH Kimura et al., J. Biol. Chem. 255 (1980) 6084.
27. DJ Thornton, HG Morris, GH Cokin, TN Huckerby, IA Nieduszynski, I Carlstedt et al., Biochem. J. 260 (1989) 277.
28. GH Tai, GM Brown, HG Morris, TN Huckerby and IA Nieduszynski, Biochem. J. 273 (1991) 307.
29. JE Silbert, J. Biol. Chem. 238 (1963) 3542.
30. JE Silbert, J. Biol. Chem. 242 (1967) 5146.
31. K Lidholt, J Riesenfeld, KG Jacobsson, DS Feingold and U Lindahl, Biochem. J. 254 (1988) 571.
32. J Riesenfeld, M Höök and U Lindahl, J. Biol. Chem. 255 (1980) 922.
33. I Jacobsson, U Lindahl, JW Jensen, L Rodén, H Prihar and DS Feingold, J. Biol. Chem. 259 (1984) 1056.
34. K Lidholt, L Kjellen and U Lindahl, Biochem. J. 261 (1989) 999.
35. JT Gallagher, M Lyon and WP Steward, Biochem. J. 236 (1986) 313.
36. U Lindahl and L Kjellen in *The Biology of the Extracelluar Matrix*, TN Wight and RP Mecham (eds), Academic Press, New York (1987) p. 59.
37. G Pejler, G Backstrom, U Lindahl, M Paulsson, M Dziadek, S Fujiwara and R Timple, J. Biol. Chem. 262 (1987) 5036.
38. ASB Edge and RG Spiro, J. Biol. Chem. 265 (1990) 15874.
39. NS Fedarko and HE Conrad, J. Cell Biol. 102 (1986) 587.
40. JT Gallagher and A Walker, Biochem. J. 230 (1985) 665.
41. IA Niedduszynski, TN Huckerby, JM Dickenson, GM Brown, GH Tai and MT Bayliss, Biochem. Soc. Trans. 18 (1990) 792.
42. M Klagsbrun, Prog. Growth Factor Res. 1 (1989) 207.
43. M Klagsbrun, Curr. Opin. Cell Biol. 2 (1990) 857.
44. C Basilico and D Moscatelli, Adv. Cancer Res. 59 (1992) 115.
45. A Tanaka, K Miyamoto, N Minamino, M Takeda, B Sato, H Matsuo et al., Proc. Natl. Acad. Sci. USA 89 (1992) 8928.
46. M Miyamoto, K Naruo, C Seko, S Matsumoto, T Kondo and T Kurokawa, Mol. Cell Biol. 13 (1993) 4251.
47. D Gospodarowicz, J. Biol. Chem. 250 (1975) 2515.
48. D Gospodarowicz, J Weseman and J Moran, Nature 256 (1975) 216.
49. I Vlodavsky, R Bar-Shavit, R Ishai-Michaeli, P Bashkin and Z Fuks, Trends Biochem. Sci. 16 (1991) 268.
50. A Yayon, M Klagsbrun, JD Esko, P Leder and DM Ornitz, Cell 64 (1991) 841.
51. DM Ornitz, A Yayon, JG Flanagan, CM Svahn, E Levi and P. Leader, Mol. Cell Biol. 12 (1992) 240.
52. M Jaye, J Schlessinger and CA Dionne, Biochim. Biophys. Acta 1135 (1992) 185.
53. X Zhu, H Komiya, A Chirino, S Faham, GM Fox, T Arakawa et al., Science 251 (1991) 90.
54. J Zhang, LS Cousens, PJ Barr and SR Sprang, Proc. Natl. Acad. Sci. USA 88 (1991) 3446.
55. M Klagsbrun and A Baird, Cell 67 (1991) 229.
56. E Amaya, TJ Musci and MW Kirshner, Cell 66 (1991) 257.
57. WJ Fantl, DE Johnson and LT Williams, Ann. Rev. Biochem. 62 (1993) 453.
58. H Gitay-Goren, S Soker, I Vlodavsky and G Neufeld, J. Biol. Chem. 267 (1992) 6093.
59. BB Olwin and A Rapraeger, J. Cell Biol. 118 (1992) 631.
60. M Maccarana, B Casu and U Lindahl, J. Biol. Chem. 268 (1993) 23898.

61. S Guimond, M Maccarana, BB Olwin, U Lindahl and AC Rapraeger, J. Biol. Chem. 268 (1993) 23906.
62. JE Turnbull, DE Fernig, Y Ke, MC Wilkinson and JT Gallagher, J. Biol. Chem. 267 (1992) 10337.
63. K Mizuno and T Nakamura in *Hepatocyte Growth Factor-Scatter Factor and the C-MET Receptor*, ID Goldberg and EM Rosen (eds), Birkhäuser Verlag, Basel, Switzerland (1993), p. 1.
64. DP Bottaro, JS Rubin, DL Faletto, AML Chan, TE Kmiecek et al., Science 251 (1991) 802.
65. L Naldini, KM Weidner, E Vigina, G Gaudino, A Bardeli, C Ponzetteo et al., EMBO J. 10 (1991) 2867.
66. O Higuchi, K Mizuno, GF Vande Woude and T Nakamura, FEBS Lett. 301 (1992) 282.
67. AML Chan, JS Rubin, DP Bottaro, DW Hirschfield, M Chedid and SA Aaronson, Science 254 (1991) 1382.
68. K Matsumoto, T Takehara, H Inoue, M Hagiya, S Shimizu and T Nakamura, Biochim. Biophys. Res. Comm. 181 (1991) 691.
69. G Hartmann, L Naldini, KM Weidner, M Sachs, E Vigna, PM Comoglio et al., Proc. Natl. Acad. Sci. USA 89 (1992) 11574.
70. NA Lokker, MR Mark, EA Luis, GL Bennett, KA Robbins, JB Baker et al., EMBO J. 11 (1992) 2503.
71. M Lyon, JA Deakin, K Mizuno, T Nakamura and JT Gallagher, J. Biol. Chem. 269 (1994) 11216.
72. B Furie and BC Furie, Cell 33 (1988) 505.
73. B Dahlback, Thromb. Haemostasis 66 (1991) 49.
74. I Björk and A Danielsson in *Proteinase Inhibitors*, AJ Barrett and G Salvesen (eds), Elsevier, Amsterdam (1986) pp. 489.
75. B Casu, P Oreste, G Torri, G Zoppetti, J Choay, JC Lormeau et al., Biochem. J. 197 (1981) 599.
76. U Lindahl, L Thunberg, G Bäckström, J Riesenfeld, K Nordling and I Bjork, J. Biol. Chem. 259 (1984) 12368.
77. L Thunberg, G Bäckström and U Lindahl, Carbohydr. Res. 100 (1982) 393.
78. DH Atha, JC Lormeau, M Petitou, RD Rosenberg and J Choay, Biochemistry 24 (1985) 6723.
79. J Riesenfeld, L Thunberg, M Hook and U Lindahl, J. Biol. Chem. 256 (1981) 2389.
80. Mpetitou, P Duchaussoy, I Lederman, J Choay and P Sinay, Carbohydr. Res. 179 (1988) 163.
81. U Lindahl, G Bäckström and L Thunberg, J. Biol. Chem. 258 (1983) 9826.
82. M Kusche and U Lindahl, J. Biol. Chem. 265 (1990) 15403.
83. ST Olsen, KR Srinivasan, I Björk and JD Shore, J. Biol. Chem. 256 (1981) 11073.
84. CB Peterson and MN Blackburn, J. Biol. Chem. 262 (1987) 7559.
85. RE Jordan, GM Oosta, WT Gardner and RD Rosenberg, J. Biol. Chem. 255 (1980) 10081.
86. RW Carrell, PB Christy and DR Boswell in *Thrombosis and Haemostasis*, M Verstaaete, J Vermylen, HR Lijnen and J Arnout (eds), Leuven University Press, Leuven, (1987) pp. 1.
87. J Choay, M Petitou, JC Lormeau, P Sinaÿ, B Casu and G Gatti, Biochem. Biophys. Res. Commun. 116 (1983) 492.
88. JD Shore, ST Olsen, PA Craig, J Choay and I Björk, Ann. N. Y. Acad. Sci. 556 (1989) 75.
89. TC Laurent, A Tengblad, L Thumberg, M Höök and U Lindahl, Biochem. J. 175 (1978) 691.
90. MW Pomerantz and WG Owen, Biochim. Biophys. Acta 535 (1978) 66.
91. E Holmer, G Soderstrom and LO Andersson, Eur. J. Biochem. 93 (1979) 1.
92. GM Oosta, WT Gardner, DL Beeler and RD Rosenberg, Proc. Natl. Acad. Sci. USA 78 (1981) 829.
93. MJ Griffith, J. Biol. Chem. 257 (1982) 7360.
94. ME Nesheim, J. Biol. Chem. 258 (1983) 14708.
95. M Hoylaerts, WG Owen and D Collen, J. Biol. Chem. 259 (1984) 5670.
96. DM Tollefson, CA Petska and WJ Monafo, J. Biol. Chem. 258 (1983) 6713.
97. MF Scully, V Ellis, N Seno and VV Kakkar, Biochem. Biophys. Res. Commun. 137 (1986) 15.

98. MA Blinder, TR Andersson, U Abildgaard and DM Tollefsen, J. Biol. Chem. 264 (1989) 5128.
99. MA Blinder and DM Tollefsen, J. Biol. Chem. 265 (1990) 286.
100. H Rag, T Ulshofer and J Gerewitz, J. Biol. Chem. 265 (1990) 22386.
101. HC Whinna, MA Blinder, M Szewczyk, DM Tollefson and FC Church, J. Biol. Chem. 266 (1991) 8129.
102. RE Hurst, MC Poon and MJ Griffith, J. Clin, Invest. 72 (1983) 1042.
103. MM Maimone and DM Tollefsen, Biochem. Biophys. Res. Commun. 152 (1988) 1052.
104. B Bray, DA Lane, JM Freyssinet, G Pejler and U Lindahl, Biochem. J. 262 (1989) 225.
105. DM Tollefsen, T Sugimori and MM Maimone, Semin. Thromb. Haemostasis 16 (1990) 66.
106. MF Scully, V Ellis, N Seno and VV Kakkar, Biochem. J. 254 (1988) 547.
107. DM Tollefsen, ME Peacock and WJ Monafo, J. Biol. Chem. 261 (1986) 8854.
108. MM Maimone and DM Tollefsen, J. Biol. Chem. 265 (1990) 18263.
109. VMD Van Deerlin and DM Tollefsen, J. Biol. Chem. 266 (1991) 20223.
110. JE Shively and HE Conrad, Biochemistry 15 (1976) 3932.
111. JJ Hopwood and H Elliot, Carbohydr. Res. 117 (1983) 263.
112. RJ Linhardt, JE Turnbull, HM Wang, DK Loganathan and JT Gallagher, Biochemistry 29 (1990) 2611.
113. T Yamagata, H Saito, O Habuchi and S Suzuki, J. Biol. Chem. 243 (1968) 1523.
114. S Saito, T Yamagata and S Suzuki, J. Biol. Chem. 243 (1968) 1536.
115. S Suzuki, Methods Enzymol. 28 (1972) 911.
116. K Nakazawa and S Suzuki, J. Biol. Chem. 250 (1975) 912.
117. K Nakazawa, M Ito, T Yamagata and S Suzuki in *Keratan Sulphate: Chemistry, Biology, Chemical Pathology*, H Greiling and JE Scott (eds), The Biochemical Society, London (1989) pp. 99.
118. GM Brown, TN Huckerby, HG Morris, BL Abram and IA Nieduszyski, Biochemistry 33 (1994) 4836.
119. RJ Linhardt, KG Rice, YS Kim, DL Lohse, HM Wang and DK Loganathan, Biochem. J. 254 (1988) 781.
120. NK Karamanos, P Vanky, A Syrokou and A Hjerpe, Anal. Biochem. 225 (1995) 220.
121. UR Desai, HM Wang, SA Ampofo and RJ Linhardt, Anal. Biochem. 213 (1993) 120.
122. JE Turnbull and JT Gallagher, Biochem. J. 251 (1988) 597.
123. DK Loganathan, HM Wang, LM Mallis and RJ Linhardt, Biochemistry 29 (1990) 4362.
124. GH Tai, TN Huckerby and IA Nieduszynski, Biochem. J. 291 (1993) 889.
125. LM Mallis, HM Wang, DK Loganathan and RJ Linhardt, Anal. Chem. 61 (1989) 1453.
126. LB Jaques, Science 206 (1979) 528.
127. R Langer, RJ Linhardt, CL Cooney, M Klein, D Tapper, SM Hoffeberg and A Larson, Science 217 (1982) 261.
128. PDJ Grootenhuis, P Westerduin, D Meuleman, M Petitou and CAA van Boeckel, Nature Medicine (1996) in press.
129. M Ishihara, OJ Tyrrell, GB Stauber, S Brown, LS Cousens and RJ Stack, J. Biol. Chem. 268 (1993) 4675.
130. J Fareed, JM Walenga, D Hoppensteadt, X Huan and R Nonn, Proc. N. Y. Acad. Sci. 556 (1989) 333.
131. LF Farjardo and JR Stewart, Lab. Invest. 29 (1973) 244.
132. DA Pye, S Kumar, MJ Wang and RD Hunter, Biochim. Biophys. Acta 1220 (1994) 266.
133. EJ-M Thonar, ME Lenz, GK Klintworth, B Caterson, LM Pachman, P Glickman et al., Arthritis Rheum. 28 (1985) 1367.
134. EJ-M Thonar, LM Pachmann, ME Lenz, J Hayford, P Lynch and KE Kuettner, J. Clin. Chem. Clin. Biochem. 26 (1988) 57.
135. JM Williams, C Downey and EJ-M Thonar, Arthritis Rheum. 31 (1988) 557.

Appendix A. Glossary

Ac	acetyl
Fuc	fucose
Gal	galactose
GalNAc	N-acetylgalactosamine
GalNAc-ol	N-acetylgalactosaminitol
GlcA	glucuronic acid
GlcNAc	N-acetylglucosamine
$GlcNSO_3$	N-sulphated glucosamine
HexA	hexuronic acid
IdoA	iduronic acid
NeuAc	N-acetylneuraminic acid (sialic acid)
(2S)	O-ester sulphate group of C-2
UA	uronic acid
GAG	glycosaminoglycan
PG	proteoglycan
HS	heparan sulphate
CS	chondroitin sulphate
DS	dermatan sulphate
KS	keratan sulphate
HA	hyaluronic acid
FGF	fibroblast growth factor
FGFR	fibroblast growth factor receptor
HGF	hepatocyte growth factor
AT	antithrombin
HCII	heparin cofactor II
HSPG	heparan sulphate proteoglycan

Subject index

MIX
Papier aus verantwortungsvollen Quellen
Paper from responsible sources
FSC® C105338

If you have any concerns about our products,
you can contact us on
ProductSafety@springernature.com

In case Publisher is established outside the EU,
the EU authorized representative is:
Springer Nature Customer Service Center GmbH
Europaplatz 3, 69115 Heidelberg, Germany

Printed by Libri Plureos GmbH
in Hamburg, Germany